Lecture Notes in Mathematics

Edited by J.-M. Morel, F. Takens and B. Teissier

Editorial Policy for Multi-Author Publications: Summer Schools / Intensive Courses

1. Lecture Notes aim to report new developments in all areas of mathematics and their applications – quickly, informally and at a high level. Mathematical texts analysing new developments in modelling and numerical simulation are welcome. Manuscripts should be reasonably self-contained and rounded off. Thus they may, and often will, present not only results of the author but also related work by other people. They should provide sufficient motivation, examples and applications. There should also be an introduction making the text comprehensible to a wider audience. This clearly distinguishes Lecture Notes from journal articles or technical reports which normally are very concise. Articles intended for a journal but too long to be accepted by most journals, usually do not have this „lecture notes" character.

2. In general SUMMER SCHOOLS and other similar INTENSIVE COURSES are held to present mathematical topics that are close to the frontiers of recent research to an audience at the beginning or intermediate graduate level, who may want to continue with this area of work, for a thesis or later. This makes demands on the didactic aspects of the presentation. Because the subjects of such schools are advanced, there often exists no textbook, and so ideally, the publication resulting from such a school could be a first approximation to such a textbook.

   Usually several authors are involved in the writing, so it is not always simple to obtain a unified approach to the presentation.

   For prospective publication in LNM, the resulting manuscript should not be just a collection of course notes, each of which has been developed by an individual author with little or no co-ordination with the others, and with little or no common concept. The subject matter should dictate the structure of the book, and the authorship of each part or chapter should take secondary importance. Of course the choice of authors is crucial to the quality of the material at the school and in the book, and the intention here is not to belittle their impact, but simply to say that the book should be planned to be written by these authors jointly, and not just assembled as a result of what these authors happen to submit.

   This represents considerable preparatory work (as it is imperative to ensure that the authors know these criteria before they invest work on a manuscript), and also considerable editing work afterwards, to get the book into final shape. Still it is the form that holds the most promise of a successful book that will be used by its intended audience, rather than yet another volume of proceedings for the library shelf.

3. Manuscripts should be submitted (preferably in duplicate) either to Springer's mathematics editorial in Heidelberg, or to one of the series editors (with a copy to Springer). Volume editors are expected to arrange for the refereeing, to the usual scientific standards, of the individual contributions. If the resulting reports can be forwarded to us (series editors or Springer) this is very helpful. If no reports are forwarded or if other questions remain unclear in respect of homogeneity etc, the series editors may wish to consult external referees for an overall evaluation of the volume. A final decision to publish can be made only on the basis of the complete manuscript; however a preliminary decision can be based on a pre-final or incomplete manuscript. The strict minimum amount of material that will be considered should include a detailed outline describing the planned contents of each chapter.

   Volume editors and authors should be aware that incomplete or insufficiently close to final manuscripts almost always result in longer evaluation times. They should also be aware that parallel submission of their manuscript to another publisher while under consideration for LNM will in general lead to immediate rejection.

Continued on inside back-cover

Lecture Notes in Mathematics 1866

Ole E. Barndorff-Nielsen · Uwe Franz · Rolf Gohm
Burkhard Kümmerer · Steen Thorbjørnsen

Quantum Independent Increment Processes II

Structure of Quantum Lévy Processes, Classical Probability, and Physics

Editors:

Michael Schüermann
Uwe Franz

Springer

Editors and Authors

Ole E. Barndorff-Nielsen
Department of Mathematical Sciences
University of Aarhus
Ny Munkegade, Bldg. 350
8000 Aarhus
Denmark
e-mail: oebn@imf.au.dk

Burkhard Kümmerer
Fachbereich Mathematik
Technische Universität Darmstadt
Schlossgartenstr. 7
64289 Darmstadt
Germany
e-mail: kuemmerer@mathematik.tu-darmstadt.de

Michael Schuermann
Rolf Gohm
Uwe Franz
Institut für Mathematik und Informatik
Universität Greifswald
Friedrich-Ludwig-Jahn-Str. 15a
17487 Greifswald
Germany
e-mail: schurman@uni-greifswald.de
gohm@uni-greifswald.de
franz@uni-greifswald.de

Steen Thorbjørnsen
Department of Mathematics and
Computer Science
University of Southern Denmark
Campusvej 55
5230 Odense
Denmark
e-mail: steenth@imada.sdu.dk

Library of Congress Control Number: 2005934035

Mathematics Subject Classification (2000): 60G51, 81S25, 46L60, 58B32, 47A20, 16W30

ISSN print edition: 0075-8434
ISSN electronic edition: 1617-9692
ISBN-10 3-540-24407-7 Springer Berlin Heidelberg New York
ISBN-13 978-3-540-24407-3 Springer Berlin Heidelberg New York

DOI 10.1007/11376637

Springer is a part of Springer Science+Business Media
springer.com

Printed in The Netherlands

Typesetting: by the authors and Techbooks using a Springer LaTeX package
Cover design: *design & production* GmbH, Heidelberg

Printed on acid-free paper SPIN: 11376637 41/TechBooks 5 4 3 2 1 0

Preface

This volume is the second of two volumes containing the lectures given at the School "Quantum Independent Increment Processes: Structure and Applications to Physics". This school was held at the Alfried Krupp Wissenschaftskolleg in Greifswald during the period March 9–22, 2003. We thank the lecturers for all the hard work they accomplished. Their lectures give an introduction to current research in their domains that is accessible to Ph. D. students. We hope that the two volumes will help to bring researchers from the areas of classical and quantum probability, operator algebras and mathematical physics together and contribute to developing the subject of quantum independent increment processes.

We are greatly indebted to the Volkswagen Foundation for their financial support, without which the school would not have been possible. We also acknowledge the support by the European Community for the Research Training Network "QP-Applications: Quantum Probability with Applications to Physics, Information Theory and Biology" under contract HPRN-CT-2002-00279.

Special thanks go to Mrs. Zeidler who helped with the preparation and organisation of the school and who took care of all of the logistics.

Finally, we would like to thank all the students for coming to Greifswald and helping to make the school a success.

Neuherberg and Greifswald,
August 2005

Uwe Franz
Michael Schürmann

Contents

Contents of Volume I

List of Contributors

David Applebaum
Probability and Statistics Dept.
University of Sheffield
Hicks Building
Hounsfield Road
Sheffield, S3 7RH, UK
D.Applebaum@sheffield.ac.uk

Ole E. Barndorff-Nielsen
Dept. of Mathematical Sciences
University of Aarhus
Ny Munkegade
DK-8000 Århus, Denmark
oebn@imf.au.dk

B. V. Rajarama Bhat
Indian Statistical Institute
Bangalore, India
bhat@isibang.ac.in

Uwe Franz
GSF - Forschungszentrum für
Umwelt und Gesundheit
Institut für Biomathematik und
Biometrie
Ingolstädter Landstraße 1
85764 Neuherberg, Germany
uwe.franz@gsf.de

Rolf Gohm
Universität Greifswald
Friedrich-Ludwig-Jahnstrasse 15 A
D-17487 Greifswald, Germany
gohm@uni-greifswald.de

Burkhard Kümmerer
Fachbereich Mathematik
Technische Universität Darmstadt
Schloßgartenstraße 7
64289 Darmstadt, Germany
kuemmerer@mathematik.
tu-darmstadt.de

Johan Kustermans
KU Leuven
Departement Wiskunde
Celestijnenlaan 200B
3001 Heverlee, Belgium
johan.kustermans@wis.kuleuven.
ac.be

J. Martin Lindsay
School of Mathematical Sciences
University of Nottingham
University Park
Nottingham, NG7 2RD, UK
martin.lindsay@nottingham.ac.
uk

Steen Thorbjørnsen
Dept. of Mathematics & Computer
Science
University of Southern Denmark
Campusvej 55
DK-5230 Odense, Denmark
steenth@imada.sdu.dk

Introduction

In the seventies and eighties of the last century, non-commutative probability or quantum probability arose as an independent field of research that generalised the classical theory of probability formulated by Kolmorogov. It follows von Neumann's approach to quantum mechanics [vN96] and its subsequent operator algebraic formulation, cf. [BR87, BR97, Emc72]. Since its initiation quantum probability has steadily grown and now covers a wide span of research from the foundations of quantum mechanics and probability theory to applications in quantum information and the study of open quantum systems. For general introductions to the subject see [AL03a, AL03b, Mey95, Bia93, Par92].

Formally, quantum probability is related to classical probability in a similar way as non-commutative geometry to differential geometry or the theory of quantum groups to its classical counterpart. The classical theory is formulated in terms of function algebras and then these algebras are allowed to be non-commutative. The motivation for this generalisation is that examples of the new theory play an important role in quantum physics.

Some parts of quantum probability resemble classical probability, but there are also many significant differences. One is the notion of independence. Unlike in classical probability, there exist several notions of independence in quantum probability. In Uwe Franz's lecture, *Lévy processes on quantum groups and dual groups*, we will see that from an axiomatic point of view, independence should be understood as a product in the category of probability spaces having certain nice properties. It turns out to be possible to classify all possible notions of independence and to develop a theory of stochastic processes with independent and stationary increments for each of them.

The lecture *Classical and Free Infinite Divisibility and Lévy Processes* by O.E. Barndorff-Nielsen and S. Thorbjørnsen focuses on the similarities and differences between two of these notions, namely classical independence and free independence. The authors show that many important concepts of infinite divisibility and Lévy processes have interesting analogues in free probability.

In particular, the Υ-mappings provide a direct connection between the Lévy-Khintchine formula in free and in classical probability.

Another important concept in classical probability is the notion of Markovianity. In classical probability the class of Markov processes contains the class of processes with independent and stationary processes, i.e. Lévy processes. In quantum probability this is true for free independence [Bia98], tensor independence [Fra99], and for monotone independence [FM04], but neither for boolean nor for anti-monotone independence. See also the lecture *Random Walks on Finite Quantum Groups* by Uwe Franz and Rolf Gohm, where random walks on quantum groups, i.e. the discrete-time analogue of Lévy processes, are studied with special emphasis on their Markov structure.

Burkhard Kümmerer's lecture *Quantum Markov Processes and Application in Physics* gives a detailed introduction to quantum Markov processes. In particular, Kümmerer shows how these processes can be constructed from independent noises and how they arise in physics in the description of open quantum systems. The micro-maser and a spin-$\frac{1}{2}$-particle in a stochastic magnetic field can be naturally described by discrete-time quantum Markov processes. Repeated measurement is also a kind of Markov process, but of a different type.

References

[AL03a] S. Attal and J.M. Lindsay, editors. *Quantum Probability Communications.* QP-PQ, XI. World Sci. Publishing, Singapore, 2003. Lecture notes from a Summer School on Quantum Probability held at the University of Grenoble.

[AL03b] S. Attal and J.M. Lindsay, editors. *Quantum Probability Communications.* QP-PQ, XII. World Sci. Publishing, Singapore, 2003. Lecture notes from a Summer School on Quantum Probability held at the University of Grenoble.

[Bia93] P. Biane. *Ecole d'été de Probabilités de Saint-Flour*, volume 1608 of *Lecture Notes in Math.*, chapter Calcul stochastique non-commutatif. Springer-Verlag, Berlin, 1993.

[Bia98] P. Biane. Processes with free increments. *Math. Z.*, 227(1):143–174, 1998.

[BR87] O. Bratteli and D.W. Robinson. *Operator algebras and quantum statistical mechanics. 1. C^*- and W^*-algebras, symmetry groups, decomposition of states. 2nd ed.* Texts and Monographs in Physics. New York, NY: Springer, 1987.

[BR97] O. Bratteli and D.W. Robinson. *Operator algebras and quantum statistical mechanics. 2: Equilibrium states. Models in quantum statistical mechanics. 2nd ed.* Texts and Monographs in Physics. Berlin: Springer., 1997.

[Emc72] G.G. Emch. *Algebraic methods in statistical mechanics and quantum field theory.* Interscience Monographs and Texts in Physics and Astronomy. Vol. XXVI. New York etc.: Wiley-Interscience, 1972.

[FM04] U. Franz and N. Muraki. Markov structure on monotone Lévy processes. preprint math.PR/0401390, 2004.

[Fra99] U. Franz. Classical Markov processes from quantum Lévy processes. *Inf. Dim. Anal., Quantum Prob., and Rel. Topics*, 2(1):105–129, 1999.

[Mey95] P.-A. Meyer. *Quantum Probability for Probabilists*, volume 1538 of *Lecture Notes in Math.* Springer-Verlag, Berlin, 2nd edition, 1995.

[Par92] K.R. Parthasarathy. *An Introduction to Quantum Stochastic Calculus.* Birkhäuser, 1992.

[vN96] J. von Neumann. *Mathematical foundations of quantum mechanics.* Princeton Landmarks in Mathematics. Princeton University Press, Princeton, 1996. Translated from the German, with preface by R.T. Beyer.

Random Walks on Finite Quantum Groups

Uwe Franz[1] and Rolf Gohm[2]

[1] GSF - Forschungszentrum für Umwelt und Gesundheit
Institut für Biomathematik und Biometrie
Ingolstädter Landstraße 1
85764 Neuherberg
uwe.franz@gsf.de

[2] Ernst-Moritz-Arndt-Universität Greifswald
Institut für Mathematik und Informatik
Friedrich-Ludwig-Jahnstrasse 15 A
D-17487 Greifswald, Germany
gohm@uni-greifswald.de

Introduction

We present here the theory of quantum stochastic processes with independent increments with special emphasis on their structure as Markov processes. To avoid all technical difficulties we restrict ourselves to discrete time and finite quantum groups, i.e. finite-dimensional C^*-Hopf algebras, see Appendix A. More details can be found in the lectures of Kümmerer and Franz in this volume.

U. Franz and R. Gohm: *Random Walks on Finite Quantum Groups*,
Lect. Notes Math. **1866**, 1–32 (2006)
www.springerlink.com

Let G be a finite group. A Markov chain $(X_n)_{n\geq 0}$ with values in G is called a (left-invariant) *random walk*, if the transition probabilities are invariant under left multiplication, i.e.

$$P(X_{n+1} = g'|X_n = g) = P(X_{n+1} = hg'|X_n = hg) = p_{g^{-1}g'}$$

for all $n \geq 0$ and $g, g', h \in G$, with some probability measure $p = (p_g)_{g\in G}$ on G. Since every group element can be translated to the unit element by left multiplication with its inverse, this implies that the Markov chain looks the same everywhere in G. In many applications this is a reasonable assumption which simplifies the study of $(X_n)_{n\geq 0}$ considerably. For a survey on random walks on finite groups focusing in particular on their asymptotic behavior, see [SC04].

A quantum version of the theory of Markov processes arose in the seventies and eighties, see e.g. [AFL82, Küm88] and the references therein. The first examples of quantum random walks were constructed on duals of compact groups, see [vW90b, vW90a, Bia90, Bia91b, Bia91a, Bia92a, Bia92c, Bia92b, Bia94]. Subsequently, this work has been generalized to discrete quantum groups in general, see [Izu02, Col04, NT04, INT04]. We hope that the present lectures will also serve as an appetizer for the "quantum probabilistic potential theory" developed in these references.

It has been realized early that bialgebras and Hopf algebras are closely related to combinatorics, cf. [JR82, NS82]. Therefore it became natural to reformulate the theory of random walks in the language of bialgebras. In particular, the left-invariant Markov transition operator of some probability measure on a group G is nothing else than the left dual (or regular) action of the corresponding state on the algebra of functions on G. This leads to the algebraic approach to random walks on quantum groups in [Maj93, MRP94, Maj95, Len96, Ell04].

This lecture is organized as follows.

In Section 1, we recall the definition of random walks from classical probability. Section 2 provides a brief introduction to quantum Markov chains. For more detailed information on quantum Markov processes see, e.g., [Par03] and of course Kümmerer's lecture in this volume.

In Sections 3 and 4, we introduce the main objects of these lectures, namely quantum Markov chains that are invariant under the coaction of a finite quantum group. These constructions can also be carried out in infinite dimension, but require more careful treatment of the topological and analytical properties. For example the properties that use the Haar state become much more delicate, because discrete or locally compact quantum groups in general do not have a two-sided Haar state, but only one-sided Haar weights, cf. [Kus05].

The remainder of these lectures is devoted to three relatively independent topics.

In Section 5, we show how the coupling representation of random walks on finite quantum groups can be constructed using the multiplicative unitary.

This also gives a method to extend random walks in a natural way which is related to quantization.

In Section 6, we study the classical stochastic processes that can be obtained from random walks on finite quantum groups. There are basically two methods. Either one can restrict the random walk to some commutative subalgebra that is invariant under the transition operator, or one can look for a commutative subalgebra such that the whole process obtained by restriction is commutative. We give an explicit characterisation of the classical processes that arise in this way in several examples.

In Section 7, we study the asymptotic behavior of random walks on finite quantum groups. It is well-known that the Cesaro mean of the marginal distributions of a random walk starting at the identity on a classical group converges to an idempotent measure. These measures are Haar measures on some compact subgroup. We show that the Cesaro limit on finite quantum groups is again idempotent, but here this does not imply that it has to be a Haar state of some quantum subgroup.

Finally, we have collected some background material in the Appendix. In Section A, we summarize the basic theory of finite quantum groups, i.e. finite-dimensional C^*-Hopf algebras. The most important results are the existence of a unique two-sided Haar state and the multiplicative unitary, see Theorems A.2 and A.4. In order to illustrate the theory of random walks, we shall present explicit examples and calculations on the eight-dimensional quantum group introduced by Kac and Paljutkin in [KP66]. The defining relations of this quantum group and the formulas for its Haar state, GNS representation, dual, etc., are collected in Section B.

1 Markov Chains and Random Walks in Classical Probability

Let $(X_n)_{n\geq 0}$ be a stochastic process with values in a finite set, say $M = \{1, \ldots, d\}$. It is called *Markovian*, if the conditional probabilities onto the past of time n depend only on the value of $(X_n)_{n\geq 0}$ at time n, i.e.

$$P(X_{n+1} = i_{n+1}|X_0 = i_0, \ldots, X_n = i_n) = P(X_{n+1} = i_{n+1}|X_n = i_n)$$

for all $n \geq 0$ and all $i_0, \ldots, i_{n+1} \in \{1, \ldots, d\}$ with

$$P(X_0 = i_0, \ldots, X_n = i_n) > 0.$$

It follows that the distribution of $(X_n)_{n\geq 0}$ is uniquely determined by the *initial distribution* $(\lambda_i)_{1\leq i\leq d}$ and *transition matrices* $(p_{ij}^{(n)})_{1\leq i,j\leq d}$, $n \geq 1$, defined by

$$\lambda_i = P(X_0 = i) \qquad \text{and} \qquad p_{ij}^{(n)} = P(X_{n+1} = j|X_n = i).$$

In the following we will only consider the case, where the transition probabilities $p_{ij}^{(n)} = P(X_{n+1} = j|X_n = i)$ do not depend on n.

Definition 1.1. *A stochastic process* $(X_n)_{n\geq 0}$ *with values in* $M = \{1,\ldots,d\}$ *is called a* Markov chain *on* M *with initial distribution* $(\lambda_i)_{1\leq i\leq d}$ *and transition matrix* $(p_{ij})_{1\leq i,j\leq d}$, *if*

1. $P(X_0 = i) = \lambda_i$ *for* $i = 1,\ldots,d$,
2. $P(X_{n+1} = i_{n+1}|X_0 = i_0,\ldots,X_n = i_n) = p_{i_n i_{n+1}}$ *for all* $n \geq 0$ *and all* $i_0,\ldots,i_{n+1} \in M$ *s.t.* $P(X_0 = i_0,\ldots,X_n = i_n) > 0$.

The transition matrix of a Markov chain is a *stochastic matrix*, i.e. it has non-negative entries and the sum over a row is equal to one,

$$\sum_{j=1}^{d} p_{ij} = 1, \qquad \text{for all } 1 \leq i \leq d.$$

The following gives an equivalent characterisation of Markov chains, cf. [Nor97, Theorem 1.1.1.].

Proposition 1.2. *A stochastic process* $(X_n)_{n\geq 0}$ *is a Markov chain with initial distribution* $(\lambda_i)_{1\leq i\leq d}$ *and transition matrix* $(p_{ij})_{1\leq i,j\leq d}$ *if and only if*

$$P(X_0 = i_0, X_1 = i_1,\ldots,X_n = i_n) = \lambda_{i_0} p_{i_0 i_1} \cdots p_{i_{n-1} i_n}$$

for all $n \geq 0$ *and all* $i_0, i_1,\ldots,i_n \in M$.

If a group G is acting on the state space M of a Markov chain $(X_n)_{n\geq 0}$, then we can get a family of Markov chains $(g.X_n)_{n\geq 0}$ indexed by group elements $g \in G$. If all these Markov chains have the same transition matrices, then we call $(X_n)_{n\geq 0}$ a left-invariant *random walk* on M (w.r.t. to the action of G). This is the case if and only if the transition probabilities satisfy

$$P(X_{n+1} = h.y|X_n = h.x) = P(X_{n+1} = y|X_n = x)$$

for all $x, y \in M$, $h \in G$, and $n \geq 0$. If the state space is itself a group, then we consider the action defined by left multiplication. More precisely, we call a Markov chain $(X_n)_{n\geq 0}$ on a finite group G a *random walk* on G, if

$$P(X_{n+1} = hg'|X_n = hg) = P(X_{n+1} = g'|X_n = g)$$

for all $g, g', h \in G$, $n \geq 0$.

Example 1.3. We describe a binary message that is transmitted in a network. During each transmission one of the bits may be flipped with a small probability $p > 0$ and all bits have the same probability to be flipped. But we assume here that two or more errors can not occur during a single transmission.

If the message has length d, then the state space for the Markov chain $(X_n)_{n\geq 0}$ describing the message after n transmissions is equal to the d-dimensional hypercube $M = \{0,1\}^d \cong \mathbb{Z}_2^d$. The transition matrix is given by

$$p_{ij} = \begin{cases} 1-p & \text{if } i=j, \\ p/d & \text{if } i,j \text{ differ only in one bit,} \\ 0 & \text{if } i,j \text{ differ in more that one bit.} \end{cases}$$

This random walk is invariant for the group structure of $\mathbb{Z}_2^d$ and also for the action of the symmetry group of the hypercube.

2 Quantum Markov Chains

To motivate the definition of quantum Markov chains let us start with a reformulation of the classical situation. Let M, G be (finite) sets. Any map $b : M \times G \to M$ may be called an action of G on M. (Later we shall be interested in the case that G is a group but for the moment it is enough to have a set.) Let $\mathbb{C}^M$ respectively $\mathbb{C}^G$ be the $*$-algebra of complex functions on M respectively G. For all $g \in G$ we have unital $*$-homomorphisms $\alpha_g : \mathbb{C}^M \to \mathbb{C}^M$ given by $\alpha_g(f)(x) := f(b(x,g))$. They can be put together into a single unital $*$-homomorphism

$$\beta : \mathbb{C}^M \to \mathbb{C}^M \otimes \mathbb{C}^G, \quad f \mapsto \sum_{g \in G} \alpha_g(f) \otimes \mathbf{1}_{\{g\}},$$

where $\mathbf{1}_{\{g\}}$ denotes the indicator function of g. A nice representation of such a structure can be given by a directed labeled multigraph. For example, the graph

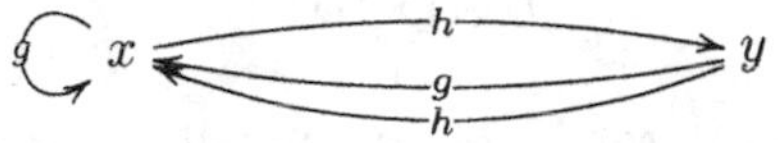

with set of vertices $M = \{x, y\}$ and set of labels $G = \{g, h\}$ represents the map $b : M \times G \to M$ with $b(x,g) = x$, $b(x,h) = y$, $b(y,g) = x = b(y,h)$. We get a natural noncommutative generalization just by allowing the algebras to become noncommutative. In [GKL04] the resulting structure is called a transition and is further analyzed. For us it is interesting to check that this is enough to construct a noncommutative or quantum Markov chain.

Let $\mathcal{B}$ and $\mathcal{A}$ be unital C^*-algebras and $\beta : \mathcal{B} \to \mathcal{B} \otimes \mathcal{A}$ a unital $*$-homomorphism. Here $\mathcal{B} \otimes \mathcal{A}$ is the minimal C^*-tensor product [Sak71]. Then we can build up the following iterative scheme ($n \geq 0$).

$$j_0 : \mathcal{B} \to \mathcal{B}, \quad b \mapsto b$$
$$j_1 : \mathcal{B} \to \mathcal{B} \otimes \mathcal{A}, \quad b \mapsto \beta(b) = b_{(0)} \otimes b_{(1)}$$

(Sweedler's notation $b_{(0)} \otimes b_{(1)}$ stands for $\sum_i b_{0i} \otimes b_{1i}$ and is very convenient in writing formulas.)

$$j_n : \mathcal{B} \to \mathcal{B} \otimes \bigotimes_1^n \mathcal{A}, \qquad j_n = (j_{n-1} \otimes \mathrm{id}_\mathcal{A}) \circ \beta,$$

$$b \mapsto j_{n-1}(b_{(0)}) \otimes b_{(1)} \in \left(\mathcal{B} \otimes \bigotimes_1^{n-1} \mathcal{A} \right) \otimes \mathcal{A}.$$

Clearly all the j_n are unital $*$-homomorphisms. If we want to have an algebra $\hat{\mathcal{B}}$ which includes all their ranges we can form the infinite tensor product $\hat{\mathcal{A}} := \bigotimes_1^\infty \mathcal{A}$ (the closure of the union of all $\bigotimes_1^n \mathcal{A}$ with the natural inclusions $x \mapsto x \otimes \mathbf{1}$) and then $\hat{\mathcal{B}} := \mathcal{B} \otimes \hat{\mathcal{A}}$.

Denote by σ the right shift on $\hat{\mathcal{A}}$, i.e., $\sigma(a_1 \otimes a_2 \otimes \ldots) = \mathbf{1} \otimes a_1 \otimes a_2 \otimes \ldots$ Using this we can also write

$$j_n : \mathcal{B} \to \hat{\mathcal{B}}, \quad b \mapsto \hat{\beta}^n(b \otimes \mathbf{1}),$$

where $\hat{\beta}$ is a unital *-homomorphism given by

$$\hat{\beta} : \hat{\mathcal{B}} \to \hat{\mathcal{B}}, \quad b \otimes a \mapsto \beta \circ (\mathrm{id}_\mathcal{B} \otimes \sigma)(b \otimes a) = \beta(b) \otimes a,$$

i.e., by applying the shift we first obtain $b \otimes \mathbf{1} \otimes a \in \hat{\mathcal{B}}$ and then interpret "$\beta\circ$" as the operation which replaces $b \otimes \mathbf{1}$ by $\beta(b)$. We may interpret $\hat{\beta}$ as a kind of time evolution producing $j_1, j_2 \ldots$

To do probability theory, consider states ψ, ϕ on $\mathcal{B}, \mathcal{A}$ and form product states

$$\psi \otimes \bigotimes_1^n \phi$$

for $\mathcal{B} \otimes \bigotimes_1^n \mathcal{A}$ (in particular for $n = \infty$ the infinite product state on $\hat{\mathcal{B}}$, which we call Ψ). Now we can think of the j_n as noncommutative random variables with distributions $\Psi \circ j_n$, and $(j_n)_{n \geq 0}$ is a noncommutative stochastic process [AFL82]. We call ψ the *initial state* and ϕ the *transition state*.

In order to analyze this process, we define for $n \geq 1$ linear maps

$$Q_{[0,n-1]} : \mathcal{B} \otimes \bigotimes_1^n \mathcal{A} \to \mathcal{B} \otimes \bigotimes_1^{n-1} \mathcal{A},$$

$$b \otimes a_1 \otimes \ldots \otimes a_{n-1} \otimes a_n \mapsto b \otimes a_1 \otimes \ldots \otimes a_{n-1}\, \phi(a_n)$$

In particular $Q := Q_{[0,0]} = \mathrm{id} \otimes \phi : \mathcal{B} \otimes \mathcal{A} \to \mathcal{B}$, $b \otimes a \mapsto b\phi(a)$.

Such maps are often called slice maps. From a probabilistic point of view, it is common to refer to idempotent norm-one (completely) positive maps onto a C^*-subalgebra as (noncommutative) conditional expectations [Sak71]. Clearly the slice map $Q_{[0,n-1]}$ is a conditional expectation (with its range embedded by $x \mapsto x \otimes \mathbf{1}$) and it has the additional property of preserving the state, i.e., $\Psi \circ Q_{[0,n-1]} = \Psi$.

Proposition 2.1. *(Markov property)*

$$Q_{[0,n-1]} \circ j_n = j_{n-1} \circ T_\phi$$

where $T_\phi : \mathcal{B} \to \mathcal{B}, \quad b \mapsto Q\,\beta(b) = (\mathrm{id} \otimes \phi) \circ \beta(b) = b_{(0)}\, \phi(b_{(1)})$.

Proof.

$$Q_{[0,n-1]} j_n(b) = Q_{[0,n-1]}\big(j_{n-1}(b_{(0)}) \otimes b_{(1)}\big) = j_{n-1}(b_{(0)})\phi(b_{(1)}) = j_{n-1}T_\phi(b).$$

□

We interpret this as a Markov property of the process $(j_n)_{n\geq 0}$. Note that if there are state-preserving conditional expectations P_{n-1} onto $j_{n-1}(\mathcal{B})$ and $P_{[0,n-1]}$ onto the algebraic span of $j_0(\mathcal{B}), \ldots, j_{n-1}(\mathcal{B})$, then because P_{n-1} is dominated by $P_{[0,n-1]}$ and $P_{[0,n-1]}$ is dominated by $Q_{[0,n-1]}$, we get

$$P_{[0,n-1]} \circ j_n = j_{n-1} \circ T_\phi \quad (Markov\ property)$$

The reader should check that for commutative algebras this is the usual Markov property of classical probability. Thus in the general case, we say that $(j_n)_{n\geq 0}$ is a *quantum Markov chain* on $\mathcal{B}$. The map T_ϕ is called the *transition operator* of the Markov chain. In the classical case as discussed in Section 1 it can be identified with the transition matrix by choosing indicator functions of single points as a basis, i.e., $T_\phi(\mathbf{1}_{\{j\}}) = \sum_{i=1}^{d} p_{ij}\mathbf{1}_{\{i\}}$. It is an instructive exercise to start with a given transition matrix (p_{ij}) and to realize the classical Markov chain with the construction above.

Analogous to the classical formula in Proposition 1.2 we can also derive the following semigroup property for transition operators from the Markov property. It is one of the main reasons why Markov chains are easier than more general processes.

Corollary 2.2. *(Semigroup property)*

$$Q\, j_n = T_\phi^n$$

Finally we note that if $(\psi \otimes \phi) \circ \beta = \psi$ then $\Psi \circ \hat{\beta} = \Psi$. This implies that the Markov chain is stationary, i.e., correlations between the random variables depend only on time differences. In particular, the state ψ is then preserved by T_ϕ, i.e., $\psi \circ T_\phi = \psi$.

The construction above is called *coupling to a shift*, and similar structures are typical for quantum Markov processes, see [Küm88, Go04].

3 Random Walks on Comodule Algebras

Let us return to the map $b : M \times G \to M$ considered in the beginning of the previous section. If G is group, then $b : M \times G \to M$ is called a (left) action of G on M, if it satisfies the following axioms expressing associativity and unit,

$$b(b(x,g),h) = b(x,hg), \quad b(x,e) = x$$

for all $x \in M$, $g,h \in G$, $e \in G$ the unit of G. In Section 1, we wrote $g.x$ instead of $b(x,g)$. As before we have the unital $*$-homomorphisms $\alpha_g : \mathbb{C}^M \to \mathbb{C}^M$. Actually, in order to get a representation of G on $\mathbb{C}^M$, i.e., $\alpha_g \alpha_h = \alpha_{gh}$ for all $g,h \in G$ we must modify the definition and use $\alpha_g(f)(x) := f(b(x,g^{-1}))$. (Otherwise we get an anti-representation. But this is a minor point at the moment.) In the associated coaction $\beta : \mathbb{C}^M \to \mathbb{C}^M \otimes \mathbb{C}^G$ the axioms above are turned into the coassociativity and counit properties. These make perfect sense not only for groups but also for quantum groups and we state them at once in this more general setting. We are rewarded with a particular interesting class of quantum Markov chains associated to quantum groups which we call random walks and which are the subject of this lecture.

Let $\mathcal{A}$ be a finite quantum group with comultiplication Δ and counit ε (see Appendix A). A C^*-algebra $\mathcal{B}$ is called an *$\mathcal{A}$-comodule algebra* if there exists a unital $*$-algebra homomorphism $\beta : \mathcal{B} \to \mathcal{B} \otimes \mathcal{A}$ such that

$$(\beta \otimes \mathrm{id}) \circ \beta = (\mathrm{id} \otimes \Delta) \circ \beta, \quad (\mathrm{id} \otimes \varepsilon) \circ \beta = \mathrm{id}.$$

Such a map β is called a *coaction*. In Sweedler's notation, the first equation applied to $b \in \mathcal{B}$ reads

$$b_{(0)(0)} \otimes b_{(0)(1)} \otimes b_{(1)} = b_{(0)} \otimes b_{(1)(1)} \otimes b_{(1)(2)},$$

which thus can safely be written as $b_{(0)} \otimes b_{(1)} \otimes b_{(2)}$.

If we start with such a coaction β then we can look at the quantum Markov chain constructed in the previous section in a different way. Define for $n \geq 1$

$$\begin{aligned} k_n : \mathcal{A} &\to \mathcal{B} \otimes \hat{\mathcal{A}} \\ a &\mapsto \mathbf{1}_{\mathcal{B}} \otimes \mathbf{1} \otimes \ldots \mathbf{1} \otimes a \otimes \mathbf{1} \otimes \ldots, \end{aligned}$$

where a is inserted at the n-th copy of $\mathcal{A}$. We can interpret the k_n as (noncommutative) random variables. Note that the k_n are identically distributed. Further, the sequence $j_0, k_1, k_2, \ldots$ is a sequence of *tensor independent* random variables, i.e., their ranges commute and the state acts as a product state on them. The convolution $j_0 \star k_1$ is defined by

$$j_0 \star k_1(b) := j_0(b_{(0)})\, k_1(b_{(1)})$$

and it is again a random variable. (Check that tensor independence is needed to get the homomorphism property.) In a similar way we can form the convolution of the k_n among each other. By induction we can prove the following formulas for the random variables j_n of the chain.

Proposition 3.1.

$$\begin{aligned} j_n &= (\beta \otimes \mathrm{id} \otimes \ldots \otimes \mathrm{id}) \ldots (\beta \otimes \mathrm{id} \otimes \mathrm{id})(\beta \otimes \mathrm{id})\beta \\ &= (\mathrm{id} \otimes \mathrm{id} \otimes \ldots \otimes \Delta) \ldots (\mathrm{id} \otimes \mathrm{id} \otimes \Delta)(\mathrm{id} \otimes \Delta)\beta \\ &= j_0 \star k_1 \star \ldots \star k_n \end{aligned}$$

Note that by the properties of coactions and comultiplications the convolution is associative and we do not need to insert brackets. The statement $j_n = j_0 \star k_1 \star \ldots \star k_n$ can be put into words by saying that the Markov chain associated to a coaction is a chain with (tensor-)independent and stationary increments. Using the convolution of states we can write the distribution of $j_n = j_0 \star k_1 \star \ldots \star k_n$ as $\psi \star \phi^{\star n}$. For all $b \in \mathcal{B}$ and $n \geq 1$ the transition operator T_ϕ satisfies

$$\psi(T_\phi^n(b)) = \Psi(j_n(b)) = \psi \star \phi^{\star n}(b),$$

and from this we can verify that

$$T_\phi^n = (\mathrm{id} \otimes \phi^{\star n}) \circ \beta,$$

i.e., given β the semigroup of transition operators (T_ϕ^n) and the semigroup $(\phi^{\star n})$ of convolution powers of the transition state are essentially the same thing.

A quantum Markov chain associated to such a coaction is called a *random walk on the $\mathcal{A}$-comodule algebra $\mathcal{B}$*. We have seen that in the commutative case this construction describes an action of a group on a set and the random walk derived from it. Because of this background, some authors call an action of a quantum group what we called a coaction. But this should always become clear from the context.

Concerning stationarity we get

Proposition 3.2. *For a state ψ on $\mathcal{B}$ the following assertions are equivalent:*

(a) $(\psi \otimes \mathrm{id}) \circ \beta = \psi(\cdot)\mathbf{1}$.
(b) $(\psi \otimes \phi) \circ \beta = \psi$ *for all states ϕ on $\mathcal{A}$.*
(c) $(\psi \otimes \eta) \circ \beta = \psi$, *where η is the Haar state on $\mathcal{A}$ (see Appendix A).*

Proof. (a)⇔(b) and (b)⇒(c) is clear. Assuming (c) and using the invariance properties of η we get for all states ϕ on $\mathcal{A}$

$$\psi = (\psi \otimes \eta)\beta = (\psi \otimes \eta \otimes \phi)(\mathrm{id} \otimes \Delta)\beta = (\psi \otimes \eta \otimes \phi)(\beta \otimes \mathrm{id})\beta = (\psi \otimes \phi)\beta,$$

which is (b). □

Such states are often called invariant for the coaction β. Of course for special states ϕ on $\mathcal{A}$ there may be other states ψ on $\mathcal{B}$ which also lead to stationary walks.

Example 3.3. For explicit examples we will use the eight-dimensional finite quantum group introduced by Kac and Paljutkin [KP66], see Appendix B.

Consider the commutative algebra $\mathcal{B} = \mathbb{C}^4$ with standard basis $v_1 = (1,0,0,0), \ldots, v_4 = (0,0,0,1)$ (and component-wise multiplication). Defining an $\mathcal{A}$-coaction by

$$\begin{aligned}\beta(v_1) &= v_1 \otimes (e_1+e_3) + v_2 \otimes (e_2+e_4)\\ &+v_3 \otimes \frac{1}{2}\left(a_{11} + \frac{1-i}{\sqrt{2}}a_{12} + \frac{1+i}{\sqrt{2}}a_{21} + a_{22}\right)\\ &+v_4 \otimes \frac{1}{2}\left(a_{11} - \frac{1-i}{\sqrt{2}}a_{12} - \frac{1+i}{\sqrt{2}}a_{21} + a_{22}\right),\end{aligned}$$

$$\begin{aligned}\beta(v_2) &= v_1 \otimes (e_2+e_4) + v_2 \otimes (e_1+e_3)\\ &+v_3 \otimes \frac{1}{2}\left(a_{11} - \frac{1-i}{\sqrt{2}}a_{12} - \frac{1+i}{\sqrt{2}}a_{21} + a_{22}\right)\\ &+v_4 \otimes \frac{1}{2}\left(a_{11} + \frac{1-i}{\sqrt{2}}a_{12} + \frac{1+i}{\sqrt{2}}a_{21} + a_{22}\right),\end{aligned}$$

$$\begin{aligned}\beta(v_3) &= v_1 \otimes \frac{1}{2}\left(a_{11} + \frac{1+i}{\sqrt{2}}a_{12} + \frac{1-i}{\sqrt{2}}a_{21} + a_{22}\right)\\ &+v_2 \otimes \frac{1}{2}\left(a_{11} - \frac{1+i}{\sqrt{2}}a_{12} - \frac{1-i}{\sqrt{2}}a_{21} + a_{22}\right)\\ &+v_3 \otimes (e_1+e_2) + v_4 \otimes (e_3+e_4),\end{aligned}$$

$$\begin{aligned}\beta(v_4) &= v_1 \otimes \frac{1}{2}\left(a_{11} - \frac{1+i}{\sqrt{2}}a_{12} - \frac{1-i}{\sqrt{2}}a_{21} + a_{22}\right)\\ &+v_2 \otimes \frac{1}{2}\left(a_{11} + \frac{1+i}{\sqrt{2}}a_{12} + \frac{1-i}{\sqrt{2}}a_{21} + a_{22}\right)\\ &+v_3 \otimes (e_3+e_4) + v_4 \otimes (e_1+e_2),\end{aligned}$$

$\mathbb{C}^4$ becomes an $\mathcal{A}$-comodule algebra.

Let ϕ be an arbitrary state on $\mathcal{A}$. It can be parametrized by $\mu_1, \mu_2, \mu_3, \mu_4, \mu_5 \geq 0$ and $x, y, z \in \mathbb{R}$ with $\mu_1+\mu_2+\mu_3+\mu_4+\mu_5 = 1$ and $x^2+y^2+z^2 \leq 1$, cf. Subsection B.3 in the Appendix. Then the transition operator $T_\phi = (\mathrm{id}\otimes\phi)\circ\Delta$ on $\mathbb{C}^4$ becomes

$$T_\phi = \begin{pmatrix} \mu_1+\mu_3 & \mu_2+\mu_4 & \frac{\mu_5}{2}\left(1+\frac{x+y}{\sqrt{2}}\right) & \frac{\mu_5}{2}\left(1-\frac{x+y}{\sqrt{2}}\right)\\ \mu_2+\mu_4 & \mu_1+\mu_3 & \frac{\mu_5}{2}\left(1-\frac{x+y}{\sqrt{2}}\right) & \frac{\mu_5}{2}\left(1+\frac{x+y}{\sqrt{2}}\right)\\ \frac{\mu_5}{2}\left(1+\frac{x-y}{\sqrt{2}}\right) & \frac{\mu_5}{2}\left(1-\frac{x-y}{\sqrt{2}}\right) & \mu_1+\mu_2 & \mu_3+\mu_4\\ \frac{\mu_5}{2}\left(1-\frac{x-y}{\sqrt{2}}\right) & \frac{\mu_5}{2}\left(1+\frac{x-y}{\sqrt{2}}\right) & \mu_3+\mu_4 & \mu_1+\mu_2 \end{pmatrix} \tag{3.1}$$

w.r.t. to the basis v_1, v_2, v_3, v_4.

The state $\psi_0 : \mathcal{B} \to \mathbb{C}$ defined by $\psi_0(v_1) = \psi_0(v_2) = \psi_0(v_3) = \psi_0(v_4) = \frac{1}{4}$ is invariant, i.e. we have

$$\psi_0 \star \phi = (\psi_0 \otimes \phi)\circ\beta = \psi_0$$

for any state ϕ on $\mathcal{A}$.

4 Random Walks on Finite Quantum Groups

The most important special case of the construction in the previous section is obtained when we choose $\mathcal{B} = \mathcal{A}$ and $\beta = \Delta$. Then we have a *random walk on the finite quantum group* $\mathcal{A}$. Let us first show that this is indeed a generalization of a left invariant random walk as discussed in the Introduction and in Section 1. Using the coassociativity of Δ we see that the transition operator $T_\phi = (\mathrm{id} \otimes \phi) \circ \Delta$ satisfies the formula

$$\Delta \circ T_\phi = (\mathrm{id} \otimes T_\phi) \circ \Delta.$$

Suppose now that $\mathcal{B} = \mathcal{A}$ consists of functions on a finite group G and $\beta = \Delta$ is the comultiplication which encodes the group multiplication, i.e.

$$\Delta(\mathbf{1}_{\{g'\}}) = \sum_{h \in G} \mathbf{1}_{\{g'h^{-1}\}} \otimes \mathbf{1}_{\{h\}} = \sum_{h \in G} \mathbf{1}_{\{h^{-1}\}} \otimes \mathbf{1}_{\{hg'\}},$$

where $\mathbf{1}_{\{g\}}$ denotes the indicator function of g. We also have

$$T_\phi(\mathbf{1}_{\{g'\}}) = \sum_{g \in G} p_{g,g'} \mathbf{1}_{\{g\}},$$

where $(p_{g,g'})$ is the transition matrix. Compare Sections 1 and 2. Inserting these formulas yields

$$(\Delta \circ T_\phi)\, \mathbf{1}_{\{g'\}} = \Delta(\sum_{g \in G} p_{g,g'} \mathbf{1}_{\{g\}}) = \sum_{h \in G} \mathbf{1}_{\{h^{-1}\}} \otimes \sum_{g \in G} p_{g,g'} \mathbf{1}_{\{hg\}},$$

$$\begin{aligned}\big[(\mathrm{id} \otimes T_\phi) \circ \Delta\big]\, \mathbf{1}_{\{g'\}} &= (\mathrm{id} \otimes T_\phi) \sum_{h \in G} \mathbf{1}_{\{h^{-1}\}} \otimes \mathbf{1}_{\{hg'\}} \\ &= \sum_{h \in G} \mathbf{1}_{\{h^{-1}\}} \otimes \sum_{g \in G} p_{hg,hg'} \mathbf{1}_{\{hg\}}.\end{aligned}$$

We conclude that $p_{g,g'} = p_{hg,hg'}$ for all $g, g', h \in G$. This is the left invariance of the random walk which was already stated in the introduction in a more probabilistic language.

For random walks on a finite quantum group there are some natural special choices for the initial distribution ψ. On the one hand, one may choose $\psi = \varepsilon$ (the counit) which in the commutative case (i.e., for a group) corresponds to starting in the unit element of the group. Then the time evolution of the distributions is given by $\varepsilon \star \phi^{\star n} = \phi^{\star n}$. In other words, we get a convolution semigroup of states.

On the other hand, stationarity of the random walk can be obtained if ψ is chosen such that

$$(\psi \otimes \phi) \circ \Delta = \psi.$$

(Note that stationarity of a random walk must be clearly distinguished from stationarity of the increments which for our definition of a random walk is automatic.) In particular we may choose the unique Haar state η of the finite quantum group $\mathcal{A}$ (see Appendix A).

Proposition 4.1. *The random walks on a finite quantum group are stationary for all choices of ϕ if and only if $\psi = \eta$.*

Proof. This follows by Proposition 3.2 together with the fact that the Haar state is characterized by its right invariance (see Appendix A). □

5 Spatial Implementation

In this section we want to represent the algebras on Hilbert spaces and obtain spatial implementations for the random walks. On a finite quantum group $\mathcal{A}$ we can introduce an inner product

$$\langle a, b \rangle = \eta(a^* b),$$

where $a, b \in \mathcal{A}$ and η is the Haar state. Because the Haar state is faithful (see Appendix A) we can think of $\mathcal{A}$ as a finite dimensional Hilbert space which we denote by $\mathcal{H}$. Further we denote by $\|\cdot\|$ the norm associated to this inner product. We consider the linear operator

$$W : \mathcal{H} \otimes \mathcal{H} \to \mathcal{H} \otimes \mathcal{H}, \quad b \otimes a \mapsto \Delta(b)(\mathbf{1} \otimes a).$$

It turns out that this operator contains all information about the quantum group and thus it is called its *fundamental operator*. We discuss some of its properties.

(a) W is unitary.

Proof. Using $(\eta \otimes \mathrm{id}) \circ \Delta = \eta(\cdot)\mathbf{1}$ it follows that

$$\begin{aligned} \|W\, b \otimes a\|^2 &= \|\Delta(b)(\mathbf{1} \otimes a)\|^2 = \eta \otimes \eta\big((\mathbf{1} \otimes a^*)\Delta(b^* b)(\mathbf{1} \otimes a)\big) \\ &= \eta\big(a^* [(\eta \otimes \mathrm{id})\Delta(b^* b)] a\big) = \eta(a^* \eta(b^* b) a) = \eta(b^* b)\, \eta(a^* a) \\ &= \eta \otimes \eta(b^* b \otimes a^* a) = \|b \otimes a\|^2. \end{aligned}$$

A similar computation works for $\sum_i b_i \otimes a_i$ instead of $b \otimes a$. Thus W is isometric and, because $\mathcal{H}$ is finite dimensional, also unitary. It can be easily checked using Sweedler's notation that with the antipode S the inverse $W^{-1} = W^*$ can be written explicitly as

$$W^{-1}(b \otimes a) = [(\mathrm{id} \otimes S)\Delta(b)](\mathbf{1} \otimes a).$$

□

(b) W satisfies the Pentagon Equation $W_{12}W_{13}W_{23} = W_{23}W_{12}$.

This is an equation on $\mathcal{H} \otimes \mathcal{H} \otimes \mathcal{H}$ and we have used the *leg notation* $W_{12} = W \otimes \mathbf{1}$, $W_{23} = \mathbf{1} \otimes W$, $W_{13} = (\mathbf{1} \otimes \tau) \circ W_{12} \circ (\mathbf{1} \otimes \tau)$, where τ is the *flip*, $\tau : \mathcal{H} \otimes \mathcal{H} \to \mathcal{H} \otimes \mathcal{H}$, $\tau(a \otimes b) = b \otimes a$.

Proof.

$$\begin{aligned} W_{12}W_{13}W_{23}\ a \otimes b \otimes c &= W_{12}W_{13}\ a \otimes b_{(1)} \otimes b_{(2)}c = W_{12}\ a_{(1)} \otimes b_{(1)} \otimes a_{(2)}b_{(2)}c \\ &= a_{(1)} \otimes a_{(2)}b_{(1)} \otimes a_{(3)}b_{(2)}c = W_{23}\ a_{(1)} \otimes a_{(2)}b \otimes c = W_{23}W_{12}\ a \otimes b \otimes c. \end{aligned}$$

□

Remark 5.1. The pentagon equation expresses the coassociativity of the comultiplication Δ. Unitaries satisfying the pentagon equation have been called *multiplicative unitaries* in [BS93].

The operator L_a of left multiplication by $a \in \mathcal{A}$ on $\mathcal{H}$

$$L_a : \mathcal{H} \to \mathcal{H}, \quad c \mapsto a\,c$$

will often simply be written as a in the following. It is always clear from the context whether $a \in \mathcal{A}$ or $a : \mathcal{H} \to \mathcal{H}$ is meant. We can also look at left multiplication as a faithful representation L of the C^*-algebra $\mathcal{A}$ on $\mathcal{H}$. In this sense we have

(c) $\quad \Delta(a) = W\ (a \otimes \mathbf{1})\ W^* \quad$ for all $a \in \mathcal{A}$

Proof. Here $\Delta(a)$ and $a \otimes \mathbf{1}$ are left multiplication operators on $\mathcal{H} \otimes \mathcal{H}$. The formula can be checked as follows.

$$\begin{aligned} W\ (a \otimes \mathbf{1})\ W^*\ b \otimes c &= W\ (a \otimes \mathbf{1})\ b_{(1)} \otimes (Sb_{(2)})c = W\ ab_{(1)} \otimes (Sb_{(2)})c \\ &= a_{(1)}b_{(1)} \otimes a_{(2)}b_{(2)}(Sb_{(3)})c = a_{(1)}b_{(1)} \otimes a_{(2)}\varepsilon(b_{(2)})c \\ &= a_{(1)}b \otimes a_{(2)}c = \Delta(a)(b \otimes c) \end{aligned}$$

□

By left multiplication we can also represent a random walk on a finite quantum group $\mathcal{A}$. Then $j_n(a)$ becomes an operator on an $(n+1)$-fold tensor product of $\mathcal{H}$. To get used to it let us show how the pentagon equation is related to our Proposition 3.1 above.

Theorem 5.2.

$$j_n(a) = W_{01}W_{02} \dots W_{0n}\ (a \otimes \mathbf{1} \otimes \dots \otimes \mathbf{1})\ W_{0n}^* \dots W_{02}^* W_{01}^*.$$

$$W_{01}W_{02} \dots W_{0n}|_{\mathcal{H}} = W_{n-1,n}W_{n-2,n-1} \dots W_{01}|_{\mathcal{H}},$$

where $|_{\mathcal{H}}$ *means restriction to* $\mathcal{H} \otimes \mathbf{1} \otimes \dots \otimes \mathbf{1}$ *and this left position gets the number zero.*

Proof. A comparison makes clear that this is nothing but Proposition 3.1 written in terms of the fundamental operator W. Alternatively, we prove the second equality by using the pentagon equation. For $n = 1$ or $n = 2$ the equation is clearly valid. Assume that it is valid for some $n \geq 2$. Then

$$W_{01} W_{02} \ldots W_{0,n-1} W_{0n} W_{0,n+1}|_{\mathcal{H}} = W_{01} W_{02} \ldots W_{0,n-1} W_{n,n+1} W_{0n}|_{\mathcal{H}}$$
$$= W_{n,n+1} W_{01} W_{02} \ldots W_{0,n-1} W_{0n}|_{\mathcal{H}} = W_{n,n+1} W_{n-1,n} \ldots W_{01}|_{\mathcal{H}}.$$

In the first line we used the pentagon equation for positions $0, n, n+1$ together with $W_{n,n+1}(\mathbf{1}\otimes\mathbf{1}) = \mathbf{1}\otimes\mathbf{1}$. In the second line we applied the fact that disjoint subscripts yield commuting operators and finally we inserted the assumption. □

It is an immediate but remarkable consequence of this representation that we have a canonical way of extending our random walk to $\mathcal{B}(\mathcal{H})$, the C^*-algebra of all (bounded) linear operators on $\mathcal{H}$. Namely, we can for $n \geq 0$ define the random variables

$$J_n : \mathcal{B}(\mathcal{H}) \to \mathcal{B}(\bigotimes_0^n \mathcal{H}) \simeq \bigotimes_0^n \mathcal{B}(\mathcal{H}),$$
$$x \mapsto W_{01} W_{02} \ldots W_{0n} \, (x \otimes \mathbf{1} \otimes \ldots \otimes \mathbf{1}) \, W_{0n}^* \ldots W_{02}^* W_{01}^*,$$

i.e., we simply insert an arbitrary operator x instead of the left multiplication operator a.

Theorem 5.3. *$(J_n)_{n\geq 0}$ is a random walk on the $\mathcal{A}$-comodule algebra $\mathcal{B}(\mathcal{H})$.*

Proof. First we show that $W \in \mathcal{B}(\mathcal{H}) \otimes \mathcal{A}$. In fact, if $x' \in \mathcal{B}(\mathcal{H})$ commutes with $\mathcal{A}$ then

$$W(\mathbf{1} \otimes x')(b \otimes a) = W(b \otimes x'a) = \Delta(b)(\mathbf{1} \otimes x'a) = \Delta(b)(\mathbf{1} \otimes x')(\mathbf{1} \otimes a)$$
$$= (\mathbf{1} \otimes x')\Delta(b)(\mathbf{1} \otimes a) = (\mathbf{1} \otimes x')W(b \otimes a).$$

Because W commutes with all $\mathbf{1}\otimes x'$ it must be contained in $\mathcal{B}(\mathcal{H}) \otimes \mathcal{A}$. (This is a special case of von Neumann's bicommutant theorem but of course the finite dimensional version used here is older and purely algebraic.) We can now define

$$\gamma : \mathcal{B}(\mathcal{H}) \to \mathcal{B}(\mathcal{H}) \otimes \mathcal{A}, \quad x \mapsto W \, (x \otimes \mathbf{1}) \, W^*,$$

and check that it is a coaction. The property $(\gamma \otimes \mathrm{id}) \circ \gamma = (\mathrm{id} \otimes \Delta) \circ \gamma$ is a consequence of the pentagon equation. It corresponds to

$$W_{01} W_{02} (x \otimes \mathbf{1} \otimes \ldots \otimes \mathbf{1}) W_{02}^* W_{01}^* = W_{01} W_{02} W_{12} (x \otimes \mathbf{1} \otimes \ldots \otimes \mathbf{1}) W_{12}^* W_{02}^* W_{01}^*$$
$$= W_{12} W_{01} \, (x \otimes \mathbf{1} \otimes \ldots \otimes \mathbf{1}) \, W_{01}^* W_{12}^*.$$

Finally we check that $(\mathrm{id} \otimes \varepsilon) \circ \gamma = \mathrm{id}$. In fact,

$$\gamma(x)(b \otimes a) = W(x \otimes \mathbf{1})W^*(b \otimes a) = W(x \otimes \mathbf{1}) \, b_{(1)} \otimes (S b_{(2)}) \, a$$

$$= [x(b_{(1)})]_{(1)} \otimes [x(b_{(1)})]_{(2)} \, (Sb_{(2)}) \, a$$

and thus

$$[(\mathrm{id} \otimes \varepsilon)\gamma(x)](b) = [x(b_{(1)})]_{(1)} \, \varepsilon([x(b_{(1)})]_{(2)}) \, \varepsilon(Sb_{(2)})$$
$$= x(b_{(1)}) \, \varepsilon(b_{(2)}) = x(b_{(1)} \, \varepsilon(b_{(2)})) = x(b),$$

i.e., $(\mathrm{id} \otimes \varepsilon)\gamma(x) = x$. Here we used $(\mathrm{id} \otimes \varepsilon) \circ \Delta = \mathrm{id}$ and the fact that $\varepsilon \circ S = \varepsilon$. □

Remark 5.4. The Haar state η on $\mathcal{A}$ is extended to a vector state on $\mathcal{B}(\mathcal{H})$ given by $\mathbf{1} \in \mathcal{H}$. Thus we have also an extension of the probabilistic features of the random walk. Note further that arbitrary states on $\mathcal{A}$ can always be extended to vector states on $\mathcal{B}(\mathcal{H})$ (see Appendix A). This means that we also find the random walks with arbitrary initial state ψ and arbitrary transition state ϕ represented on tensor products of the Hilbert space $\mathcal{H}$ and we have extensions also for them. This is an important remark because for many random walks of interest we would like to start in $\psi = \varepsilon$ and all the possible steps of the walk are small, i.e., ϕ is not a faithful state.

Remark 5.5. It is not possible to give $\mathcal{B}(\mathcal{H})$ the structure of a quantum group. For example, there cannot be a counit because $\mathcal{B}(\mathcal{H})$ as a simple algebra does not have nontrivial multiplicative linear functionals. Thus $\mathcal{B}(\mathcal{H})$ must be treated here as a $\mathcal{A}$-comodule algebra.

In fact, it is possible to generalize all these results and to work with coactions on $\mathcal{A}$-comodule algebras from the beginning. Let $\beta : \mathcal{B} \to \mathcal{B} \otimes \mathcal{A}$ be such a coaction. For convenience we continue to use the Haar state η on $\mathcal{A}$ and assume that there is a faithful stationary state ψ on $\mathcal{B}$. As before we can consider $\mathcal{A}$ as a Hilbert space $\mathcal{H}$ and additionally we have on $\mathcal{B}$ an inner product induced by ψ which yields a Hilbert space $\mathcal{K}$. By modifying the arguments above the reader should have no problems to verify the following assertions. Their proof is thus left as an exercise.

Define $V : \mathcal{K} \otimes \mathcal{H} \to \mathcal{K} \otimes \mathcal{H}$ by $b \otimes a \mapsto \beta(b)(\mathbf{1} \otimes a)$. Using Proposition 3.2, one can show that the stationarity of ψ implies that V is unitary. The map V satisfies $V_{12}V_{13}W_{23} = W_{23}V_{12}$ (with leg notation on $\mathcal{K} \otimes \mathcal{H} \otimes \mathcal{H}$) and the inverse can be written explicitly as $V^{-1}(b \otimes a) = [(\mathrm{id} \otimes S)\beta(b)](\mathbf{1} \otimes a)$. In [Wo96] such a unitary V is called *adapted* to W. We have $\beta(b) = V\,(b \otimes \mathbf{1})\,V^*$ for all $b \in \mathcal{B}$. The associated random walk $(j_n)_{n \geq 0}$ on $\mathcal{B}$ can be implemented by

$$j_n(b) = V_{01}V_{02} \ldots V_{0n} \; (b \otimes \mathbf{1} \otimes \ldots \otimes \mathbf{1}) \; V_{0n}^* \ldots V_{02}^* V_{01}^*$$

with

$$V_{01}V_{02} \ldots V_{0n}|_{\mathcal{K}} = W_{n-1,n}W_{n-2,n-1} \ldots W_{12}V_{01}|_{\mathcal{K}}.$$

These formulas can be used to extend this random walk to a random walk $(J_n)_{n \geq 0}$ on $\mathcal{B}(\mathcal{K})$.

Remark 5.6. There is an extended transition operator $Z : \mathcal{B}(\mathcal{K}) \to \mathcal{B}(\mathcal{K})$ corresponding to the extension of the random walk. It can be described explicitly as follows. Define an isometry

$$v : \mathcal{K} \to \mathcal{K} \otimes \mathcal{H}, \quad b \mapsto V^*(b \otimes \mathbf{1}) = b_{(0)} \otimes Sb_{(1)}.$$

Then we have

$$Z : \mathcal{B}(\mathcal{K}) \to \mathcal{B}(\mathcal{K}), \quad x \mapsto v^* \, x \otimes \mathbf{1} \, v.$$

Because v is isometric, Z is a unital completely positive map which extends T_η. Such extended transition operators are discussed in the general frame of quantum Markov chains in [Go04]. See also [GKL04] for applications in noncommutative coding.

What is the meaning of these extensions? We think that this is an interesting question which leads to a promising direction of research. Let us indicate an interpretation in terms of quantization.

First we quickly review some facts which are discussed in more detail for example in [Maj95]. On $\mathcal{A}$ we have an action T of its dual $\mathcal{A}^*$ which sends $\phi \in \mathcal{A}^*$ to

$$T_\phi : \mathcal{A} \to \mathcal{A}, \quad a \mapsto a_{(0)} \, \phi(a_{(1)}).$$

Note that if ϕ is a state then T_ϕ is nothing but the transition operator considered earlier. It is also possible to consider T as a representation of the (convolution) algebra $\mathcal{A}^*$ on $\mathcal{H}$ which is called the regular representation. We can now form the crossed product $\mathcal{A} \rtimes \mathcal{A}^*$ which as a vector space is $\mathcal{A} \otimes \mathcal{A}^*$ and becomes an algebra with the multiplication

$$(c \otimes \phi)(d \otimes \psi) = c \, T_{\phi_{(1)}}(d) \ \otimes \ \phi_{(2)} \star \psi,$$

where $\Delta\phi = \phi_{(1)} \otimes \phi_{(2)} \in \mathcal{A}^* \otimes \mathcal{A}^* \cong (\mathcal{A} \otimes \mathcal{A})^*$ is defined by $\Delta\phi(a \otimes b) = \phi(ab)$ for $a, b \in \mathcal{A}$.

There is a representation $\mathcal{S}$ of $\mathcal{A} \rtimes \mathcal{A}^*$ on $\mathcal{H}$ called the Schrödinger representation and given by

$$\mathcal{S}(c \otimes \phi) = L_c \, T_\phi.$$

Note further that the representations L and T are contained in $\mathcal{S}$ by choosing $c \otimes \varepsilon$ and $\mathbf{1} \otimes \phi$.

Theorem 5.7.

$$\mathcal{S}(\mathcal{A} \otimes \mathcal{A}^*) = \mathcal{B}(\mathcal{H}).$$

If $(c_i), (\phi_i)$ are dual bases in $\mathcal{A}, \mathcal{A}^*$, then the fundamental operator W can be written as

$$W = \sum_i T_{\phi_i} \otimes L_{c_i}$$

Proof. See [Maj95], 6.1.6. Note that this once more implies $W \in \mathcal{B}(\mathcal{H}) \otimes \mathcal{A}$ which was used earlier. □

We consider an example. For a finite group G both $\mathcal{A}$ and $\mathcal{A}^*$ can be realized by the vector space of complex functions on G, but in the first case we have pointwise multiplication while in the second case we need convolution, i.e., indicator functions $\mathbf{1}_{\{g\}}$ for $g \in G$ are multiplied according to the group rule and for general functions the multiplication is obtained by linear extension. These indicator functions provide dual bases as occurring in the theorem and we obtain

$$W = \sum_{g \in G} T_g \otimes L_g,$$

where

$$\begin{aligned} L_g &:= L_{\mathbf{1}_{\{g\}}} : \mathbf{1}_{\{h\}} \mapsto \delta_{g,h}\, \mathbf{1}_{\{h\}}, \\ T_g &:= T_{\mathbf{1}_{\{g\}}} : \mathbf{1}_{\{h\}} \mapsto \mathbf{1}_{\{hg^{-1}\}}. \end{aligned}$$

The reader may rediscover here the map $b : M \times G \to M$ (for $M = G$) discussed in the beginning of the Sections 2 and 3. It is also instructive to check the pentagon equation directly.

$$\begin{aligned} W_{12}W_{13}W_{23} &= \sum_{a,b,c} (T_a \otimes L_a \otimes \mathbf{1})(T_b \otimes \mathbf{1} \otimes L_b)(\mathbf{1} \otimes T_c \otimes L_c) \\ &= \sum_{a,b,c} T_a T_b \otimes L_a T_c \otimes L_b L_c = \sum_{a,c} T_a T_c \otimes L_a T_c \otimes L_c \\ &= \sum_{a,c} T_{ac} \otimes L_a T_c \otimes L_c = \sum_{a,c} T_a \otimes L_{ac^{-1}} T_c \otimes L_c, \end{aligned}$$

where the last equality is obtained by the substitution $a \leftrightarrow ac^{-1}$. This coincides with

$$W_{23}W_{12} = \sum_{a,c} (\mathbf{1} \otimes T_c \otimes L_c)(T_a \otimes L_a \otimes \mathbf{1}) = \sum_{a,c} T_a \otimes T_c L_a \otimes L_c$$

precisely because of the relations

$$T_c\, L_a = L_{ac^{-1}}\, T_c \quad \text{for all } a, c \in G.$$

This is a version of the *canonical commutation relations*. In quantum mechanics, for $G = \mathbb{R}$, they encode Heisenberg's uncertainty principle. This explains why $\mathcal{S}$ is called a Schrödinger representation. Its irreducibility in the case $G = \mathbb{R}$ is a well-known theorem. For more details see [Maj95, Chapter 6.1].

Thus Theorem 5.7 may be interpreted as a generalization of these facts to quantum groups. Our purpose here has been to give an interpretation of the extension of random walks to $\mathcal{B}(\mathcal{H})$ in terms of quantization. Indeed, we see that $\mathcal{B}(\mathcal{H})$ can be obtained as a crossed product, and similarly as in Heisenberg's situation where the algebra $\mathcal{B}(\mathcal{H})$ occurs by appending to the observable of position a noncommuting observable of momentum, in our case we get $\mathcal{B}(\mathcal{H})$ by appending to the original algebra of observables all the transition operators of potential random walks.

6 Classical Versions

In this section we will show how one can recover a classical Markov chain from a quantum Markov chain. We will apply a folklore theorem that says that one gets a classical Markov process, if a quantum Markov process can be restricted to a commutative algebra, cf. [AFL82, Küm88, BP95, Bia98, BKS97].

For random walks on quantum groups we have the following result.

Theorem 6.1. *Let $\mathcal{A}$ be a finite quantum group, $(j_n)_{n\geq 0}$ a random walk on a finite dimensional $\mathcal{A}$-comodule algebra $\mathcal{B}$, and $\mathcal{B}_0$ a unital abelian sub-*-algebra of $\mathcal{B}$. The algebra $\mathcal{B}_0$ is isomorphic to the algebra of functions on a finite set, say $\mathcal{B}_0 \cong \mathbb{C}^{\{1,\dots,d\}}$.*

If the transition operator T_ϕ of $(j_n)_{n\geq 0}$ leaves $\mathcal{B}_0$ invariant, then there exists a classical Markov chain $(X_n)_{n\geq 0}$ with values in $\{1,\dots,d\}$, whose probabilities can be computed as time-ordered moments of $(j_n)_{n\in\mathbb{N}}$, i.e.,

$$P(X_0 = i_0, \dots, X_\ell = i_\ell) = \Psi\big(j_0(\mathbf{1}_{\{i_0\}}) \cdots j_\ell(\mathbf{1}_{\{i_\ell\}})\big) \tag{6.1}$$

for all $\ell \geq 0$ and $i_0, \dots, i_\ell \in \{1, \dots, d\}$.

Proof. We use the indicator functions $\mathbf{1}_{\{1\}}, \dots, \mathbf{1}_{\{d\}}$,

$$\mathbf{1}_{\{i\}}(j) = \delta_{ij}, \qquad 1 \leq i, j, \leq d,$$

as a basis for $\mathcal{B}_0 \subseteq \mathcal{B}$. They are positive, therefore $\lambda_1 = \Psi\big(j_0(\mathbf{1}_{\{1\}})\big), \dots, \lambda_d = \Psi\big(j_0(\mathbf{1}_{\{d\}})\big)$ are non-negative. Since furthermore

$$\lambda_1 + \dots + \lambda_d = \Psi\big(j_0(\mathbf{1}_{\{1\}})\big) + \dots + \Psi\big(j_0(\mathbf{1}_{\{d\}})\big) = \Psi\big(j_0(\mathbf{1})\big) = \Psi(\mathbf{1}) = 1,$$

these numbers define a probability measure on $\{1, \dots, d\}$.

Define now $(p_{ij})_{1\leq i,j\leq d}$ by

$$T_\phi(\mathbf{1}_{\{j\}}) = \sum_{i=1}^{d} p_{ij} \mathbf{1}_{\{i\}}.$$

Since $T_\phi = (\mathrm{id}\otimes\phi)\circ\beta$ is positive, we have $p_{ij} \geq 0$ for $1 \leq i, j \leq d$. Furthermore, $T_\phi(\mathbf{1}) = \mathbf{1}$ implies

$$\mathbf{1} = T_\phi(\mathbf{1}) = T_\phi\left(\sum_{j=1}^{d} \mathbf{1}_{\{j\}}\right) = \sum_{j=1}^{d}\sum_{i=1}^{d} p_{ij}\mathbf{1}_{\{i\}}$$

i.e. $\sum_{j=1}^{d} p_{ij} = 1$ and so $(p_{ij})_{1\leq i,j\leq d}$ is a stochastic matrix.

Therefore there exists a unique Markov chain $(X_n)_{n\geq 0}$ with initial distribution $(\lambda_i)_{1\leq i\leq d}$ and transition matrix $(p_{ij})_{1\leq i,j\leq d}$.

We show by induction that Equation (6.1) holds.

For $\ell = 0$ this is clear by definition of $\lambda_1, \ldots, \lambda_d$. Let now $\ell \geq 1$ and $i_0, \ldots, i_\ell \in \{1, \ldots, d\}$. Then we have

$$\begin{aligned}
\Psi\big(j_0(\mathbf{1}_{\{i_0\}})\cdots j_\ell(\mathbf{1}_{\{i_\ell\}})\big) &= \Psi\big(j_0(\mathbf{1}_{\{i_0\}})\cdots j_{\ell-1}(\mathbf{1}_{\{i_{\ell-1}\}})j_{\ell-1}(\mathbf{1}_{\{i_\ell\}(1)})k_\ell(\mathbf{1}_{\{i_\ell\}(2)})\big) \\
&= \Psi\big(j_0(\mathbf{1}_{\{i_0\}})\cdots j_{\ell-1}(\mathbf{1}_{\{i_{\ell-1}\}}\mathbf{1}_{\{i_\ell\}(1)})\big)\phi(\mathbf{1}_{\{i_\ell\}(2)}) \\
&= \Psi\Big(j_0(\mathbf{1}_{\{i_0\}})\cdots j_{\ell-1}\big(\mathbf{1}_{\{i_{\ell-1}\}}T_\phi(\mathbf{1}_{\{i_\ell\}})\big)\Big) \\
&= \Psi\big(j_0(\mathbf{1}_{\{i_0\}})\cdots j_{\ell-1}(\mathbf{1}_{\{i_{\ell-1}\}})\big)p_{i_{\ell-1},i_\ell} \\
&= \lambda_{i_0}p_{i_0 i_1}\cdots p_{i_{\ell-1}i_\ell} \\
&= P(X_0 = i_0, \ldots, X_\ell = i_\ell),
\end{aligned}$$

by Proposition 1.2. □

Remark 6.2. If the condition that T_ϕ leaves $\mathcal{A}_0$ invariant is dropped, then one can still compute the "probabilities"

$$\begin{aligned}
\text{“}P(X_0 = i_0, \ldots, X_\ell = i_\ell)\text{”} &= \Psi\big(j_0(\mathbf{1}_{\{i_0\}})\cdots j_\ell(\mathbf{1}_{\{i_\ell\}})\big) \\
&= \Psi\Big(P_{[0,\ell-1]}\big(j_0(\mathbf{1}_{\{i_0\}})\cdots j_\ell(\mathbf{1}_{\{i_\ell\}})\big)\Big) \\
&= \Psi\Big(j_0(\mathbf{1}_{\{i_0\}})\cdots j_{\ell-1}(\mathbf{1}_{\{i_{\ell-1}\}})j_{\ell-1}\big(T_\phi(\mathbf{1}_{\{i_\ell\}})\big)\Big) \\
&= \Psi\Big(j_0(\mathbf{1}_{\{i_0\}})\cdots j_{\ell-1}\big(\mathbf{1}_{\{i_{\ell-1}\}}T_\phi(\mathbf{1}_{\{i_\ell\}})\big)\Big) \\
&= \cdots \\
&= \psi\Big(\mathbf{1}_{\{i_0\}}T_\phi\big(\mathbf{1}_{\{i_1\}}T_\phi(\cdots \mathbf{1}_{\{i_{\ell-1}\}}T_\phi(\mathbf{1}_{\{i_\ell\}})\cdots)\big)\Big),
\end{aligned}$$

but in general they are no longer positive or even real, and so it is impossible to construct a classical stochastic process $(X_n)_{n\geq 0}$ from them. We give an example where no classical process exists in Example 6.4.

Example 6.3. The comodule algebra $\mathcal{B} = \mathbb{C}^4$ that we considered in Example 3.3 is abelian, so we can take $\mathcal{B}_0 = \mathcal{B}$. For any pair of a state ψ on $\mathcal{B}$ and a state ϕ on $\mathcal{A}$, we get a random walk on $\mathcal{B}$ and a corresponding Markov chain $(X_n)_{n\geq 0}$ on $\{1,2,3,4\}$. We identify $\mathbb{C}^{\{1,2,3,4\}}$ with $\mathcal{B}$ by $v_i \equiv \mathbf{1}_{\{i\}}$ for $i = 1,2,3,4$.

The initial distribution of $(X_n)_{n\geq 0}$ is given by $\lambda_i = \psi(v_i)$ and the transition matrix is given in Equation (3.1).

Example 6.4. . Let us now consider random walks on the Kac-Paljutkin quantum group $\mathcal{A}$ itself. For the defining relations, the calculation of the dual of $\mathcal{A}$ and a parametrization of all states on $\mathcal{A}$, see Appendix B. Let us consider here transition states of the form

$$\phi = \mu_1\eta_1 + \mu_2\eta_2 + \mu_3\eta_3 + \mu_4\eta_4,$$

with $\mu_1, \mu_2, \mu_3, \mu_4 \in [0,1]$, $\mu_1 + \mu_2 + \mu_3 + \mu_4 = 1$.

The transition operators $T_\phi = (\mathrm{id}\otimes\phi)\circ\Delta$ of these states leave the abelian subalgebra $\mathcal{A}_0 = \mathrm{span}\,\{e_1, e_2, e_3, e_4\} \cong \mathbb{C}^4$ invariant. The transition matrix of the associated classical Markov chain on $\{1,2,3,4\}$ that arises by identifying $e_i \equiv \mathbf{1}_{\{i\}}$ for $i = 1,2,3,4$ has the form

$$\begin{pmatrix} \mu_1 & \mu_2 & \mu_3 & \mu_4 \\ \mu_2 & \mu_1 & \mu_4 & \mu_3 \\ \mu_3 & \mu_4 & \mu_1 & \mu_2 \\ \mu_4 & \mu_3 & \mu_2 & \mu_1 \end{pmatrix}.$$

This is actually the transition matrix of a random walk on the group $\mathbb{Z}_2 \times \mathbb{Z}_2$.

The subalgebra $\mathrm{span}\,\{a_{11}, a_{12}, a_{21}, a_{22}\} \cong M_2$ is also invariant under these states, T_ϕ acts on it by

$$T_\phi(X) = \mu_1 X + \mu_2 V_2^* X V_2 + \mu_3 V_3^* X V_3 + \mu_4 V_4^* X V_4$$

for $X = aa_{11} + ba_{12} + ca_{21} + da_{22} \cong \begin{pmatrix} a & b \\ c & d \end{pmatrix}$, $a,b,c,d \in \mathbb{C}$, with

$$V_2 = \begin{pmatrix} 0 & i \\ 1 & 0 \end{pmatrix}, \qquad V_3 = \begin{pmatrix} 0 & -i \\ 1 & 0 \end{pmatrix}, \qquad V_4 = \begin{pmatrix} 1 & 0 \\ 0 & -1 \end{pmatrix}.$$

Let $u = \begin{pmatrix} \cos\vartheta \\ e^{i\delta}\sin\vartheta \end{pmatrix}$ be a unit vector in $\mathbb{C}$ and denote by p_u the orthogonal projection onto u. The maximal abelian subalgebra $\mathcal{A}_u = \mathrm{span}\,\{p_u, \mathbf{1} - p_u\}$ in $M_2 \subset \mathcal{A}$ is in general not invariant under T_ϕ.

E.g., for $u = \frac{1}{\sqrt{2}}\begin{pmatrix} 1 \\ 1 \end{pmatrix}$ we get the algebra $\mathcal{A}_u = \mathrm{span}\left\{\begin{pmatrix} a & b \\ b & a \end{pmatrix}\middle| a,b\in\mathbb{C}\right\}$. It can be identified with $\mathbb{C}^{\{1,2\}}$ via $\begin{pmatrix} a & b \\ b & a \end{pmatrix} \equiv (a+b)\mathbf{1}_{\{1\}} + (a-b)\mathbf{1}_{\{2\}}$.

Specializing to the transition state $\phi = \eta_2$ and starting from the Haar measure $\psi = \eta$, we see that the time-ordered joint moment

$$\begin{aligned} \Psi\big(j_0(\mathbf{1}_{\{1\}})j_1(\mathbf{1}_{\{1\}})j_2(\mathbf{1}_{\{2\}})j_3(\mathbf{1}_{\{2\}})\big) &= \eta\Big(\mathbf{1}_{\{1\}}T_{\eta_2}\big(\mathbf{1}_{\{1\}}T_{\eta_2}\big(\mathbf{1}_{\{2\}}T_{\eta_2}(\mathbf{1}_{\{2\}})\big)\big)\Big) \\ &= \frac{1}{4}\mathrm{Tr}\left(\begin{pmatrix} \frac12 & \frac12 \\ \frac12 & \frac12 \end{pmatrix} V_2^* \begin{pmatrix} \frac12 & \frac12 \\ \frac12 & \frac12 \end{pmatrix} V_2^* \begin{pmatrix} \frac12 & -\frac12 \\ -\frac12 & \frac12 \end{pmatrix} V_2^* \begin{pmatrix} \frac12 & -\frac12 \\ -\frac12 & \frac12 \end{pmatrix} V_2^3\right) \\ &= \frac{1}{4}\mathrm{Tr}\begin{pmatrix} -\frac{1+i}{8} & \frac{-1+i}{8} \\ -\frac{1+i}{8} & \frac{-1+i}{8} \end{pmatrix} = -\frac{1}{16} \end{aligned}$$

is negative and can not be obtained from a classical Markov chain.

Example 6.5. For states in $\mathrm{span}\,\{\eta_1, \eta_2, \eta_3, \eta_4, \alpha_{11} + \alpha_{22}\}$, the center $Z(\mathcal{A}) = \mathrm{span}\,\{e_1, e_2, e_3, e_4, a_{11} + a_{22}\}$ of $\mathcal{A}$ is invariant under T_ϕ, see also [NT04, Proposition 2.1]. A state on $\mathcal{A}$, parametrized as in Equation (B.1), belongs to this set if and only if $x = y = z = 0$. With respect to the basis $e_1, e_2, e_3, e_4, a_{11} + a_{22}$ of $Z(\mathcal{A})$ we get

$$T_\phi|_{Z(\mathcal{A})} = \begin{pmatrix} \mu_1 & \mu_2 & \mu_3 & \mu_4 & \mu_5 \\ \mu_2 & \mu_1 & \mu_4 & \mu_3 & \mu_5 \\ \mu_3 & \mu_4 & \mu_1 & \mu_2 & \mu_5 \\ \mu_4 & \mu_3 & \mu_2 & \mu_1 & \mu_5 \\ \frac{\mu_5}{4} & \frac{\mu_5}{4} & \frac{\mu_5}{4} & \frac{\mu_5}{4} & 1-\mu_5 \end{pmatrix}$$

for the transition matrix of the classical Markov process that has the same time-ordered joint moments.

For Lévy processes or random walks on quantum groups there exists another way to prove the existence of a classical version that does not use the Markov property. We will illustrate this on an example.

Example 6.6. We consider restrictions to the center $Z(\mathcal{A})$ of $\mathcal{A}$. If $a \in Z(\mathcal{A})$, then $a \otimes \mathbf{1} \in Z(\mathcal{A} \otimes \mathcal{A})$ and therefore

$$[a \otimes \mathbf{1}, \Delta(b)] = 0 \qquad \text{for all } a, b \in Z(\mathcal{A}).$$

This implies that the range of the restriction $(j_n|_{Z(\mathcal{A})})_{n\geq 0}$ of any random walk on $\mathcal{A}$ to $Z(\mathcal{A})$ is commutative, i.e.

$$\begin{aligned} &\big[j_\ell(a), j_n(b)\big] \\ &= \big[(j_0 \star k_1 \star \cdots \star k_\ell)(a), (j_0 \star k_1 \star \cdots \star k_n)(b)\big] \\ &= \big[(j_0 \star k_1 \star \cdots \star k_\ell)(a), (j_0 \star k_1 \star \cdots \star k_\ell)(b_{(1)})(k_{\ell+1} \star \cdots k_n)(b_{(2)})\big] \\ &= m\big(j_\ell \otimes (k_{\ell+1} \star \cdots \star k_n)([a \otimes \mathbf{1}, \Delta(b)])\big) = 0 \end{aligned}$$

for all $0 \leq \ell \leq n$ and $a, b \in Z(\mathcal{A})$. Here m denotes the multiplication, $m : \mathcal{A} \otimes \mathcal{A} \to \mathcal{A}$, $m(a \otimes b) = ab$ for $a, b \in \mathcal{A}$. Therefore the restriction $(j_n|_{Z(\mathcal{A})})_{n\geq 0}$ corresponds to a classical process, see also [Sch93, Proposition 4.2.3] and [Fra99, Theorem 2.1].

Let us now take states for which T_ϕ does not leave the center of $\mathcal{A}$ invariant, e.g. $\mu_1 = \mu_2 = \mu_3 = \mu_4 = x = y = 0$, $\mu_5 = 1$, $z \in [-1, 1]$, i.e.

$$\phi_z = \frac{1+z}{2}\alpha_{11} + \frac{1-z}{2}\alpha_{22}.$$

In this particular case we have the invariant commutative subalgebra $\mathcal{A}_0 = \text{span}\{e_1, e_2, e_3, e_4, a_{11}, a_{22}\}$ which contains the center $Z(\mathcal{A})$. If we identity $\mathcal{A}_0$ with $\mathbb{C}^{\{1,\ldots,6\}}$ via $e_1 \equiv \mathbf{1}_{\{1\}}, \ldots, e_4 \equiv \mathbf{1}_{\{4\}}, a_{11} \equiv \mathbf{1}_{\{5\}}, a_{22} \equiv \mathbf{1}_{\{6\}}$, then the transition matrix of the associated classical Markov chain is

$$\begin{pmatrix} 0 & 0 & 0 & 0 & \frac{1+z}{2} & \frac{1-z}{2} \\ 0 & 0 & 0 & 0 & \frac{1-z}{2} & \frac{1+z}{2} \\ 0 & 0 & 0 & 0 & \frac{1-z}{2} & \frac{1+z}{2} \\ 0 & 0 & 0 & 0 & \frac{1+z}{2} & \frac{1-z}{2} \\ \frac{1+z}{4} & \frac{1-z}{4} & \frac{1-z}{4} & \frac{1+z}{4} & 0 & 0 \\ \frac{1-z}{4} & \frac{1+z}{4} & \frac{1+z}{4} & \frac{1-z}{4} & 0 & 0 \end{pmatrix}.$$

The classical process corresponding to the center $Z(\mathcal{A})$ arises from this Markov chain by "gluing" the two states 5 and 6 into one. More precisely, if $(X_n)_{n\geq 0}$ is a Markov chain that has the same time-ordered moments as $(j_n)_{n\geq 0}$ restricted to $\mathcal{A}_0$, and if $g : \{1,\ldots,6\} \to \{1,\ldots,5\}$ is the mapping defined by $g(i) = i$ for $i = 1,\ldots,5$ and $g(6) = 5$, then $(Y_n)_{n\geq 0}$ with $Y_n = g(X_n)$, for $n \geq 0$, has the same joint moments as $(j_n)_{n\geq 0}$ restricted to the center $Z(\mathcal{A})$ of $\mathcal{A}$. Note that $(Y_n)_{n\geq 0}$ is not a Markov process.

7 Asymptotic Behavior

Theorem 7.1. *Let ϕ be a state on a finite quantum group $\mathcal{A}$. Then the Cesaro mean*

$$\phi_n = \frac{1}{n}\sum_{k=1}^{n} \phi^{\star n}, \qquad n \in \mathbb{N}$$

converges to an idempotent state on $\mathcal{A}$, i.e. to a state ϕ_∞ such that $\phi_\infty \star \phi_\infty = \phi_\infty$.

Proof. Let ϕ' be an accumulation point of $(\phi_n)_{n\geq 0}$, this exists since the states on $\mathcal{A}$ form a compact set. We have

$$||\phi_n - \phi \star \phi_n|| = \frac{1}{n}||\phi - \phi^{\star n+1}|| \leq \frac{2}{n}.$$

and choosing a sequence $(n_k)_{k\geq 0}$ such that $\phi_{n_k} \to \phi'$, we get $\phi \star \phi' = \phi'$ and similarly $\phi' \star \phi = \phi'$. By linearity this implies $\phi_n \star \phi' = \phi' = \phi' \star \phi_n$. If ϕ'' is another accumulation point of (ϕ_n) and $(m_\ell)_{\ell\geq 0}$ a sequence such that $\phi_{m_\ell} \to \phi''$, then we get $\phi'' \star \phi' = \phi' = \phi' \star \phi''$ and thus $\phi' = \phi''$ by symmetry. Therefore the sequence (ϕ_n) has a unique accumulation point, i.e., it converges. □

Remark 7.2. If ϕ is faithful, then the Cesaro limit ϕ_∞ is the Haar state on $\mathcal{A}$.

Remark 7.3. Due to "cyclicity" the sequence $(\phi^{\star n})_{n\in\mathbb{N}}$ does not converge in general. Take, e.g., the state $\phi = \eta_2$ on the Kac-Paljutkin quantum group $\mathcal{A}$, then we have

$$\eta_2^{\star n} = \begin{cases} \eta_2 & \text{if } n \text{ is odd,} \\ \varepsilon & \text{if } n \text{ is even,} \end{cases}$$

but

$$\lim_{n\to\infty} \frac{1}{n}\sum_{k=1}^{n} \eta_2^k = \frac{\varepsilon + \eta_2}{2}.$$

Example 7.4. Pal[Pal96] has shown that there exist exactly the following eight idempotent states on the Kac-Paljutkin quantum group [KP66],

$$\begin{aligned}
\rho_1 &= \eta_1 = \varepsilon, \\
\rho_2 &= \frac{1}{2}(\eta_1 + \eta_2), \\
\rho_3 &= \frac{1}{2}(\eta_1 + \eta_3), \\
\rho_4 &= \frac{1}{2}(\eta_1 + \eta_4), \\
\rho_5 &= \frac{1}{4}(\eta_1 + \eta_2 + \eta_3 + \eta_4), \\
\rho_6 &= \frac{1}{4}(\eta_1 + \eta_4) + \frac{1}{2}\alpha_{11}, \\
\rho_7 &= \frac{1}{4}(\eta_1 + \eta_4) + \frac{1}{2}\alpha_{22}, \\
\rho_8 &= \frac{1}{8}(\eta_1 + \eta_2 + \eta_3 + \eta_4) + \frac{1}{4}(\alpha_{11} + \alpha_{22}) = \eta.
\end{aligned}$$

On locally compact groups idempotent probability measures are Haar measures on some compact subgroup, cf. [Hey77, 1.5.6]. But Pal has shown that ρ_6 and ρ_7 are not Haar states on some "quantum sub-group" of $\mathcal{A}$.

To understand this, we compute the null spaces $\mathcal{N}_\rho = \{a | \rho(a^*a) = 0\}$ for the idempotent states. We get

$$\begin{aligned}
\mathcal{N}_\varepsilon &= \text{span}\{e_2, e_3, e_4, a_{11}, a_{12}, a_{21}, a_{22}\}, \\
\mathcal{N}_{\rho_2} &= \text{span}\{e_3, e_4, a_{11}, a_{12}, a_{21}, a_{22}\}, \\
\mathcal{N}_{\rho_3} &= \text{span}\{e_2, e_4, a_{11}, a_{12}, a_{21}, a_{22}\}, \\
\mathcal{N}_{\rho_4} &= \text{span}\{e_2, e_3, a_{11}, a_{12}, a_{21}, a_{22}\}, \\
\mathcal{N}_{\rho_5} &= \text{span}\{a_{11}, a_{12}, a_{21}, a_{22}\}, \\
\mathcal{N}_{\rho_6} &= \text{span}\{e_2, e_3, a_{12}, a_{22}\}, \\
\mathcal{N}_{\rho_7} &= \text{span}\{e_2, e_3, a_{11}, a_{21}\}, \\
\mathcal{N}_\eta &= \{0\}.
\end{aligned}$$

All null spaces of idempotent states are coideals. $\mathcal{N}_\varepsilon, \mathcal{N}_{\rho_2}, \mathcal{N}_{\rho_3}, \mathcal{N}_{\rho_4}, \mathcal{N}_{\rho_5}, \mathcal{N}_\eta$ are even Hopf ideals, so that we can obtain new quantum groups by dividing out these null spaces. The idempotent states $\varepsilon, \rho_2, \rho_3, \rho_4, \rho_5, \eta$ are equal to the composition of the canonical projection onto this quotient and the Haar state of the quotient. In this sense they can be understood as Haar states on quantum subgroups of $\mathcal{A}$. We obtain the following quantum groups,

$$\begin{aligned}
\mathcal{A}/\mathcal{N}_\varepsilon &\cong \mathbb{C} \cong \text{functions on the trivial group}, \\
\mathcal{A}/\mathcal{N}_{\rho_2} \cong \mathcal{A}/\mathcal{N}_{\rho_3} \cong \mathcal{A}/\mathcal{N}_{\rho_4} &\cong \text{functions on the group } \mathbb{Z}_2, \\
\mathcal{A}/\mathcal{N}_{\rho_5} &\cong \text{functions on the group } \mathbb{Z}_2 \times \mathbb{Z}_2, \\
\mathcal{A}/\mathcal{N}_\eta &\cong \mathcal{A}.
\end{aligned}$$

But the null spaces of ρ_6 and ρ_7 are only coideals and left ideals. Therefore the quotients $\mathcal{A}/\mathcal{N}_{\rho_6}$ and $\mathcal{A}/\mathcal{N}_{\rho_7}$ inherit only a $\mathcal{A}$-module coalgebra structure, but no quantum group structure, and ρ_6, ρ_7 can not be interpreted as Haar states on some quantum subgroup of $\mathcal{A}$, cf. [Pal96].

We define an order for states on $\mathcal{A}$ by

$$\phi_1 \preceq \phi_2 \qquad \Leftrightarrow \qquad \mathcal{N}_{\phi_1} \subseteq \mathcal{N}_{\phi_2}.$$

The resulting lattice structure for the idempotent states on $\mathcal{A}$ can be represented by the following Hasse diagram,

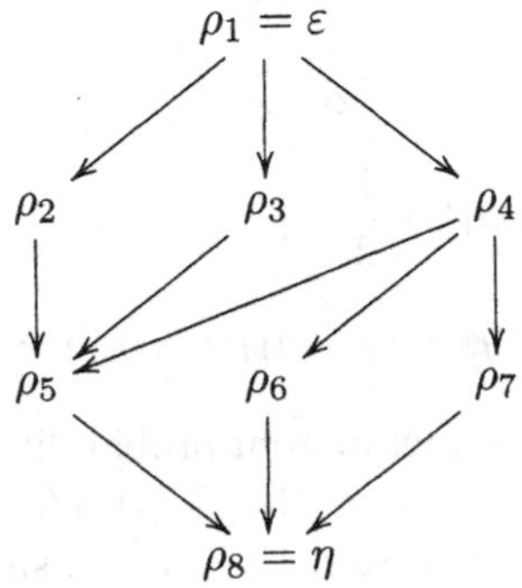

Note that the convolution product of two idempotent states is equal to their greatest lower bound in this lattice, $\rho_i \star \rho_j = \rho_i \wedge \rho_j$ for $i, j, \in \{1, \ldots, 8\}$.

A Finite Quantum Groups

In this section we briefly summarize the facts on finite quantum groups that are used throughout the main text. For proofs and more details, see [KP66, Maj95, VD97].

Recall that a *bialgebra* is a unital associative algebra $\mathcal{A}$ equipped with two unital algebra homomorphisms $\varepsilon : \mathcal{A} \to \mathbb{C}$ and $\Delta : \mathcal{A} \to \mathcal{A} \otimes \mathcal{A}$ such that

$$(\mathrm{id} \otimes \Delta) \circ \Delta = (\Delta \otimes \mathrm{id}) \circ \Delta$$
$$(\mathrm{id} \otimes \varepsilon) \circ \Delta = \mathrm{id} = (\varepsilon \otimes \mathrm{id}) \circ \Delta.$$

We call ε and Δ the *counit* and the *comultiplication* or *coproduct* of $\mathcal{A}$.

For the coproduct $\Delta(a) = \sum_i a_{(1)i} \otimes a_{(2)i} \in \mathcal{A} \otimes \mathcal{A}$ we will often suppress the summation symbol and use the shorthand notation $\Delta(a) = a_{(1)} \otimes a_{(2)}$ introduced by Sweedler[Swe69].

If $\mathcal{A}$ has an involution $* : \mathcal{A} \to \mathcal{A}$ such that ε and Δ are $*$-algebra homomorphisms, then we call $\mathcal{A}$ a *$*$-bialgebra* or an *involutive bialgebra*.

If there exists furthermore a linear map $S : \mathcal{A} \to \mathcal{A}$ (called *antipode*) satisfying

$$a_{(1)} S(a_{(2)}) = \varepsilon(a)\mathbf{1} = S(a_{(1)}) a_{(2)}$$

for all $a \in \mathcal{A}$, then we call $\mathcal{A}$ a *$*$-Hopf algebra* or an *involutive Hopf algebra*.

Definition A.1. *A* finite quantum group *is a finite dimensional C^*-Hopf algebra, i.e. a $*$-Hopf algebra $\mathcal{A}$, whose algebra is a finite dimensional C^*-algebra.*

Note that finite dimensional C^*-algebras are very concrete objects, namely they are multi-matrix algebras $\bigoplus_{n=1}^{N} M_{k_n}$, where M_k denotes the algebra of $k \times k$-matrices. Not every multi-matrix algebra carries a Hopf algebra structure. For example, the direct sum must contain a one-dimensional summand to make possible the existence of a counit.

First examples are of course the group algebras of finite groups. Another example is examined in detail in Appendix B.

Theorem A.2. *Let $\mathcal{A}$ be a finite quantum group. Then there exists a unique state η on $\mathcal{A}$ such that*

$$(\mathrm{id} \otimes \eta) \circ \Delta(a) = \eta(a)\mathbf{1} \tag{A.1}$$

for all $a \in \mathcal{A}$.

The state η is called the *Haar state* of $\mathcal{A}$. The defining property (A.1) is called left invariance. On finite (and more generally on compact) quantum groups left invariance is equivalent to right invariance, i.e. the Haar state satisfies also

$$(\eta \otimes \mathrm{id}) \circ \Delta(a) = \eta(a)\mathbf{1}.$$

One can show that it is even a faithful trace, i.e. $\eta(a^*a) = 0$ implies $a = 0$ and

$$\eta(ab) = \eta(ba)$$

for all $a, b \in \mathcal{A}$.

This is a nontrivial result. See [VD97] for a careful discussion of it. Using the unique Haar state we also get a distinguished inner product on $\mathcal{A}$, namely for $a, b \in \mathcal{A}$

$$\langle a, b \rangle = \eta(a^*b).$$

The corresponding Hilbert space is denoted by $\mathcal{H}$.

Proposition A.3. *Every state on $\mathcal{A}$ can be realized as a vector state in $\mathcal{H}$.*

Proof. Because $\mathcal{A}$ is finite dimensional every linear functional can be written in the form

$$\phi_a : b \mapsto \eta(a^*b) = \langle a, b \rangle.$$

Such a functional is positive iff $a \in \mathcal{A}$ is positive. In fact, since η is a trace, it is clear that $a \geq 0$ implies $\phi_a \geq 0$. Conversely, assume $\phi_a \geq 0$. Convince yourself that it is enough to consider $a, b \in M_k$ where M_k is one of the summands of the multi-matrix algebra $\mathcal{A}$. The restriction of η is a multiple of the usual trace. Inserting the one-dimensional projections for b shows that a is positive.

Because a is positive there is a unique positive square root. We can now write $\phi_a = \langle a^{\frac{1}{2}}, \cdot\, a^{\frac{1}{2}} \rangle$ and if ϕ_a is a state then $a^{\frac{1}{2}}$ is a unit vector in $\mathcal{H}$. □

Note that an equation $\phi = \langle d, \cdot\, d\rangle$ does not determine d uniquely. But the vector constructed in the proof is unique and all these vectors together generate a positive cone associated to η.

The following result was already introduced and used in Section 5.

Theorem A.4. *Let $\mathcal{A}$ be a finite quantum group with Haar state η. Then the map $W : \mathcal{A}\otimes\mathcal{A} \to \mathcal{A}\otimes\mathcal{A}$ defined by*

$$W(b\otimes a) = \Delta(b)(\mathbf{1}\otimes a), \qquad a, b \in \mathcal{A},$$

is unitary with respect to the inner product defined by

$$\langle b\otimes a, d\otimes c\rangle = \eta(b^*d)\,\eta(a^*c),$$

for $a, b, c, d \in \mathcal{A}$.

Furthermore, it satisfies the pentagon equation

$$W_{12}W_{13}W_{23} = W_{23}W_{12}.$$

We used the leg notation $W_{12} = W\otimes \mathrm{id}$, $W_{23} = \mathrm{id}\otimes W$, $W_{13} = (\mathrm{id}\otimes\tau)\circ W_{12}\circ(\mathrm{id}\otimes\tau)$, *where τ is the* flip, $\tau : \mathcal{A}\otimes\mathcal{A}\to\mathcal{A}\otimes\mathcal{A}$, $\tau(a\otimes b) = b\otimes a$.

Remark A.5. The operator $W : \mathcal{A}\otimes\mathcal{A} \to \mathcal{A}\otimes\mathcal{A}$ is called the *fundamental operator* or *multiplicative unitary* of $\mathcal{A}$, cf. [BS93, BBS99].

B The Eight-Dimensional Kac-Paljutkin Quantum Group

In this section we give the defining relations and the main structure of an eight-dimensional quantum group introduced by Kac and Paljutkin [KP66]. This is actually the smallest finite quantum group that does not come from a group as the group algebra or the algebra of functions on the group. In other words, it is the C^*-Hopf algebra with the smallest dimension, which is neither commutative nor cocommutative.

Consider the multi-matrix algebra $\mathcal{A} = \mathbb{C}\oplus\mathbb{C}\oplus\mathbb{C}\oplus\mathbb{C}\oplus\mathcal{M}_2(\mathbb{C})$, with the usual multiplication and involution. We shall use the basis

$$\begin{aligned}
e_1 &= 1\oplus 0\oplus 0\oplus 0\oplus 0, & a_{11} &= 0\oplus 0\oplus 0\oplus 0\oplus \begin{pmatrix}1&0\\0&0\end{pmatrix},\\
e_2 &= 0\oplus 1\oplus 0\oplus 0\oplus 0, & a_{12} &= 0\oplus 0\oplus 0\oplus 0\oplus \begin{pmatrix}0&1\\0&0\end{pmatrix},\\
e_3 &= 0\oplus 0\oplus 1\oplus 0\oplus 0, & a_{21} &= 0\oplus 0\oplus 0\oplus 0\oplus \begin{pmatrix}0&0\\1&0\end{pmatrix},\\
e_4 &= 0\oplus 0\oplus 0\oplus 1\oplus 0, & a_{22} &= 0\oplus 0\oplus 0\oplus 0\oplus \begin{pmatrix}0&0\\0&1\end{pmatrix}.
\end{aligned}$$

The algebra $\mathcal{A}$ is an eight-dimensional C*-algebra. Its unit is of course $\mathbf{1} = e_1 + e_2 + e_3 + e_4 + a_{11} + a_{22}$. We shall need the trace Tr on $\mathcal{A}$,

$$\mathrm{Tr}\left(x_1 \oplus x_2 \oplus x_3 \oplus x_4 \oplus \begin{pmatrix} c_{11} & c_{12} \\ c_{21} & c_{22} \end{pmatrix}\right) = x_1 + x_2 + x_3 + x_4 + c_{11} + c_{22}.$$

Note that Tr is normalized to be equal to one on minimal projections.

The following defines a coproduct on $\mathcal{A}$,

$$\begin{aligned}
\Delta(e_1) &= e_1 \otimes e_1 + e_2 \otimes e_2 + e_3 \otimes e_3 + e_4 \otimes e_4 \\
&\quad + \frac{1}{2} a_{11} \otimes a_{11} + \frac{1}{2} a_{12} \otimes a_{12} + \frac{1}{2} a_{21} \otimes a_{21} + \frac{1}{2} a_{22} \otimes a_{22}, \\
\Delta(e_2) &= e_1 \otimes e_2 + e_2 \otimes e_1 + e_3 \otimes e_4 + e_4 \otimes e_3 \\
&\quad + \frac{1}{2} a_{11} \otimes a_{22} + \frac{1}{2} a_{22} \otimes a_{11} + \frac{i}{2} a_{21} \otimes a_{12} - \frac{i}{2} a_{12} \otimes a_{21}, \\
\Delta(e_3) &= e_1 \otimes e_3 + e_3 \otimes e_1 + e_2 \otimes e_4 + e_4 \otimes e_2 \\
&\quad + \frac{1}{2} a_{11} \otimes a_{22} + \frac{1}{2} a_{22} \otimes a_{11} - \frac{i}{2} a_{21} \otimes a_{12} + \frac{i}{2} a_{12} \otimes a_{21}, \\
\Delta(e_4) &= e_1 \otimes e_4 + e_4 \otimes e_1 + e_2 \otimes e_3 + e_3 \otimes e_2 \\
&\quad + \frac{1}{2} a_{11} \otimes a_{11} + \frac{1}{2} a_{22} \otimes a_{22} - \frac{1}{2} a_{12} \otimes a_{12} - \frac{1}{2} a_{21} \otimes a_{21}, \\
\Delta(a_{11}) &= e_1 \otimes a_{11} + a_{11} \otimes e_1 + e_2 \otimes a_{22} + a_{22} \otimes e_2 \\
&\quad + e_3 \otimes a_{22} + a_{22} \otimes e_3 + e_4 \otimes a_{11} + a_{11} \otimes e_4, \\
\Delta(a_{12}) &= e_1 \otimes a_{12} + a_{12} \otimes e_1 + i e_2 \otimes a_{21} - i a_{21} \otimes e_2 \\
&\quad - i e_3 \otimes a_{21} + i a_{21} \otimes e_3 - e_4 \otimes a_{12} - a_{12} \otimes e_4, \\
\Delta(a_{21}) &= e_1 \otimes a_{21} + a_{21} \otimes e_1 - i e_2 \otimes a_{12} + i a_{12} \otimes e_2 \\
&\quad + i e_3 \otimes a_{12} - i a_{12} \otimes e_3 - e_4 \otimes a_{21} - a_{21} \otimes e_4, \\
\Delta(a_{22}) &= e_1 \otimes a_{22} + a_{22} \otimes e_1 + e_2 \otimes a_{11} + a_{11} \otimes e_2 \\
&\quad e_3 \otimes a_{11} + a_{11} \otimes e_3 + e_4 \otimes a_{22} + a_{22} \otimes e_4.
\end{aligned}$$

The counit is given by

$$\varepsilon\left(x_1 \oplus x_2 \oplus x_3 \oplus x_4 \oplus \begin{pmatrix} c_{11} & c_{12} \\ c_{21} & c_{22} \end{pmatrix}\right) = x_1$$

The antipode is the transpose map, i.e.

$$S(e_i) = e_i, \qquad S(a_{jk}) = a_{kj},$$

for $i = 1, 2, 3, 4$, $j, k = 1, 2$.

B.1 The Haar State

Finite quantum groups have unique Haar elements h satisfying $h^* = h = h^2$, $\varepsilon(h) = 1$, and

$$ah = \varepsilon(a)h = ha \qquad \text{for all } a \in \mathcal{A},$$

cf. [VD97]. For the Kac-Paljutkin quantum group it is given by $h = e_1$. An invariant functional is given by $\phi(a) = \mathrm{Tr}(aK^{-1})$, with $K = (\mathrm{Tr} \otimes \mathrm{id})\Delta(h) = e_1 + e_2 + e_3 + e_4 + \frac{1}{2}(a_{11} + a_{22})$ and $K^{-1} = e_1 + e_2 + e_3 + e_4 + 2(a_{11} + a_{22})$. On an arbitrary element of $\mathcal{A}$ the action of ϕ is given by

$$\phi\left(x_1 \oplus x_2 \oplus x_3 \oplus x_4 \oplus \begin{pmatrix} c_{11} & c_{12} \\ c_{21} & c_{22} \end{pmatrix}\right) = x_1 + x_2 + x_3 + x_3 + 2c_{11} + 2c_{22}.$$

Normalizing η so that $\eta(1) = 1$, we get the *Haar state* $\eta = \frac{1}{8}\phi$.

B.2 The Dual of $\mathcal{A}$

The dual $\mathcal{A}^*$ of a finite quantum groups $\mathcal{A}$ is again a finite quantum group, see [VD97]. Its morphisms are the duals of the morphisms of $\mathcal{A}$, e.g.

$$m_{\mathcal{A}^*} = \Delta_{\mathcal{A}}^* : \mathcal{A}^* \otimes \mathcal{A}^* \cong (\mathcal{A} \otimes \mathcal{A})^* \to \mathcal{A}^*, \quad m_{\mathcal{A}^*}(\phi_1 \otimes \phi_2) = (\phi_1 \otimes \phi_2) \circ \Delta$$

and

$$\Delta_{\mathcal{A}^*} = m_{\mathcal{A}}^* : \mathcal{A}^* \to \mathcal{A}^* \otimes \mathcal{A}^* \cong (\mathcal{A} \otimes \mathcal{A})^*, \quad \Delta_{\mathcal{A}^*}\phi = \phi \circ m_{\mathcal{A}}.$$

The involution of $\mathcal{A}^*$ is given by $\phi^*(a) = \overline{\phi\big((Sa)^*\big)}$ for $\phi \in \mathcal{A}^*$, $a \in \mathcal{A}$. To show that $\mathcal{A}^*$ is indeed a C^*-algebra, one can show that the dual regular action of $\mathcal{A}^*$ on $\mathcal{A}$ defined by $T_\phi a = \phi(a_{(2)})a_{(1)}$ for $\phi \in \mathcal{A}^*$, $a \in \mathcal{A}$, is a faithful $*$-representation of $\mathcal{A}^*$ w.r.t. the inner product on $\mathcal{A}$ defined by

$$\langle a, b \rangle = \eta(a^* b)$$

for $a, b \in \mathcal{A}$, cf. [VD97, Proposition 2.3].

For the Kac-Paljutkin quantum group $\mathcal{A}$ the dual $\mathcal{A}^*$ actually turns out to be isomorphic to $\mathcal{A}$ itself.

Denote by $\{\eta_1, \eta_2, \eta_3, \eta_4, \alpha_{11}, \alpha_{12}, \alpha_{21}, \alpha_{22}\}$ the basis of $\mathcal{A}^*$ that is dual to $\{e_1, e_2, e_3, e_4, a_{11}, a_{12}, a_{21}, a_{22}\}$, i.e. the functionals on $\mathcal{A}$ defined by

$$\begin{aligned} \eta_i(e_j) &= \delta_{ij}, & \eta_i(a_{rs}) &= 0, \\ \alpha_{k\ell}(e_j) &= 0, & \alpha_{k\ell}(a_{rs}) &= \delta_{kr}\delta_{\ell s}, \end{aligned}$$

for $i, j = 1, 2, 3, 4$, $k, \ell, r, s = 1, 2$.

We leave the verification of the following as an exercise.

The functionals

$$\begin{aligned} f_1 &= \frac{1}{8}(\eta_1 + \eta_2 + \eta_3 + \eta_4 + 2\alpha_{11} + 2\alpha_{22}), \\ f_2 &= \frac{1}{8}(\eta_1 - \eta_2 - \eta_3 + \eta_4 - 2\alpha_{11} + 2\alpha_{22}), \\ f_3 &= \frac{1}{8}(\eta_1 - \eta_2 - \eta_3 + \eta_4 + 2\alpha_{11} - 2\alpha_{22}), \\ f_4 &= \frac{1}{8}(\eta_1 + \eta_2 + \eta_3 + \eta_4 - 2\alpha_{11} - 2\alpha_{22}), \end{aligned}$$

are minimal projections in $\mathcal{A}^*$. Furthermore

$$\begin{aligned} b_{11} &= \frac{1}{4}(\eta_1 + \eta_2 - \eta_3 - \eta_4), \\ b_{12} &= \frac{1-i}{2\sqrt{2}}(\alpha_{12} + i\alpha_{21}), \\ b_{21} &= \frac{1+i}{2\sqrt{2}}(\alpha_{12} - i\alpha_{21}), \\ b_{22} &= \frac{1}{4}(\eta_1 - \eta_2 + \eta_3 - \eta_4), \end{aligned}$$

are matrix units, i.e. satisfy the relations

$$b_{ij}b_{k\ell} = \delta_{jk}b_{i\ell} \quad \text{and} \quad (b_{ij})^* = b_{ji},$$

and the "mixed" products vanish,

$$f_i b_{jk} = 0 = b_{jk} f_i, \qquad i = 1,2,3,4, \quad j,k = 1,2.$$

Therefore $\mathcal{A}^* \cong \mathbb{C}^4 \oplus \mathcal{M}_2(\mathbb{C}) \cong \mathcal{A}$ as an algebra. But actually, $e_i \mapsto f_i$ and $a_{ij} \mapsto b_{ij}$ defines even a C*-Hopf algebra isomorphism from $\mathcal{A}$ to $\mathcal{A}^*$.

B.3 The States on $\mathcal{A}$

On $\mathbb{C}$ there exists only one state, the identity map. States on $\mathcal{M}_2(\mathbb{C})$ are given by density matrices, i.e., positive semi-definite matrices with trace one. More precisely, for any state ϕ on $\mathcal{M}_2(\mathbb{C})$ there exists a unique density matrix $\rho \in \mathcal{M}_2(\mathbb{C})$ such that

$$\phi(A) = Tr(\rho A),$$

for all $A \in \mathcal{M}_2(\mathbb{C})$. The 2×2 density matrices can be parametrized by the unit ball $B_1 = \{(x,y,z) \in \mathbb{R}^3 | x^2 + y^2 + z^2 \leq 1\}$,

$$\rho(x,y,z) = \frac{1}{2}\begin{pmatrix} 1+z & x+iy \\ x-iy & 1-z \end{pmatrix}$$

A state on $\mathcal{A}$ is a convex combination of states on the four copies of $\mathbb{C}$ and a state on $\mathcal{M}_2(\mathbb{C})$. All states on $\mathcal{A}$ can therefore be parametrized by the set $\{(\mu_1, \mu_2, \mu_3, \mu_4, \mu_5, x, y, z) \in \mathbb{R}^8 | x^2 + y^2 + z^2 = 1; \mu_1 + \mu_2 + \mu_3 + \mu_4 + \mu_5 = 1; \mu_1, \mu_2, \mu_3, \mu_4, \mu_5 \geq 0\}$. They are given by

$$\phi = \mathrm{Tr}(\mu \,\cdot\,) = 8\eta(K\mu \,\cdot\,)$$

where

$$\mu = \mu_1 \oplus \mu_2 \oplus \mu_3 \oplus \mu_4 \oplus \frac{\mu_5}{2}\begin{pmatrix} 1+z & x+iy \\ x-iy & 1-z \end{pmatrix}.$$

With respect to the dual basis, the state ϕ can be written as

$$\phi = \mu_1\eta_1 + \mu_2\eta_2 + \mu_3\eta_3 + \mu_4\eta_4 \\ +\frac{\mu_5}{2}\big((1+z)\alpha_{11} + (x-iy)\alpha_{12} + (x+iy)\alpha_{21} + (1-z)\alpha_{22}\big). \tag{B.1}$$

The regular representation $T_\phi = (\mathrm{id}\otimes\phi)\circ\Delta$ of ϕ on $\mathcal{A}$ has the matrix

$$\begin{pmatrix}
\mu_1 & \mu_2 & \mu_3 & \mu_4 & \frac{1+z}{2\sqrt{2}}\mu_5 & \frac{x-iy}{2\sqrt{2}}\mu_5 & \frac{x+iy}{2\sqrt{2}}\mu_5 & \frac{1-z}{2\sqrt{2}}\mu_5 \\
\mu_2 & \mu_1 & \mu_4 & \mu_3 & \frac{1-z}{2\sqrt{2}}\mu_5 & -\frac{ix-y}{2\sqrt{2}}\mu_5 & \frac{ix+y}{2\sqrt{2}}\mu_5 & \frac{1+z}{2\sqrt{2}}\mu_5 \\
\mu_3 & \mu_4 & \mu_1 & \mu_2 & \frac{1-z}{2\sqrt{2}}\mu_5 & \frac{ix-y}{2\sqrt{2}}\mu_5 & -\frac{ix+y}{2\sqrt{2}}\mu_5 & \frac{1+z}{2\sqrt{2}}\mu_5 \\
\mu_4 & \mu_3 & \mu_2 & \mu_1 & \frac{1+z}{2\sqrt{2}}\mu_5 & -\frac{x-iy}{2\sqrt{2}}\mu_5 & -\frac{x+iy}{2\sqrt{2}}\mu_5 & \frac{1-z}{2\sqrt{2}}\mu_5 \\
\frac{1+z}{2\sqrt{2}}\mu_5 & \frac{1-z}{2\sqrt{2}}\mu_5 & \frac{1-z}{2\sqrt{2}}\mu_5 & \frac{1+z}{2\sqrt{2}}\mu_5 & \mu_1+\mu_4 & 0 & 0 & \mu_2+\mu_3 \\
\frac{x-iy}{2\sqrt{2}}\mu_5 & \frac{ix-y}{2\sqrt{2}}\mu_5 & -\frac{ix-y}{2\sqrt{2}}\mu_5 & -\frac{x-iy}{2\sqrt{2}}\mu_5 & 0 & \mu_1-\mu_4 & -i\mu_2+i\mu_3 & 0 \\
\frac{x+iy}{2\sqrt{2}}\mu_5 & -\frac{ix+y}{2\sqrt{2}}\mu_5 & \frac{ix+y}{2\sqrt{2}}\mu_5 & -\frac{x+iy}{2\sqrt{2}}\mu_5 & 0 & i\mu_2-i\mu_3 & \mu_1-\mu_4 & 0 \\
\frac{1-z}{2\sqrt{2}}\mu_5 & \frac{1+z}{2\sqrt{2}}\mu_5 & \frac{1+z}{2\sqrt{2}}\mu_5 & \frac{1-z}{2\sqrt{2}}\mu_5 & \mu_2+\mu_3 & 0 & 0 & \mu_1+\mu_4
\end{pmatrix}.$$

with respect to the basis $(2\sqrt{2}e_1, 2\sqrt{2}e_2, 2\sqrt{2}e_3, 2\sqrt{2}e_4, 2a_{11}, 2a_{12}, 2a_{21}, 2a_{22})$.

In terms of the basis of matrix units of $\mathcal{A}^*$, ϕ takes the form

$$\begin{aligned}
\phi &= (\mu_1+\mu_2+\mu_3+\mu_4+\mu_5)f_1 + (\mu_1-\mu_2-\mu_3+\mu_4-z\mu_5)f_2 \\
&\quad +(\mu_1-\mu_2-\mu_3+\mu_4+z\mu_5)f_3 + (\mu_1+\mu_2+\mu_3+\mu_4-\mu_5)f_4 \\
&\quad +(\mu_1+\mu_2-\mu_3-\mu_4)b_{11} + (\mu_1-\mu_2+\mu_3-\mu_4)b_{22} \\
&\quad +\frac{x+y}{\sqrt{2}}\mu_5 b_{12} + \frac{x-y}{\sqrt{2}}\mu_5 b_{21}
\end{aligned}$$

or

$$\begin{aligned}
\phi &= (\mu_1+\mu_2+\mu_3+\mu_4+\mu_5)\oplus(\mu_1-\mu_2-\mu_3+\mu_4-z\mu_5)\oplus \\
&\quad \oplus(\mu_1-\mu_2-\mu_3+\mu_4+z\mu_5)\oplus(\mu_1+\mu_2+\mu_3+\mu_4-\mu_5)\oplus \\
&\quad \oplus\begin{pmatrix} \mu_1+\mu_2-\mu_3-\mu_4 & \frac{x+y}{\sqrt{2}}\mu_5 \\ \frac{x-y}{\sqrt{2}}\mu_5 & \mu_1-\mu_2+\mu_3-\mu_4 \end{pmatrix}
\end{aligned}$$

in matrix form.

Remark: Note that the states on $\mathcal{A}$ are in general not positive for the $*$-algebra structure of $\mathcal{A}^*$.

If $\phi \in \mathcal{A}^*$ is positive for the $*$-algebra structure of $\mathcal{A}^*$, then T_ϕ is positive definite on the GNS Hilbert space $\mathcal{H} \cong \mathcal{A}$ of the Haar state η, since the regular representation is a $*$-representation, cf. [VD97].

On the other hand, if $\phi \in \mathcal{A}^*$ is positive as a functional on $\mathcal{A}$, then $T_\phi = (\mathrm{id}\otimes\phi)\circ\Delta$ is completely positive as a map from the C^*-algebra $\mathcal{A}$ to itself.

References

[AFL82] L. Accardi, A. Frigerio, and J.T. Lewis. Quantum stochastic processes. *Publ. RIMS*, 18:97–133, 1982.

[BBS99] S. Baaj, E. Blanchard, and G. Skandalis. Unitaires multiplicatifs en dimension finie et leurs sous-objets. *Ann. Inst. Fourier (Grenoble)*, 49(4):1305–1344, 1999.

[Bia90] P. Biane. Marches de Bernoulli quantiques. In *Séminaire de Probabilités, XXIV, 1988/89, Lecture Notes in Math.*, Vol. 1426, pp. 329–344. Springer, Berlin, 1990.

[Bia91a] P. Biane. Quantum random walk on the dual of SU(n). *Probab. Theory Related Fields*, 89(1):117–129, 1991.

[Bia91b] P. Biane. Some properties of quantum Bernoulli random walks. In *Quantum probability & related topics*, QP-PQ, VI, pages 193–203. World Sci. Publishing, River Edge, NJ, 1991.

[Bia92a] P. Biane. Équation de Choquet-Deny sur le dual d'un groupe compact. *Probab. Theory Related Fields*, 94(1):39–51, 1992.

[Bia92b] P. Biane. Frontière de Martin du dual de SU(2). In *Séminaire de Probabilités, XXVI, Lecture Notes in Math.*, Vol. 1526, pp. 225–233. Springer, Berlin, 1992.

[Bia92c] P. Biane. Minuscule weights and random walks on lattices. In *Quantum probability & related topics*, QP-PQ, VII, pages 51–65. World Sci. Publishing, River Edge, NJ, 1992.

[Bia94] P. Biane. Théorème de Ney-Spitzer sur le dual de SU(2). *Trans. Amer. Math. Soc.*, 345(1):179–194, 1994.

[Bia98] P. Biane. Processes with free increments. *Math. Z.*, 227(1):143–174, 1998.

[BKS97] M. Bożejko, B. Kümmerer, and R. Speicher. q-Gaussian processes: Noncommutative and classical aspects. *Commun. Math. Phys.*, 185(1):129–154, 1997.

[BP95] B.V.R. Bhat and K.R. Parthasarathy. Markov dilations of nonconservative dynamical semigroups and a quantum boundary theory. *Ann. Inst. H. Poincaré Probab. Statist.*, 31(4):601–651, 1995.

[BS93] S. Baaj and G. Skandalis. Unitaires multiplicatifs et dualité pour les produits croisés de C^*-algèbres. *Ann. Sci. École Norm. Sup. (4)*, 26(4):425–488, 1993.

[Col04] B. Collins. Martin boundary theory of some quantum random walks. *Ann. Inst. H. Poincaré Probab. Statist.*, 40(3):367–384, 2004.

[Ell04] D. Ellinas. On algebraic and quantum random walks. In: *Quantum Probability and Infinite Dimensional Analysis: From Foundations to Applications*, Quantum Probability and White Noise Calculus, Vol. XVIII, U. Franz and M. Schürmann (eds.), World Scientific, 2005.

[Fra99] U. Franz. Classical Markov Processes from Quantum Lévy Processes. Infin. Dim. Anal., Quantum Prob. and Rel. Topics, 2(1):105-129, 1999.

[Go04] R. Gohm. *Noncommutative Stationary Processes.* Lecture Notes in Math., Vol. 1839, Springer, 2004.

[GKL04] R. Gohm, B. Kümmerer and T. Lang. Noncommutative symbolic coding. Preprint, 2004

[Hey77] H. Heyer. *Probability measures on locally compact groups.* Springer-Verlag, Berlin, 1977.

[INT04] M. Izumi, S. Neshveyev, and L. Tuset. Poisson boundary of the dual of SUq(n). Preprint math.OA/0402074, 2004.

[Izu02] M. Izumi. Non-commutative Poisson boundaries and compact quantum group actions. *Adv. Math.*, 169(1):1–57, 2002.

[JR82] S.A. Joni and G.-C. Rota. Coalgebras and bialgebras in combinatorics. *Contemporary Mathematics*, 6:1–47, 1982.

[KP66] G.I. Kac and V.G. Paljutkin. Finite ring groups. *Trudy Moskov. Mat. Obšč.*, 15:224–261, 1966. Translated in Trans. Moscow Math. Soc. (1967), 251-284.

[Küm88] B. Kümmerer. Survey on a theory of non-commutativ stationary Markov processes. In L. Accardi and W.v. Waldenfels, editors, *Quantum Probability and Applications III*, pages 228–244. Springer-Verlag, 1988.

[Kus05] J. Kustermans. Locally compact quantum groups. In: D. Applebaum, B.V.R. Bhat, J. Kustermans, J.M. Lindsay. *Quantum Independent Increment Processes I: From Classical Probability to Quantum Stochastic Calculus* U. Franz, M. Schürmann (eds.), Lecture Notes in Math., Vol. 1865, pp. 99-180, Springer, 2005.

[Len96] R. Lenczewski. Quantum random walk for $U_q(\mathrm{su}(2))$ and a new example of quantum noise. *J. Math. Phys.*, 37(5):2260–2278, 1996.

[Maj93] S. Majid. Quantum random walks and time-reversal. *Int. J. Mod. Phys.*, 8:4521–4545, 1993.

[Maj95] S. Majid. *Foundations of quantum group theory.* Cambridge University Press, 1995.

[MRP94] S. Majid and M. J. Rodríguez-Plaza. Random walk and the heat equation on superspace and anyspace. *J. Math. Phys.*, 35:3753–3760, 1994.

[NS82] W. Nichols and M. Sweedler. Hopf algebras and combinatorics. *Contemporary Mathematics*, 6:49–84, 1982.

[NT04] S. Neshveyev and L. Tuset. The Martin boundary of a discrete quantum group. *J. Reine Angew. Math.*, 568:23–70, 2004.

[Nor97] J.R. Norris. *Markov Chains.* Cambridge University Press, 1997.

[Pal96] A. Pal. A counterexample on idempotent states on a compact quantum group. *Lett. Math. Phys.*, 37(1):75–77, 1996.

[Par03] K.R. Parthasarathy. Quantum probability and strong quantum Markov processes. In: *Quantum Probability Communications, Vol. XII (Grenoble, 1998)*, World Scientific, pp. 59–138, 2003.

[Sak71] S. Sakai. *C^*-Algebras and W^*-Algebras.* Springer, Berlin 1971.

[SC04] L. Saloff-Coste. Random walks on finite groups. In *Probability on discrete structures, Encyclopaedia Math. Sci.*, Vol. 110, pp. 263–346. Springer, Berlin, 2004.

[Sch93] M. Schürmann. *White Noise on Bialgebras.* Lecture Notes in Math., Vol. 1544, Springer, Berlin, 1993.

[Swe69] M. E. Sweedler. *Hopf Algebras.* Benjamin, New York, 1969.

[VD97] A. Van Daele. The Haar measure on finite quantum groups. *Proc. Amer. Math. Soc.*, 125(12):3489–3500, 1997.

[vW90a] W. von Waldenfels. Illustration of the quantum central limit theorem by independent addition of spins. In *Séminaire de Probabilités, XXIV, 1988/89, Lecture Notes in Math.*, Vol. 1426, pp. 349–356. Springer, Berlin, 1990.

[vW90b] W. von Waldenfels. The Markov process of total spins. In *Séminaire de Probabilités, XXIV, 1988/89, Lecture Notes in Math.*, Vol. 1426, pp. 357–361. Springer, Berlin, 1990.

[Wo96] S.L. Woronowicz. From multiplicative unitaries to quantum groups. *Int. J. Math.*, 7(1):127-149, 1996.

Classical and Free Infinite Divisibility and Lévy Processes

Ole E. Barndorff-Nielsen[1] and Steen Thorbjørnsen[2]

[1] Dept. of Mathematical Sciences
University of Aarhus
Ny Munkegade
DK-8000 Århus, Denmark
oebn@imf.au.dk

[2] Dept. of Mathematics & Computer Science
University of Southern Denmark
Campusvej 55
DK-5230 Odense, Denmark
steenth@imada.sdu.dk

O.E. Barndorff-Nielsen and S. Thorbjørnsen: *Classical and Free Infinite Divisibility and Lévy Processes*, Lect. Notes Math. **1866**, 33–159 (2006)
www.springerlink.com

1 Introduction

The present lecture notes have grown out of a wish to understand whether certain important concepts of classical infinite divisibility and Lévy processes, such as selfdecomposability and the Lévy-Itô decomposition, have natural and interesting analogues in free probability. The study of this question has led to new links between classical and free Lévy theory, and to some new results in the classical setting, that seem of independent interest. The new concept of Upsilon mappings have a key role in both respects. These are regularizing mappings from the set of Lévy measures into itself or, otherwise interpreted, mappings of the class of infinitely divisible laws into itself. One of these mappings, Υ, provides a direct connection to the Lévy-Khintchine formula of free probability.

The next Section recalls a number of concepts and results from the classical framework, and in Section 3 the basic Upsilon mappings Υ_0 and Υ are introduced and studied. They are shown to be smooth, injective and regularizing, and their relation to important subclasses of infinitely divisible laws is discussed. Subsequently Υ_0 and Υ are generalized to one-parameter families of mappings $(\Upsilon_0^\alpha)_{\alpha\in[0,1]}$ and $(\Upsilon^\alpha)_{\alpha\in[0,1]}$ with similar properties, and which interpolate between Υ_0 (resp. Υ) and the identity mapping on the set of Lévy measures (resp. the class of infinitely divisible laws). Other types of Upsilon mappings are also considered, including some generalizations to higher dimensions. Section 4 gives an introduction to non-commutative probability, particularly free infinite divisibility, and then takes up some of the above-mentioned questions concerning links between classical and free Lévy theory. The discussion of such links is continued in Section 5, centered around the Upsilon mapping Υ and the closely associated Bercovici-Pata mapping Λ. The final Section 6 discusses free stochastic integration and establishes a free analogue of the Lévy-Ito representation.

The material presented in these lecture notes is based on the authors' papers [BaTh02a], [BaTh02b], [BaTh02c], [BaTh04a], [BaTh04b] and [BaTh05].

2 Classical Infinite Divisibility and Lévy Processes

The classical theory of infinite divisibility and Lévy processes was founded by Kolmogorov, Lévy and Khintchine in the Nineteen Thirties. The monographs [Sa99] and [Be96],[Be97] are main sources for information on this theory. For some more recent results, including various types of applications, see [BaMiRe01].

Here we recall some of the most basic facts of the theory, and we discuss a hierarchy of important subclasses of the space of infinitely divisible distributions.

2.1 Basics of Infinite Divisibility

The class of infinitely divisible probability measures on the real line will here be denoted by $\mathcal{ID}(*)$. A probability measure μ on $\mathbb{R}$ belongs to $\mathcal{ID}(*)$ if there exists, for each positive integer n, a probability measure μ_n, such that

$$\mu = \underbrace{\mu_n * \mu_n * \cdots * \mu_n}_{n \text{ terms}},$$

where $*$ denotes the usual convolution of probability measures.

We recall that a probability measure μ on $\mathbb{R}$ is infinitely divisible if and only if its characteristic function (or Fourier transform) f_μ has the Lévy-Khintchine representation:

$$\log f_\mu(u) = \mathrm{i}\gamma u + \int_{\mathbb{R}} \Big(\mathrm{e}^{\mathrm{i}ut} - 1 - \frac{\mathrm{i}ut}{1+t^2}\Big)\frac{1+t^2}{t^2}\,\sigma(\mathrm{d}t), \quad (u \in \mathbb{R}), \qquad (2.1)$$

where γ is a real constant and σ is a finite measure on $\mathbb{R}$. In that case, the pair (γ, σ) is uniquely determined, and is termed the generating pair for μ.

The function $\log f_\mu$ is called the *cumulant transform* for μ and is also denoted by C_μ, as we shall do often in the sequel.

In the literature, there are several alternative ways of writing the above representation. In recent literature, the following version seems to be preferred (see e.g. [Sa99]):

$$\log f_\mu(u) = \mathrm{i}\eta u - \tfrac{1}{2}au^2 + \int_{\mathbb{R}} \left(\mathrm{e}^{\mathrm{i}ut} - 1 - \mathrm{i}ut1_{[-1,1]}(t)\right) \rho(\mathrm{d}t), \quad (u \in \mathbb{R}), \quad (2.2)$$

where η is a real constant, a is a non-negative constant and ρ is a Lévy measure on $\mathbb{R}$ according to Definition 2.1 below. Again, a, ρ and η are uniquely determined by μ and the triplet (a, ρ, η) is called the *characteristic triplet* for μ.

Definition 2.1. *A Borel measure ρ on $\mathbb{R}$ is called a Lévy measure, if it satisfies the following conditions:*

$$\rho(\{0\}) = 0 \quad \textit{and} \quad \int_{\mathbb{R}} \min\{1, t^2\} \, \rho(\mathrm{d}t) < \infty.$$

The relationship between the two representations (2.1) and (2.2) is as follows:

$$\begin{aligned} a &= \sigma(\{0\}), \\ \rho(\mathrm{d}t) &= \frac{1+t^2}{t^2} \cdot 1_{\mathbb{R}\setminus\{0\}}(t)\, \sigma(\mathrm{d}t), \\ \eta &= \gamma + \int_{\mathbb{R}} t\Big(1_{[-1,1]}(t) - \frac{1}{1+t^2}\Big)\, \rho(\mathrm{d}t). \end{aligned} \quad (2.3)$$

2.2 Classical Lévy Processes

For a (real-valued) random variable X defined on a probability space $(\Omega, \mathcal{F}, P)$, we denote by $L\{X\}$ the distribution[1] of X.

Definition 2.2. *A real valued stochastic process $(X_t)_{t\geq 0}$, defined on a probability space $(\Omega, \mathcal{F}, P)$, is called a Lévy process, if it satisfies the following conditions:*

(i) *whenever $n \in \mathbb{N}$ and $0 \leq t_0 < t_1 < \cdots < t_n$, the increments*

$$X_{t_0}, X_{t_1} - X_{t_0}, X_{t_2} - X_{t_1}, \ldots, X_{t_n} - X_{t_{n-1}},$$

are independent random variables.

(ii) *$X_0 = 0$, almost surely.*

(iii) *for any s, t in $[0, \infty[$, the distribution of $X_{s+t} - X_s$ does not depend on s.*

(iv) *(X_t) is stochastically continuous, i.e. for any s in $[0, \infty[$ and any positive ϵ, we have: $\lim_{t\to 0} P(|X_{s+t} - X_s| > \epsilon) = 0$.*

(v) *for almost all ω in Ω, the sample path $t \mapsto X_t(\omega)$ is right continuous (in $t \geq 0$) and has left limits (in $t > 0$).*

---

[1] L stands for "the law of".

If a stochastic process $(X_t)_{t\geq 0}$ satisfies conditions (i)-(iv) in the definition above, we say that (X_t) is a *Lévy process in law*. If (X_t) satisfies conditions (i), (ii), (iv) and (v) (respectively (i), (ii) and (iv)) it is called an *additive process* (respectively an *additive process in law*). Any Lévy process in law (X_t) has a modification which is a Lévy process, i.e. there exists a Lévy process (Y_t), defined on the same probability space as (X_t), and such that $X_t = Y_t$ with probability one, for all t. Similarly any additive process in law has a modification which is a genuine additive process. These assertions can be found in [Sa99, Theorem 11.5].

Note that condition (iv) is equivalent to the condition that $X_{s+t} - X_s \to 0$ in distribution, as $t \to 0$. Note also that under the assumption of (ii) and (iii), this condition is equivalent to saying that $X_t \to 0$ in distribution, as $t \searrow 0$.

The concepts of infinitely divisible probability measures and of Lévy processes are closely connected, since there is a one-to-one correspondance between them. Indeed, if (X_t) is a Lévy process, then $L\{X_t\}$ is infinitely divisible for all t in $[0, \infty[$, since for any positive integer n

$$X_t = \sum_{j=1}^{n}(X_{jt/n} - X_{(j-1)t/n}),$$

and hence, by (i) and (iii) of Definition 2.2,

$$L\{X_t\} = \underbrace{L\{X_{t/n}\} * L\{X_{t/n}\} * \cdots * L\{X_{t/n}\}}_{n \text{ terms}}.$$

Moreover, for each t, $L\{X_t\}$ is uniquely determined by $L\{X_1\}$ via the relation $L\{X_t\} = L\{X_1\}^t$ (see [Sa99, Theorem 7.10]). Conversely, for any infinitely divisible distribution μ on $\mathbb{R}$, there exists a Lévy process (X_t) (on some probability space $(\Omega, \mathcal{F}, P)$), such that $L\{X_1\} = \mu$ (cf. [Sa99, Theorem 7.10 and Corollary 11.6]).

2.3 Integration with Respect to Lévy Processes

We start with a general discussion of the existence of stochastic integrals w.r.t. (classical) Lévy processes and their associated cumulant functions. Some related results are given in [ChSh02] and [Sa00], but they do not fully cover the situation considered below.

Throughout, we shall use the notation $C\{u \ddagger X\}$ to denote the cumulant function of (the distribution of) a random variable X, evaluated at the real number u.

Recall that a sequence (σ_n) of *finite* measures on $\mathbb{R}$ is said to converge weakly to a finite measure σ on $\mathbb{R}$, if

$$\int_{\mathbb{R}} f(t)\,\sigma_n(\mathrm{d}t) \to \int_{\mathbb{R}} f(t)\,\sigma(\mathrm{d}t), \quad \text{as } n \to \infty, \tag{2.4}$$

for any bounded continuous function $f\colon \mathbb{R} \to \mathbb{C}$. In that case, we write $\sigma_n \xrightarrow{\text{w}} \sigma$, as $n \to \infty$.

Remark 2.3. Recall that a sequence (x_n) of points in a metric space (M, d) converges to a point x in M, if and only if every subsequence $(x_{n'})$ has a subsequence $(x_{n''})$ converging to x. Taking $M = \mathbb{R}$ it is an immediate consequence of (2.4) that $\sigma_n \xrightarrow{\text{w}} \sigma$ if and only if any subsequence $(\sigma_{n'})$ has a subsequence $(\sigma_{n''})$ which converges weakly to σ. This observation, which we shall make use of in the folowing, follows also from the fact, that weak convergence can be viewed as convergence w.r.t. a certain metric on the set of bounded measures on $\mathbb{R}$ (the Lévy metric).

Lemma 2.4. *Let $(X_{n,m})_{n,m\in\mathbb{N}}$ be a family of random variables indexed by $\mathbb{N}\times\mathbb{N}$ and all defined on the same probability space $(\Omega, \mathcal{F}, P)$. Assume that*

$$\forall u \in \mathbb{R}\colon \int_{\mathbb{R}} \mathrm{e}^{\mathrm{i}tu}\, L\{X_{n,m}\}(\mathrm{d}t) \to 1, \quad \textit{as } n, m \to \infty. \tag{2.5}$$

Then $X_{n,m} \xrightarrow{\text{P}} 0$, as $n, m \to \infty$, in the sense that

$$\forall \epsilon > 0\colon P(|X_{n,m}| > \epsilon) \to 0, \quad \textit{as } n, m \to \infty. \tag{2.6}$$

Proof. This is, of course, a variant of the usual continuity theorem for characteristic functions. For completeness, we include a proof.

To prove (2.6), it suffices, by a standard argument, to prove that $L\{X_{n,m}\} \xrightarrow{\text{w}} \delta_0$, as $n, m \to \infty$, i.e. that

$$\forall f \in C_b(\mathbb{R})\colon \int_{\mathbb{R}} f(t)\, L\{X_{n,m}\}(\mathrm{d}t) \longrightarrow \int_{\mathbb{R}} f(t)\, \delta_0(\mathrm{d}t) = f(0), \quad \text{as } n, m \to \infty, \tag{2.7}$$

where $C_b(\mathbb{R})$ denotes the space of continuous bounded functions $f\colon \mathbb{R} \to \mathbb{R}$.

So assume that (2.7) is not satisfied. Then we may choose f in $C_b(\mathbb{R})$ and ϵ in $]0, \infty[$ such that

$$\forall N \in \mathbb{N}\ \exists n, m \geq N\colon \Big| \int_{\mathbb{R}} f(t)\, L\{X_{n,m}\}(\mathrm{d}t) - f(0) \Big| \geq \epsilon.$$

By an inductive argument, we may choose a sequence $n_1 \leq n_2 < n_3 \leq n_4 < \cdots$, of positive integers, such that

$$\forall k \in \mathbb{N}\colon \Big| \int_{\mathbb{R}} f(t)\, L\{X_{n_{2k}, n_{2k-1}}\}(\mathrm{d}t) - f(0) \Big| \geq \epsilon. \tag{2.8}$$

On the other hand, it follows from (2.5) that

$$\forall u \in \mathbb{R}\colon \int_{\mathbb{R}} \mathrm{e}^{\mathrm{i}tu}\, L\{X_{n_{2k}, n_{2k-1}}\}(\mathrm{d}t) \to 1, \quad \text{as } k \to \infty,$$

so by the usual continuity theorem for characteristic functions, we find that $L\{X_{n_{2k}, n_{2k-1}}\} \xrightarrow{\text{w}} \delta_0$. But this contradicts (2.8). □

Lemma 2.5. *Assume that $0 \le a < b < \infty$, and let $f\colon [a,b] \to \mathbb{R}$ be a continuous function. Let, further, $(X_t)_{t\ge 0}$ be a (classical) Lévy process, and put $\mu = L\{X_1\}$. Then the stochastic integral $\int_a^b f(t)\,\mathrm{d}X_t$ exists as the limit, in probability, of approximating Riemann sums. Furthermore, $L\{\int_a^b f(t)\,\mathrm{d}X_t\} \in \mathcal{ID}(*)$, and*

$$C\{u \ddagger \textstyle\int_a^b f(t)\,\mathrm{d}X_t\} = \int_a^b C_\mu(uf(t))\,\mathrm{d}t,$$

for all u in $\mathbb{R}$.

Proof. This is well-known, but, for completeness, we sketch the proof: By definition (cf. [Lu75]), $\int_a^b f(t)\,\mathrm{d}X_t$ is the limit in probability of the Riemann sums:

$$R_n := \sum_{j=1}^n f(t_j^{(n)})\big(X_{t_j^{(n)}} - X_{t_{j-1}^{(n)}}\big),$$

where, for each n, $a = t_0^{(n)} < t_1^{(n)} < \cdots < t_n^{(n)} = b$ is a subdivision of $[a,b]$, such that $\max_{j=1,2,\dots,n}(t_j^{(n)} - t_{j-1}^{(n)}) \to 0$ as $n \to \infty$. Since (X_t) has stationary, independent increments, it follows that for any u in $\mathbb{R}$,

$$\begin{aligned}
C\{u \ddagger R_n\} &= \sum_{j=1}^n C\{f(t_j^{(n)})u \ddagger (X_{t_j^{(n)}} - X_{t_{j-1}^{(n)}})\} \\
&= \sum_{j=1}^n C\{f(t_j^{(n)})u \ddagger X_{t_j^{(n)} - t_{j-1}^{(n)}}\} \\
&= \sum_{j=1}^n C_\mu(f(t_j^{(n)})u)\cdot(t_j^{(n)} - t_{j-1}^{(n)}),
\end{aligned}$$

where, in the last equality, we used [Sa99, Theorem 7.10]. Since C_μ and f are both continuous, it follows that

$$C\{u \ddagger \textstyle\int_a^b f(t)\,\mathrm{d}X_t\} = \lim_{n\to\infty} \sum_{j=1}^n C_\mu(f(t_j^{(n)})u)\cdot(t_j^{(n)} - t_{j-1}^{(n)}) = \int_a^b C_\mu(f(t)u)\,\mathrm{d}t,$$

for any u in $\mathbb{R}$. □

Proposition 2.6. *Assume that $0 \le a < b \le \infty$, and let $f\colon]a,b[\to \mathbb{R}$ be a continuous function. Let, further, $(X_t)_{t\ge 0}$ be a classical Lévy process, and put $\mu = L\{X_1\}$. Assume that*

$$\forall u \in \mathbb{R}\colon \int_a^b \big|C_\mu(uf(t))\big|\,\mathrm{d}t < \infty.$$

Then the stochastic integral $\int_a^b f(t)\,\mathrm{d}X_t$ *exists as the limit, in probability, of the sequence* $(\int_{a_n}^{b_n} f(t)\,\mathrm{d}X_t)_{n\in\mathbb{N}}$*, where* (a_n) *and* (b_n) *are arbitrary sequences in* $]a,b[$ *such that* $a_n \le b_n$ *for all* n *and* $a_n \searrow a$ *and* $b_n \nearrow b$ *as* $n \to \infty$.

Furthermore, $L\{\int_a^b f(t)\,\mathrm{d}X_t\} \in \mathcal{ID}(*)$ *and*

$$C\{u \ddagger \textstyle\int_a^b f(t)\,\mathrm{d}X_t\} = \displaystyle\int_a^b C_\mu(uf(t))\,\mathrm{d}t, \tag{2.9}$$

for all u *in* $\mathbb{R}$.

Proof. Let (a_n) and (b_n) be arbitrary sequences in $]a,b[$, such that $a_n \le b_n$ for all n and $a_n \searrow a$ and $b_n \nearrow b$ as $n \to \infty$. Then, for each n, consider the stochastic integral $\int_{a_n}^{b_n} f(t)\,\mathrm{d}X_t$. Since the topology corresponding to convergence in probability is complete, the convergence of the sequence $(\int_{a_n}^{b_n} f(t)\,\mathrm{d}X_t)_{n\in\mathbb{N}}$ will follow, once we have verified that it is a Cauchy sequence. Towards this end, note that whenever $n > m$ we have that

$$\int_{a_n}^{b_n} f(t)\,\mathrm{d}X_t - \int_{a_m}^{b_m} f(t)\,\mathrm{d}X_t = \int_{a_n}^{a_m} f(t)\,\mathrm{d}X_t + \int_{b_m}^{b_n} f(t)\,\mathrm{d}X_t,$$

so it suffices to show that

$$\int_{a_n}^{a_m} f(t)\,\mathrm{d}X_t \xrightarrow{\ \mathrm{P}\ } 0 \quad \text{and} \quad \int_{b_m}^{b_n} f(t)\,\mathrm{d}X_t \xrightarrow{\ \mathrm{P}\ } 0, \quad \text{as } n,m \to \infty.$$

By Lemma 2.4, this, in turn, will follow if we prove that

$$\forall u \in \mathbb{R}\colon C\{u \ddagger \textstyle\int_{a_n}^{a_m} f(t)\,\mathrm{d}X_t\} \longrightarrow 0, \qquad \text{as } n,m \to \infty,$$

and

$$\forall u \in \mathbb{R}\colon C\{u \ddagger \textstyle\int_{b_m}^{b_n} f(t)\,\mathrm{d}X_t\} \longrightarrow 0, \qquad \text{as } n,m \to \infty. \tag{2.10}$$

But for n,m in $\mathbb{N}$, $m < n$, it follows from Lemma 2.5 that

$$\left|C\{u \ddagger \textstyle\int_{a_n}^{a_m} f(t)\,\mathrm{d}X_t\}\right| \le \displaystyle\int_{a_n}^{a_m} \left|C_\mu(uf(t))\right|\mathrm{d}t, \tag{2.11}$$

and since $\int_a^b |C_\mu(uf(t))|\,\mathrm{d}t < \infty$, the right hand side of (2.11) tends to 0 as $n,m \to \infty$. Statement (2.10) follows similarly.

To prove that $\lim_{n\to\infty} \int_{a_n}^{b_n} f(t)\,\mathrm{d}X_t$ does not depend on the choice of sequences (a_n) and (b_n), let (a_n') and (b_n') be sequences in $]a,b[$, also satisfying that $a_n' \le b_n'$ for all n, and that $a_n' \searrow a$ and $b_n' \nearrow b$ as $n \to \infty$. We may then, by an inductive argument, choose sequences $n_1 < n_2 < n_3 < \cdots$ and $m_1 < m_2 < m_3 \cdots$ of positive integers, such that

$$a_{n_1} > a'_{m_1} > a_{n_2} > a'_{m_2} > \cdots, \quad \text{and} \quad b_{n_1} < b'_{m_1} < b_{n_2} < b'_{m_2} < \cdots.$$

Consider then the sequences (a''_k) and (b''_k) given by:

$$a''_{2k-1} = a_{n_k},\ a''_{2k} = a'_{m_k}, \quad \text{and} \quad b''_{2k-1} = b_{n_k},\ b''_{2k} = b'_{m_k}, \qquad (k \in \mathbb{N}).$$

Then $a''_k \leq b''_k$ for all k, and $a''_k \searrow a$ and $b''_k \nearrow b$ as $k \to \infty$. Thus, by the argument given above, all of the following limits exist (in probability), and, by "sub-sequence considerations", they have to be equal:

$$\begin{aligned}
\lim_{n\to\infty} \int_{a_n}^{b_n} f(t)\,\mathrm{d}X_t &= \lim_{k\to\infty} \int_{a_{n_k}}^{b_{n_k}} f(t)\,\mathrm{d}X_t = \lim_{k\to\infty} \int_{a''_{2k-1}}^{b''_{2k-1}} f(t)\,\mathrm{d}X_t \\
&= \lim_{k\to\infty} \int_{a''_k}^{b''_k} f(t)\,\mathrm{d}X_t = \lim_{k\to\infty} \int_{a''_{2k}}^{b''_{2k}} f(t)\,\mathrm{d}X_t \\
&= \lim_{k\to\infty} \int_{a'_{m_k}}^{b'_{m_k}} f(t)\,\mathrm{d}X_t = \lim_{n\to\infty} \int_{a'_n}^{b'_n} f(t)\,\mathrm{d}X_t,
\end{aligned}$$

as desired.

To verify, finally, the last statements of the proposition, let (a_n) and (b_n) be sequences as above, so that, by definition, $\int_a^b f(t)\,\mathrm{d}X_t = \lim_{n\to\infty} \int_{a_n}^{b_n} f(t)\,\mathrm{d}X_t$ in probability. Since $\mathcal{ID}(*)$ is closed under weak convergence, this implies that $L\{\int_a^b f(t)\,\mathrm{d}X_t\} \in \mathcal{ID}(*)$. To prove (2.9), we find next, using Gnedenko's theorem (cf. [GnKo68, §19, Theorem 1] and Lemma 2.5, that

$$\begin{aligned}
C\{u \ddagger \textstyle\int_a^b f(t)\,\mathrm{d}X_t\} &= \lim_{n\to\infty} C\{u \ddagger \textstyle\int_{a_n}^{b_n} f(t)\,\mathrm{d}X_t\} \\
&= \lim_{n\to\infty} \int_{a_n}^{b_n} C_\mu(uf(t))\,\mathrm{d}t = \int_a^b C_\mu(uf(t))\,\mathrm{d}t,
\end{aligned}$$

for any u in $\mathbb{R}$, and where the last equality follows from the assumption that $\int_a^b |C_\mu(uf(t))|\,\mathrm{d}t < \infty$. This concludes the proof. □

2.4 The Classical Lévy-Itô Decomposition

The Lévy-Itô decomposition represents a (classical) Lévy process (X_t) as the sum of two independent Lévy processes, the first of which is continuous (and hence a Brownian motion) and the second of which is, loosely speaking, the sum of the jumps of (X_t). In order to rigorously describe the sum of jumps part, one needs to introduce the notion of Poisson random measures. Before doing so, we introduce some notation: For any λ in $[0, \infty]$ we denote by $\mathrm{Poiss}^*(\lambda)$ the (classical) Poisson distribution with mean λ. In particular, $\mathrm{Poiss}^*(0) = \delta_0$ and $\mathrm{Poiss}^*(\infty) = \delta_\infty$.

Definition 2.7. *Let $(\Theta, \mathcal{E}, \nu)$ be a σ-finite measure space and let $(\Omega, \mathcal{F}, P)$ be a probability space. A Poisson random measure on $(\Theta, \mathcal{E}, \nu)$ and defined on $(\Omega, \mathcal{F}, P)$ is a mapping $N: \mathcal{E} \times \Omega \to [0, \infty]$, satisfying the following conditions:*

(i) *For each* E *in* $\mathcal{E}$, $N(E) = N(E, \cdot)$ *is a random variable on* $(\Omega, \mathcal{F}, P)$.
(ii) *For each* E *in* $\mathcal{E}$, $L\{N(E)\} = \text{Poiss}^*(\nu(E))$.
(iii) *If* $E_1, \dots, E_n$ *are disjoint sets from* $\mathcal{E}$, *then* $N(E_1), \dots, N(E_n)$ *are independent random variables.*
(iv) *For each fixed* ω *in* Ω, *the mapping* $E \mapsto N(E, \omega)$ *is a (positive) measure on* $\mathcal{E}$.

In the setting of Definition 2.7, the measure ν is called the *intensity measure* for the Poisson random measure N. Let $(\Theta, \mathcal{E}, \nu)$ be a σ-finite measure space, and let N be a Poisson random measure on it (defined on some probability space $(\Omega, \mathcal{F}, P)$). Then for any $\mathcal{E}$-measurable function $f\colon \Theta \to [0, \infty]$, we may, for all ω in Ω, consider the integral $\int_\Theta f(\theta)\, N(\mathrm{d}\theta, \omega)$. We obtain, thus, an everywhere defined mapping on Ω, given by: $\omega \mapsto \int_\Theta f(\theta)\, N(\mathrm{d}\theta, \omega)$. This observation is the starting point for the theory of integration with respect to Poisson random measures, from which we shall need the following basic properties:

Proposition 2.8. *Let* N *be a Poisson random measure on the* σ*-finite measure space* $(\Theta, \mathcal{E}, \nu)$, *defined on the probability space* $(\Omega, \mathcal{F}, P)$.

(i) *For any positive* $\mathcal{E}$*-measurable function* $f\colon \Theta \to [0, \infty]$, $\int_\Theta f(\theta)\, N(\mathrm{d}\theta)$ *is an* $\mathcal{F}$*-measurable positive function, and*

$$\mathbb{E}\Big\{\int_\Theta f(\theta)\, N(\mathrm{d}\theta)\Big\} = \int_\Theta f\, \mathrm{d}\nu.$$

(ii) *If* f *is a real-valued function in* $\mathcal{L}^1(\Theta, \mathcal{E}, \nu)$, *then* $f \in \mathcal{L}^1(\Theta, \mathcal{E}, N(\cdot, \omega))$ *for almost all* ω *in* Ω, $\int_\Theta f(\theta)\, N(\mathrm{d}\theta) \in \mathcal{L}^1(\Omega, \mathcal{F}, P)$ *and*

$$\mathbb{E}\Big\{\int_\Theta f(\theta)\, N(\mathrm{d}\theta)\Big\} = \int_\Theta f\, \mathrm{d}\nu.$$

The proof of the above proposition follows the usual pattern, proving it first for simple (positive) $\mathcal{E}$-measurable functions and then, via an approximation argument, obtaining the results in general. We shall adapt the same method in developing integration theory with respect to free Poisson random measures in Section 6.4 below.

We are now in a position to state the Lévy-Itô decomposition for classical Lévy processes. We denote the Lebesgue measure on $\mathbb{R}$ by Leb.

Theorem 2.9 (Lévy-Itô Decomposition). *Let* (X_t) *be a classical (genuine) Lévy process, defined on a probability space* $(\Omega, \mathcal{F}, P)$, *and let* ρ *be the Lévy measure appearing in the generating triplet for* $L\{X_1\}$.

(i) *Assume that* $\int_{-1}^{1} |x|\, \rho(\mathrm{d}x) < \infty$. *Then* (X_t) *has a representation in the form:*

$$X_t \overset{\text{a.s.}}{=} \gamma t + \sqrt{a} B_t + \int_{]0,t] \times \mathbb{R}} x\, N(\mathrm{d}s, \mathrm{d}x), \tag{2.12}$$

where $\gamma \in \mathbb{R}$, $a \geq 0$, (B_t) *is a Brownian motion and* N *is a Poisson random measure on* $(]0,\infty[\times\mathbb{R}, \mathrm{Leb}\otimes\rho)$. *Furthermore, the last two terms on the right hand side of* (2.12) *are independent Lévy processes on* $(\Omega, \mathcal{F}, P)$.

(ii) *If* $\int_{-1}^{1} |x|\,\rho(\mathrm{d}x) = \infty$, *then we still have a decomposition like* (2.12), *but the integral* $\int_{]0,t]\times\mathbb{R}} x\, N(\mathrm{d}s, \mathrm{d}x)$ *no longer makes sense and has to be replaced by the limit:*

$$Y_t = \lim_{\epsilon\searrow 0}\Big[\int_{]0,t]\times(\mathbb{R}\setminus[-\epsilon,\epsilon])} xN(\mathrm{d}u,\mathrm{d}x) - \int_{]0,t]\times([-1,1]\setminus[-\epsilon,\epsilon])} x\mathrm{Leb}\otimes\rho(\mathrm{d}u,\mathrm{d}x)\Big].$$

The process (Y_t) *is, again, a Lévy process, which is independent of* (B_t).

The symbol $\overset{\text{a.s.}}{=}$ in (2.12) means that the two random variables are equal with probability 1 (a.s. stands for "almost surely"). The Poisson random measure N appearing in the right hand side of (2.12) is, specifically, given by

$$N(E,\omega) = \#\{s \in]0,\infty[\,\big|\, (s, \Delta X_s(\omega)) \in E\},$$

for any Borel subset E of $]0,\infty[\times(\mathbb{R}\setminus\{0\})$, and where $\Delta X_s = X_s - \lim_{u\nearrow s} X_u$. Consequently, the integral in the right hand side of (2.12) is, indeed, the sum of the jumps of X_t until time t: $\int_{]0,t]\times\mathbb{R}} x\, N(\mathrm{d}s, \mathrm{d}x) = \sum_{s\leq t} \Delta X_s$. The condition $\int_{-1}^{1} |x|\,\rho(\mathrm{d}x) < \infty$ ensures that this sum converges. Without that condition, one has to consider the "compensated sums of jumps" given by the process (Y_t). For a proof of Theorem 2.9 we refer to [Sa99].

2.5 Classes of Infinitely Divisible Probability Measures

In the following, we study, in various connections, dilations of Borel measures by constants. If ρ is a Borel measure on $\mathbb{R}$ and c is a non-zero real constant, then the dilation of ρ by c is the measure $D_c\rho$ given by

$$D_c\rho(B) = \rho(c^{-1}B),$$

for any Borel set B. Furthermore, we put $D_0\rho = \delta_0$ (the Dirac measure at 0). We shall also make use of terminology like

$$D_c\rho(\mathrm{d}x) = \rho(c^{-1}\mathrm{d}x),$$

whenever $c \neq 0$. With this notation at hand, we now introduce several important classes of infinitely divisible probability measures on $\mathbb{R}$.

In classical probability theory, we have the following fundamental hierarchy:

$$\mathcal{G}(*) \subset \mathcal{S}(*) \subset \mathcal{R}(*) \subset \mathcal{T}(*) \subset \begin{Bmatrix} \mathcal{L}(*) \\ \mathcal{B}(*) \end{Bmatrix} \subset \mathcal{ID}(*) \subset \mathcal{P}, \tag{2.13}$$

where

(i) $\mathcal{P}$ is the class of all probability measures on $\mathbb{R}$.
(ii) $\mathcal{ID}(*)$ is the class of infinitely divisible probability measures on $\mathbb{R}$ (as defined above).
(iii) $\mathcal{L}(*)$ is the class of selfdecomposable probability measures on $\mathbb{R}$, i.e.

$$\mu \in \mathcal{L}(*) \iff \forall c \in \,]0,1[\; \exists \mu_c \in \mathcal{P}\colon \mu = D_c\mu * \mu_c.$$

(iv) $\mathcal{B}(*)$ is the *Goldie-Steutel-Bondesson class*, i.e. the smallest subclass of $\mathcal{ID}(*)$, which contains all mixtures of positive and negative exponential distributions[2] and is closed under convolution and weak limits.
(v) $\mathcal{T}(*)$ is the *Thorin Class*, i.e. the smallest subclass of $\mathcal{ID}(*)$, which contains all positive and negative Gamma distributions[2] and is closed under convolution and weak limits.
(vi) $\mathcal{R}(*)$ is the class of tempered stable distributions, which will defined below in terms of the Lévy-Khintchine representation.
(vii) $\mathcal{S}(*)$ is the class of stable probability measures on $\mathbb{R}$, i.e.

$$\mu \in \mathcal{S}(*) \iff \{\psi(\mu) \mid \psi\colon \mathbb{R} \to \mathbb{R},\ \text{increasing affine transformation}\}$$
$$\text{is closed under convolution } * .$$

(viii) $\mathcal{G}(*)$ is the class of Gaussian (or normal) distributions on $\mathbb{R}$.

The classes of probability measures, defined above, are all of considerable importance in classical probability and are of major applied interest. In particular the classes $\mathcal{S}(*)$ and $\mathcal{L}(*)$ have received a lot of attention. This is, partly, explained by their characterizations as limit distributions of certain types of sums of independent random variables. Briefly, the stable laws are those that occur as limiting distributions for $n \to \infty$ of affine transformations of sums $X_1 + \cdots + X_n$ of independent identically distributed random variables (subject to the assumption of uniform asymptotic neglibility). Dropping the assumption of identical distribution one arrives at the class $\mathcal{L}(*)$. Finally, the class $\mathcal{ID}(*)$ of all infinitely divisible distributions consists of the limiting laws for sums of independent random variables of the form $X_{n1} + \cdots + X_{nk_n}$ (again subject to the assumption of uniform asymptotic neglibility).

An alternative characterization of selfdecomposability says that (the distribution of) a random variable Y is selfdecomposable if and only if for all c in $]0,1[$ the characteristic function f of Y can be factorised as

$$f(\zeta) = f(c\zeta) f_c(\zeta), \tag{2.14}$$

for some characteristic function f_c (which then, as can be proved, necessarily corresponds to an infinitely divisible random variable Y_c). In other words, considering Y_c as independent of Y we have a representation in law

[2] A negative exponential (resp. Gamma) distribution is of the form $D_{-1}\mu$, where μ is a positive exponential (resp. Gamma) distribution.

$$Y \stackrel{\mathrm{d}}{=} cY + Y_c$$

(where the symbol $\stackrel{\mathrm{d}}{=}$ means that the random variables on the left and right hand side have the same distribution). This latter formulation makes the idea of selfdecomposability of immediate appeal from the viewpoint of mathematical modeling. Yet another key characterization is given by the following result which was first proved by Wolfe in [Wo82] and later generalized and strengthened by Jurek and Verwaat ([JuVe83], cf. also Jurek and Mason, [JuMa93, Theorem 3.6.6]): A random variable Y has law in $\mathcal{L}(*)$ if and only if Y has a representation of the form

$$Y \stackrel{\mathrm{d}}{=} \int_0^\infty \mathrm{e}^{-t}\, \mathrm{d}X_t, \tag{2.15}$$

where X_t is a Lévy process satisfying $\mathbb{E}\{\log(1+|X_1|)\} < \infty$. The process $X = (X_t)_{t\geq 0}$ is termed the *background driving Lévy process* or the BDLP corresponding to Y.

There is a very extensive literature on the theory and applications of stable laws. A standard reference for the theoretical properties is [SaTa94], but see also [Fe71] and [BaMiRe01]. In comparison, work on selfdecomposability has up till recently been somewhat limited. However, a comprehensive account of the theoretical aspects of selfdecomposability, and indeed of infinite divisibility in general, is now available in [Sa99]. Applications of selfdecomposability are discussed, inter alia, in [BrReTw82], [Ba98], [BaSh01a] and [BaSh01b].

The class $\mathcal{R}(*)$, its d-dimensional version $\mathcal{R}^d(*)$, and the associated Lévy processes and Ornstein-Uhlenbeck type processes were introduced and studied extensively by Rosinski (see [Ros04]), following earlier works by other authors on special instances of this kind of stochastic objects (see references in [Ros04]). These processes are of considerable interest as they exhibit stable like behaviour over short time spans and - in the Lévy process case - Gaussian behaviour for long lags. That paper also develops powerful series representations of shot noise type for the processes.

By $\mathcal{ID}^+(*)$ we denote the class of infinitely divisible probability measures, which are concentrated on $[0,\infty[$. The classes $\mathcal{S}^+(*), \mathcal{R}^+(*), \mathcal{T}^+(*), \mathcal{B}^+(*)$ and $\mathcal{L}^+(*)$ are defined similarly. The class $\mathcal{T}^+(*)$, in particular, is the class of measures which was originally studied by O. Thorin in [Th77]. He introduced it as the smallest subclass of $\mathcal{ID}(*)$, which contains the Gamma distributions and is closed under convolution and weak limits. This group of distributions is also referred to as generalized gamma convolutions and have been extensively studied by Bondesson in [Bo92]. (It is noteworthy, in the present context, that Bondesson uses Pick functions, which are essentially Cauchy transforms, as a main tool in his investigations. The Cauchy transform also occur as a key tool in the study of free infinite divisibility; see Section 4.4).

Example 2.10. An important class of generalized Gamma convolutions are the generalized inverse Gaussian distributions: Assume that λ in $\mathbb{R}$ and γ, δ in

$[0, \infty[$ satisfy the conditions: $\lambda < 0 \Rightarrow \delta > 0$, $\lambda = 0 \Rightarrow \gamma, \delta > 0$ and $\lambda > 0 \Rightarrow \gamma > 0$. Then the *generalized inverse Gaussian distribution* $\mathrm{GIG}(\lambda, \delta, \gamma)$ is the distribution on $\mathbb{R}_+$ with density (w.r.t. Lebesgue measure) given by

$$g(t; \lambda, \delta, \gamma) = \frac{(\gamma/\delta)^\lambda}{2K_\lambda(\delta\gamma)} t^{\lambda-1} \exp\{ - \tfrac{1}{2}(\delta^2 t^{-1} + \gamma^2 t)\}, \quad t \geq 0,$$

where K_λ is the modified Bessel function of the third kind and with index λ. For all λ, δ, γ (subject to the above restrictions) $\mathrm{GIG}(\lambda, \delta, \gamma)$ belongs to $\mathcal{T}^+(*)$, and it is not stable unless $\lambda = -\frac{1}{2}$ and $\gamma = 0$. For special choices of the parameters, one obtains the gamma distributions (and hence the exponential and χ^2 distributions), the inverse Gaussian distributions, the reciprocal inverse Gaussian distributions[3] and the reciprocal gamma distributions.

Example 2.11. A particularly important group of examples of selfdecomposable laws, supported on the whole real line, are the marginal laws of subordinated Brownian motion with drift, when the subordinator process is generated by one of the generalized gamma convolutions. The induced selfdecomposability of the marginals follows from a result due to Sato (cf. [Sa00]).

We introduce next some notation that will be convenient in Section 3.3 below. There, we shall also consider translations of the measures in the classes $\mathcal{T}^+(*)$, $\mathcal{L}^+(*)$ and $\mathcal{ID}^+(*)$. For a real constant c, we consider the mapping $\tau_c \colon \mathbb{R} \to \mathbb{R}$ given by

$$\tau_c(x) = x + c, \qquad (x \in \mathbb{R}),$$

i.e. τ_c is translation by c. For a Borel measure μ on $\mathbb{R}$, we may then consider the translated measure $\tau_c(\mu)$ given by

$$\tau_c(\mu)(B) = \mu(B - c),$$

for any Borel set B in $\mathbb{R}$. Note, in particular, that if μ is infinitely divisible with characteristic triplet (a, ρ, η), then $\tau_c(\mu)$ is infinitely divisible with characteristic triplet $(a, \rho, \eta + c)$.

Definition 2.12. *We introduce the following notation:*

$$\mathcal{ID}^+_\tau(*) = \{\mu \in \mathcal{ID}(*) \mid \exists c \in \mathbb{R} \colon \tau_c(\mu) \in \mathcal{ID}^+(*)\}$$

$$\mathcal{L}^+_\tau(*) = \{\mu \in \mathcal{ID}(*) \mid \exists c \in \mathbb{R} \colon \tau_c(\mu) \in \mathcal{L}^+(*)\} = \mathcal{ID}^+_\tau \cap \mathcal{L}(*)$$

$$\mathcal{T}^+_\tau(*) = \{\mu \in \mathcal{ID}(*) \mid \exists c \in \mathbb{R} \colon \tau_c(\mu) \in \mathcal{T}^+(*)\} = \mathcal{ID}^+_\tau \cap \mathcal{T}(*).$$

---

[3]The inverse Gaussian distributions and the reciprocal inverse Gaussian distributions are, respectively, the first and the last passage time distributions to a constant level by a Brownian motion with drift.

Remark 2.13. The probability measures in $\mathcal{ID}^+(*)$ are characterized among the measures in $\mathcal{ID}(*)$ as those with characteristic triplets in the form $(0,\rho,\eta)$, where ρ is concentrated on $[0,\infty[$, $\int_{[0,1]} t\,\rho(\mathrm{d}t) < \infty$ and $\eta \geq \int_{[0,1]} t\,\rho(\mathrm{d}t)$ (cf. [Sa99, Theorem 24.11]). Consequently, the class $\mathcal{ID}^+_\tau(*)$ can be characterized as that of measures in $\mathcal{ID}(*)$ with generating triplets in the form $(0,\eta,\rho)$, where ρ is concentrated on $[0,\infty[$ and $\int_{[0,1]} t\,\rho(\mathrm{d}t) < \infty$.

Characterization in Terms of Lévy Measures

We shall say that a nonnegative function k with domain $\mathbb{R}\backslash\{0\}$ is *monotone on* $\mathbb{R}\backslash\{0\}$ if k is increasing on $(-\infty,0)$ and decreasing on $(0,\infty)$. And we say that k is *completely monotone on* $\mathbb{R}\backslash\{0\}$ if k is of the form

$$k(t) = \begin{cases} \int_0^\infty \mathrm{e}^{-ts}\nu(\mathrm{d}s), & \text{for } t > 0 \\ \int_{-\infty}^0 \mathrm{e}^{-ts}\nu(\mathrm{d}s), & \text{for } t < 0 \end{cases} \tag{2.16}$$

for some Borel measure ν on $\mathbb{R}\backslash\{0\}$. Note in this case that ν is necessarily a Radon measure on $\mathbb{R}\setminus\{0\}$. Indeed, for any compact subset K of $]0,\infty[$, we may consider the strictly positive number $m := \inf_{s\in K} \mathrm{e}^{-s}$. Then,

$$\nu(K) \leq m^{-1}\int_K \mathrm{e}^{-s}\,\nu(\mathrm{d}s) \leq m^{-1}\int_0^\infty \mathrm{e}^{-s}\,\nu(\mathrm{d}s) = m^{-1}k(1) < \infty.$$

Similarly, $\nu(K) < \infty$ for any compact subset of K of $]-\infty,0[$.

With the notation just introduced, we can now state simple characterizations of the Lévy measures of each of the classes $\mathcal{S}(*), \mathcal{T}(*), \mathcal{R}(*), \mathcal{L}(*), \mathcal{B}(*)$ as follows. In all cases the Lévy measure has a density r of the form

$$r(t) = \begin{cases} c_+ t^{-a_+} k(t), & \text{for } t > 0, \\ c_- |t|^{-a_-} k(t), & \text{for } t < 0, \end{cases} \tag{2.17}$$

where a_+, a_-, c_+, c_- are non-negative constants and where $k \geq 0$ is monotone on $\mathbb{R}\backslash\{0\}$.

- The Lévy measures of $\mathcal{S}(*)$ are characterized by having densities r of the form (2.17) with $a_\pm = 1+\alpha$, $\alpha \in]0,2[$, and k constant on $\mathbb{R}_{<0}$ and on $\mathbb{R}_{>0}$.
- The Lévy measures of $\mathcal{R}(*)$ are characterized by having densities r of the form (2.17) with $a_\pm = 1+\alpha$, $\alpha \in]0,2[$, and k completely monotone on $\mathbb{R}\backslash\{0\}$ with $k(0+) = k(0-) = 1$.
- The Lévy measures of $\mathcal{T}(*)$ are characterized by having densities r of the form (2.17) with $a_\pm = 1$ and k completely monotone on $\mathbb{R}\backslash\{0\}$.
- The Lévy measures of $\mathcal{L}(*)$ are characterized by having densities r of the form (2.17) with $a_\pm = 1$ and k monotone on $\mathbb{R}\backslash\{0\}$.
- The Lévy measures of $\mathcal{B}(*)$ are characterized by having densities r of the form (2.17) with $a_\pm = 0$ and k completely monotone on $\mathbb{R}\backslash\{0\}$.

In the case of $\mathcal{S}(*)$ and $\mathcal{L}(*)$ these characterizations are well known, see for instance [Sa99]. For $\mathcal{T}(*)$, $\mathcal{R}(*)$ and $\mathcal{B}(*)$ we indicate the proofs in Section 3.

3 Upsilon Mappings

The term Upsilon mappings is used to indicate a class of one-to-one regularizing mappings from the set of Lévy measures into itself or, equivalently, from the set of infinitely divisible distributions into itself. They are defined as deterministic integrals but have a third interpretation in terms of stochastic integrals with respect to Lévy processes. In addition to the regularizing effect, the mappings have simple relations to the classes of infinitely divisible laws discussed in the foregoing section. Some extensions to multivariate settings are briefly discussed at the end of the section.

3.1 The Mapping Υ_0

Let ρ be a Borel measure on $\mathbb{R}$, and consider the family $(D_x\rho)_{x>0}$ of Borel measures on $\mathbb{R}$. Assume that ρ has density r w.r.t. some σ-finite Borel measure σ on $\mathbb{R}$: $\rho(\mathrm{d}t) = r(t)\,\sigma(\mathrm{d}t)$. Then $(D_x\rho)_{x>0}$ is a Markov kernel, i.e. for any Borel subset B of $\mathbb{R}$, the mapping $x \mapsto D_x\rho(B)$ is Borel measurable. Indeed, for any x in $]0,\infty[$ we have

$$D_x\rho(B) = \rho(x^{-1}B) = \int_{\mathbb{R}} 1_{x^{-1}B}(t)r(t)\,\sigma(\mathrm{d}t) = \int_{\mathbb{R}} 1_B(xt)r(t)\,\sigma(\mathrm{d}t).$$

Since the function $(t,x) \mapsto 1_B(tx)r(t)$ is a Borel function of two variables, and since σ is σ-finite, it follows from Tonelli's theorem that the function $x \mapsto \int_{\mathbb{R}} 1_B(xt)r(t)\,\sigma(\mathrm{d}t)$ is a Borel function, as claimed.

Assume now that ρ is Borel measure on $\mathbb{R}$, which has a density r w.r.t. some σ-finite Borel measure on $\mathbb{R}$. Then the above considerations allow us to define a new Borel measure $\tilde{\rho}$ on $\mathbb{R}$ by:

$$\tilde{\rho} = \int_0^\infty (D_x\rho)\mathrm{e}^{-x}\,\mathrm{d}x, \tag{3.1}$$

or more precisely:

$$\tilde{\rho}(B) = \int_0^\infty D_x\rho(B)\mathrm{e}^{-x}\,\mathrm{d}x,$$

for any Borel subset B of $\mathbb{R}$. In the following we usually assume that ρ is a σ-finite, although many of the results are actually valid in the slightly more general situation, where ρ is only assumed to have a (possibly infinite) density w.r.t. a σ-finite measure. In fact, we are mainly interested in the case where ρ is a Lévy measure (recall that Lévy measures are automatically σ-finite).

Definition 3.1. *Let $\mathfrak{M}(\mathbb{R})$ denote the class of all positive Borel measure on $\mathbb{R}$ and let $\mathfrak{M}_L(\mathbb{R})$ denote the subclass of all Lévy measure on $\mathbb{R}$. We then define a mapping $\Upsilon_0\colon \mathfrak{M}_L(\mathbb{R}) \to \mathfrak{M}(\mathbb{R})$ by*

$$\Upsilon_0(\rho) = \int_0^\infty (D_x\rho)\mathrm{e}^{-x}\,\mathrm{d}x, \qquad (\rho \in \mathfrak{M}_L(\mathbb{R})).$$

As we shall see at the end of this section, the range of Υ_0 is actually a genuine subset of $\mathfrak{M}_L(\mathbb{R})$ (cf. Corollary 3.10 below).

In the following we consider further, for a measure ρ on $\mathbb{R}$, the transformation of $\rho_{|\mathbb{R}\setminus\{0\}}$ by the mapping $x \mapsto x^{-1} : \mathbb{R}\setminus\{0\} \to \mathbb{R}\setminus\{0\}$ (here $\rho_{|\mathbb{R}\setminus\{0\}}$ denotes the restriction of ρ to $\mathbb{R}\setminus\{0\}$). The transformed measure will be denoted by ω and occasionally also by $\underset{\leftarrow}{\rho}$. Note that ω is σ-finite if ρ is, and that ρ is a Lévy measure if and only if $\rho(\{0\}) = 0$ and ω satisfies the property:

$$\int_{\mathbb{R}} \min\{1, s^{-2}\}\, \omega(\mathrm{d}s) < \infty. \tag{3.2}$$

Theorem 3.2. *Let ρ be a σ-finite Borel measure on $\mathbb{R}$, and consider the Borel function $\tilde{r} : \mathbb{R}\setminus\{0\} \to [0, \infty]$, given by*

$$\tilde{r}(t) = \begin{cases} \int_{]0,\infty[} s\mathrm{e}^{-ts}\, \omega(\mathrm{d}s), & \text{if } t > 0, \\ \int_{]-\infty,0[} |s|\mathrm{e}^{-ts}\, \omega(\mathrm{d}s), & \text{if } t < 0, \end{cases} \tag{3.3}$$

where ω is the transformation of $\rho_{|\mathbb{R}\setminus\{0\}}$ by the mapping $x \mapsto x^{-1} : \mathbb{R}\setminus\{0\} \to \mathbb{R}\setminus\{0\}$.

Then the measure $\tilde{\rho}$, defined in (3.1), is given by:

$$\tilde{\rho}(\mathrm{d}t) = \rho(\{0\})\delta_0(\mathrm{d}t) + \tilde{r}(t)\,\mathrm{d}t.$$

Proof. We have to show that

$$\tilde{\rho}(B) = \rho(\{0\})\delta_0(B) + \int_{B\setminus\{0\}} \tilde{r}(t)\,\mathrm{d}t, \tag{3.4}$$

for any Borel set B of $\mathbb{R}$. Clearly, it suffices to verify (3.4) in the two cases $B \subseteq [0, \infty[$ and $B \subseteq]-\infty, 0]$. If $B \subseteq [0, \infty[$, we find that

$$\begin{aligned} \tilde{\rho}(B) &= \int_0^\infty \Big(\int_{[0,\infty[} 1_B(s)\, D_x\rho(\mathrm{d}s) \Big) \mathrm{e}^{-x}\,\mathrm{d}x \\ &= \int_0^\infty \Big(\int_{[0,\infty[} 1_B(sx)\, \rho(\mathrm{d}s) \Big) \mathrm{e}^{-x}\,\mathrm{d}x \\ &= \int_{[0,\infty[} \Big(\int_0^\infty 1_B(sx)\mathrm{e}^{-x}\,\mathrm{d}x \Big) \rho(\mathrm{d}s). \end{aligned}$$

Using, for $s > 0$, the change of variable $u = sx$, we find that

$$\begin{aligned} \tilde{\rho}(B) &= \Big(1_B(0) \int_0^\infty \mathrm{e}^{-x}\,\mathrm{d}x \Big) \rho(\{0\}) + \int_{]0,\infty[} \Big(\int_0^\infty 1_B(u)\mathrm{e}^{-u/s}s^{-1}\,\mathrm{d}u \Big) \rho(\mathrm{d}s) \\ &= \rho(\{0\})\delta_0(B) + \int_0^\infty 1_B(u) \Big(\int_{]0,\infty[} s^{-1}\mathrm{e}^{-u/s}\, \rho(\mathrm{d}s) \Big)\,\mathrm{d}u \\ &= \rho(\{0\})\delta_0(B) + \int_0^\infty 1_B(u) \Big(\int_{]0,\infty[} s\mathrm{e}^{-us}\, \omega(\mathrm{d}s) \Big)\,\mathrm{d}u, \end{aligned}$$

as desired. The case $B \subseteq\,]-\infty, 0]$ is proved similarly or by applying, what we have just established, to the set $-B$ and the measure $D_{-1}\rho$. □

Corollary 3.3. *Let ρ be a σ-finite Borel measure on $\mathbb{R}$ and consider the measure $\tilde{\rho}$ given by (3.1). Then*

$$\tilde{\rho}(\{t\}) = \begin{cases} 0, & \text{if } t \in \mathbb{R} \setminus \{0\}, \\ \rho(\{0\}), & \text{if } t = 0. \end{cases}$$

Corollary 3.4. *Let $r\colon \mathbb{R} \to [0, \infty[$ be a non-negative Borel function and let ρ be the measure on $\mathbb{R}$ with density r w.r.t. Lebesgue measure: $\rho(\mathrm{d}t) = r(t)\,\mathrm{d}t$. Consider further the measure $\tilde{\rho}$ given by (3.1). Then $\tilde{\rho}$ is absolutely continuous w.r.t. Lebesgue measure and the density, $\tilde{r}$, is given by*

$$\tilde{r}(t) = \begin{cases} \int_0^\infty y^{-1} r(y^{-1}) \mathrm{e}^{-ty}\,\mathrm{d}y, & \text{if } t > 0, \\ \int_{-\infty}^0 -y^{-1} r(y^{-1}) \mathrm{e}^{-ty}\,\mathrm{d}y, & \text{if } t < 0. \end{cases}$$

Proof. This follows immediately from Theorem 3.2 together with the fact that the measure ω has density

$$s \mapsto s^{-2} r(s^{-1}), \qquad (s \in \mathbb{R} \setminus \{0\}),$$

w.r.t. Lebesgue measure. □

Corollary 3.5. *Let ρ be a Lévy measure on $\mathbb{R}$. Then the measure $\Upsilon_0(\rho)$ is absolutely continuous w.r.t. Lebesgue measure. The density, $\tilde{r}$, is given by* (3.3) *and is a C^∞-function on $\mathbb{R} \setminus \{0\}$.*

Proof. We only have to verify that $\tilde{r}$ is a C^∞-function on $\mathbb{R} \setminus \{0\}$. But this follows from the usual theorem on differentiation under the integral sign, since, by (3.2),

$$\int_{]0,\infty[} s^p \mathrm{e}^{-ts}\,\omega(\mathrm{d}s) < \infty \quad \text{and} \quad \int_{]-\infty,0[} |s|^p \mathrm{e}^{ts}\,\omega(\mathrm{d}s) < \infty,$$

for any t in $]0, \infty[$ and any p in $\mathbb{N}$. □

Proposition 3.6. *Let ρ be a σ-finite measure on $\mathbb{R}$, let $\tilde{\rho}$ be the measure given by* (3.1) *and let ω be the transformation of $\rho_{|\mathbb{R}\setminus\{0\}}$ under the mapping $t \mapsto t^{-1}$. We then have*

$$\tilde{\rho}([t, \infty[) = \int_0^\infty \mathrm{e}^{-ts}\,\omega(\mathrm{d}s), \qquad (t \in\,]0, \infty[), \tag{3.5}$$

and

$$\tilde{\rho}(]-\infty, t]) = \int_{-\infty}^0 \mathrm{e}^{-ts}\,\omega(\mathrm{d}s), \qquad (t \in\,]-\infty, 0[). \tag{3.6}$$

Proof. Using Theorem 3.2 we find, for $t > 0$, that

$$\tilde{\rho}([t,\infty[) = \int_t^\infty \Big(\int_{]0,\infty[} s\mathrm{e}^{-us}\,\omega(\mathrm{d}s)\Big)\,\mathrm{d}u = \int_{]0,\infty[} \Big(\int_t^\infty \mathrm{e}^{-us} s\,\mathrm{d}u\Big)\,\omega(\mathrm{d}s)$$
$$= \int_{]0,\infty[} \Big(\int_{ts}^\infty \mathrm{e}^{-x}\,\mathrm{d}x\Big)\,\omega(\mathrm{d}s) = \int_{]0,\infty[} \mathrm{e}^{-ts}\,\omega(\mathrm{d}s),$$

where we have used the change of variable $x = us$. Formula (3.6) is proved similarly. □

Corollary 3.7. *The mapping* $\Upsilon_0\colon \mathfrak{M}_L(\mathbb{R}) \to \mathfrak{M}(\mathbb{R})$ *is injective.*

Proof. Suppose $\rho \in \mathfrak{M}_L(\mathbb{R})$ and let ω be the transformation of $\rho_{|\mathbb{R}\setminus\{0\}}$ be the mapping $t \mapsto t^{-1}$. Let, further, ω_+ and ω_- denote the restrictions of ω to $]0,\infty[$ and $]-\infty,0[$, respectively. By (3.2) it follows then that the Laplace transform for ω_+ is well-defined on all of $]0,\infty[$. Furthermore, (3.5) shows that this Laplace transform is uniquely determined by $\tilde{\rho}$. Hence, by uniqueness of Laplace transforms (cf. [Fe71, Theorem 1a, Chapter XIII.1]), ω_+ is uniquely determined by $\tilde{\rho}$. Arguing similarly for the measure $D_{-1}\omega_-$, it follows that $D_{-1}\omega_-$ (and hence ω_-) is uniquely determined by $\tilde{\rho}$. Altogether, ω (and hence ρ) is uniquely determined by $\tilde{\rho}$. □

Proposition 3.8. *Let ρ be a σ-finite measure on $\mathbb{R}$ and let $\tilde{\rho}$ be the measure given by* (3.1)*. Then for any p in $[0,\infty[$, we have that*

$$\int_{\mathbb{R}} |t|^p\,\tilde{\rho}(\mathrm{d}t) = \Gamma(p+1)\int_{\mathbb{R}} |t|^p\,\rho(\mathrm{d}t).$$

In particular, the p'th moment of $\tilde{\rho}$ and ρ exist simultaneously, in which case

$$\int_{\mathbb{R}} t^p\,\tilde{\rho}(\mathrm{d}t) = \Gamma(p+1)\int_{\mathbb{R}} t^p\,\rho(\mathrm{d}t). \tag{3.7}$$

Proof. Let p from $[0,\infty[$ be given. Then

$$\int_{\mathbb{R}} |t|^p\,\tilde{\rho}(\mathrm{d}t) = \int_0^\infty \Big(\int_{\mathbb{R}} |t|^p\,D_x\rho(\mathrm{d}t)\Big)\mathrm{e}^{-x}\,\mathrm{d}x \int_0^\infty \Big(\int_{\mathbb{R}} |tx|^p\,\rho(\mathrm{d}t)\Big)\mathrm{e}^{-x}\,\mathrm{d}x$$
$$= \int_{\mathbb{R}} |t|^p \Big(\int_0^\infty x^p\mathrm{e}^{-x}\,\mathrm{d}x\Big)\,\rho(\mathrm{d}t) = \Gamma(p+1)\int_{\mathbb{R}} |t|^p\,\rho(\mathrm{d}t).$$

If the integrals above are finite, we can perform the same calculation without taking absolute values, and this establishes (3.7). □

Proposition 3.9. *Let ρ be a σ-finite Borel measure on $\mathbb{R}$ and let $\tilde{\rho}$ be the measure given by* (3.1)*. We then have*

$$\int_{\mathbb{R}\setminus[-1,1]} 1\,\tilde{\rho}(\mathrm{d}t) = \int_{\mathbb{R}\setminus\{0\}} \mathrm{e}^{-1/|t|}\,\rho(\mathrm{d}t) \tag{3.8}$$

$$\int_{[-1,1]} t^2\,\tilde{\rho}(\mathrm{d}t) = \int_{\mathbb{R}\setminus\{0\}} 2t^2 - \mathrm{e}^{-1/|t|}(1+2|t|+2t^2)\,\rho(\mathrm{d}t). \tag{3.9}$$

In particular

$$\int_{\mathbb{R}} \min\{1,t^2\}\,\tilde{\rho}(\mathrm{d}t) = \int_{\mathbb{R}\setminus\{0\}} 2t^2\big(1-\mathrm{e}^{-1/|t|}(|t|^{-1}+1)\big)\,\rho(\mathrm{d}t), \tag{3.10}$$

and consequently

$$\int_{\mathbb{R}} \min\{1,t^2\}\,\tilde{\rho}(\mathrm{d}t) < \infty \iff \int_{\mathbb{R}} \min\{1,t^2\}\,\rho(\mathrm{d}t) < \infty. \tag{3.11}$$

Proof. We note first that

$$\begin{aligned}
\int_{\mathbb{R}\setminus[-1,1]} 1\,\tilde{\rho}(\mathrm{d}t) &= \int_0^\infty \Big(\int_{\mathbb{R}} 1_{]1,\infty[}(|t|)\,D_x\rho(\mathrm{d}t)\Big)\mathrm{e}^{-x}\,\mathrm{d}x \\
&= \int_0^\infty \Big(\int_{\mathbb{R}} 1_{]1,\infty[}(|tx|)\,\rho(\mathrm{d}t)\Big)\mathrm{e}^{-x}\,\mathrm{d}x \\
&= \int_{\mathbb{R}\setminus\{0\}} \Big(\int_{1/|t|}^\infty \mathrm{e}^{-x}\,\mathrm{d}x\Big)\rho(\mathrm{d}t) \\
&= \int_{\mathbb{R}\setminus\{0\}} \mathrm{e}^{-1/|t|}\,\rho(\mathrm{d}t),
\end{aligned}$$

which proves (3.8). Regarding (3.9) we find that

$$\begin{aligned}
\int_{[-1,1]} t^2\,\tilde{\rho}(\mathrm{d}t) &= \int_0^\infty \Big(\int_{\mathbb{R}} 1_{[0,1]}(|t|)t^2\,D_x\rho(\mathrm{d}t)\Big)\mathrm{e}^{-x}\,\mathrm{d}x \\
&= \int_0^\infty \Big(\int_{\mathbb{R}} 1_{[0,1]}(|tx|)t^2x^2\,\rho(\mathrm{d}t)\Big)\mathrm{e}^{-x}\,\mathrm{d}x \\
&= \int_{\mathbb{R}\setminus\{0\}} \Big(\int_0^{1/|t|} x^2\mathrm{e}^{-x}\,\mathrm{d}x\Big)t^2\,\rho(\mathrm{d}t) \\
&= \int_{\mathbb{R}\setminus\{0\}} (2-\mathrm{e}^{-1/|t|}(t^{-2}+2|t|^{-1}+2))t^2\,\rho(\mathrm{d}t) \\
&= \int_{\mathbb{R}\setminus\{0\}} 2t^2 - \mathrm{e}^{-1/|t|}(1+2|t|+2t^2)\,\rho(\mathrm{d}t),
\end{aligned}$$

as claimed. Combining (3.8) and (3.9), we immediately get (3.10). To deduce finally (3.11), note first that for any positive u, we have by second order Taylor expansion

$$\frac{2}{u^2}\big(1 - \mathrm{e}^{-u}(u+1)\big) = \frac{2\mathrm{e}^{-u}}{u^2}\big(\mathrm{e}^u - u + 1\big) = \mathrm{e}^{\xi - u}, \tag{3.12}$$

for some number ξ in $]0, u[$. It follows thus that

$$\forall t \in \mathbb{R} \setminus \{0\} \colon 0 < 2t^2\big(1 - \mathrm{e}^{-1/|t|}(|t|^{-1} + 1)\big) \leq 1, \tag{3.13}$$

and from the upper bound together with (3.10), the implication "$\Leftarrow$" in (3.11) follows readily. Regarding the converse implication, note that (3.12) also shows that

$$\lim_{|t| \to \infty} 2t^2\big(1 - \mathrm{e}^{-1/|t|}(|t|^{-1} + 1)\big) = 1,$$

and together with the lower bound in (3.13), this implies that

$$\inf_{t \in \mathbb{R} \setminus [-1,1]} 2t^2\big(1 - \mathrm{e}^{-1/|t|}(|t|^{-1} + 1)\big) > 0. \tag{3.14}$$

Note also that

$$\lim_{t \to 0} 2\big(1 - \mathrm{e}^{-1/|t|}(|t|^{-1} + 1)\big) = 2 \lim_{u \to \infty} \big(1 - \mathrm{e}^{-u}(u+1)\big) = 2,$$

so that

$$\inf_{t \in [-1,1] \setminus \{0\}} 2\big(1 - \mathrm{e}^{-1/|t|}(|t|^{-1} + 1)\big) > 0. \tag{3.15}$$

Combining (3.14),(3.15) and (3.10), the implication "$\Rightarrow$" in (3.11) follows. This completes the proof. □

Corollary 3.10. *For any Lévy measure ρ on $\mathbb{R}$, $\Upsilon_0(\rho)$ is again a Lévy measure on $\mathbb{R}$. Moreover, a Lévy measure υ on $\mathbb{R}$ is in the range of Υ_0 if and only if the function $F_\upsilon \colon \mathbb{R} \setminus \{0\} \to [0, \infty[$ given by*

$$F_\upsilon(t) = \begin{cases} \upsilon(]-\infty, t]), & \text{if } t < 0, \\ \upsilon([t, \infty[), & \text{if } t > 0, \end{cases}$$

is completely monotone (cf. (2.16)*).*

Proof. It follows immediately from (3.11) that $\Upsilon(\rho)$ is a Lévy measure if ρ is.

Regarding the second statement of the corollary, we already saw in Proposition 3.6 that $F_{\Upsilon(\rho)}$ is completely monotone for any Lévy measure ρ on $\mathbb{R}$. Assume conversely that υ is a Lévy measure on $\mathbb{R}$, such that F_υ is completely monotone, i.e.

$$\upsilon([t, \infty[) = \int_0^\infty \mathrm{e}^{-ts}\, \omega(\mathrm{d}s), \qquad (t \in]0, \infty[),$$

and

$$\upsilon(]-\infty,t]) = \int_{-\infty}^{0} \mathrm{e}^{-ts}\,\omega(\mathrm{d}s), \qquad (t \in]-\infty,0[).$$

for some Radon measure ω on $\mathbb{R}\setminus\{0\}$. Now let ρ be the transformation of ω by the mapping $t \mapsto t^{-1}\colon \mathbb{R}\setminus\{0\} \to \mathbb{R}\setminus\{0\}$. Then ρ is clearly a Radon measure on $\mathbb{R}\setminus\{0\}$, too. Setting $\rho(\{0\}) = 0$, we may thus consider ρ as a σ-finite measure on $\mathbb{R}$. Applying then Proposition 3.6 to ρ, it follows that $\tilde{\rho}$ and υ coincide on all intervals in the form $]-\infty,-t]$ or $[t,\infty[$ for $t > 0$. Since also $\tilde{\rho}(\{0\} = 0 = \upsilon(\{0\})$ by Corollary 2.3, we conclude that $\tilde{\rho} = \upsilon$. Combining this with formula (3.11), it follows finally that ρ is a Lévy measure and that $\upsilon = \tilde{\rho} = \Upsilon_0(\rho)$. □

Proposition 3.11. *Let ρ be a σ-finite measure concentrated on $[0,\infty[$ and let $\tilde{\rho}$ be the measure given by* (3.1). *We then have*

$$\int_{]1,\infty[} 1\,\tilde{\rho}(\mathrm{d}t) = \int_{]0,\infty[} \mathrm{e}^{-1/t}\,\rho(\mathrm{d}t), \tag{3.16}$$

$$\int_{[0,1]} t\,\tilde{\rho}(\mathrm{d}t) = \int_{]0,\infty[} t(1-\mathrm{e}^{-1/t}) - \mathrm{e}^{-1/t}\,\rho(\mathrm{d}t). \tag{3.17}$$

In particular

$$\int_{[0,\infty[} \min\{1,t\}\,\tilde{\rho}(\mathrm{d}t) = \int_{]0,\infty[} t(1-\mathrm{e}^{-1/t})\,\rho(\mathrm{d}t), \tag{3.18}$$

and therefore

$$\int_{[0,\infty[} \min\{1,t\}\,\tilde{\rho}(\mathrm{d}t) < \infty \iff \int_{[0,\infty[} \min\{1,t\}\,\rho(\mathrm{d}t) < \infty. \tag{3.19}$$

Proof. Note first that (3.18) follows immediately from (3.16) and (3.17). To prove (3.16), note that by definition of $\tilde{\rho}$, we have

$$\begin{aligned}
\int_{]1,\infty[} 1\,\tilde{\rho}(\mathrm{d}t) &= \int_0^\infty \Big(\int_{[0,\infty[} 1_{]1,\infty[}(t)\,D_x\rho(\mathrm{d}t)\Big)\mathrm{e}^{-x}\,\mathrm{d}x \\
&= \int_0^\infty \Big(\int_{[0,\infty[} 1_{]1,\infty[}(tx)\,\rho(\mathrm{d}t)\Big)\mathrm{e}^{-x}\,\mathrm{d}x \\
&= \int_{]0,\infty[} \Big(\int_{1/t}^\infty \mathrm{e}^{-x}\,\mathrm{d}x\Big)\,\rho(\mathrm{d}t) \\
&= \int_{]0,\infty[} \mathrm{e}^{-1/t}\,\rho(\mathrm{d}t).
\end{aligned}$$

Regarding (3.17), we find similarly that

$$\begin{aligned}
\int_{[0,1]} t\,\tilde{\rho}(\mathrm{d}t) &= \int_0^\infty \Big(\int_{[0,1]} t\,D_x\rho(\mathrm{d}t)\Big)\mathrm{e}^{-x}\,\mathrm{d}x \\
&= \int_0^\infty \Big(\int_{[0,\infty[} tx1_{[0,1]}(tx)\,\rho(\mathrm{d}t)\Big)\mathrm{e}^{-x}\,\mathrm{d}x \\
&= \int_{]0,\infty[} t\Big(\int_0^{1/t} x\mathrm{e}^{-x}\,\mathrm{d}x\Big)\,\rho(\mathrm{d}t) \\
&= \int_{]0,\infty[} t(1-\mathrm{e}^{-1/t}(\tfrac{1}{t}+1))\,\rho(\mathrm{d}t) \\
&= \int_{]0,\infty[} t(1-\mathrm{e}^{-1/t})-\mathrm{e}^{-1/t}\,\rho(\mathrm{d}t).
\end{aligned}$$

Finally, (3.19) follows from (3.18) by noting that

$$0 \le t(1-\mathrm{e}^{-1/t}) = -\frac{\mathrm{e}^{-1/t}-1}{1/t} \le 1, \qquad \text{whenever } t>0,$$

and that

$$\lim_{t\searrow 0}(1-\mathrm{e}^{-1/t}) = 1 = \lim_{t\to\infty} t(1-\mathrm{e}^{-1/t}).$$

This concludes the proof. □

3.2 The Mapping Υ

We now extend the mapping Υ_0 to a mapping Υ from $\mathcal{ID}(*)$ into $\mathcal{ID}(*)$.

Definition 3.12. *For any μ in $\mathcal{ID}(*)$, with characteristic triplet (a,ρ,η), we take $\Upsilon(\mu)$ to be the element of $\mathcal{ID}(*)$ whose characteristic triplet is $(2a,\tilde{\rho},\tilde{\eta})$ where*

$$\tilde{\eta} = \eta + \int_0^\infty \Big(\int_{\mathbb{R}} t\,\big(1_{[-1,1]}(t) - 1_{[-x,x]}(t)\big)\,D_x\rho(\mathrm{d}t)\Big)\mathrm{e}^{-x}\,\mathrm{d}x \tag{3.20}$$

and

$$\tilde{\rho} = \Upsilon_0(\rho) = \int_0^\infty (D_x\rho)\mathrm{e}^{-x}\mathrm{d}x. \tag{3.21}$$

Note that it is an immediate consequence of Proposition 3.9 that the measure $\tilde{\rho}$ in Definition 3.12 is indeed a Lévy measure. We verify next that the integral in (3.20) is well-defined.

Lemma 3.13. *Let ρ be a Lévy measure on $\mathbb{R}$. Then for any x in $]0,\infty[$, we have that*

$$\int_{\mathbb{R}} \big|ux\cdot\big(1_{[-1,1]}(ux) - 1_{[-x,x]}(ux)\big)\big|\,\rho(\mathrm{d}u) < \infty.$$

Furthermore,

$$\int_0^\infty \Big(\int_{\mathbb{R}} \big| ux \cdot \big(1_{[-1,1]}(ux) - 1_{[-x,x]}(ux)\big)\big| \, \rho(\mathrm{d}u) \Big) \mathrm{e}^{-x} \, \mathrm{d}x < \infty.$$

Proof. Note first that for any x in $]0, \infty[$ we have that

$$\begin{aligned}
\int_{\mathbb{R}} \big| ux \cdot \big(1_{[-1,1]}(ux) - 1_{[-x,x]}(ux)\big)\big| \, \rho(\mathrm{d}u) \\
&= \int_{\mathbb{R}} \big| ux \cdot \big(1_{[-x^{-1},x^{-1}]}(u) - 1_{[-1,1]}(u)\big)\big| \, \rho(\mathrm{d}u) \\
&= \begin{cases} x \int_{\mathbb{R}} |u| \cdot 1_{[-x^{-1},x^{-1}] \setminus [-1,1]}(u) \, \rho(\mathrm{d}u), & \text{if } x \le 1, \\ x \int_{\mathbb{R}} |u| \cdot 1_{[-1,1] \setminus [-x^{-1},x^{-1}]}(u) \, \rho(\mathrm{d}u), & \text{if } x > 1. \end{cases}
\end{aligned}$$

Note then that whenever $0 < \epsilon < K$, we have that

$$|u| \cdot 1_{[-K,K] \setminus [-\epsilon,\epsilon]}(u) \le \min\{K, \tfrac{u^2}{\epsilon}\} \le \max\{K, \epsilon^{-1}\} \min\{u^2, 1\},$$

for any u in $\mathbb{R}$. Hence, if $0 < x \le 1$, we find that

$$\begin{aligned}
x \int_{\mathbb{R}} \big| u \cdot \big(1_{[-x^{-1},x^{-1}]}(u) - 1_{[-1,1]}(u)\big)\big| \, \rho(\mathrm{d}u) \\
&\le x \max\{x^{-1}, 1\} \int_{\mathbb{R}} \min\{u^2, 1\} \, \rho(\mathrm{d}u) = \int_{\mathbb{R}} \min\{u^2, 1\} \, \rho(\mathrm{d}u) < \infty,
\end{aligned}$$

since ρ is a Lévy measure. Similarly, if $x \ge 1$,

$$\begin{aligned}
x \int_{\mathbb{R}} \big| u \cdot \big(1_{[-1,1]}(u) - 1_{[-x^{-1},x^{-1}]}(u)\big)\big| \, \rho(\mathrm{d}u) \\
&\le x \max\{1, x\} \int_{\mathbb{R}} \min\{u^2, 1\} \, \rho(\mathrm{d}u) = x^2 \int_{\mathbb{R}} \min\{u^2, 1\} \, \rho(\mathrm{d}u) < \infty.
\end{aligned}$$

Altogether, we find that

$$\begin{aligned}
\int_0^\infty \Big(\int_{\mathbb{R}} \big| ux \cdot \big(1_{[-1,1]}(ux) - 1_{[-x,x]}(ux)\big)\big| \, \rho(\mathrm{d}u) \Big) \mathrm{e}^{-x} \, \mathrm{d}x \\
&\le \int_{\mathbb{R}} \min\{u^2, 1\} \, \rho(\mathrm{d}u) \cdot \Big(\int_0^1 \mathrm{e}^{-x} \, \mathrm{d}x + \int_1^\infty x^2 \mathrm{e}^{-x} \, \mathrm{d}x \Big) < \infty,
\end{aligned}$$

as asserted. □

Remark 3.14. In connection with (3.20), note that it follows from Lemma 3.13 above that the integral

$$\int_0^\infty \Big(\int_{\mathbb{R}} u\big(1_{[-1,1]}(u) - 1_{[-x,x]}(u)\big)\, D_x\rho(\mathrm{d}u)\Big) \mathrm{e}^{-x}\,\mathrm{d}x,$$

is well-defined. Indeed,

$$\int_0^\infty \Big(\int_{\mathbb{R}} \big|u\big(1_{[-1,1]}(u) - 1_{[-x,x]}(u)\big)\big|\, D_x\rho(\mathrm{d}u)\Big) \mathrm{e}^{-x}\,\mathrm{d}x$$
$$= \int_0^\infty \Big(\int_{\mathbb{R}} \big|ux\big(1_{[-1,1]}(ux) - 1_{[-x,x]}(ux)\big)\big|\, \rho(\mathrm{d}u)\Big) \mathrm{e}^{-x}\,\mathrm{d}x.$$

Having established that the definition of Υ is meaningful, we prove next a key formula for the cumulant transform of $\Upsilon(\mu)$ (Theorem 3.17 below). From that formula we derive subsequently a number of important properties of Υ. We start with the following technical result.

Lemma 3.15. *Let ρ be a Lévy measure on $\mathbb{R}$. Then for any number ζ in $]-\infty, 0[$, we have that*

$$\int_0^\infty \Big(\int_{\mathbb{R}} \big|\mathrm{e}^{\mathrm{i}\zeta tx} - 1 - \mathrm{i}\zeta tx 1_{[-1,1]}(t)\big|\, \rho(\mathrm{d}t)\Big) \mathrm{e}^{-x}\,\mathrm{d}x < \infty.$$

Proof. Let ζ from $]-\infty, 0[$ and x in $[0,\infty[$ be given. Note first that

$$\int_{\mathbb{R}\setminus[-1,1]} \big|\mathrm{e}^{\mathrm{i}\zeta tx} - 1 - \mathrm{i}\zeta tx 1_{[-1,1]}(t)\big|\, \rho(\mathrm{d}t) = \int_{\mathbb{R}\setminus[-1,1]} \big|\mathrm{e}^{\mathrm{i}\zeta tx} - 1\big|\, \rho(\mathrm{d}t)$$
$$\le 2\int_{\mathbb{R}\setminus[-1,1]} \min\{1, t^2\}\rho(\mathrm{d}t)$$
$$\le 2\int_{\mathbb{R}} \min\{1, t^2\}\rho(\mathrm{d}t).$$

To estimate $\int_{-1}^1 |\mathrm{e}^{\mathrm{i}\zeta tx} - 1 - \mathrm{i}\zeta tx|\, \rho(\mathrm{d}t)$, we note that for any real number t, it follows by standard second order Taylor expansion that

$$\big|\mathrm{e}^{\mathrm{i}\zeta tx} - 1 - \mathrm{i}\zeta tx\big| \le \frac{1}{\sqrt{2}}(\zeta tx)^2,$$

and hence

$$\int_{-1}^1 \big|\mathrm{e}^{\mathrm{i}\zeta tx} - 1 - \mathrm{i}\zeta tx\big|\, \rho(\mathrm{d}t) \le \frac{1}{\sqrt{2}}(\zeta x)^2 \int_{-1}^1 t^2\, \rho(\mathrm{d}t)$$
$$\le \frac{1}{\sqrt{2}}(\zeta x)^2 \int_{\mathbb{R}} \min\{1, t^2\}\, \rho(\mathrm{d}t).$$

Altogether, we find that for any number x in $[0,\infty[$,

$$\int_{\mathbb{R}} \left| e^{i\zeta tx} - 1 - i\zeta tx 1_{[-1,1]}(t) \right| \rho(dt) \le \Big(2 + \frac{1}{\sqrt{2}}(\zeta x)^2\Big) \int_{\mathbb{R}} \min\{1, t^2\}\, \rho(dt),$$

and therefore

$$\int_0^\infty \Big(\int_{\mathbb{R}} \left| e^{i\zeta tx} - 1 - i\zeta tx 1_{[-1,1]}(t) \right| \rho(dt) \Big) e^{-x}\, dx$$
$$\le \int_{\mathbb{R}} \min\{1, t^2\}\, \rho(dt) \int_0^\infty \Big(2 + \frac{1}{\sqrt{2}}(\zeta x)^2\Big) e^{-x}\, dx < \infty,$$

as desired. □

Theorem 3.16. *Let μ be a measure in $\mathcal{ID}(*)$ with characteristic triplet (a, ρ, η). Then the cumulant function of $\Upsilon(\mu)$ is representable as*

$$C_{\Upsilon(\mu)}(\zeta) = i\eta\zeta - a\zeta^2 + \int_{\mathbb{R}} \Big(\frac{1}{1 - i\zeta t} - 1 - i\zeta t 1_{[-1,1]}(t) \Big) \rho(dt), \tag{3.22}$$

for any ζ in $\mathbb{R}$.

Proof. Recall first that for any $z \in \mathbb{C}$ with $\mathrm{Re} z < 1$ we have

$$\frac{1}{1 - z} = \int_0^\infty e^{zx} e^{-x} dx,$$

implying that for ζ real with $\zeta \le 0$

$$\frac{1}{1 - i\zeta t} - 1 - i\zeta t 1_{[-1,1]}(t) = \int_0^\infty \big(e^{i\zeta tx} - 1 - i\zeta tx 1_{[-1,1]}(t)\big)\, e^{-x} dx. \tag{3.23}$$

Now, let μ from $\mathcal{ID}(*)$ be given and let (a, ρ, η) be the characteristic triplet for μ. Then by the above calculation

$$\int_{\mathbb{R}} \Big(\frac{1}{1 - i\zeta t} - 1 - i\zeta t 1_{[-1,1]}(t) \Big) \rho(dt)$$
$$= \int_{\mathbb{R}} \Big(\int_0^\infty \big(e^{i\zeta tx} - 1 - i\zeta tx 1_{[-1,1]}(t)\big) e^{-x}\, dx \Big) \rho(dt)$$
$$= \int_0^\infty \Big(\int_{\mathbb{R}} \big(e^{i\zeta u} - 1 - i\zeta u 1_{[-x,x]}(u)\big)\, \rho(x^{-1} du) \Big) e^{-x}\, dx$$
$$= \int_0^\infty \Big(\int_{\mathbb{R}} \big(e^{i\zeta u} - 1 - i\zeta u 1_{[-1,1]}(u)\big)\, \rho(x^{-1} du) \Big) e^{-x}\, dx$$
$$\quad + i\zeta \int_0^\infty \Big(\int_{\mathbb{R}} u\big(1_{[-1,1]}(u) - 1_{[-x,x]}(u)\big)\, \rho(x^{-1} du) \Big) e^{-x}\, dx$$
$$= \int_{\mathbb{R}} \big(e^{i\zeta u} - 1 - i\zeta u 1_{[-1,1]}(u)\big)\, \tilde{\rho}(du)$$
$$\quad + i\zeta \int_0^\infty \Big(\int_{\mathbb{R}} u\big(1_{[-1,1]}(u) - 1_{[-x,x]}(u)\big)\, \rho(x^{-1} du) \Big) e^{-x}\, dx,$$

where we have changed the order of integration in accordance with Lemma 3.15. Comparing the above calculation with Definition 3.12, the theorem follows readily. □

Theorem 3.17. *For any μ in $\mathcal{ID}(*)$ we have*

$$C_{\Upsilon(\mu)}(z) = \int_0^\infty C_\mu(zx)\mathrm{e}^{-x}\,\mathrm{d}x, \qquad (z \in \mathbb{R}).$$

Proof. Let (a, ρ, η) be the characteristic triplet for μ. For arbitrary z in $\mathbb{R}$, we then have

$$\begin{aligned}
&\int_0^\infty C_\mu(zx)\mathrm{e}^{-x}\,\mathrm{d}x \\
&= \int_0^\infty \Big(\mathrm{i}\eta zx - \frac{1}{2}az^2x^2 + \int_{\mathbb{R}} \big(\mathrm{e}^{\mathrm{i}tzx} - 1 - \mathrm{i}tzx1_{[-1,1]}(t)\big)\,\rho(\mathrm{d}t)\Big)\mathrm{e}^{-x}\,\mathrm{d}x \\
&= \mathrm{i}\eta z\int_0^\infty x\mathrm{e}^{-x}\,\mathrm{d}x - \frac{1}{2}az^2\int_0^\infty x^2\mathrm{e}^{-x}\,\mathrm{d}x \\
&\qquad + \int_{\mathbb{R}}\Big(\int_0^\infty \big(\mathrm{e}^{\mathrm{i}tzx} - 1 - \mathrm{i}tzx1_{[-1,1]}(t)\big)\mathrm{e}^{-x}\,\mathrm{d}x\Big)\,\rho(\mathrm{d}t) \\
&= \mathrm{i}\eta z - az^2 + \int_{\mathbb{R}}\Big(\frac{1}{1-\mathrm{i}zt} - 1 - \mathrm{i}zt1_{[-1,1]}(t)\Big)\,\rho(\mathrm{d}t),
\end{aligned} \tag{3.24}$$

where the last equality uses (3.23). According to Theorem 3.16, the resulting expression in (3.24) equals $C_{\Upsilon(\mu)}(z)$, and the theorem follows. □

Based on Theorem 3.17 we establish next a number of interesting properties for Υ.

Proposition 3.18. *The mapping $\Upsilon\colon \mathcal{ID}(*) \to \mathcal{ID}(*)$ has the following properties:*

(i) *Υ is injective.*
(ii) *For any measures μ, ν in $\mathcal{ID}(*)$, $\Upsilon(\mu * \nu) = \Upsilon(\mu) * \Upsilon(\nu)$.*
(iii) *For any measure μ in $\mathcal{ID}(*)$ and any constant c in $\mathbb{R}$, $\Upsilon(D_c\mu) = D_c\Upsilon(\mu)$.*
(iv) *For any constant c in $\mathbb{R}$, $\Upsilon(\delta_c) = \delta_c$.*
(v) *Υ is continuous w.r.t. weak convergence*[4].

Proof. (i) This is an immediate consequence of the definition of Υ together with the injectivity of Υ_0 (cf. Corollary 3.7).

(ii) Suppose $\mu_1, \mu_2 \in \mathcal{ID}(*)$. Then for any z in $\mathbb{R}$ we have by Proposition 3.17

[4] In fact, it can be proved that Υ is a homeomorphism onto its range with respect to weak convergence; see [BaTh04c].

$$C_{\Upsilon(\mu_1*\mu_2)}(z) = \int_0^\infty C_{\mu_1*\mu_2}(zx)\mathrm{e}^{-x}\,\mathrm{d}x = \int_0^\infty (C_{\mu_1}(zx) + C_{\mu_2}(zx))\mathrm{e}^{-x}\,\mathrm{d}x$$
$$= C_{\Upsilon(\mu_1)}(z) + C_{\Upsilon(\mu_2)}(z) = C_{\Upsilon(\mu_1)*\Upsilon(\mu_2)}(z),$$

which verifies statement (ii)

(iii) Suppose $\mu \in \mathcal{ID}(*)$ and $c \in \mathbb{R}$. Then for any z in $\mathbb{R}$,

$$C_{\Upsilon(D_c\mu)}(z) = \int_0^\infty C_{D_c\mu}(zx)\mathrm{e}^{-x}\,\mathrm{d}x = \int_0^\infty C_\mu(czx)\mathrm{e}^{-x}\,\mathrm{d}x$$
$$= C_{\Upsilon(\mu)}(cz) = C_{D_c\Upsilon(\mu)}(z),$$

which verifies (iii).

(iv) Let c from $\mathbb{R}$ be given. For z in $\mathbb{R}$ we then have

$$C_{\Upsilon(\delta_c)}(z) = \int_0^\infty C_{\delta_c}(zx)\mathrm{e}^{-x}\,\mathrm{d}x = \int_0^\infty \mathrm{i}czx\mathrm{e}^{-x}\,\mathrm{d}x = \mathrm{i}cz = C_{\delta_c}(z),$$

which verifies (iv).

(v) Although we might give a direct proof of (v) at the present stage (see the proof of Theorem 3.40), we postpone the proof to Section 5.3, where we can give an easy argument based on the continuity of the Bercovici-Pata bijection Λ (introduced in Section 5.1) and the connection between Υ and Λ (see Section 5.2).

Corollary 3.19. *The mapping $\Upsilon\colon \mathcal{ID}(*) \to \mathcal{ID}(*)$ preserves stability and selfdecomposability. More precisely, we have*

$$\Upsilon(\mathcal{S}(*)) = \mathcal{S}(*) \quad \textit{and} \quad \Upsilon(\mathcal{L}(*)) \subseteq \mathcal{L}(*).$$

Proof. Suppose $\mu \in \mathcal{S}(*)$ and that $c, c' > 0$ and $d, d' \in \mathbb{R}$. Then

$$(D_c\mu * \delta_d) * (D_{c'}\mu * \delta_{d'}) = D_{c''}\mu * \delta_{d''},$$

for suitable c'' in $]0,\infty[$ and d'' in $\mathbb{R}$. Using now (ii)-(iv) of Proposition 3.18, we find that

$$\begin{aligned}(D_c\Upsilon(\mu) * \delta_d) * (D_{c'}\Upsilon(\mu) * \delta_{d'}) &= (\Upsilon(D_c\mu) * \Upsilon(\delta_d)) * (\Upsilon(D_{c'}\mu) * \Upsilon(\delta_{d'}))\\ &= \Upsilon(D_c\mu * \delta_d) * \Upsilon(D_{c'}\mu * \delta_{d'})\\ &= \Upsilon\big((D_c\mu * \delta_d) * (D_{c'}\mu * \delta_{d'})\big)\\ &= \Upsilon\big(D_{c''}\mu * \delta_{d''}\big)\\ &= D_{c''}\Upsilon(\mu) * \delta_{d''},\end{aligned}$$

which shows that $\Upsilon(\mu) \in \mathcal{S}(*)$. This verifies the inclusion $\Upsilon(\mathcal{S}(*)) \subseteq \mathcal{S}(*)$. To prove the converse inclusion, we use Corollary 3.4 (the following argument, in fact, also shows the inclusion just verified above). As described in Section 2.5, the stable laws are characterized by having Lévy measures in the form $r(t)\,\mathrm{d}t$, where

$$r(t) = \begin{cases} c_+ t^{-1-\alpha}, & \text{for } t > 0, \\ c_- |t|^{-1-\alpha}, & \text{for } t < 0, \end{cases}$$

with $\alpha \in \,]0,2[$ and $c_+, c_- \geq 0$. Using Corollary 3.4, it follows then that for μ in $\mathcal{S}(*)$, the Lévy measure for $\Upsilon(\mu)$ takes the form $\tilde{r}(t)\,\mathrm{d}t$, with $\tilde{r}(t)$ given by

$$\begin{aligned} \tilde{r}(t) &= \begin{cases} \int_0^\infty y^{-1} r(y^{-1}) \mathrm{e}^{-ty}\,\mathrm{d}y, & \text{if } t > 0, \\ \int_{-\infty}^0 -y^{-1} r(y^{-1}) \mathrm{e}^{-ty}\,\mathrm{d}y, & \text{if } t < 0, \end{cases} \\ &= \begin{cases} c_+ \Gamma(1+\alpha) t^{-1-\alpha}, & \text{if } t > 0, \\ c_- \Gamma(1+\alpha)|t|^{-1-\alpha}, & \text{if } t < 0, \end{cases} \end{aligned} \tag{3.25}$$

where the second equality follows by a standard calculation. Formula (3.25) shows, in particular, that any measure in $\mathcal{S}(*)$ is the image by Υ of another measure in $\mathcal{S}(*)$.

Assume next that $\mu \in \mathcal{L}(*)$. Then for any c in $]0,1[$, there exists a measure μ_c in $\mathcal{ID}(*)$, such that $\mu = D_c\mu * \mu_c$. Using now (ii)-(iii) of Proposition 3.18, we find that

$$\Upsilon(\mu) = \Upsilon(D_c\mu * \mu_c) = \Upsilon(D_c\mu) * \Upsilon(\mu_c) = D_c\Upsilon(\mu) * \Upsilon(\mu_c),$$

which shows that $\Upsilon(\mu) \in \mathcal{L}(*)$. □

Remark 3.20. By the definition of Υ and Corollary 3.5 it follows that the Lévy measure for any probability measure in the range $\Upsilon(\mathcal{ID}(*))$ of Υ has a C^∞ density w.r.t. Lebesgue measure. This implies that the mapping $\Upsilon\colon \mathcal{ID}(*) \to \mathcal{ID}(*)$ *is not surjective.* In particular it is apparent that the (classical) Poisson distributions are not in the image of Υ, since the characteristic triplet for the Poisson distribution with mean $c > 0$ is $(0, c\delta_1, c)$. In [BaMaSa04], it was proved that the full range of Υ is the Goldie-Steutel-Bondesson class $\mathcal{B}(*)$. In Theorem 3.27 below, we show that $\Upsilon(\mathcal{L}(*)) = \mathcal{T}(*)$.

We end this section with some results on properties of distributions that are preserved by the mapping Υ. The first of these results is an immediate consequence of Proposition 3.11.

Corollary 3.21. *Let μ be a measure in $\mathcal{ID}(*)$. Then $\mu \in \mathcal{ID}_\tau^+(*)$ if and only if $\Upsilon(\mu) \in \mathcal{ID}_\tau^+(*)$.*

Proof. For a measure μ in $\mathcal{ID}(*)$ with Lévy measure ρ, $\Upsilon(\mu)$ has Lévy measure $\Upsilon_0(\rho) = \tilde{\rho}$. Hence, the corollary follows immediately from formula (3.19) and the characterization of $\mathcal{ID}_\tau^+(*)$ given in Remark 2.13. □

The next result shows that the mapping Υ has the same property as that of Υ_0 exhibited in Proposition 3.8.

Proposition 3.22. *For any measure μ in $\mathcal{ID}(*)$ and any positive number p, we have*

$$\mu \text{ has } p\text{'th moment} \iff \Upsilon(\mu) \text{ has } p\text{'th moment.}$$

Proof. Let μ in $\mathcal{ID}(*)$ be given and put $\nu = \Upsilon(\mu)$. Let (a, ρ, η) be the characteristic triplet for μ and $(2a, \tilde{\rho}, \tilde{\eta})$ the characteristic triplet for ν (in particular $\tilde{\rho} = \Upsilon_0(\rho)$. Now by [Sa99, Corollary 25.8] we have

$$\int_{\mathbb{R}} |x|^p \, \mu(\mathrm{d}x) < \infty \iff \int_{[-1,1]^c} |x|^p \, \rho(\mathrm{d}x) < \infty, \tag{3.26}$$

and

$$\int_{\mathbb{R}} |x|^p \, \nu(\mathrm{d}x) < \infty \iff \int_{[-1,1]^c} |x|^p \, \tilde{\rho}(\mathrm{d}x) < \infty. \tag{3.27}$$

Note next that

$$\begin{aligned} \int_{[-1,1]^c} |x|^p \tilde{\rho}(\mathrm{d}x) &= \int_0^\infty \Big(\int_{[-1,1]^c} |x|^p \, D_y\rho(\mathrm{d}x) \Big) \mathrm{e}^{-y} \, \mathrm{d}y \\ &= \int_0^\infty \Big(\int_{\mathbb{R}} |xy|^p 1_{[-1,1]^c}(xy) \, \rho(\mathrm{d}x) \Big) \mathrm{e}^{-y} \, \mathrm{d}y \qquad (3.28) \\ &= \int_{\mathbb{R}} |x|^p \Big(\int_{1/|x|}^\infty y^p \mathrm{e}^{-y} \, \mathrm{d}y \Big) \, \rho(\mathrm{d}x), \end{aligned}$$

where we interpret $\int_{1/|x|}^\infty y^p \mathrm{e}^{-y} \, \mathrm{d}y$ as 0, when $x = 0$.

Assume now that μ has p'th moment. Then by (3.26), $\int_{[-1,1]^c} |x|^p \, \rho(\mathrm{d}x) < \infty$, and by (3.28)

$$\begin{aligned} &\int_{[-1,1]^c} |x|^p \tilde{\rho}(\mathrm{d}x) \\ &\qquad \le \int_{[-1,1]} |x|^p \Big(\int_{1/|x|}^\infty y^p \mathrm{e}^{-y} \, \mathrm{d}y \Big) \, \rho(\mathrm{d}x) + \Gamma(p+1) \int_{[-1,1]^c} |x|^p \, \rho(\mathrm{d}x). \end{aligned}$$

By (3.27), it remains thus to show that

$$\int_{[-1,1]} |x|^p \Big(\int_{1/|x|}^\infty y^p \mathrm{e}^{-y} \, \mathrm{d}y \Big) \, \rho(\mathrm{d}x) < \infty. \tag{3.29}$$

If $p \ge 2$, then this is obvious:

$$\int_{[-1,1]} |x|^p \Big(\int_{1/|x|}^\infty y^p \mathrm{e}^{-y} \, \mathrm{d}y \Big) \, \rho(\mathrm{d}x) \le \Gamma(p+1) \int_{[-1,1]} |x|^p \, \rho(\mathrm{d}x) < \infty,$$

since ρ is a Lévy measure. For p in $]0,2[$ we note first that for any numbers t, q in $]0,\infty[$ we have

$$\int_t y^p e^{-y}\,dy = \int_t^\infty \frac{y^{p+q}}{y^q} e^{-y}\,dy \le t^{-q} \int_t^\infty y^{p+q} e^{-y}\,dy \le t^{-q}\Gamma(p+q+1).$$

Using this with $t = 1/|x|$, we find for any positive q that

$$\int_{[-1,1]} |x|^p \Big(\int_{1/|x|}^\infty y^p e^{-y}\,dy \Big)\,\rho(dx) \le \Gamma(p+q+1) \int_{[-1,1]} |x|^{p+q}\,\rho(dx).$$

Choosing $q = 2 - p$ we find as desired that

$$\int_{[-1,1]} |x|^p \Big(\int_{1/|x|}^\infty y^p e^{-y}\,dy \Big)\,\rho(dx) \le \Gamma(3) \int_{[-1,1]} |x|^2\,\rho(dx) < \infty,$$

since ρ is a Lévy measure.

Assume conversely that $\nu = \Upsilon(\mu)$ has p'th moment. Then by (3.27), we have $\int_{[-1,1]^c} |x|^p\,\tilde{\rho}(dx) < \infty$, and by (3.26) we have to show that $\int_{[-1,1]^c} |x|^p\,\rho(dx) < \infty$. For this, note that whenever $|x| > 1$ we have

$$\int_{1/|x|}^\infty y^p e^{-y}\,dy \ge \int_1^\infty y^p e^{-y}\,dy \in\,]0,\infty[.$$

Setting $c(p) = \int_1^\infty y^p e^{-y}\,dy$ and using (3.28) we find thus that

$$\begin{aligned}\int_{[-1,1]^c} |x|^p\,\rho(dx) &\le \frac{1}{c(p)} \int_{[-1,1]^c} |x|^p \Big(\int_{1/|x|}^\infty y^p e^{-y}\,dy \Big)\,\rho(dx) \\ &\le \frac{1}{c(p)} \int_{[-1,1]^c} |x|^p\,\tilde{\rho}(dx) < \infty,\end{aligned}$$

as desired. □

3.3 Relations between Υ_0, Υ and the Classes $\mathcal{L}(*)$, $\mathcal{T}(*)$

In this section we establish a close connection between the mapping Υ and the relationship between the classes $\mathcal{T}(*)$ and $\mathcal{L}(*)$. More precisely, we prove that $\Upsilon(\mathcal{L}(*)) = \mathcal{T}(*)$ and also that $\Upsilon(\mathcal{L}_\tau^+(*)) = \mathcal{T}_\tau^+(*)$. We consider the latter equality first.

The Positive Thorin Class

We start by establishing the following technical result on the connection between complete monotonicity and Lévy densities for measures in $\mathcal{ID}^+(*)$.

Lemma 3.23. *Let ν be a Borel measure on $[0,\infty[$ such that*

$$\forall t > 0\colon \int_{[0,\infty[} e^{-ts}\,\nu(ds) < \infty,$$

and note that ν is necessarily a Radon measure. Let $q\colon]0,\infty[\to [0,\infty[$ be the function given by:

$$q(t) = \frac{1}{t}\int_{[0,\infty[} \mathrm{e}^{-ts}\,\nu(\mathrm{d}s), \qquad (t > 0).$$

Then q satisfies the condition

$$\int_0^\infty \min\{1,t\}q(t)\,\mathrm{d}t < \infty, \tag{3.30}$$

if and only if ν satisfies the following three conditions:

(a) $\nu(\{0\}) = 0$,
(b) $\int_{]0,1]} |\log(t)|\,\nu(\mathrm{d}t) < \infty$,
(c) $\int_{[1,\infty[} \frac{1}{t}\,\nu(\mathrm{d}t) < \infty$.

Proof. We note first that

$$\begin{aligned}\int_0^1 tq(t)\,\mathrm{d}t &= \int_0^1\int_{[0,\infty[} \mathrm{e}^{-ts}\,\nu(\mathrm{d}s)\,\mathrm{d}t = \int_{[0,\infty[}\Big(\int_0^1 \mathrm{e}^{-ts}\,\mathrm{d}t\Big)\,\nu(\mathrm{d}s)\\ &= \nu(\{0\}) + \int_{]0,\infty[} \tfrac{1}{s}(1-\mathrm{e}^{-s})\,\nu(\mathrm{d}s).\end{aligned} \tag{3.31}$$

Note next that

$$\begin{aligned}\int_1^\infty q(t)\,\mathrm{d}t &= \int_1^\infty \tfrac{1}{t}\int_{[0,\infty[} \mathrm{e}^{-ts}\,\nu(\mathrm{d}s)\,\mathrm{d}t = \int_{[0,\infty[}\Big(\int_1^\infty \tfrac{1}{t}\mathrm{e}^{-ts}\,\mathrm{d}t\Big)\,\nu(\mathrm{d}s)\\ &= \int_{[0,\infty[}\Big(\int_s^\infty \tfrac{1}{t}\mathrm{e}^{-t}\,\mathrm{d}t\Big)\,\nu(\mathrm{d}s) = \int_0^\infty \tfrac{1}{t}\mathrm{e}^{-t}\Big(\int_{[0,t]} 1\,\nu(\mathrm{d}s)\Big)\,\mathrm{d}t\\ &= \int_0^\infty \tfrac{1}{t}\mathrm{e}^{-t}\nu([0,t])\,\mathrm{d}t.\end{aligned} \tag{3.32}$$

Assume now that (3.30) is satisfied. It follows then from (3.32) that

$$\infty > \int_0^1 \tfrac{1}{t}\mathrm{e}^{-t}\nu([0,t])\,\mathrm{d}t \geq \mathrm{e}^{-1}\int_0^1 \tfrac{1}{t}\nu([0,t])\,\mathrm{d}t.$$

Here, by partial (Stieltjes) integration,

$$\begin{aligned}\int_0^1 \tfrac{1}{t}\nu([0,t])\,\mathrm{d}t &= \Big[\log(t)\nu([0,t])\Big]_0^1 - \int_{]0,1]} \log(t)\,\nu(\mathrm{d}t)\\ &= \lim_{t\searrow 0} |\log(t)|\nu([0,t]) + \int_{]0,1]} |\log(t)|\,\nu(\mathrm{d}t),\end{aligned}$$

so we may conclude that

$$\lim_{t\searrow 0} |\log(t)|\nu([0,t]) < \infty \quad \text{and} \quad \int_{]0,1]} |\log(t)|\,\nu(\mathrm{d}t) < \infty,$$

and this implies that (a) and (b) are satisfied. Regarding (c), note that it follows from (3.30) and (3.31) that

$$\infty > \int_0^1 tq(t)\,\mathrm{d}t \geq \int_{[1,\infty[} \tfrac{1}{s}(1-\mathrm{e}^{-s})\,\nu(\mathrm{d}s) \geq (1-\mathrm{e}^{-1}) \int_{[1,\infty[} \tfrac{1}{s}\,\nu(\mathrm{d}s),$$

and hence (c) follows.

Assume conversely that ν satisfies conditions (a), (b) and (c). Then by (3.31) we have

$$\int_0^1 tq(t)\,\mathrm{d}t = \int_{]0,\infty[} \tfrac{1}{s}(1-\mathrm{e}^{-s})\,\nu(\mathrm{d}s) \leq \int_{]0,1[} 1\,\nu(\mathrm{d}s) + \int_{[1,\infty[} \tfrac{1}{s}\,\nu(\mathrm{d}s),$$

where we have used that $\frac{1}{s}(1-\mathrm{e}^{-s}) \leq 1$ for all positive s. Thus, by (b) and (c), $\int_0^1 tq(t)\,\mathrm{d}t < \infty$. Regarding $\int_1^\infty q(t)\,\mathrm{d}t$, note that for any s in $]0,1]$ we have (using (a))

$$\begin{aligned} 0 \leq |\log(s)|\nu([0,s]) &= \int_{]0,s]} \log(s^{-1})\,\nu(\mathrm{d}u) \leq \int_{]0,s]} \log(u^{-1})\,\nu(\mathrm{d}u) \\ &= \int_{]0,s]} |\log(u)|\,\nu(\mathrm{d}u), \end{aligned}$$

and hence it follows from (b) that $|\log(s)|\nu([0,s]) \to 0$ as $s \searrow 0$. By partial integration we obtain thus that

$$\begin{aligned} \infty > \int_{]0,1]} |\log(s)|\,\nu(\mathrm{d}s) &= \Big[|\log(s)|\nu([0,s])\Big]_0^1 + \int_0^1 \tfrac{1}{s}\nu([0,s])\,\mathrm{d}s \\ &= \int_0^1 \tfrac{1}{s}\nu([0,s])\,\mathrm{d}s \\ &\geq \int_0^1 \tfrac{1}{s}\mathrm{e}^{-s}\nu([0,s])\,\mathrm{d}s. \end{aligned}$$

By (3.32) and (b) it remains, thus, to show that $\int_1^\infty \frac{1}{s}\mathrm{e}^{-s}\nu([0,s])\,\mathrm{d}s < \infty$. For that, it obviously suffices to prove that $\frac{1}{s}\nu([0,s]) \to 0$ as $s \to \infty$. Note, towards this end, that whenever $s \geq t \geq 1$, we have

$$\tfrac{1}{s}\nu([0,s]) = \tfrac{1}{s}\nu([0,t]) + \int_{]t,s]} \tfrac{1}{s}\,\nu(\mathrm{d}u) \leq \tfrac{1}{s}\nu([0,t]) + \int_{]t,s]} \tfrac{1}{u}\,\nu(\mathrm{d}u),$$

and hence, for any t in $[1,\infty[$,

$$\limsup_{s\to\infty} \tfrac{1}{s}\nu([0,s]) \le \int_{]t,\infty[} \tfrac{1}{u}\,\nu(\mathrm{d}u).$$

Letting finally $t\to\infty$, it follows from (c) that

$$\limsup_{s\to\infty} \tfrac{1}{s}\nu([0,s]) = 0,$$

as desired. □

Theorem 3.24. *The mapping Υ maps the class $\mathcal{L}_\tau^+(*)$ onto the class $\mathcal{T}_\tau^+(*)$, i.e.*

$$\Upsilon(\mathcal{L}_\tau^+(*)) = \mathcal{T}_\tau^+(*).$$

Proof. Assume that $\mu \in \mathcal{L}_\tau^+(*)$ with generating triplet (a,ρ,η). Then, by Remark 2.13, $a=0$, ρ is concentrated on $[0,\infty[$, and $\int_0^\infty \min\{1,t\}\,\rho(\mathrm{d}t)<\infty$. Furthermore, since μ is selfdecomposable, $\rho(\mathrm{d}t) = r(t)\,\mathrm{d}t$ for some density function $r\colon [0,\infty[\to[0,\infty[$, satisfying that the function $q(t)=tr(t)$ $(t\ge 0)$ is decreasing (cf. the last paragraph in Section 2.5).

Now the measure $\Upsilon(\mu)$ has generating triplet $(0,\tilde\rho,\tilde\eta)$, where $\tilde\rho$ has density $\tilde r$ given by

$$\tilde r(t) = \int_0^\infty q(s^{-1})\mathrm{e}^{-ts}\,\mathrm{d}s, \qquad (t\ge 0),$$

(cf. Corollary 3.4). We already know from Corollary 3.21 that $\Upsilon(\mu)\in\mathcal{ID}_\tau^+(*)$, so it remains to show that the function $t\mapsto t\tilde r(t)$ is completely monotone, i.e. that

$$t\tilde r(t) = \int_{[0,\infty[} \mathrm{e}^{-ts}\,\nu(\mathrm{d}s), \qquad (t>0),$$

for some (Radon) measure ν on $[0,\infty[$. Note for this, that the function $s\mapsto q(s^{-1})$ is increasing on $]0,\infty[$. This implies, in particular, that $s\mapsto q(s^{-1})$ has only countably many points of discontinuity, and hence, by changing r on a Lebesgue null-set, we may assume that $s\mapsto q(s^{-1})$ is increasing and right continuous. Note finally that $q(s^{-1})\to 0$ as $s\searrow 0$. Indeed, since $s\mapsto q(s^{-1})$ is increasing, the limit $\beta=\lim_{s\searrow 0} q(s^{-1})$ exists and equals $\inf_{s>0} q(s^{-1})$. Since $sr(s)=q(s)\to\beta$ as $s\to\infty$ and $\int_1^\infty r(s)\,\mathrm{d}s<\infty$, we must have $\beta=0$. We may now let ν be the Stieltjes measure corresponding to the function $s\mapsto q(s^{-1})$, i.e.

$$\nu(]-\infty,s]) = \begin{cases} q(s^{-1}), & \text{if } s>0,\\ 0, & \text{if } s\le 0.\end{cases}$$

Then, whenever $t\in\,]0,\infty[$ and $0<a<b<\infty$, we have by partial integration

$$\int_a^b q(s^{-1})t\mathrm{e}^{-ts}\,\mathrm{d}s = \Big[-q(s^{-1})\mathrm{e}^{-ts}\Big]_a^b + \int_{]a,b]} \mathrm{e}^{-ts}\,\nu(\mathrm{d}s). \qquad (3.33)$$

Here $q(a^{-1})\mathrm{e}^{-ta} \to 0$ as $a \searrow 0$. Furthermore, since $\int_0^\infty q(s^{-1})t\mathrm{e}^{-ts}\,\mathrm{d}s = t\tilde{r}(t) < \infty$, it follows from (3.33) that $\gamma = \lim_{b\to\infty} q(b^{-1})\mathrm{e}^{-bt}$ exists in $[0,\infty]$. Now $sr(s)\mathrm{e}^{-t/s} = q(s)\mathrm{e}^{-t/s} \to \gamma$ as $s \searrow 0$, and since $\int_0^1 sr(s)\,\mathrm{d}s < \infty$, this implies that $\gamma = 0$. Letting, finally, $a \to 0$ and $b \to \infty$ in (3.33), we may now conclude that

$$t\tilde{r}(t) = \int_0^\infty q(s^{-1})t\mathrm{e}^{-ts}\,\mathrm{d}s = \int_{]0,\infty[} \mathrm{e}^{-ts}\nu(\mathrm{d}s), \qquad (t > 0),$$

as desired.

Assume conversely that $\tilde{\mu} \in \mathcal{T}_\tau^+(*)$ with generating triplet $(a, \tilde{\rho}, \tilde{\eta})$. Then $a = 0$, $\tilde{\rho}$ is concentrated on $[0,\infty[$ and $\int_0^\infty \min\{1,t\}\,\tilde{\rho}(\mathrm{d}t) < \infty$. Furthermore, $\tilde{\rho}$ has a density $\tilde{r}$ in the form

$$\tilde{r}(t) = \frac{1}{t}\int_{[0,\infty[} \mathrm{e}^{-ts}\,\nu(\mathrm{d}s), \qquad (t > 0),$$

for some (Radon) measure ν on $[0,\infty[$, satisfying conditions (a),(b) and (c) of Lemma 3.23.

We define next a function $r\colon]0,\infty[\to [0,\infty[$ by

$$r(s) = \tfrac{1}{s}\nu([0,\tfrac{1}{s}]), \qquad (s > 0). \tag{3.34}$$

Furthermore, we put

$$q(s) = sr(s) = \nu([0,\tfrac{1}{s}]), \qquad (s > 0),$$

and we note that q is decreasing on $]0,\infty[$ and that $q(s^{-1}) = \nu([0,s])$. Note also that, since $\nu(\{0\}) = 0$ (cf. Lemma 3.23),

$$0 \le \nu([0,s])\mathrm{e}^{-ts} \le \nu([0,s]) \to 0, \qquad \text{as } s \searrow 0,$$

for any $t > 0$. Furthermore, since $\int_{[1,\infty[} \frac{1}{s}\,\nu(\mathrm{d}s) < \infty$ (cf. Lemma 3.23), it follows as in the last part of the proof of Lemma 3.23 that $\frac{1}{s}\nu([0,s]) \to 0$ as $s \to \infty$. This implies, in particular, that $q(s^{-1})\mathrm{e}^{-ts} = \nu([0,s])\mathrm{e}^{-ts} = \frac{1}{s}\nu([0,s])s\mathrm{e}^{-ts} \to 0$ as $s \to \infty$ for any positive t. By partial integration, we now conclude that

$$\int_0^\infty q(s^{-1})t\mathrm{e}^{-ts}\,\mathrm{d}s = \Big[-q(s^{-1})\mathrm{e}^{-ts}\Big]_0^\infty + \int_{]0,\infty[} \mathrm{e}^{-ts}\,\nu(\mathrm{d}s) = t\tilde{r}(t),$$

for any positive t. Hence,

$$\tilde{r}(t) = \int_0^\infty q(s^{-1})\mathrm{e}^{-ts}\,\mathrm{d}s = \int_0^\infty s^{-1}r(s^{-1})\mathrm{e}^{-ts}\,\mathrm{d}s, \qquad (t > 0),$$

and by Corollary 3.4, this means that

$$\tilde{\rho} = \int_0^\infty (D_x\rho)\mathrm{e}^{-x}\,\mathrm{d}x,$$

where $\rho(\mathrm{d}t) = r(t)\,\mathrm{d}t$. Note that since ν is a Radon measure, r is bounded on compact subsets of $]0,\infty[$, and hence ρ is σ-finite. We may thus apply Proposition 3.11 to conclude that $\int_0^\infty \min\{1,t\}\,\rho(\mathrm{d}t) < \infty$, so in particular ρ is a Lévy measure. Now, let μ be the measure in $\mathcal{ID}(*)$ with generating triplet $(0,\rho,\eta)$, where

$$\eta = \tilde{\eta} - \int_0^\infty \Big(\int_{\mathbb{R}} t\big(1_{[-1,1]}(t) - 1_{[-x,x]}(t)\big)\,D_x\rho(\mathrm{d}t)\Big)\mathrm{e}^{-x}\,\mathrm{d}x.$$

Then $\Upsilon(\mu) = \tilde{\mu}$ and $\mu \in \mathcal{ID}_\tau^+(*)$ (cf. Corollary 3.21). Moreover, since $tr(t) = q(t)$ is a decreasing function of t, it follows that μ is selfdecomposable (cf. the last paragraph of Section 2.5). This concludes the proof. □

The General Thorin Class

We start again with some technical results on complete monotonicity.

Lemma 3.25. *Let ν be a Borel measure on $[0,\infty[$ satisfying that*

$$\forall t > 0\colon \int_{[0,\infty[} \mathrm{e}^{-ts}\,\nu(\mathrm{d}s) < \infty,$$

and note that ν is a Radon measure on $[0,\infty[$. Let further $q\colon]0,\infty[\to [0,\infty[$ be the function given by

$$q(t) = \frac{1}{t}\int_{[0,\infty[} \mathrm{e}^{-ts}\,\nu(\mathrm{d}s), \qquad (t > 0). \tag{3.35}$$

Then q is a Lévy density (i.e. $\int_0^\infty \min\{1,t^2\}q(t)\,\mathrm{d}t < \infty$) if and only if ν satisfies the following three conditions:

(a) $\nu(\{0\}) = 0$.
(b) $\int_{]0,1[} |\log(t)|\,\nu(\mathrm{d}t) < \infty$.
(c) $\int_{[1,\infty[} \frac{1}{t^2}\,\nu(\mathrm{d}t) < \infty$.

Proof. We note first that

$$\begin{aligned}\int_0^1 t^2 q(t)\,\mathrm{d}t &= \int_0^1 t\Big(\int_{[0,\infty[} \mathrm{e}^{-ts}\,\nu(\mathrm{d}s)\Big)\,\mathrm{d}t = \int_{[0,\infty[}\Big(\int_0^1 t\mathrm{e}^{-ts}\,\mathrm{d}t\Big)\,\nu(\mathrm{d}s)\\ &= \frac{1}{2}\nu(\{0\}) + \int_{]0,\infty[} \frac{1}{s^2}(1 - \mathrm{e}^{-s} - s\mathrm{e}^{-s})\,\nu(\mathrm{d}s).\end{aligned} \tag{3.36}$$

Exactly as in the proof of Lemma 3.23 we have also that

$$\int_1^\infty q(t)\,\mathrm{d}t = \int_0^\infty \frac{1}{t}\mathrm{e}^{-t}\nu([0,t])\,\mathrm{d}t. \tag{3.37}$$

Assume now that q is a Lévy density. Exactly as in the proof of Lemma 3.23, formula (3.37) then implies that ν satisfies conditions (a) and (b). Regarding (c), note that by (3.36),

$$\infty > \int_0^1 t^2 q(t)\,\mathrm{d}t \geq \int_{[1,\infty[} \frac{1}{s^2}(1-\mathrm{e}^{-s}-s\mathrm{e}^{-s})\,\nu(\mathrm{d}s) \geq (1-2\mathrm{e}^{-1})\int_{[1,\infty[} \frac{1}{s^2}\,\nu(\mathrm{d}s),$$

where we used that $s \mapsto 1-\mathrm{e}^{-s}-s\mathrm{e}^{-s}$ is an increasing function on $[0,\infty[$. It follows thus that (c) is satisfied too.

Assume conversely that ν satisfies (a),(b) and (c). Then by (3.36) we have

$$\int_0^1 t^2 q(t)\,\mathrm{d}t = \int_{]0,\infty[} \frac{1}{s^2}(1-\mathrm{e}^{-s}-s\mathrm{e}^{-s})\,\nu(\mathrm{d}s) \leq \int_{]0,1[} 1\,\nu(\mathrm{d}s) + \int_{[1,\infty[} \frac{1}{s^2}\,\nu(\mathrm{d}s),$$

where we used that $s^{-2}(1-\mathrm{e}^{-s}-s\mathrm{e}^{-s}) = \int_0^1 t\mathrm{e}^{-ts}\,\mathrm{d}t \leq 1$ for all positive s. Hence, using (c) (and the fact that ν is a Radon measure on $[0,\infty[$), we see that $\int_0^1 t^2 q(t)\,\mathrm{d}t < \infty$.

Regarding $\int_1^\infty q(t)\,\mathrm{d}t$, we find by application of (a) and (b), exactly as in the proof of Lemma 3.23, that

$$\infty > \int_{]0,1]} |\log(s)|\,\nu(\mathrm{d}s) \geq \int_0^1 \frac{1}{s}\mathrm{e}^{-s}\nu([0,s])\,\mathrm{d}s.$$

By (3.37), it remains thus to show that $\int_1^\infty \frac{1}{s}\mathrm{e}^{-s}\nu([0,s])\,\mathrm{d}s < \infty$, and this clearly follows, if we prove that $s^{-2}\nu([0,s]) \to 0$ as $s \to \infty$ (since ν is a Radon measure). The latter assertion is established similarly to the last part of the proof of Lemma 3.23: Whenever $s \geq t \geq 1$, we have

$$\frac{1}{s^2}\nu([0,s]) \leq \frac{1}{s^2}\nu([0,t]) + \int_{]t,s]} \frac{1}{u^2}\,\nu(\mathrm{d}u),$$

and hence for any t in $[1,\infty[$,

$$\limsup_{s\to\infty} \frac{1}{s^2}\nu([0,s]) \leq \int_{]t,\infty[} \frac{1}{u^2}\nu(\mathrm{d}u). \tag{3.38}$$

Letting finally $t \to \infty$ in (3.38), it follows from (c) that

$$\limsup_{s\to\infty} s^{-2}\nu([0,s]) = 0.$$

This completes the proof. □

Corollary 3.26. *Let ν be a Borel measure on $\mathbb{R}$ satisfying that*

$$\forall t \in \mathbb{R}\setminus\{0\}\colon \int_{\mathbb{R}} \mathrm{e}^{-|ts|}\,\nu(\mathrm{d}s) < \infty,$$

and note that ν is necessarily a Radon measure on $\mathbb{R}$. Let $q\colon \mathbb{R}\setminus\{0\} \to [0,\infty[$ be the function defined by:

$$q(t) = \begin{cases} \frac{1}{t}\int_{[0,\infty[} \mathrm{e}^{-ts}\,\nu(\mathrm{d}s), & \text{if } t>0,\\ \frac{1}{|t|}\int_{]-\infty,0]} \mathrm{e}^{-ts}\,\nu(\mathrm{d}s), & \text{if } t<0.\end{cases}$$

Then q is a Lévy density (i.e. $\int_{\mathbb{R}} \min\{1,t^2\}q(t)\,\mathrm{d}t < \infty$), if and only if ν satisfies the following three conditions:

(d) $\nu(\{0\}) = 0$.
(e) $\int_{[-1,1]\setminus\{0\}} \big|\log|t|\big|\,\nu(\mathrm{d}t) < \infty$.
(f) $\int_{\mathbb{R}\setminus]-1,1[} \frac{1}{t^2}\,\nu(\mathrm{d}t) < \infty$.

Proof. Let ν_+ and ν_- be the restrictions of ν to $[0,\infty[$ and $]-\infty,0]$, respectively. Let, further, $\check{\nu}_-$ be the transformation of ν_- by the mapping $s \mapsto -s$, and put $\check{q}(t) = q(-t)$. Note then that

$$\check{q}(t) = \frac{1}{t}\int_{[0,\infty[} \mathrm{e}^{-ts}\,\check{\nu}_-(\mathrm{d}s), \qquad (t>0).$$

By application of Lemma 3.25, we now have

$$\begin{aligned} q \text{ is a Lévy density on } \mathbb{R} &\iff q \text{ and } \check{q} \text{ are Lévy densities on } [0,\infty[\\ &\iff \nu_+ \text{ and } \check{\nu}_- \text{ satisfy (a),(b) and (c) of Lemma 3.25}\\ &\iff \nu \text{ satisfies (d),(e) and (f).}\end{aligned}$$

This proves the corollary. □

Theorem 3.27. *The mapping Υ maps the class of selfdecomposable distributions on $\mathbb{R}$ onto the generalized Thorin class, i.e.*

$$\Upsilon(\mathcal{L}(*)) = \mathcal{T}(*).$$

Proof. We prove first that $\Upsilon(\mathcal{L}(*)) \subseteq \mathcal{T}(*)$. So let μ be a measure in $\mathcal{L}(*)$ and consider its generating triplet (a,ρ,η). Then $a \geq 0$, $\eta \in \mathbb{R}$ and $\rho(\mathrm{d}t) = r(t)\,\mathrm{d}t$ for some density function, $r(t)$, satisfying that the function

$$q(t) := |t|r(t), \qquad (t \in \mathbb{R}),$$

is increasing on $]-\infty, 0[$ and decreasing on $]0, \infty[$. Next, let $(2a, \tilde{\rho}, \tilde{\eta})$ be the generating triplet for $\Upsilon(\mu)$. From Lemma 3.4 we know that $\tilde{\rho}$ has the following density w.r.t. Lebesgue measure:

$$\tilde{r}(t) = \begin{cases} \int_0^\infty q(y^{-1})\mathrm{e}^{-ty}\,\mathrm{d}y, & \text{if } t > 0, \\ \int_{-\infty}^0 q(y^{-1})\mathrm{e}^{-ty}\,\mathrm{d}y, & \text{if } t < 0. \end{cases}$$

Note that the function $y \mapsto q(y^{-1})$ is increasing on $]0, \infty[$. Thus, as in the proof of Theorem 3.24, we may, by changing $r(t)$ on a null-set, assume that $y \mapsto q(y^{-1})$ is increasing and right-continuous on $]0, \infty[$. Furthermore, since $\int_1^\infty \frac{1}{s}q(s)\,\mathrm{d}s = \int_1^\infty r(s)\,\mathrm{d}s < \infty$, it follows as in the proof of Theorem 3.24 that $q(y^{-1}) \to 0$ as $y \searrow 0$. Thus, we may let ν_+ be the Stieltjes measure corresponding to the function $y \mapsto q(y^{-1})$ on $]0, \infty[$, i.e.

$$\nu_+(]-\infty, y]) = \begin{cases} 0, & \text{if } y \leq 0, \\ q(y^{-1}), & \text{if } y > 0. \end{cases}$$

Now, whenever $t > 0$ and $0 < b < c < \infty$, we have by partial Stieltjes integration that

$$t\int_b^c q(s^{-1})\mathrm{e}^{-ts}\,\mathrm{d}s = \Big[-\mathrm{e}^{-ts}q(s^{-1})\Big]_b^c + \int_b^c \mathrm{e}^{-ts}\,\nu_+(\mathrm{d}s). \tag{3.39}$$

Here, $\mathrm{e}^{-tb}q(b^{-1}) \leq q(b^{-1}) \to 0$ as $b \searrow 0$. Since $\int_0^\infty q(s^{-1})\mathrm{e}^{-ts}\,\mathrm{d}s = \tilde{r}(t) < \infty$, (3.39) shows, furthermore, that the limit

$$\gamma := \lim_{c\to\infty} \mathrm{e}^{-tc}q(c^{-1}) = \lim_{s\searrow 0} \mathrm{e}^{-t/s}sr(s)$$

exists in $[0, \infty]$. Since $\int_0^\infty s^2 r(s)\,\mathrm{d}s < \infty$, it follows that we must have $\gamma = 0$. From (3.39), it follows thus that

$$t\tilde{r}(t) = t\int_0^\infty q(s^{-1})\mathrm{e}^{-ts}\,\mathrm{d}s = \int_0^\infty \mathrm{e}^{-ts}\,\nu_+(\mathrm{d}s). \tag{3.40}$$

Replacing now $r(s)$ by $r(-s)$ for s in $]0, \infty[$, the argument just given yields the existence of a measure $\check{\nu}_-$ on $[0, \infty[$, such that (after changing r on a null-set)

$$\check{\nu}_-(]-\infty, y]) = \begin{cases} 0, & \text{if } y \leq 0, \\ q(-y^{-1}), & \text{if } y > 0. \end{cases}$$

Furthermore, the measure $\check{\nu}_-$ satisfies the identity

$$t\int_0^\infty q(-s^{-1})\mathrm{e}^{-ts}\,\mathrm{d}s = \int_0^\infty \mathrm{e}^{-ts}\,\check{\nu}_-(\mathrm{d}s), \qquad (t > 0).$$

Next, let ν_- be the transformation of $\check{\nu}_-$ by the mapping $s \mapsto -s$. For t in $]-\infty, 0[$ we then have

$$\begin{aligned}|t|\tilde{r}(t) &= |t|\int_{-\infty}^{0} q(s^{-1})\mathrm{e}^{-ts}\,\mathrm{d}s = |t|\int_{0}^{\infty} q(-s^{-1})\mathrm{e}^{-|t|s}\,\mathrm{d}s \\ &= \int_{0}^{\infty} \mathrm{e}^{-|t|s}\,\check{\nu}_{-}(\mathrm{d}s) = \int_{-\infty}^{0} \mathrm{e}^{-ts}\,\nu_{-}(\mathrm{d}s).\end{aligned} \tag{3.41}$$

Putting finally $\nu = \nu_+ + \nu_-$, it follows from (3.40) and (3.41) that

$$|t|\tilde{r}(t) = \begin{cases} \int_0^\infty \mathrm{e}^{-ts}\,\nu(\mathrm{d}s), & \text{if } t > 0, \\ \int_{-\infty}^0 \mathrm{e}^{-ts}\,\nu(\mathrm{d}s), & \text{if } t < 0, \end{cases}$$

and this shows that $\Upsilon(\mu) \in \mathcal{T}(*)$, as desired (cf. the last paragraph in Section 2.5).

Consider, conversely, a measure $\tilde{\mu}$ in $\mathcal{T}(*)$ with generating triplet $(a, \tilde{\rho}, \tilde{\eta})$. Then $a \geq 0$, $\tilde{\eta} \in \mathbb{R}$ and $\tilde{\rho}$ has a density, $\tilde{r}$, w.r.t. Lebesgue measure such that

$$|t|\tilde{r}(t) = \begin{cases} \int_0^\infty \mathrm{e}^{-ts}\,\nu(\mathrm{d}s), & \text{if } t > 0, \\ \int_{-\infty}^0 \mathrm{e}^{-ts}\,\nu(\mathrm{d}s), & \text{if } t < 0, \end{cases}$$

for some (Radon) measure ν on $\mathbb{R}$ satisfying conditions (d),(e) and (f) of Corollary 3.26. Define then the function $r\colon \mathbb{R}\setminus\{0\} \to [0,\infty[$ by

$$r(s) = \begin{cases} \frac{1}{s}\nu([0,\frac{1}{s}]), & \text{if } s > 0, \\ \frac{1}{|s|}\nu([\frac{1}{s},0]), & \text{if } s < 0, \end{cases}$$

and put furthermore

$$q(t) = |s|r(s) = \begin{cases} \nu([0,\frac{1}{s}]), & \text{if } s > 0, \\ \nu([\frac{1}{s},0]), & \text{if } s < 0. \end{cases} \tag{3.42}$$

Note that since $\nu(\{0\}) = 0$ (cf. Corollary 3.26), we have

$$\forall t > 0\colon \nu([0,s])\mathrm{e}^{-ts} \leq \nu([0,s]) \to 0, \quad \text{as } s \searrow 0,$$

and

$$\forall t < 0\colon \nu([s,0])\mathrm{e}^{-ts} \leq \nu([s,0]) \to 0, \quad \text{as } s \nearrow 0.$$

Furthermore, since $\int_{\mathbb{R}\setminus[-1,1]} \frac{1}{s^2}\,\nu(\mathrm{d}s) < \infty$, it follows as in the last part of the proof of Lemma 3.25 that

$$\lim_{s\to\infty} s^{-2}\nu([0,s]) = 0 = \lim_{s\to-\infty} s^{-2}\nu([s,0]).$$

In particular it follows that

$$\forall t > 0\colon \lim_{s\to\infty} \nu([0,s])\mathrm{e}^{-ts} = 0, \quad \text{and that} \quad \forall t < 0\colon \lim_{s\to-\infty} \nu([s,0])\mathrm{e}^{-ts} = 0.$$

By partial Stieltjes integration, we find now for $t > 0$ that

$$\begin{aligned} t\int_0^\infty q(s^{-1})\mathrm{e}^{-ts}\,\mathrm{d}s &= \Big[-q(s^{-1})\mathrm{e}^{-ts}\Big]_0^\infty + \int_0^\infty \mathrm{e}^{-ts}\,\nu(\mathrm{d}s) \\ &= \int_0^\infty \mathrm{e}^{-ts}\,\nu(\mathrm{d}s) = t\tilde{r}(t). \end{aligned} \tag{3.43}$$

Denoting by $\check{\nu}$ the transformation of ν by the mapping $s \mapsto -s$, we find similarly for $t < 0$ that

$$\begin{aligned} |t|\tilde{r}(t) &= \int_{-\infty}^0 \mathrm{e}^{-ts}\,\nu(\mathrm{d}s) = \int_0^\infty \mathrm{e}^{-|t|s}\,\check{\nu}(\mathrm{d}s) \\ &= \Big[\mathrm{e}^{-|t|s}q(-s^{-1})\Big]_0^\infty + |t|\int_0^\infty \mathrm{e}^{-|t|s}q(-s^{-1})\,\mathrm{d}s = |t|\int_{-\infty}^0 \mathrm{e}^{-ts}q(s^{-1})\,\mathrm{d}s. \end{aligned} \tag{3.44}$$

Combining now (3.43) and (3.44) it follows that

$$\tilde{r}(t) = \begin{cases} \int_0^\infty q(s^{-1})\mathrm{e}^{-ts}\,\mathrm{d}s, & \text{if } t > 0, \\ \int_{-\infty}^0 q(s^{-1})\mathrm{e}^{-sy}\,\mathrm{d}s, & \text{if } t < 0. \end{cases}$$

By Corollary 3.4 we may thus conclude that $\tilde{\rho}(\mathrm{d}t) = \int_0^\infty (D_x\rho)\mathrm{e}^{-x}\,\mathrm{d}x$, where $\rho(\mathrm{d}t) = r(t)\,\mathrm{d}t$. Since ν is a Radon measure, r is bounded on compact subsets of $\mathbb{R}\setminus\{0\}$, so that ρ is, in particular, σ-finite. By Proposition 3.9, it follows then that $\int_{\mathbb{R}} \min\{1,t^2\}\,\rho(\mathrm{d}t) < \infty$, so that ρ is actually a Lévy measure and $\Upsilon_0(\rho) = \tilde{\rho}$.

Let, finally, μ be the measure in $\mathcal{ID}(*)$ with generating triplet $(\frac{1}{2}a, \rho, \eta)$, where

$$\eta = \tilde{\eta} - \int_0^\infty \Big(\int_{\mathbb{R}} t\big(1_{[-1,1]}(t) - 1_{[-x,x]}(t)\big)\,D_x\rho(\mathrm{d}t)\Big)\mathrm{e}^{-x}\,\mathrm{d}x.$$

Then $\Upsilon(\mu) = \tilde{\mu}$, and since q is increasing on $]-\infty, 0[$ and decreasing on $]0,\infty[$ (cf. (3.42)), we have that $\mu \in \mathcal{L}(*)$. This concludes the proof. □

3.4 The Mappings Υ_0^α and Υ^α, $\alpha \in [0,1]$

As announced in Section 1, we now introduce two families of mappings $\{\Upsilon_0^\alpha\}_{0\le\alpha\le1}$ and $\{\Upsilon^\alpha\}_{0\le\alpha\le1}$ that, respectively, generalize Υ_0 and Υ, with $\Upsilon_0^0 = \Upsilon_0$, $\Upsilon^0 = \Upsilon$ and with Υ_0^1 and Υ^1 the identity mappings on $\mathfrak{M}_L$ and $\mathcal{ID}(*)$, respectively. The Mittag-Leffler function takes a natural role in this.

A review of relevant properties of the Mittag-Leffler function is given. The transformation Υ_0^α is defined in terms of the associated stable law and is shown to be injective, with absolutely continuous images. Then Υ_0^α is extended to a mapping $\Upsilon^\alpha : \mathcal{ID}(*) \to \mathcal{ID}(*)$, in analogy with the extension of Υ_0 to Υ, and properties of Υ^α are discussed. Finally, stochastic representations of Υ and Υ^α are given.

The Mittag-Leffler Function

The Mittag-Leffler function of negative real argument and index $\alpha > 0$ is given by

$$E_\alpha(-t) = \sum_{k=0}^{\infty} \frac{(-t)^k}{\Gamma(\alpha k + 1)}, \qquad (t > 0). \tag{3.45}$$

In particular we have $E_1(-t) = \mathrm{e}^{-t}$, and if we define E_0 by setting $\alpha = 0$ on the right hand side of (3.45) then $E_0(-t) = (1+t)^{-1}$ (whenever $|t| < 1$).

The Mittag-Leffler function is infinitely differentiable and completely monotone if and only if $0 < \alpha \leq 1$. Hence for $0 < \alpha \leq 1$ it is representable as a Laplace transform and, in fact, for α in $]0,1[$ we have (see [Fe71, p. 453])

$$E_\alpha(-t) = \int_0^\infty \mathrm{e}^{-tx} \zeta_\alpha(x)\,\mathrm{d}x, \tag{3.46}$$

where

$$\zeta_\alpha(x) = \alpha^{-1} x^{-1-1/\alpha} \sigma_\alpha(x^{-1/\alpha}), \qquad (x > 0), \tag{3.47}$$

and σ_α denotes the density function of the positive stable law with index α and Laplace transform $\exp(-\theta^\alpha)$. Note that, for $0 < \alpha < 1$, the function $\zeta_\alpha(x)$ is simply the probability density obtained from $\sigma_\alpha(y)$ by the transformation $x = y^{-\alpha}$. In other words, if we denote the distribution functions determined by ζ_α and σ_α by Z_α and S_α, respectively, then

$$Z_\alpha(x) = 1 - S_\alpha(x^{-1/\alpha}). \tag{3.48}$$

As kindly pointed out to us by Marc Yor, ζ_α has a direct interpretation as the probability density of $l_1^{(\alpha)}$ where $l_t^{(\alpha)}$ denotes the local time of a Bessel process with dimension $2(1-\alpha)$. The law of $l_1^{(\alpha)}$ is called the *Mittag-Leffler distribution.* See [MoOs69] and [ChYo03, p. 114]; cf. also [GrRoVaYo99]. Defining $\zeta_\alpha(x)$ as e^{-x} for $\alpha = 0$ and as the Dirac density at 1 when $\alpha = 1$, formula (3.46) remains valid for all α in $[0,1]$.

For later use, we note that the probability measure $\zeta_\alpha(x)\,\mathrm{d}x$ has moments of all orders. Indeed, for α in $]0,1[$ and any p in $\mathbb{N}$ we have

$$\int_0^\infty x^p \zeta_\alpha(x)\,\mathrm{d}x = \int_0^\infty x^{-p\alpha} \sigma_\alpha(x)\,\mathrm{d}x,$$

where clearly $\int_1^\infty x^{-p\alpha}\sigma_\alpha(x)\,\mathrm{d}x < \infty$. Furthermore, by partial integration,

$$\begin{aligned} \int_0^1 x^{-p\alpha}\sigma_\alpha(x)\,\mathrm{d}x &= \left[x^{-p\alpha} S_\alpha(x)\right]_0^1 + p\alpha \int_0^1 x^{-p\alpha-1} S_\alpha(x)\,\mathrm{d}x \\ &= S_\alpha(1) + p\alpha \int_0^1 x^{-p\alpha-1} S_\alpha(x)\,\mathrm{d}x < \infty, \end{aligned}$$

where we make use (twice) of the relation

$$\mathrm{e}^{x^{-\alpha}} S_\alpha(x) \to 0, \quad \text{as } x \searrow 0,$$

(cf. [Fe71, Theorem 1, p.448]). Combining the observation just made with (3.45) and (3.46), we obtain the formula

$$\int_0^\infty x^k \zeta_\alpha(x)\,\mathrm{d}x = \frac{k!}{\Gamma(\alpha k+1)}, \qquad (k \in \mathbb{N}_0), \tag{3.49}$$

which holds for all α in $[0,1]$.

The Mapping Υ_0^α

As before, we denote by $\mathfrak{M}$ the class of all Borel measures on $\mathbb{R}$, and $\mathfrak{M}_L$ is the subclass of all Lévy measures on $\mathbb{R}$.

Definition 3.28. *For any α in $]0,1[$, we define the mapping $\Upsilon_0^\alpha \colon \mathfrak{M}_L \to \mathfrak{M}$ by the expression:*

$$\Upsilon_0^\alpha(\rho) = \int_0^\infty (D_x\rho)\zeta_\alpha(x)\,\mathrm{d}x, \qquad (\rho \in \mathfrak{M}_L). \tag{3.50}$$

We shall see, shortly, that Υ_0^α actually maps $\mathfrak{M}_L$ into itself. In the sequel, we shall often use $\tilde{\rho}_\alpha$ as shorthand notation for $\Upsilon_0^\alpha(\rho)$. Note that with the interpretation of $\zeta_\alpha(x)\mathrm{d}x$ for $\alpha = 0$ and 1, given above, the formula (3.50) specializes to $\Upsilon_0^1(\rho) = \rho$ and $\Upsilon_0^0(\rho) = \Upsilon_0(\rho)$.

Using (3.47), the formula (3.50) may be reexpressed as

$$\tilde{\rho}_\alpha(\mathrm{d}t) = \int_0^\infty \rho(x^\alpha \mathrm{d}t)\sigma_\alpha(x)\,\mathrm{d}x. \tag{3.51}$$

Note also that $\tilde{\rho}_\alpha(\mathrm{d}t)$ can be written as

$$\tilde{\rho}_\alpha(\mathrm{d}t) = \int_0^\infty \rho\big(\tfrac{1}{R_\alpha(y)}\mathrm{d}t\big)\,\mathrm{d}y,$$

where R_α denotes the inverse function of the distribution function Z_α of $\zeta_\alpha(x)\,\mathrm{d}x$.

Theorem 3.29. *The mapping Υ_0^α sends Lévy measures to Lévy measures.*

For the proof of this theorem we use the following technical result:

Lemma 3.30. *For any Lévy measure ρ on $\mathbb{R}$ and any positive x, we have*

$$\int_{\mathbb{R}\setminus[-1,1]} 1\, D_x\rho(\mathrm{d}t) \le \max\{1,x^2\} \int_{\mathbb{R}} \min\{1,t^2\}\,\rho(\mathrm{d}t), \tag{3.52}$$

and also

$$\int_{[-1,1]} t^2\, D_x\rho(\mathrm{d}t) \le \max\{1,x^2\} \int_{\mathbb{R}} \min\{1,t^2\}\,\rho(\mathrm{d}t). \tag{3.53}$$

Proof. Note first that

$$\int_{\mathbb{R}\setminus[-1,1]} 1\, D_x\rho(\mathrm{d}t) = D_x\rho(\mathbb{R}\setminus[-1,1]) = \rho(\mathbb{R}\setminus[-x^{-1},x^{-1}]).$$

If $0 < x \le 1$, then

$$\rho(\mathbb{R}\setminus[-x^{-1},x^{-1}]) \le \rho(\mathbb{R}\setminus[-1,1]) \le \int_{\mathbb{R}} \min\{1,t^2\}\,\rho(\mathrm{d}t),$$

and if $x > 1$,

$$\begin{aligned}\rho(\mathbb{R}\setminus[-x^{-1},x^{-1}]) &\le \int_{[-1,1]\setminus[-x^{-1},x^{-1}]} x^2t^2\,\rho(\mathrm{d}t) + \int_{\mathbb{R}\setminus[-1,1]} 1\,\rho(\mathrm{d}t)\\ &\le x^2 \int_{\mathbb{R}} \min\{1,t^2\}\,\rho(\mathrm{d}t).\end{aligned}$$

This verifies (3.52). Note next that

$$\int_{[-1,1]} t^2\, D_x\rho(\mathrm{d}t) = \int_{\mathbb{R}} x^2t^2 1_{[-x^{-1},x^{-1}]}(t)\rho(\mathrm{d}t).$$

If $x \ge 1$, we find that

$$\int_{\mathbb{R}} x^2t^2 1_{[-x^{-1},x^{-1}]}(t)\,\rho(\mathrm{d}t) \le x^2 \int_{\mathbb{R}} t^2 1_{[-1,1]}(t)\,\rho(\mathrm{d}t) \le x^2 \int_{\mathbb{R}} \min\{1,t^2\}\,\rho(\mathrm{d}t),$$

and, if $0 < x < 1$,

$$\begin{aligned}&\int_{\mathbb{R}} x^2t^2 1_{[-x^{-1},x^{-1}]}(t)\,\rho(\mathrm{d}t)\\ &\qquad = x^2 \int_{-1}^{1} t^2\,\rho(\mathrm{d}t) + x^2 \int_{\mathbb{R}} t^2 1_{[-x^{-1},x^{-1}]\setminus[-1,1]}(t)\,\rho(\mathrm{d}t)\\ &\qquad \le x^2 \int_{-1}^{1} t^2\,\rho(\mathrm{d}t) + x^2 \int_{\mathbb{R}} x^{-2} 1_{[-x^{-1},x^{-1}]\setminus[-1,1]}(t)\,\rho(\mathrm{d}t)\\ &\qquad \le \int_{-1}^{1} t^2\,\rho(\mathrm{d}t) + \int_{\mathbb{R}} 1_{\mathbb{R}\setminus[-1,1]}(t)\,\rho(\mathrm{d}t)\\ &\qquad = \int_{\mathbb{R}} \min\{1,t^2\}\,\rho(\mathrm{d}t).\end{aligned}$$

This verifies (3.53). □

Proof of Theorem 3.29. Let ρ be a Lévy measure on $\mathbb{R}$ and consider the measure $\tilde{\rho}_\alpha = \Upsilon^\alpha(\rho)$. Using Lemma 3.30 and (3.49) we then have

$$\int_{\mathbb{R}} \min\{1,t^2\}\,\tilde{\rho}_\alpha(\mathrm{d}t) = \int_0^\infty \Big(\int_{\mathbb{R}} \min\{1,t^2\}\,D_x\rho(\mathrm{d}t)\Big)\zeta_\alpha(x)\,\mathrm{d}x$$

$$= \int_0^\infty 2\max\{1,x^2\}\Big(\int_{\mathbb{R}} \min\{1,t^2\}\rho(dt)\Big)\zeta_\alpha(x)\,\mathrm{d}x$$

$$= 2\int_{\mathbb{R}} \min\{1,t^2\}\rho(dt)\int_0^\infty 2\max\{1,x^2\}\zeta_\alpha(x)\,\mathrm{d}x < \infty,$$

as desired. □

Absolute Continuity

As in Section 3.1, we let ω denote the transformation of the Lévy measure ρ by the mapping $x \mapsto x^{-1}$.

Theorem 3.31. *For any Lévy measure ρ the Lévy measure $\tilde{\rho}_\alpha$ given by (3.50) is absolutely continuous with respect to Lebesgue measure. The density $\tilde{r}_\alpha$ is the function on $\mathbb{R}\backslash\{0\}$ given by*

$$\tilde{r}_\alpha(t) = \begin{cases} \int_0^\infty s\zeta_\alpha(st)\,\omega(\mathrm{d}s), & \text{if } t > 0, \\ \int_{-\infty}^0 |s|\zeta_\alpha(st)\,\omega(\mathrm{d}s), & \text{if } t < 0. \end{cases}$$

Proof. It suffices to prove that the restrictions of $\tilde{\rho}_\alpha$ to $]-\infty,0[$ and $]0,\infty[$ equal those of $\tilde{r}_\alpha(t)\,dt$. For a Borel subset B of $]0,\infty[$, we find that

$$\int_B \tilde{r}_\alpha(t)\,\mathrm{d}t = \int_B \Big(\int_0^\infty s\zeta_\alpha(st)\,\omega(\mathrm{d}s)\Big)\,\mathrm{d}t = \int_0^\infty \Big(\int_0^\infty s1_B(t)\zeta_\alpha(st)\,\mathrm{d}t\Big)\,\omega(\mathrm{d}s)$$

$$= \int_0^\infty \Big(\int_0^\infty 1_B(s^{-1}u)\zeta_\alpha(u)\,\mathrm{d}u\Big)\,\omega(\mathrm{d}s),$$

where we have used the change of variable $u = st$. Changing again the order of integration, we have

$$\int_B \tilde{r}_\alpha(t)\,\mathrm{d}t = \int_0^\infty \Big(\int_0^\infty 1_B(s^{-1}u)\,\omega(\mathrm{d}s)\Big)\zeta_\alpha(u)\,\mathrm{d}u$$

$$= \int_0^\infty \Big(\int_0^\infty 1_B(su)\,\rho(\mathrm{d}s)\Big)\zeta_\alpha(u)\,\mathrm{d}u$$

$$= \int_0^\infty \rho(u^{-1}B)\zeta_\alpha(u)\,\mathrm{d}u = \tilde{\rho}_\alpha(B).$$

One proves similarly that the restriction to $]-\infty,0[$ of $\tilde{\rho}_\alpha$ equals that of $\tilde{r}_\alpha(t)\,\mathrm{d}t$. □

Corollary 3.32. *Letting, as above, Z_α denote the distribution function for the probability measure $\zeta_\alpha(t)\,\mathrm{d}t$, we have*

$$\tilde{\rho}_\alpha([t,\infty[) = \int_0^\infty (1 - Z_\alpha(st))\,\omega(\mathrm{d}s) = \int_0^\infty S_\alpha((ts)^{-1/\alpha})\,\omega(\mathrm{d}s) \tag{3.54}$$

for t in $]0,\infty[$, and

$$\tilde{\rho}_\alpha(]-\infty,t]) = \int_{-\infty}^0 (1 - Z_\alpha(st))\,\omega(\mathrm{d}s) = \int_{-\infty}^0 S_\alpha((ts)^{-1/\alpha})\,\omega(\mathrm{d}s) \tag{3.55}$$

for t in $]-\infty,0[$.

Proof. For t in $[0,\infty[$ we find that

$$\begin{aligned}
\tilde{\rho}_\alpha([t,\infty[) &= \int_t^\infty \Big(\int_0^\infty s\zeta_\alpha(su)\,\omega(\mathrm{d}s)\Big)\,\mathrm{d}u \\
&= \int_0^\infty \Big(\int_0^\infty s\zeta_\alpha(su)1_{[t,\infty[}(u)\,\mathrm{d}u\Big)\,\omega(\mathrm{d}s) \\
&= \int_0^\infty \Big(\int_0^\infty \zeta_\alpha(w)1_{[t,\infty[}(s^{-1}w)\,\mathrm{d}w\Big)\,\omega(\mathrm{d}s) \\
&= \int_0^\infty \Big(\int_0^\infty \zeta_\alpha(w)1_{[st,\infty[}(w)\,\mathrm{d}w\Big)\,\omega(\mathrm{d}s) \\
&= \int_0^\infty (1 - Z_\alpha(st))\,\omega(\mathrm{d}s) \\
&= \int_0^\infty S_\alpha((st)^{-1/\alpha})\,\omega(\mathrm{d}s),
\end{aligned}$$

where the last equality follows from (3.48). Formula (3.55) is proved similarly. □

Injectivity of Υ_0^α

In order to show that the mappings $\Upsilon_\alpha\colon \mathcal{ID}(*) \to \mathcal{ID}(*)$ are injective, we first introduce a Laplace like transform: Let ρ be a Lévy measure on $\mathbb{R}$, and as above let ω be the transformation of ρ by the mapping $t \mapsto t^{-1}\colon \mathbb{R}\setminus\{0\} \to \mathbb{R}\setminus\{0\}$. Then ω satisfies

$$\omega(\{0\}) = 0 \quad \text{and} \quad \int_{\mathbb{R}} \min\{1,t^{-2}\}\,\omega(\mathrm{d}t) < \infty. \tag{3.56}$$

For any $\theta,\beta > 0$ we then define

$$\mathcal{L}_\beta(\theta \ddagger \omega) = \int_{\mathbb{R}} \mathrm{e}^{-\theta|t|^\beta}\,\omega(\mathrm{d}t).$$

It follows immediately from (3.56) that $\mathcal{L}_\beta(\theta \ddagger \omega)$ is a finite, positive number for all $\theta, \beta > 0$. For $\beta = 1$, we recover the usual Laplace transform.

Proposition 3.33. *Let α be a fixed number in $]0,1[$, let ρ be a Lévy measure on $\mathbb{R}$, and put $\tilde{\rho}_\alpha = \Upsilon_0^\alpha(\rho)$. Let further ω and $\tilde{\omega}_\alpha$ denote, respectively, the transformations of ρ and $\tilde{\rho}_\alpha$ by the mapping $t \mapsto t^{-1}\colon \mathbb{R}\setminus\{0\} \to \mathbb{R}\setminus\{0\}$. We then have*

$$\mathcal{L}_{1/\alpha}(\theta^{1/\alpha} \ddagger \tilde{\omega}_\alpha) = \mathcal{L}_1(\theta \ddagger \omega), \qquad (\theta \in]0,\infty[).$$

Proof. Recall first from Theorem 3.31 that $\tilde{\rho}_\alpha(\mathrm{d}t) = \tilde{r}_\alpha(t)\,\mathrm{d}t$, where

$$\tilde{r}_\alpha(t) = \begin{cases} \int_0^\infty s\zeta_\alpha(st)\,\omega(\mathrm{d}s), & \text{if } t > 0,\\ \int_{-\infty}^0 |s|\zeta_\alpha(st)\,\omega(\mathrm{d}s), & \text{if } t < 0. \end{cases}$$

Consequently, $\tilde{\omega}_\alpha$ has the following density w.r.t. Lebesgue measure:

$$\tilde{r}_\alpha(t^{-1})t^{-2} = \begin{cases} \int_0^\infty st^{-2}\zeta_\alpha(st^{-1})\,\omega(\mathrm{d}s), & \text{if } t > 0,\\ \int_{-\infty}^0 |s|t^{-2}\zeta_\alpha(st^{-1})\,\omega(\mathrm{d}s), & \text{if } t < 0. \end{cases}$$

For any positive θ, we then find

$$\begin{aligned}
&\int_0^\infty \mathrm{e}^{-\theta t^{1/\alpha}}\,\tilde{\omega}_\alpha(\mathrm{d}t)\\
&= \int_0^\infty \mathrm{e}^{-\theta t^{1/\alpha}}\Big(\int_0^\infty st^{-2}\zeta_\alpha(st^{-1})\,\omega(\mathrm{d}s)\Big)\,\mathrm{d}t\\
&= \int_0^\infty \Big(\int_0^\infty \mathrm{e}^{-\theta t^{1/\alpha}} t^{-2}\zeta_\alpha(st^{-1})\,\mathrm{d}t\Big)\, s\omega(\mathrm{d}s)\\
&= \int_0^\infty \Big(\int_0^\infty \mathrm{e}^{-\theta t^{1/\alpha}} t^{-2}\big[\alpha^{-1}(st^{-1})^{-1-1/\alpha}\sigma_\alpha((st^{-1})^{-1/\alpha})\big]\,\mathrm{d}t\Big)\, s\omega(\mathrm{d}s)\\
&= \frac{1}{\alpha}\int_0^\infty \Big(\int_0^\infty \mathrm{e}^{-\theta t^{1/\alpha}} t^{-1+1/\alpha}\sigma_\alpha(s^{-1/\alpha}t^{1/\alpha})\,\mathrm{d}t\Big)\, s^{-1/\alpha}\omega(\mathrm{d}s),
\end{aligned}$$

where we have used (3.47). Applying now the change of variable: $u = s^{-1/\alpha}t^{1/\alpha}$, we find that

$$\begin{aligned}
\int_0^\infty \mathrm{e}^{-\theta t^{1/\alpha}}\,\tilde{\omega}_\alpha(\mathrm{d}t) &= \int_0^\infty \Big(\int_0^\infty \mathrm{e}^{-\theta s^{1/\alpha}u}\sigma_\alpha(u)\,\mathrm{d}u\Big)\,\omega(\mathrm{d}s)\\
&= \int_0^\infty \mathrm{e}^{-(\theta s^{1/\alpha})^\alpha}\,\omega(\mathrm{d}s) \qquad (3.57)\\
&= \int_0^\infty \mathrm{e}^{-\theta^\alpha s}\,\omega(\mathrm{d}s),
\end{aligned}$$

where we used that the Laplace transform of $\sigma_\alpha(t)\,dt$ is given by

$$\int_0^\infty e^{-\eta t}\sigma_\alpha(t)\,dt = e^{-\eta^\alpha}, \qquad (\eta > 0),$$

(cf. [Fe71, Theorem 1, p. 448]). Applying next the above calculation to the measure $\breve{\omega} := D_{-1}\omega$, we find for any positive θ that

$$\begin{aligned}\int_{-\infty}^0 e^{-\theta|t|^{1/\alpha}}\,\tilde{\omega}_\alpha(dt) &= \int_{-\infty}^0 e^{-\theta|t|^{1/\alpha}}\Big(\int_{-\infty}^0 |s|t^{-2}\zeta_\alpha(st^{-1})\,\omega(ds)\Big)\,dt \\ &= \int_0^\infty e^{-\theta t^{1/\alpha}}\Big(\int_0^\infty st^{-2}\zeta_\alpha(st^{-1})\,\breve{\omega}(ds)\Big)\,dt \\ &= \int_0^\infty e^{-\theta^\alpha s}\,\breve{\omega}(ds) \\ &= \int_{-\infty}^0 e^{-\theta^\alpha|s|}\,\omega(ds).\end{aligned} \tag{3.58}$$

Combining formulae (3.57) and (3.58), it follows immediately that $\mathcal{L}_{1/\alpha}(\theta \ddagger \tilde{\omega}_\alpha) = \mathcal{L}_1(\theta^\alpha \ddagger \omega)$, for any positive θ. □

Corollary 3.34. *For each α in $]0,1[$, the mapping $\Upsilon_0^\alpha\colon \mathfrak{M}_L \to \mathfrak{M}_L$ is injective.*

Proof. With notation as in Proposition 3.33, it follows immediately from that same proposition that the (usual) Laplace transform of ω is uniquely determined by $\tilde{\rho}_\alpha = \Upsilon_0^\alpha(\rho)$. As in the proof of Corollary 3.7, this implies that ω, and hence ρ, is uniquely determined by $\Upsilon_0^\alpha(\rho)$. □

The Mapping Υ^α

Our next objective is to "extend" Υ_0^α to a mapping $\Upsilon^\alpha\colon \mathcal{ID}(*) \to \mathcal{ID}(*)$.

Definition 3.35. *For a probability measure μ in $\mathcal{ID}(*)$ with generating triplet (a,ρ,η), we let $\Upsilon^\alpha(\mu)$ denote the measure in $\mathcal{ID}(*)$ with generating triplet $(c_\alpha a, \tilde{\rho}_\alpha, \eta_\alpha)$, where $\tilde{\rho}_\alpha = \Upsilon_0^\alpha(\rho)$ is defined by (3.50) while*

$$c_\alpha = \frac{2}{\Gamma(2\alpha+1)} \qquad \textit{for} \quad 0 \le \alpha \le 1$$

and

$$\eta_\alpha = \frac{\eta}{\Gamma(\alpha+1)} + \int_0^\infty \Big(\int_{\mathbb{R}} t\big(1_{[-1,1]}(t) - 1_{[-x^{-1},x^{-1}]}(t)\big)\rho(x^{-1}dt)\Big)\zeta_\alpha(x)\,dx. \tag{3.59}$$

To see that the integral in (3.59) is well-defined, we note that it was shown, although not explicitly stated, in the proof of Lemma 3.13 that

$$\int_{\mathbb{R}} |ux|\big|1_{[-1,1]}(ux) - 1_{[-x,x]}(ux)\big|\,\rho(\mathrm{d}x) \le \max\{1,x^2\}\int_0^\infty \min\{1,u^2\}\,\rho(\mathrm{d}u).$$

Together with (3.49), this verifies that η_α is well-defined. Note also that since Υ_0^α is injective (cf. Corollary 3.34), it follows immediately from the definition above that so is Υ^α. The choice of the constants c_α and η_α is motivated by the following two results, which should be seen as analogues of Theorems 3.16 and 3.17. In addition, the choice of c_α and η_α is essential to the stochastic interpretation of Υ^α given in Theorem 3.44 below. Note that for $\alpha = 0$, we recover the mapping Υ, whereas putting $\alpha = 1$ produces the identity mapping on $\mathcal{ID}(*)$.

Theorem 3.36. *Let μ be a measure in $\mathcal{ID}(*)$ with characteristic triplet (a,ρ,η). Then the cumulant function of $\Upsilon^\alpha(\mu)$ is representable as*

$$C_{\Upsilon^\alpha(\mu)}(\zeta) = \frac{\mathrm{i}\eta\zeta}{\Gamma(\alpha+1)} - \tfrac{1}{2}c_\alpha a\zeta^2 + \int_{\mathbb{R}} \big(E_\alpha(\mathrm{i}\zeta t) - 1 - \mathrm{i}\zeta\tfrac{t}{\Gamma(\alpha+1)}1_{[-1,1]}(t)\big)\,\rho(\mathrm{d}t), \tag{3.60}$$

for any ζ in $\mathbb{R}$, and where E_α is the Mittag-Leffler function.

Proof. For every $0 \le \alpha \le 1$ we note first that for any ζ in $\mathbb{R}$,

$$E_\alpha(\mathrm{i}\zeta t) - 1 - \mathrm{i}\zeta\frac{t}{\Gamma(\alpha+1)}1_{[-1,1]}(t) = \int_0^\infty (\mathrm{e}^{\mathrm{i}\zeta tx} - 1 - \mathrm{i}\zeta tx1_{[-1,1]}(t))\zeta_\alpha(x)\,\mathrm{d}x, \tag{3.61}$$

which follows immediately from the above-mentioned properties of E_α and the probability density ζ_α (including the interpretation of $\zeta_\alpha(x)\mathrm{d}x$ for $\alpha = 0$ or 1). Note in particular that $\int_0^\infty x\zeta_\alpha(x)\mathrm{d}x = \frac{1}{\Gamma(\alpha+1)}$ (cf. (3.49)).

We note next that it was established in the proof of Lemma 3.15 that

$$\int_0^\infty \big|\mathrm{e}^{\mathrm{i}\zeta tx} - 1 - \mathrm{i}\zeta tx1_{[-1,1]}(t)\big|\,\rho(\mathrm{d}t) \le \Big(2 + \frac{1}{\sqrt{2}}(\zeta x)^2\Big)\int_{\mathbb{R}} \min\{1,t^2\}\,\rho(\mathrm{d}t).$$

Together with Tonelli's theorem, (3.61) and (3.49), this verifies that the integral in (3.60) is well-defined, and that it is permissible to change the order of integration in the following calculation:

$$\begin{aligned}
&\int_{\mathbb{R}} \big(E_\alpha(\mathrm{i}\zeta t) - 1 - \mathrm{i}\zeta \tfrac{t}{\Gamma(\alpha+1)} 1_{[-1,1]}(t)\big)\, \rho(\mathrm{d}t) \\
&\quad = \int_{\mathbb{R}} \Big(\int_0^\infty (\mathrm{e}^{\mathrm{i}\zeta t x} - 1 - \mathrm{i}\zeta t x 1_{[-1,1]}(t)) \zeta_\alpha(x)\, \mathrm{d}x \Big) \rho(\mathrm{d}t) \\
&\quad = \int_0^\infty \Big(\int_{\mathbb{R}} (\mathrm{e}^{\mathrm{i}\zeta u} - 1 - \mathrm{i}\zeta u 1_{[-x^{-1},x^{-1}]}(u))\, \rho(x^{-1}\mathrm{d}u) \Big) \zeta_\alpha(x)\, \mathrm{d}x \\
&\quad = \int_0^\infty \Big(\int_{\mathbb{R}} (\mathrm{e}^{\mathrm{i}\zeta u} - 1 - \mathrm{i}\zeta u 1_{[-1,1]}(u))\, \rho(x^{-1}\mathrm{d}u) \Big) \zeta_\alpha(x)\, \mathrm{d}x \\
&\qquad\qquad + \mathrm{i}\zeta \int_0^\infty \Big(\int_{\mathbb{R}} u\big(1_{[-1,1]}(u) - 1_{[-x^{-1},x^{-1}]}(u)\big)\, \rho(x^{-1}\mathrm{d}u) \Big) \zeta_\alpha(x)\, \mathrm{d}x \\
&\quad = \int_{\mathbb{R}} (\mathrm{e}^{\mathrm{i}\zeta u} - 1 - \mathrm{i}\zeta u 1_{[-1,1]}(u))\, \tilde{\rho}_\alpha(\mathrm{d}u) \\
&\qquad\qquad + \mathrm{i}\zeta \int_0^\infty \Big(\int_{\mathbb{R}} u\big(1_{[-1,1]}(u) - 1_{[-x^{-1},x^{-1}]}(u)\big)\, \rho(x^{-1}\mathrm{d}u) \Big) \zeta_\alpha(x)\, \mathrm{d}x.
\end{aligned}$$

Comparing the above calculation with Definition 3.35, the theorem follows readily. □

Proposition 3.37. *For any* α *in* $]0,1[$ *and any measure* μ *in* $\mathcal{ID}(*)$ *we have*

$$C_{\Upsilon^\alpha(\mu)}(z) = \int_0^\infty C_\mu(zx)\zeta_\alpha(x)\, \mathrm{d}x, \qquad (z \in \mathbb{R}).$$

Proof. Let (a, ρ, η) be the characteristic triplet for μ. For arbitrary z in $\mathbb{R}$, we then have

$$\begin{aligned}
&\int_0^\infty C_\mu(zx)\zeta_\alpha(x)\, \mathrm{d}x \\
&= \int_0^\infty \Big(\mathrm{i}\eta z x - \frac{1}{2} a z^2 x^2 + \int_{\mathbb{R}} (\mathrm{e}^{\mathrm{i}tzx} - 1 - \mathrm{i}tzx 1_{[-1,1]}(t))\, \rho(\mathrm{d}t) \Big) \zeta_\alpha(x)\, \mathrm{d}x \\
&= \mathrm{i}\eta z \int_0^\infty x\zeta_\alpha(x)\, \mathrm{d}x - \frac{1}{2} a z^2 \int_0^\infty x^2 \zeta_\alpha(x)\, \mathrm{d}x \\
&\qquad\qquad + \int_{\mathbb{R}} \Big(\int_0^\infty (\mathrm{e}^{\mathrm{i}tzx} - 1 - \mathrm{i}tzx 1_{[-1,1]}(t)) \zeta_\alpha(x)\, \mathrm{d}x \Big) \rho(\mathrm{d}t) \\
&= \frac{\mathrm{i}\eta z}{\Gamma(\alpha+1)} - \frac{a z^2}{\Gamma(2\alpha+1)} + \int_{\mathbb{R}} \big(E_\alpha(\mathrm{i}zt) - 1 - \mathrm{i}z \tfrac{t}{\Gamma(\alpha+1)} 1_{[-1,1]}(t)\big)\, \rho(\mathrm{d}t),
\end{aligned} \tag{3.62}$$

where the last equality uses (3.49) as well as (3.61). According to Theorem 3.36, the resulting expression in (3.62) equals $C_{\Upsilon^\alpha(\mu)}(z)$, and the proposition follows. □

Properties of Υ^α

We prove next that the mappings Υ^α posses properties similar to those of Υ established in Proposition 3.18.

Proposition 3.38. *For each α in $]0,1[$, the mapping $\Upsilon^\alpha\colon \mathcal{ID}(*) \to \mathcal{ID}(*)$ has the following algebraic properties:*

(i) *For any μ_1, μ_2 in $\mathcal{ID}(*)$, $\Upsilon^\alpha(\mu_1 * \mu_2) = \Upsilon^\alpha(\mu_1) * \Upsilon^\alpha(\mu_2)$.*
(ii) *For any μ in $\mathcal{ID}(*)$ and any c in $\mathbb{R}$, $\Upsilon^\alpha(D_c\mu) = D_c\Upsilon^\alpha(\mu)$.*
(iii) *For any c in $\mathbb{R}$, $\Upsilon^\alpha(\delta_c) = \delta_c$.*

Proof. Suppose $\mu_1, \mu_2 \in \mathcal{ID}(*)$. Then for any z in $\mathbb{R}$ we have by Proposition 3.37

$$\begin{aligned} C_{\Upsilon^\alpha(\mu_1*\mu_2)}(z) &= \int_0^\infty C_{\mu_1*\mu_2}(zx)\zeta_\alpha(x)\,\mathrm{d}x \\ &= \int_0^\infty \big(C_{\mu_1}(zx) + C_{\mu_2}(zx)\big)\zeta_\alpha(x)\,\mathrm{d}x \\ &= C_{\Upsilon^\alpha(\mu_1)}(z) + C_{\Upsilon^\alpha(\mu_2)}(z) = C_{\Upsilon^\alpha(\mu_1)*\Upsilon^\alpha(\mu_2)}(z), \end{aligned}$$

which verifies statement (i). Statements (ii) and (iii) follow similarly by applications of Proposition 3.37. □

Corollary 3.39. *For each α in $[0,1]$, the mapping $\Upsilon^\alpha\colon \mathcal{ID}(*) \to \mathcal{ID}(*)$ preserves the notions of stability and selfdecomposability, i.e.*

$$\Upsilon^\alpha(\mathcal{S}(*)) \subseteq \mathcal{S}(*) \quad \textit{and} \quad \Upsilon^\alpha(\mathcal{L}(*)) \subseteq \mathcal{L}(*).$$

Proof. This follows as in the proof of Corollary 3.19. □

Theorem 3.40. *For each α in $]0,1[$, the mapping $\Upsilon^\alpha\colon \mathcal{ID}(*) \to \mathcal{ID}(*)$ is continuous with respect to weak convergence*[5].

For the proof of this theorem we use the following

Lemma 3.41. *For any real numbers ζ and t we have*

$$\left|\mathrm{e}^{\mathrm{i}\zeta t} - 1 - \frac{\mathrm{i}\zeta t}{1+t^2}\right| \frac{1+t^2}{t^2} \le 5\max\{1, |\zeta|^2\}. \tag{3.63}$$

Proof. For $t = 0$ the left hand side of (3.63) is interpreted as $\frac{1}{2}\zeta^2$, and the inequality holds trivially. Thus, we assume that $t \neq 0$, and clearly we may assume that $\zeta \neq 0$ too.

For t in $\mathbb{R} \setminus [-1,1]$, note that $\frac{1+t^2}{t^2} \le 2$, and hence

[5] In fact, it can be proved that Υ^α is a homeomorphism onto its range with respect to weak convergence; see [BaTh04c].

$$\Big|e^{i\zeta t} - 1 - \frac{i\zeta t}{1+t^2}\Big|\frac{1+t^2}{t^2} \le (1+1)\frac{1+t^2}{t^2} + \Big|\frac{i\zeta}{t}\Big| \le 4 + |\zeta| \le 5\max\{1, |\zeta|^2\}.$$

For t in $[-1,1] \setminus \{0\}$, note first that

$$\begin{aligned}\Big(e^{i\zeta t} - 1 - \frac{i\zeta t}{1+t^2}\Big)\frac{1+t^2}{t^2} &= \Big(e^{i\zeta t} - 1 - i\zeta t + i\zeta t\frac{t^2}{1+t^2}\Big)\frac{1+t^2}{t^2} \\ &= \Big(\big(\cos(\zeta t) - 1\big) + i\big(\sin(\zeta t) - \zeta t\big)\Big)\frac{1+t^2}{t^2} + i\zeta t. \end{aligned} \tag{3.64}$$

Using the mean value theorem, there is a real number ξ_1 strictly between 0 and t, such that

$$\frac{\cos(\zeta t) - 1}{t^2} = \frac{1}{t}\Big(\frac{\cos(\zeta t) - 1}{t}\Big) = -\frac{1}{t}\sin(\zeta\xi_1)\zeta,$$

and hence

$$\Big|\frac{\cos(\zeta t) - 1}{t^2}\Big| = \Big|\zeta^2 \cdot \frac{\xi_1}{t} \cdot \frac{\sin(\zeta\xi_1)}{\zeta\xi_1}\Big| \le |\zeta|^2. \tag{3.65}$$

Appealing once more to the mean value theorem, there are, for any non-zero real number x, real numbers ξ_2 between 0 and x and ξ_3 between 0 and ξ_2, such that

$$\frac{\sin(x)}{x} - 1 = \cos(\xi_2) - 1 = -\xi_2\sin(\xi_3), \qquad \text{and hence} \qquad \Big|\frac{\sin(x)}{x} - 1\Big| \le |x|.$$

As a consequence

$$\frac{1}{t^2} \cdot \big|\sin(\zeta t) - \zeta t\big| = \frac{1}{t^2} \cdot |\zeta t| \cdot \Big|\frac{\sin(\zeta t)}{\zeta t} - 1\Big| \le \frac{1}{t^2} \cdot |\zeta t|^2 = |\zeta|^2. \tag{3.66}$$

Combining (3.64)-(3.66), it follows for t in $[-1,1] \setminus \{0\}$ that

$$\Big|e^{i\zeta t} - 1 - \frac{i\zeta t}{1+t^2}\Big|\frac{1+t^2}{t^2} \le (|\zeta|^2 + |\zeta|^2) \cdot 2 + |\zeta| \le 5\max\{1, |\zeta|^2\}.$$

This completes the proof. □

Corollary 3.42. *Let μ be an infinitely divisible probability measure on $\mathbb{R}$ with generating pair (γ, σ) (see Section 2.1). Then for any real number ζ we have*

$$|C_\mu(\zeta)| \le (|\gamma| + 5\sigma(\mathbb{R}))\max\{1, |\zeta|^2\}.$$

Proof. This follows immediately from Lemma 3.41 and the representation:

$$C_\mu(\zeta) = i\gamma\zeta + \int_{\mathbb{R}}\Big(e^{i\zeta t} - 1 - \frac{i\zeta t}{1+t^2}\Big)\frac{1+t^2}{t^2}\,\sigma(dt). \qquad ■$$

Proof of Theorem 3.40. Let (μ_n) be a sequence of measures from $\mathcal{ID}(*)$, and suppose that $\mu_n \xrightarrow{\text{w}} \mu$ for some measure μ in $\mathcal{ID}(*)$. We need to show that $\Upsilon^\alpha(\mu_n) \xrightarrow{\text{w}} \Upsilon_\alpha(\mu)$. For this, it suffices to show that

$$C_{\Upsilon^\alpha(\mu_n)}(z) \longrightarrow C_{\Upsilon^\alpha(\mu)}(z), \qquad (z \in \mathbb{R}). \tag{3.67}$$

By Proposition 3.37,

$$C_{\Upsilon^\alpha(\mu_n)}(z) = \int_0^\infty C_{\mu_n}(zx)\zeta_\alpha(x)\,\mathrm{d}x \quad \text{and} \quad C_{\Upsilon^\alpha(\mu)}(z) = \int_0^\infty C_\mu(zx)\zeta_\alpha(x)\,\mathrm{d}x,$$

for all n in $\mathbb{N}$ and z in $\mathbb{R}$. According to [Sa99, Lemma 7.7],

$$C_{\mu_n}(y) \longrightarrow C_\mu(y), \quad \text{for all } y \text{ in } \mathbb{R},$$

so by the dominated convergence theorem, (3.67) follows, if, for each z in $\mathbb{R}$, we find a Borel function $h_z\colon [0,\infty[\to [0,\infty[$, such that

$$\forall n \in \mathbb{N}\ \forall x \in [0,\infty[\colon \left|C_{\mu_n}(zx)\zeta_\alpha(x)\right| \le h_z(x) \qquad \text{and} \qquad \int_0^\infty h_z(x)\,\mathrm{d}x < \infty. \tag{3.68}$$

Towards that end, let, for each n in $\mathbb{N}$, (γ_n, σ_n) denote the generating pair for μ_n. Since $\mu_n \xrightarrow{\text{w}} \mu$, Gnedenko's theorem (cf. [GnKo68, Theorem 1, p.87]) asserts that

$$S := \sup_{n\in\mathbb{N}} \sigma_n(\mathbb{R}) < \infty \quad \text{and} \quad G := \sup_{n\in\mathbb{N}} |\gamma_n| < \infty.$$

Now, by Corollary 3.42, for any n in $\mathbb{N}$, z in $\mathbb{R}$ and x in $[0,\infty[$ we have

$$\left|C_{\mu_n}(zx)\zeta_\alpha(x)\right| \le (G+5S)\max\{1, z^2x^2\}\zeta_\alpha(x),$$

and here, by formula (3.49),

$$\begin{aligned}\int_0^\infty (G+5S)\max\{1, z^2x^2\}\zeta_\alpha(x)\,\mathrm{d}x &\le (G+5S)\int_{\mathbb{R}} (1+z^2x^2)\zeta_\alpha(x)\,\mathrm{d}x \\ &= (G+5S) + (G+5S)z^2\tfrac{2}{\Gamma(2\alpha+1)} < \infty.\end{aligned}$$

Thus, for any z in $\mathbb{R}$, the Borel function

$$h_z(x) = (G+5S)\max\{1, z^2x^2\}\zeta_\alpha(x), \qquad (x \in [0,\infty[),$$

satisfies (3.68). This concludes the proof. □

We close this section by mentioning that a replacement of e^{-y} by $\zeta_\alpha(y)$ in the proof of Proposition 3.22 produces a proof of the following assertion:

$$\forall \mu \in \mathcal{ID}(*)\ \forall \alpha \in [0,1]\colon \mu \text{ has } p\text{'th moment} \iff \Upsilon^\alpha(\mu) \text{ has } p\text{'th moment}.$$

3.5 Stochastic Interpretation of Υ and Υ^α

The purpose of this section is to show that for any measure μ in $\mathcal{ID}(*)$, the measure $\Upsilon(\mu)$ can be realized as the distribution of a stochastic integral w.r.t. to the (classical) Lévy process corresponding to μ. We establish also a similar stochastic interpretation of $\Upsilon^\alpha(\mu)$ for any α in $]0,1[$. The main tool in this is Proposition 2.6.

Theorem 3.43. *Let μ be an arbitrary measure in $\mathcal{ID}(*)$, and let (X_t) be a (classical) Lévy process (in law), such that $L\{X_1\} = \mu$. Then the stochastic integral*

$$Z = \int_0^1 -\log(1-t)\,\mathrm{d}X_t$$

exists, as the limit in probability, of the stochastic integrals $\int_0^{1-1/n} -\log(1-t)\,\mathrm{d}X_t$, as $n \to \infty$. Furthermore, the distribution of Z is exactly $\Upsilon(\mu)$.

Proof. The existence of the stochastic integral $\int_0^1 -\log(1-t)\,\mathrm{d}X_t$ follows from Proposition 2.6, once we have verified that $\int_0^1 |C_\mu(-u\log(1-t))|\,\mathrm{d}t < \infty$, for any u in $\mathbb{R}$. Using the change of variable: $t = 1 - \mathrm{e}^{-x}$, $x \in \mathbb{R}$, we find that

$$\int_0^1 \big|C_\mu(-u\log(1-t))\big|\,\mathrm{d}t = \int_0^\infty \big|C_\mu(ux)\big|\mathrm{e}^{-x}\,\mathrm{d}x,$$

and here the right hand side is finite, according to Lemma 3.15.

Combining next Proposition 2.6 and Theorem 3.17 we find for any u in $\mathbb{R}$ that

$$C_{L\{Z\}}(u) = \int_0^1 C_\mu(-u\log(1-t))\,\mathrm{d}t = \int_0^\infty C_\mu(ux)\mathrm{e}^{-x}\,\mathrm{d}x = C_{\Upsilon(\mu)}(u),$$

which implies that $L\{Z\} = \Upsilon(\mu)$, as desired. □

Before proving the analog of Theorem 3.43 for Υ^α, recall that R_α denotes the inverse of the distribution function Z_α of the probability measure $\zeta_\alpha(x)\,\mathrm{d}x$.

Theorem 3.44. *Let μ be an arbitrary measure in $\mathcal{ID}(*)$, and let (X_t) be a (classical) Lévy process (in law), such that $L\{X_1\} = \mu$. For each $\alpha \in\,]0,1[$, the stochastic integral*

$$Y = \int_0^1 R_\alpha(s)\,\mathrm{d}X_s \tag{3.69}$$

exists, as a limit in probability, and the law of Y is $\Upsilon^\alpha(\mu)$.

Proof. It suffices to consider α in $]0,1[$. In order to ensure the existence of the stochastic integral in (3.69) , it suffices, by Proposition 2.6, to verify that $\int_0^1 |C_\mu(zR_\alpha(t))|\,\mathrm{d}t < \infty$ for all z in $\mathbb{R}$. Denoting by λ the Lebesgue measure on $[0,1]$, note that $Z_\alpha(\zeta_\alpha(x)\,\mathrm{d}x) = \lambda$, so that $R_\alpha(\lambda) = \zeta_\alpha(x)\,\mathrm{d}x$. Hence, we find that

$$\begin{aligned}\int_0^1 \left|C_\mu(zR_\alpha(t))\right| \mathrm{d}t &= \int_0^\infty \left|C_\mu(zu)\right| R_\alpha(\lambda)(\mathrm{d}u) \\ &= \int_0^\infty \left|C_\mu(zu)\right| \cdot \zeta_\alpha(u)\,\mathrm{d}u \\ &\leq \int_0^\infty \big(|\gamma| + 5\nu(\mathbb{R})\big) \max\{1, z^2u^2\}\zeta_\alpha(u)\,\mathrm{d}u < \infty,\end{aligned}$$

where (γ,ν) is the generating pair for μ (cf. Corollary 3.42). Thus, by Proposition 2.6, the stochastic integral $Y = \int_0^1 R_\alpha(t)\,\mathrm{d}X_t$ makes sense, and the cumulant function of Y is given by

$$C\{z \ddagger Y\} = \int_0^1 C_\mu(zR_\alpha(t))\,\mathrm{d}t = \int_0^1 C_\mu(zu)\zeta_\alpha(u)\,\mathrm{d}u = C_{\Upsilon^\alpha(\mu)}(z),$$

where we have used Theorem 3.37. This completes the proof. □

3.6 Mappings of Upsilon-Type: Further Results

We now summarize several pieces of recent work that extend some of the results presented in the previous part of the present section.

We start by considering a general concept of Upsilon transformations, that has the transformations Υ_0 and Υ_0^α as special cases. Another special case, denoted $\Upsilon_0^{(q)}$ $(q > -2)$ is briefly discussed; this is related to the tempered stable distributions. Further, extensions of the mappings Υ_0 and Υ_0^α to multivariate infinitely divisible distributions are discussed, and applications of these to the construction of Lévy copulas with desirable properties is indicated. Finally, a generalization of $\Upsilon_0^{(q)}$ to transformations of the class $\mathfrak{M}_L(\mathrm{M}_m^+)$ of Lévy measures on the cone of positive definite $m \times m$ matrices is mentioned.

General Upsilon Transformations

The collaborative work discussed in the subsequent parts of the present Section have led to taking up a systematic study of *generalized Upsilon transformations*. Here we mention some first results of this, based on unpublished notes by V. Pérez-Abreu, J. Rosinski, K. Sato and the authors. Detailed expositions will appear elsewhere.

Let ρ be a Lévy measure on $\mathbb{R}$, let τ be a measure on $\mathbb{R}_{>0}$ and introduce the measure ρ_τ on $\mathbb{R}$ by

$$\rho_\tau(\mathrm{d}x) = \int_0^\infty \rho(y^{-1}\mathrm{d}x)\tau(\mathrm{d}y). \tag{3.70}$$

Note here that if X is an infinitely divisible random variable with Lévy measure $\rho(\mathrm{d}x)$ then yX has Lévy measure $\rho(y^{-1}\mathrm{d}x)$.

Definition 3.45. *Given a measure τ on $\mathbb{R}_{>0}$ we define Υ_0^τ as the mapping $\Upsilon_0^\tau : \rho \mapsto \rho_\tau$ where ρ_τ is given by (3.70) and the domain of Υ_0^τ is*

$$\mathrm{dom}_L \Upsilon_0^\tau = \left\{ \rho \in \mathfrak{M}_L(\mathbb{R}) \,\middle|\, \rho_\tau \in \mathfrak{M}_L(\mathbb{R}) \right\}.$$

We have $\mathrm{dom}_L \Upsilon_0^\tau = \mathfrak{M}_L(\mathbb{R})$ if and only if

$$\int_0^\infty (1+y^2)\,\tau(\mathrm{d}y) < \infty.$$

Furthermore, letting

$$\mathfrak{M}_0(\mathbb{R}) = \left\{ \rho \in \mathfrak{M}(\mathbb{R}) \,\middle|\, \int_0^\infty (1+|t|)\,\rho(\mathrm{d}t) < \infty \right\}$$

(finite variation case) we have $\Upsilon_0^\tau : \mathfrak{M}_0(\mathbb{R}) \to \mathfrak{M}_0(\mathbb{R})$ if and only if

$$\int_0^\infty (1+|y|)\,\tau(\mathrm{d}y) < \infty.$$

Mappings of type Υ_0^τ have the important property of being commutative under composition. Under rather weak conditions the mappings are one-to-one, and the image Lévy measures possess densities with respect to Lebesgue measure. This is true, in particular, of the examples considered below.

Now, suppose that τ has a density h that is a continuous function on $\mathbb{R}_{>0}$. Then writing ρ_h for ρ_τ we have

$$\rho_h(\mathrm{d}x) = \int_0^\infty \rho(y^{-1}\mathrm{d}x)h(y)\mathrm{d}y. \tag{3.71}$$

Clearly, the mappings Υ_0 and Υ_0^α are special instances of (3.71).

Example 3.46. Φ_0 transformation. The Υ_0^h transformation obtained by letting

$$h(y) = 1_{[-1,1]}(y)y^{-1}$$

is denoted by Φ_0. Its domain is

$$\mathrm{dom}_L \Phi_0 = \left\{ \rho \in \mathfrak{M}_L(\mathbb{R}) \,\middle|\, \int_{\mathbb{R}\setminus[-1,1]} \log|y|\,\rho(\mathrm{d}y) < \infty \right\}.$$

As is well known, this transformation maps $\mathrm{dom}_L \Phi_0$ onto the class of selfdecomposable Lévy measures.

Example 3.47. $\Upsilon_0^{(q)}$ transformations. The special version of Υ_0^h obtained by taking

$$h(y) = y^q e^{-y}$$

is denoted $\Upsilon_0^{(q)}$. For each $q > -1$, $\mathrm{dom}_L \Upsilon_0^{(q)} = \mathfrak{M}_L(\mathbb{R})$, for $q = -1$ the domain equals $\mathrm{dom}_L \Phi_0$, while, for $q \in (-2, -1)$, $\Upsilon_0^{(q)}$ has domain

$$\mathrm{dom}_L \Upsilon_0^{(q)} = \left\{ \rho \in \mathfrak{M}_L(\mathbb{R}) \;\middle|\; \int_{\mathbb{R}\setminus[-1,1]} |y|^{-q-1} \rho(\mathrm{d}y) < \infty \right\}.$$

These transformations are closely related to the tempered stable laws. In fact, let $\sigma(\mathrm{d}x) = c_{\pm}\alpha x^{-1-\alpha} k(x)\mathrm{d}x$ with

$$k(x) = \int_0^\infty e^{-xc} \nu(\mathrm{d}c)$$

be the Lévy measure of an element in $\mathcal{R}(*)$. Then σ is the image under $\Upsilon_0^{(-1-\alpha)}$ of the Lévy measure

$$\rho(\mathrm{d}x) = x^{-\alpha} \underset{\leftarrow}{\nu}(\mathrm{d}x), \tag{3.72}$$

where $\underset{\leftarrow}{\nu}$ is the image of the measure ν under the mapping $x \mapsto x^{-1}$.

Interestingly, $\Upsilon_0 \Phi_0 = \Phi_0 \Upsilon_0 = \Upsilon_0^{(-1)}$. The transformations Υ_0^h may in wide generality be characterized in terms of stochastic integrals, as follows. Let

$$H(\xi) = \int_\xi^\infty h(y)\,\mathrm{d}y,$$

set $s = H(\xi)$ and let K, with derivative k, be the inverse function of H, so that $K(H(\xi)) = \xi$ and hence, by differentiation, $k(s)h(\xi) = 1$. Let ρ be an arbitrary element of $\mathfrak{M}_L(\mathbb{R})$ and let L be a Lévy process such that L_1 has Lévy measure ρ. Then, under mild regularity conditions, the integral

$$Y = \int_0^{H(0)} K(s)\,\mathrm{d}L_s \tag{3.73}$$

exists and the random variable Y is infinitely divisible with Lévy measure $\rho_h = \Upsilon_0^h(\rho)$.

Upsilon Transformations of $\mathcal{ID}^d(*)$

The present subsection is based on the paper [BaMaSa04] to which we refer for proofs, additional results, details and references.

We denote the class of infinitely divisible probability laws on $\mathbb{R}^d$ by $\mathcal{ID}^d(*)$. Let h be a function as in the previous subsection and let L be a d-dimensional Lévy process. Then, under a mild regularity condition on h, a d-dimensional random vector Y is determined by

$$Y = \int_0^{H(0)} K(s)\,\mathrm{d}L_s$$

cf. the previous subsection.

If h is the density determining Υ_0 then each of the components of Y belongs to class $\mathcal{B}(*)$ and Y is said to be of class $\mathcal{B}^d(*)$, the d-dimensional Goldie-Steutel-Bondesson class. Similarly, the d-dimensional Thorin class $\mathcal{T}^d(*)$ is defined by taking the components of L_1 to be in $\mathcal{L}(*)$. In [BaMaSa04], probabilistic characterizations of $\mathcal{B}^d(*)$ and $\mathcal{T}^d(*)$ are given, and relations to selfdecomposability and to iterations of Υ_0 and Φ_0 are studied in considerable detail.

Application to Lévy Copulas

We proceed to indicate some applications of Υ_0 and Φ_0 and of the above-mentioned results to the construction of Lévy copulas for which the associated probability measures have prescribed marginals in the Goldie-Steutel-Bondesson or Thorin class or Lévy class (the class of selfdecomposable laws). For proofs and details, see [BaLi04].

The concept of copulas for multivariate probability distributions has an analogue for multivariate Lévy measures, termed Lévy copulas. Similar to probabilistic copulas, a Lévy copula describes the dependence structure of a multivariate Lévy measure. The Lévy measure, ρ say, is then completely characterized by knowledge of the Lévy copula and the m one-dimensional margins which are obtained as projections of ρ onto the coordinate axes. An advantage of modeling dependence via Lévy copulas rather that distributional copulas is that the resulting probability laws are automatically infinitely divisible.

For simplicity, we consider only Lévy measures and Lévy copulas living on $\mathbb{R}^m_{>0}$. Suppose that $\mu_1, \dots, \mu_m$ are one-dimensional infinitely divisible distributions, all of which are in the Goldie-Steutel-Bondesson class or the Thorin class or the Lévy class. Using any Lévy copula gives an infinitely divisible distribution μ with margins $\mu_1, \dots, \mu_m$. But μ itself does not necessarily belong to the Bondesson class or the Thorin class or the Lévy class, i.e. not every Lévy copula gives rise to such distributions. However, that can be achieved by the use of Upsilon transformations. For the Goldie-Steutel-Bondesson class and the Lévy class this is done with the help of the mappings Υ_0 and Φ_0, respectively, and combining the mappings Φ_0 and Υ_0 one can construct multivariate distributions in the Thorin class with prescribed margins in the Thorin class.

Upsilon Transformations for Matrix Subordinators

The present subsection is based on the paper [BaPA05] to which we refer for proofs, additional results, details and references.

An extension of Υ_0 to a one-to-one mapping of the class of d-dimensional Lévy measures into itself was considered in the previous subsection. Here we shall briefly discuss another type of generalization, to one-to-one mappings of $\mathcal{ID}_+^{m\times m}(*)$, the set of infinitely divisible positive semidefinite $m \times m$ matrices, into itself. This class of mappings constitutes an extension to the positive definite matrix setting of the class $\{\Upsilon_0^{(q)}\}_{-1<q<\infty}$ considered above, and we shall use the same notation $\Upsilon_0^{(q)}$ in the general matrix case.

We begin by reviewing several facts about infinitely divisible matrices with values in the cone $\overline{\mathbb{M}}_m^+$ of symmetric nonnegative definite $m \times m$ matrices.

Let $\mathbb{M}_{m\times m}$ denote the linear space of $m \times m$ real matrices, $\mathbb{M}_m$ the linear subspace of symmetric matrices, $\overline{\mathbb{M}}_m^+$ the closed cone of non-negative definite matrices in $\mathbb{M}_m$, $\mathbb{M}_m^+$ and $\{X > 0\}$ the open cone of positive definite matrices in $\mathbb{M}_m$.

For $X \in \mathbb{M}_{m\times m}$, $X^\top$ is the transpose of X and $\mathrm{tr}(X)$ the trace of X. For X in $\overline{\mathbb{M}}_m^+$, $X^{1/2}$ is the unique symmetric matrix in $\overline{\mathbb{M}}_m^+$ such that $X = X^{1/2}X^{1/2}$. Given a nonsingular matrix X in $\mathbb{M}_{m\times m}$, X^{-1} denotes its inverse, $|X|$ its determinant and $X^{-\top}$ the inverse of its transpose. When X is in $\mathbb{M}_m^+$ we simply write $X > 0$.

The cone $\overline{\mathbb{M}}_m^+$ is not a linear subspace of the linear space $\mathbb{M}_{m\times m}$ of $m \times m$ matrices and the theory of infinite divisibility on Euclidean spaces does not apply immediately to $\overline{\mathbb{M}}_m^+$. In general, the study of infinitely divisible random elements in closed cones requires separate work.

A random matrix M is *infinitely divisible in* $\overline{\mathbb{M}}_m^+$ if and only if for each integer $p \geq 1$ there exist p independent identically distributed random matrices $M_1, ..., M_p$ in $\overline{\mathbb{M}}_m^+$ such that $M \stackrel{\mathrm{d}}{=} M_1 + \cdots + M_p$. In this case, the Lévy-Khintchine representation has the following special form, which is obtained from [Sk91] p.156-157.

Proposition 3.48. *An infinitely divisible random matrix M is infinitely divisible in $\overline{\mathbb{M}}_m^+$ if and only if its cumulant transform is of the form*

$$\mathcal{C}(\Theta; M) = \mathrm{itr}(\Psi^0\Theta) + \int_{\overline{\mathbb{M}}_m^+} (\mathrm{e}^{\mathrm{itr}(X\Theta)} - 1)\rho(\mathrm{d}X), \quad \Theta \in \mathbb{M}_m^+, \tag{3.74}$$

where $\Psi^0 \in \overline{\mathbb{M}}_m^+$ and the Lévy measure ρ satisfies $\rho(\mathbb{M}_m \backslash \overline{\mathbb{M}}_m^+) = 0$ and has order of singularity

$$\int_{\overline{\mathbb{M}}_m^+} \min(1, \|X\|)\rho(\mathrm{d}X) < \infty. \tag{3.75}$$

Moreover, the Laplace transform of M is given by

$$\mathcal{L}_M(\Theta) = \exp\{-\mathcal{K}(\Theta; M)\}, \quad \Theta \in \mathbb{M}_m^+, \tag{3.76}$$

where $\mathcal{K}$ is the Laplace exponent

$$\mathcal{K}(\Theta; M) = \mathrm{tr}(\Psi^0\Theta) + \int_{\overline{\mathbb{M}}_m^+} (1 - \mathrm{e}^{-\mathrm{tr}(X\Theta)})\rho(\mathrm{d}X). \tag{3.77}$$

For ρ in $\mathfrak{M}_L(\mathrm{M}_m^+)$ and $q > -1$ consider the mapping $\Upsilon_0^{(q)} : \rho \mapsto \rho_q$ given by

$$\rho_q(\mathrm{d}Z) = \int_{X>0} \rho(\overline{X}^{-\top}\mathrm{d}Z\overline{X}^{-1})\,|X|^q\,\mathrm{e}^{-\mathrm{tr}(X)}\mathrm{d}X. \tag{3.78}$$

The measure ρ_q is a Lévy measure on $\overline{\mathrm{M}}_m^+$.

To establish that for each $q > -1$ the mapping $\Upsilon_0^{(q)}$ is one-to-one the following type of Laplace transform of elements $\rho \in \mathfrak{M}_L(\mathrm{M}_m^+)$ is introduced:

$$\mathcal{L}^p\rho(\Theta) = \int_{X>0} \mathrm{e}^{-\mathrm{tr}(X\Theta)}\,|X|^p\,\rho(\mathrm{d}X). \tag{3.79}$$

For any $p \geq 1$ and ρ in $\mathfrak{M}_L(\mathrm{M}_m^+)$, the transform (3.79) is finite for any $\Theta \in \mathrm{M}_m^+$, and the following theorem implies the bijectivity.

Theorem 3.49. *Let $p \geq 1$ and $p + q \geq 1$. Then*

$$\mathcal{L}^p\rho_q(\Theta) = |\Theta|^{-\frac{1}{2}(m+1)-(p+q)} \int_{V>0} \mathcal{L}^p\rho(\mathbf{V})\,|V|^{p+q}\,\mathrm{e}^{-\mathrm{tr}(\mathbf{\Theta}^{-1}V)}\mathrm{d}V. \tag{3.80}$$

for $\Theta \in \mathrm{M}_m^+$

As in the one-dimensional case, the transformed Lévy measure determined by the mapping $\Upsilon_0^{(q)}$ is absolutely continuous (with respect to Lebesgue measure on M_m^+) and the density possesses an integral representation, showing in particular that the density is a completely monotone function on M_m^+.

Theorem 3.50. *For each $q > -1$ the Lévy measure ρ_q is absolutely continuous with Lévy density r_q given by*

$$r_q(X) = |\mathbf{X}|^q \int_{Y>0} |Y|^{-\frac{1}{2}(m+1)-q}\,\mathrm{e}^{-\mathrm{tr}(\mathbf{X}\mathbf{Y}^{-1})}\rho(\mathrm{d}Y) \tag{3.81}$$

$$= |\mathbf{X}|^q \int_{Y>0} |Y|^{\frac{1}{2}(m+1)+q}\,\mathrm{e}^{-\mathrm{tr}(\mathbf{X}\mathbf{Y})}\,\underleftarrow{\rho}(\mathrm{d}Y). \tag{3.82}$$

4 Free Infinite Divisibility and Lévy Processes

Free probability is a subject in the theory of non-commutative probability. It was originated by Voiculescu in the Nineteen Eighties and has since been extensively studied, see e.g. [VoDyNi92], [Vo98] and [Bi03]. The present section provides an introduction to the area, somewhat in parallel to the exposition of the classical case in Section 2.5. Analogues of some of the subclasses of $\mathcal{ID}(*)$ discussed in that section are introduced. Finally, a discussion of free Lévy processes is given.

4.1 Non-Commutative Probability and Operator Theory

In classical probability, one might say that the basic objects of study are random variables, represented as measurable functions from a probability space $(\Omega, \mathcal{F}, P)$ into the real numbers $\mathbb{R}$ equipped with the Borel σ-algebra $\mathcal{B}$. To any such random variable $X: \Omega \to \mathbb{R}$ the distribution μ_X of X is determined by the equation:

$$\int_{\mathbb{R}} f(t)\ \mu_X(\mathrm{d}t) = \mathbb{E}(f(X)),$$

for any bounded Borel function $f: \mathbb{R} \to \mathbb{R}$, and where $\mathbb{E}$ denotes expectation (or integration) w.r.t. P. We shall also use the notation $L\{X\}$ for μ_X.

In *non-commutative* probability, one replaces the random variables by (self-adjoint) operators on a Hilbert space $\mathcal{H}$. These operators are then referred to as "non–commutative random variables". The term non-commutative refers to the fact that, in this setting, the multiplication of "random variables" (i.e. composition of operators) is no longer commutative, as opposed to the usual multiplication of classical random variables. The non-commutative situation is often remarkably different from the classical one, and most often more complicated.

By $\mathcal{B}(\mathcal{H})$ we denote the vector space of all bounded operators on $\mathcal{H}$, i.e. linear mappings $a: \mathcal{H} \to \mathcal{H}$, which are continuous, or, equivalently, which satisfy that

$$\|a\| := \sup\{\|a\xi\| \mid \xi \in \mathcal{H},\ \|\xi\| \leq 1\} < \infty.$$

The mapping $a \mapsto \|a\|$ is a norm on $\mathcal{B}(\mathcal{H})$, called the operator norm, and $\mathcal{B}(\mathcal{H})$ is complete in the operator norm. Composition of operators form a (non-commutative) multiplication on $\mathcal{B}(\mathcal{H})$, which, together with the linear operations, turns $\mathcal{B}(\mathcal{H})$ into an algebra.

Recall next that $\mathcal{B}(\mathcal{H})$ is equipped with an involution (the adjoint operation) $a \mapsto a^*: \mathcal{B}(\mathcal{H}) \to \mathcal{B}(\mathcal{H})$, which is given by:

$$\langle a\xi, \eta\rangle = \langle \xi, a^*\eta\rangle, \quad (a \in \mathcal{B}(\mathcal{H}),\ \xi, \eta \in \mathcal{H}).$$

Instead of working with the whole algebra $\mathcal{B}(\mathcal{H})$ as the set of "random variables" under consideration, it is, for most purposes, natural to restrict attention to certain subalgebras of $\mathcal{B}(\mathcal{H})$.

A (unital) *C^*-algebra* acting on a Hilbert space $\mathcal{H}$ is a subalgebra of $\mathcal{B}(\mathcal{H})$, which contains the multiplicative unit $\mathbf{1}$ of $\mathcal{B}(\mathcal{H})$ (i.e. $\mathbf{1}$ is the identity mapping on $\mathcal{H}$), and which is closed under the adjoint operation and topologically closed w.r.t. the operator norm.

A *von Neumann algebra*, acting on $\mathcal{H}$, is a unital C^*-algebra acting on $\mathcal{H}$, which is even closed in the weak operator topology on $\mathcal{B}(\mathcal{H})$ (i.e. the weak topology on $\mathcal{B}(\mathcal{H})$ induced by the linear functionals: $a \mapsto \langle a\xi, \eta\rangle$, $\xi, \eta \in \mathcal{H}$).

A *state* on the (unital) C^*-algebra $\mathcal{A}$ is a positive linear functional $\tau: \mathcal{A} \to \mathbb{C}$, taking the value 1 at the identity operator $\mathbf{1}$ on $\mathcal{H}$. If τ satisfies, in addition, the trace property:

$$\tau(ab) = \tau(ba), \quad (a, b \in \mathcal{A}),$$

then τ is called a *tracial state*[6]. A tracial state τ on a von Neumann algebra $\mathcal{A}$ is called *normal*, if its restriction to the unit ball of $\mathcal{A}$ (w.r.t. the operator norm) is continuous in the weak operator topology.

Definition 4.1. (i) *A C^*-probability space is a pair $(\mathcal{A}, \tau)$, where $\mathcal{A}$ is a unital C^*-algebra and τ is a faithful state on $\mathcal{A}$.*
(ii) *A W^*-probability space is a pair $(\mathcal{A}, \tau)$, where $\mathcal{A}$ is a von Neumann algebra and τ is a faithful, normal tracial state on $\mathcal{A}$.*

The assumed faithfulness of τ in Definition 4.1 means that τ does not annihilate any non-zero positive operator. It implies that $\mathcal{A}$ is finite in the sense of F. Murray and J. von Neumann.

In the following, we shall mostly be dealing with W^*-probability spaces. So suppose that $(\mathcal{A}, \tau)$ is a W^*-probability space and that a is a selfadjoint operator (i.e. $a^* = a$) in $\mathcal{A}$. Then, as in the classical case, we can associate a (spectral) distribution to a in a natural way: Indeed, by the Riesz representation theorem, there exists a unique probability measure μ_a on $(\mathbb{R}, \mathcal{B})$, satisfying that

$$\int_{\mathbb{R}} f(t)\ \mu_a(\mathrm{d}t) = \tau(f(a)), \tag{4.1}$$

for any bounded Borel function $f\colon \mathbb{R} \to \mathbb{R}$. In formula (4.1), $f(a)$ has the obvious meaning if f is a polynomial. For general Borel functions f, $f(a)$ is defined in terms of spectral theory (see e.g. [Ru91]).

The (spectral) distribution μ_a of a selfadjoint operator a in $\mathcal{A}$ is automatically concentrated on the spectrum $\mathrm{sp}(a)$, and is thus, in particular, compactly supported. If one wants to be able to consider any probability measure μ on $\mathbb{R}$ as the spectral distribution of some selfadjoint operator, then it is necessary to take unbounded (i.e. non-continuous) operators into account. Such an operator a is, generally, not defined on all of $\mathcal{H}$, but only on a subspace $\mathcal{D}(a)$ of $\mathcal{H}$, called the domain of a. We say then that a is an operator *in* $\mathcal{H}$ rather than *on* $\mathcal{H}$. For most of the interesting examples, $\mathcal{D}(a)$ is a dense subspace of $\mathcal{H}$, in which case a is said to be densely defined. We have included a detailed discussion on unbounded operators in the Appendix (Section A), from which we extract the following brief discussion.

If $(\mathcal{A}, \tau)$ is a W^*-probability space acting on $\mathcal{H}$ and a is an unbounded operator in $\mathcal{H}$, a cannot be an element of $\mathcal{A}$. The closest a can get to $\mathcal{A}$ is to be *affiliated* with $\mathcal{A}$, which means that a commutes with any unitary operator u, that commutes with all elements of $\mathcal{A}$. If a is selfadjoint, a is affiliated with $\mathcal{A}$ if and only if $f(a) \in \mathcal{A}$ for any bounded Borel function $f\colon \mathbb{R} \to \mathbb{R}$. In this case,

[6] In quantum physics, τ is of the form $\tau(a) = \mathrm{tr}(\rho a)$, where ρ is a trace class selfadjoint operator on $\mathcal{H}$ with trace 1, that expresses the state of a quantum system, and a would be an observable, i.e. a selfadjoint operator on $\mathcal{H}$, the mean value of the outcome of observing a being $\mathrm{tr}(\rho a)$.

(4.1) determines, again, a unique probability measure μ_a on $\mathbb{R}$, which we also refer to as the (spectral) distribution of a, and which generally has unbounded support. Furthermore, any probability measure on $\mathbb{R}$ can be realized as the (spectral) distribution of some selfadjoint operator affiliated with some W^*-probability space. In the following we shall also use the notation $L\{a\}$ for the distribution of a (possibly unbounded) operator a affiliated with $(\mathcal{A}, \tau)$. By $\overline{\mathcal{A}}$ we denote the set of operators in $\mathcal{H}$ which are affiliated with $\mathcal{A}$.

4.2 Free Independence

The key concept on relations between classical random variables X and Y is *independence*. One way of defining that X and Y (defined on the same probability space $(\Omega, \mathcal{F}, P)$) are independent is to ask that all compositions of X and Y with bounded Borel functions be uncorrelated:

$$\mathbb{E}\{[f(X) - \mathbb{E}\{f(X)\}] \cdot [g(Y) - \mathbb{E}\{g(Y)\}]\} = 0,$$

for any bounded Borel functions $f, g \colon \mathbb{R} \to \mathbb{R}$.

In the early 1980's, D.V. Voiculescu introduced the notion of *free independence* among non-commutative random variables:

Definition 4.2. *Let $a_1, a_2, \dots, a_r$ be selfadjoint operators affiliated with a W^*-probability space $(\mathcal{A}, \tau)$. We say then that $a_1, a_2, \dots, a_r$ are freely independent w.r.t. τ, if*

$$\tau\{[f_1(a_{i_1}) - \tau(f_1(a_{i_1}))][f_2(a_{i_2}) - \tau(f_2(a_{i_2}))] \cdots [f_p(a_{i_p}) - \tau(f_p(a_{i_p}))]\} = 0,$$

for any p in $\mathbb{N}$, any bounded Borel functions $f_1, f_2, \dots, f_p \colon \mathbb{R} \to \mathbb{R}$ and any indices $i_1, i_2, \dots, i_p$ in $\{1, 2, \dots, r\}$ satisfying that $i_1 \neq i_2, i_2 \neq i_3, \dots, i_{p-1} \neq i_p$.

At a first glance, the definition of free independence looks, perhaps, quite similar to the definition of classical independence given above, and indeed, in many respects free independence is conceptually similar to classical independence. For example, if $a_1, a_2, \dots, a_r$ are freely independent selfadjoint operators affiliated with $(\mathcal{A}, \tau)$, then all numbers of the form $\tau\{f_1(a_{i_1}) f_2(a_{i_2}) \cdots f_p(a_{i_p})\}$ (where $i_1, i_2, \dots, i_p \in \{1, 2, \dots, r\}$ and $f_1, f_2, \dots, f_p \colon \mathbb{R} \to \mathbb{R}$ are bounded Borel functions), are uniquely determined by the distributions $L\{a_i\}$, $i = 1, 2, \dots, r$. On the other hand, free independence is a truly non-commutative notion, which can be seen, for instance, from the easily checked fact that two classical random variables are never freely independent, unless one of them is trivial, i.e. constant with probability one (see e.g. [Vo98]).

Voiculescu originally introduced free independence in connection with his deep studies of the von Neumann algebras associated to the free group factors (see [Vo85], [Vo91], [Vo90]). We prefer in these notes, however, to indicate the significance of free independence by explaining its connection with random matrices. In the 1950's, the phycicist E.P. Wigner showed that the spectral distribution of large selfadjoint random matrices with independent complex Gaussian entries is, approximately, the semi-circle distribution, i.e. the distribution on $\mathbb{R}$ with density $s \mapsto \sqrt{4-s^2} \cdot 1_{[-2,2]}(s)$ w.r.t. Lebesgue measure. More precisely, for each n in $\mathbb{N}$, let $X^{(n)}$ be a selfadjoint complex Gaussian random matrix of the kind considered by Wigner (and suitably normalized), and let tr_n denote the (usual) tracial state on the $n \times n$ matrices $M_n(\mathbb{C})$. Then for any positive integer p, Wigner showed that

$$\mathbb{E}\{\mathrm{tr}_n\big[(X^{(n)})^p\big]\} \underset{n\to\infty}{\longrightarrow} \int_{-2}^{2} s^p\sqrt{4-s^2}\,\mathrm{d}s.$$

In the late 1980's, Voiculescu generalized Wigner's result to families of independent selfadjoint Gaussian random matrices (cf. [Vo91]): For each n in $\mathbb{N}$, let $X_1^{(n)}, X_2^{(n)}, \ldots, X_r^{(n)}$ be independent[7] random matrices of the kind considered by Wigner. Then for any indices $i_1, i_2, \ldots, i_p$ in $\{1, 2, \ldots, r\}$,

$$\mathbb{E}\{\mathrm{tr}_n\big[X_{i_1}^{(n)}X_{i_2}^{(n)}\cdots X_{i_p}^{(n)}\big]\} \underset{n\to\infty}{\longrightarrow} \tau\{x_{i_1}x_{i_2}\cdots x_{i_p}\},$$

where $x_1, x_2, \ldots, x_r$ are freely independent selfadjoint operators in a W^*-probability space $(\mathcal{A}, \tau)$, and such that $L\{x_i\}$ is the semi-circle distribution for each i.

By Voiculescu's result, free independence describes what the assumed classical independence between the random matrices is turned into, as $n \to \infty$. Also, from a classical probabilistic point of view, free probability theory may be considered as (an aspect of) the probability theory of large random matrices.

Voiculescu's result reveals another general fact in free probability, namely that the role of the Gaussian distribution in classical probability is taken over by the semi-circle distribution in free probability. In particular, as also proved by Voiculescu, the limit distribution appearing in the free version of the central limit theorem is the semi-circle distribution (see e.g. [VoDyNi92]).

4.3 Free Independence and Convergence in Probability

In this section, we study the relationship between convergence in probability and free independence. The results will be used in the proof of the free Lévy-Itô decomposition in Section 6.5 below. We start by defining the notion of convergence in probability in the non-commutative setting:

Definition 4.3. *Let $(\mathcal{A}, \tau)$ be a W^*-probability space and let a and a_n, $n \in \mathbb{N}$, be operators in $\overline{\mathcal{A}}$. We say then that $a_n \to a$ in probability, as $n \to \infty$, if $|a_n - a| \to 0$ in distribution, i.e. if $L\{|a_n - a|\} \to \delta_0$ weakly.*

[7] in the classical sense; at the level of the entries.

Convergence in probability, as defined above, corresponds to the so-called *measure topology*, which is discussed in detail in the Appendix (Section A). As mentioned there, if we assume that the operators a_n and a are all selfadjoint, then convergence in probability is equivalent to the condition:

$$L\{a_n - a\} \xrightarrow{\text{w}} \delta_0.$$

Lemma 4.4. *Let (b_n) be a sequence of (not necessarily selfadjoint) operators in a W^*-probability space $(\mathcal{A}, \tau)$, and assume that $\|b_n\| \le 1$ for all n. Assume, further, that $b_n \to b$ in probability as $n \to \infty$ for some operator b in $\mathcal{A}$. Then also $\|b\| \le 1$ and $\tau(b_n) \to \tau(b)$, as $n \to \infty$.*

Proof. To see that $\|b\| \le 1$, note first that $b_n^* b_n \to b^* b$ in probability as $n \to \infty$, since operator multiplication and the adjoint operation are both continuous operations in the measure topology. This implies that $b_n^* b_n \to b^* b$ in distribution, i.e. that $L\{b_n^* b_n\} \xrightarrow{\text{w}} L\{b^* b\}$ as $n \to \infty$ (cf. Proposition A.9). Since $\text{supp}(L\{b_n^* b_n\}) = \text{sp}(b_n^* b_n) \subseteq [0,1]$ for all n (recall that τ is faithful), a standard argument shows that also $[0,1] \supseteq \text{supp}(L\{b^* b\}) = \text{sp}(b^* b)$, whence $\|b\| \le 1$.

To prove the second statement, consider, for each n in $\mathbb{N}$, $b_n' = \frac{1}{2}(b_n + b_n^*)$ and $b_n'' = \frac{1}{2\text{i}}(b_n - b_n^*)$, and define b', b'' similarly from b. Then b_n', b_n'', b', b'' are all selfadjoint operators in $\mathcal{A}$ of norm less than or equal to 1. Since addition, scalar-multiplication and the adjoint operation are all continuous operations in the measure topology, it follows, furthermore, that $b_n' \to b'$ and $b_n'' \to b''$ in probability as $n \to \infty$. As above, this implies that $L\{b_n'\} \xrightarrow{\text{w}} L\{b'\}$ and $L\{b_n''\} \xrightarrow{\text{w}} L\{b''\}$ as $n \to \infty$.

Now, choose a continuous bounded function $f \colon \mathbb{R} \to \mathbb{R}$, such that $f(x) = x$ for all x in $[-1,1]$. Then, since $\text{sp}(b_n'), \text{sp}(b')$ are contained in $[-1,1]$, we find that

$$\begin{aligned} \tau(b_n') = \tau(f(b_n')) &= \int_{\mathbb{R}} f(x)\, L\{b_n'\}(\text{d}x) \underset{n\to\infty}{\longrightarrow} \int_{\mathbb{R}} f(x)\, L\{b'\}(\text{d}x) \\ &= \tau(f(b')) = \tau(b'). \end{aligned}$$

Similarly, $\tau(b_n'') \to \tau(b'')$ as $n \to \infty$, and hence also $\tau(b_n) = \tau(b_n' + \text{i}b_n'') \to \tau(b' + \text{i}b'') = \tau(b)$, as $n \to \infty$. □

Lemma 4.5. *Let r be a positive integer, and let $(b_{1,n})_{n\in\mathbb{N}}, \dots, (b_{r,n})_{n\in\mathbb{N}}$ be sequences of bounded (not necessarily selfadjoint) operators in the W^*-probability space $(\mathcal{A}, \tau)$. Assume, for each j, that $\|b_{j,n}\| \le 1$ for all n and that $b_{j,n} \to b_j$ in probability as $n \to \infty$, for some operator b_j in $\mathcal{A}$. If $b_{1,n}, b_{2,n}, \dots, b_{r,n}$ are freely independent for each n, then the operators $b_1, b_2, \dots, b_r$ are also freely independent.*

Proof. Assume that $b_{1,n}, b_{2,n}, \dots, b_{r,n}$ are freely independent for all n, and let $i_1, i_2, \dots, i_p$ in $\{1, 2, \dots, r\}$ be given. Then there is a universal polynomial

$P_{i_1,\dots,i_p}$ in rp complex variables, depending only on $i_1,\dots,i_p$, such that for all n in $\mathbb{N}$,

$$\tau(b_{i_1,n}b_{i_2,n}\cdots b_{i_p,n}) = P_{i_1,\dots,i_p}\Big[\big\{\tau(b_{1,n}^{\ell})\big\}_{1\le\ell\le p},\dots,\big\{\tau(b_{r,n}^{\ell})\big\}_{1\le\ell\le p}\Big]. \quad (4.2)$$

Now, since operator multiplication is a continuous operation with respect to the measure topology, $b_{i_1,n}b_{i_2,n}\cdots b_{i_p,n} \to b_{i_1}b_{i_2}\cdots b_{i_p}$ in probability as $n\to\infty$. Furthermore, $\|b_{i_1,n}b_{i_2,n}\cdots b_{i_p,n}\| \le 1$ for all n, so by Lemma 4.4 we have

$$\tau\big(b_{i_1,n}b_{i_2,n}\cdots b_{i_p,n}\big) \underset{n\to\infty}{\longrightarrow} \tau\big(b_{i_1}b_{i_2}\cdots b_{i_p}\big).$$

Similarly,

$$\tau(b_{j,n}^{\ell}) \underset{n\to\infty}{\longrightarrow} \tau(b_j^{\ell}), \quad \text{for any } j \text{ in } \{1,2,\dots,r\} \text{ and } \ell \text{ in } \mathbb{N}.$$

Combining these observations with (4.2), we conclude that also

$$\tau(b_{i_1}b_{i_2}\cdots b_{i_p}) = P_{i_1,\dots,i_p}\Big[\big\{\tau(b_1^{\ell})\big\}_{1\le\ell\le p},\dots,\big\{\tau(b_r^{\ell})\big\}_{1\le\ell\le p}\Big],$$

and since this holds for arbitrary $i_1,\dots,i_p$ in $\{1,2,\dots,r\}$, it follows that $b_1,\dots,b_r$ are freely independent, as desired. □

For a selfadjoint operator a affiliated with a W^*-probability space $(\mathcal{A},\tau)$, we denote by $\kappa(a)$ the *Cayley transform* of a, i.e.

$$\kappa(a) = (a - \mathrm{i}\mathbf{1}_{\mathcal{A}})(a + \mathrm{i}\mathbf{1}_{\mathcal{A}})^{-1}.$$

Recall that even though a may be an unbounded operator, $\kappa(a)$ is a unitary operator in $\mathcal{A}$.

Lemma 4.6. *Let $a_1,a_2,\dots,a_r$ be selfadjoint operators affiliated with the W^*-probability space $(\mathcal{A},\tau)$. Then $a_1,a_2,\dots,a_r$ are freely independent if and only if $\kappa(a_1),\kappa(a_2),\dots,\kappa(a_r)$ are freely independent.*

Proof. This is an immediate consequence of the fact that a_j and $\kappa(a_j)$ generate the same von Neumann subalgebra of $\mathcal{A}$ for each j (cf. [Pe89, Lemma 5.2.8]). □

Proposition 4.7. *Suppose $r\in\mathbb{N}$ and that $(a_{1,n})_{n\in\mathbb{N}},\dots,(a_{r,n})_{n\in\mathbb{N}}$ are sequences of selfadjoint operators affiliated with the W^*-probability space $(\mathcal{A},\tau)$. Assume, further, that for each j in $\{1,2,\dots,r\}$, $a_{j,n}\to a_j$ in probability as $n\to\infty$, for some selfadjoint operator a_j affiliated with $(\mathcal{A},\tau)$. If the operators $a_{1,n},a_{2,n},\dots,a_{r,n}$ are freely independent for each n, then the operators $a_1,a_2,\dots,a_r$ are also freely independent.*

Proof. Assume that $a_{1,n}, a_{2,n}, \dots, a_{r,n}$ are freely independent for all n. Then, by Lemma 4.6, the unitaries $\kappa(a_{1,n}), \dots, \kappa(a_{r,n})$ are freely independent for each n in $\mathbb{N}$. Moreover, since the Cayley transform is continuous in the measure topology (cf. [St59, Lemma 5.3]), we have

$$\kappa(a_{j,n}) \underset{n\to\infty}{\longrightarrow} \kappa(a_j), \qquad \text{in probability,}$$

for each j. Hence, by Lemma 4.5, the unitaries $\kappa(a_1), \dots, \kappa(a_r)$ are freely independent, and, appealing once more to Lemma 4.6, this means that $a_1, \dots, a_r$ themselves are freely independent. □

Remark 4.8. Let $\mathcal{B}$ and $\mathcal{C}$ be two freely independent von Neumann subalgebras of a W^*-probability space $(\mathcal{A}, \tau)$. Let, further, (b_n) and (c_n) be two sequences of selfadjoint operators, which are affiliated with $\mathcal{B}$ and $\mathcal{C}$, respectively, in the sense that $f(b_n) \in \mathcal{B}$ and $g(c_n) \in \mathcal{C}$ for any n in $\mathbb{N}$ and any bounded Borel functions $f, g\colon \mathbb{R} \to \mathbb{R}$. Assume that $b_n \to b$ and $c_n \to c$ in probability as $n \to \infty$. Then b and c are also freely independent. This follows, of course, from Proposition 4.7, but it is also an immediate consequence of the fact that the set $\overline{\mathcal{B}}$ of closed, densely defined operators, affiliated with $\mathcal{B}$, is complete (and hence closed) in the measure topology. Indeed, the restriction to $\overline{\mathcal{B}}$ of the measure topology on $\overline{\mathcal{A}}$ is the measure topology on $\overline{\mathcal{B}}$ (induced by $\tau_{|\mathcal{B}}$). Thus, b is affiliated with $\mathcal{B}$ and similarly c is affiliated with $\mathcal{C}$, so that, in particular, b and c are freely independent.

4.4 Free Additive Convolution

From a probabilistic point of view, free additive convolution may be considered merely as a new type of convolution on the set of probability measures on $\mathbb{R}$. Let a and b be selfadjoint operators in a W^*-probability space $(\mathcal{A}, \tau)$, and note that $a + b$ is selfadjoint too. Denote then the (spectral) distributions of a, b and $a + b$ by μ_a, μ_b and μ_{a+b}. If a and b are freely independent, it is not hard to see that the moments of μ_{a+b} (and hence μ_{a+b} itself) is uniquely determined by μ_a and μ_b. Hence we may write $\mu_a \boxplus \mu_b$ instead of μ_{a+b}, and we say that $\mu_a \boxplus \mu_b$ is the *free additive*[8] *convolution* of μ_a and μ_b.

Since the distribution μ_a of a selfadjoint operator a in $\mathcal{A}$ is a compactly supported probability measure on $\mathbb{R}$, the definition of free additive convolution, stated above, works at most for all compactly supported probability measures on $\mathbb{R}$. On the other hand, given any two compactly supported probability measures μ_1 and μ_2 on $\mathbb{R}$, it follows from a free product construction (see [VoDyNi92]), that it is always possible to find a W^*-probability space

[8]The reason for the term additive is that there exists another convolution operation called *free multiplicative convolution*, which arises naturally out of the non-commutative setting (i.e. the non-commutative multiplication of operators). In the present notes we do not consider free multiplicative convolution.

$(\mathcal{A}, \tau)$ and free selfadjoint operators a, b in $\mathcal{A}$, such that a and b have distributions μ_1 and μ_2 respectively. Thus, the operation $\boxplus$ introduced above is, in fact, defined on all compactly supported probability measures on $\mathbb{R}$. To extend this operation to all probability measures on $\mathbb{R}$, one needs, as indicated above, to consider unbounded selfadjoint operators in a Hilbert space, and then to proceed with a construction similar to that described above. We postpone a detailed discussion of this matter to the Appendix (see Remark A.3), since, for our present purposes, it is possible to study free additive convolution by virtue of the Voiculescu transform, which we introduce next.

By $\mathbb{C}^+$ (respectively $\mathbb{C}^-$) we denote the set of complex numbers with strictly positive (respectively strictly negative) imaginary part.

Let μ be a probability measure on $\mathbb{R}$, and consider its Cauchy (or Stieltjes) transform $G_\mu\colon \mathbb{C}^+ \to \mathbb{C}^-$ given by:

$$G_\mu(z) = \int_{\mathbb{R}} \frac{1}{z-t}\, \mu(\mathrm{d}t), \quad (z \in \mathbb{C}^+).$$

Then define the mapping $F_\mu\colon \mathbb{C}^+ \to \mathbb{C}^+$ by:

$$F_\mu(z) = \frac{1}{G_\mu(z)}, \quad (z \in \mathbb{C}^+),$$

and note that F_μ is analytic on $\mathbb{C}^+$. It was proved by Bercovici and Voiculescu in [BeVo93, Proposition 5.4 and Corollary 5.5] that there exist positive numbers η and M, such that F_μ has an (analytic) right inverse F_μ^{-1} defined on the region

$$\Gamma_{\eta,M} := \{z \in \mathbb{C} \mid |\mathrm{Re}(z)| < \eta \mathrm{Im}(z),\ \mathrm{Im}(z) > M\}.$$

In other words, there exists an open subset $G_{\eta,M}$ of $\mathbb{C}^+$ such that F_μ is injective on $G_{\eta,M}$ and such that $F_\mu(G_{\eta,M}) = \Gamma_{\eta,M}$.

Now the *Voiculescu transform* ϕ_μ of μ is defined by

$$\phi_\mu(z) = F_\mu^{-1}(z) - z,$$

on any region of the form $\Gamma_{\eta,M}$, where F_μ^{-1} is defined. It follows from [BeVo93, Corollary 5.3] that $\mathrm{Im}(F_\mu^{-1}(z)) \le \mathrm{Im}(z)$ and hence $\mathrm{Im}(\phi_\mu(z)) \le 0$ for all z in $\Gamma_{\eta,M}$.

The Voiculescu transform ϕ_μ should be viewed as a modification of Voiculescu's $\mathcal{R}$-transform (see e.g. [VoDyNi92]), since we have the correspondence:

$$\phi_\mu(z) = \mathcal{R}_\mu(\tfrac{1}{z}).$$

A third variant, which we shall also make use of is the *free cumulant transform*, given by:

$$\mathcal{C}_\mu(z) = z\mathcal{R}_\mu(z) = z\phi_\mu(\tfrac{1}{z}). \tag{4.3}$$

The key property of the Voiculescu transform is the following important result, which shows that the Voiculescu transform (and its variants) can be viewed as the free analogue of the classical cumulant function (the logarithm of the characteristic function). The result was first proved by Voiculescu for probability measures μ with compact support, and then by Maassen in the case where μ has variance. Finally Bercovici and Voiculescu proved the general case.

Theorem 4.9 ([Vo86],[Ma92],[BeVo93]). *Let μ_1 and μ_2 be probability measures on $\mathbb{R}$, and consider their free additive convolution $\mu_1 \boxplus \mu_2$. Then*

$$\phi_{\mu_1 \boxplus \mu_2}(z) = \phi_{\mu_1}(z) + \phi_{\mu_2}(z),$$

for all z in any region $\Gamma_{\eta,M}$, where all three functions are defined.

Remark 4.10. We shall need the fact that a probability measure on $\mathbb{R}$ is uniquely determined by its Voiculescu transform. To see this, suppose μ and μ' are probability measures on $\mathbb{R}$, such that $\phi_\mu = \phi_{\mu'}$, on a region $\Gamma_{\eta,M}$. It follows then that also $F_\mu = F_{\mu'}$ on some open subset of $\mathbb{C}^+$, and hence (by analytic continuation), $F_\mu = F_{\mu'}$ on all of $\mathbb{C}^+$. Consequently μ and μ' have the same Cauchy (or Stieltjes) transform, and by the Stieltjes Inversion Formula (cf. e.g. [Ch78, page 90]), this means that $\mu = \mu'$.

In [BeVo93, Proposition 5.6], Bercovici and Voiculescu proved the following characterization of Voiculescu transforms:

Theorem 4.11 ([BeVo93]). *Let ϕ be an analytic function defined on a region $\Gamma_{\eta,M}$, for some positive numbers η and M. Then the following assertions are equivalent:*

(i) *There exists a probability measure μ on $\mathbb{R}$, such that $\phi(z) = \phi_\mu(z)$ for all z in a domain $\Gamma_{\eta,M'}$, where $M' \geq M$.*
(ii) *There exists a number M' greater than or equal to M, such that*
  (a) *$Im(\phi(z)) \leq 0$ for all z in $\Gamma_{\eta,M'}$.*
  (b) *$\phi(z)/z \to 0$, as $|z| \to \infty$, $z \in \Gamma_{\eta,M'}$.*
  (c) *For any positive integer n and any points $z_1, \dots, z_n$ in $\Gamma_{\eta,M'}$, the $n \times n$ matrix*

$$\left[\frac{z_j - \overline{z_k}}{z_j + \phi(z_j) - \overline{z_k} - \overline{\phi(z_k)}}\right]_{1 \leq j,k \leq n},$$

  is positive definite.

The relationship between weak convergence of probability measures and the Voiculescu transform was settled in [BeVo93, Proposition 5.7] and [BePa96, Proposition 1]:

Proposition 4.12 ([BeVo93],[BePa96]). *Let (μ_n) be a sequence of probability measures on $\mathbb{R}$. Then the following assertions are equivalent:*

(a) *The sequence (μ_n) converges weakly to a probability measure μ on $\mathbb{R}$.*

(b) *There exist positive numbers* η *and* M*, and a function* ϕ*, such that all the functions* ϕ*,* ϕ_{μ_n} *are defined on* $\Gamma_{\eta,M}$*, and such that*
  (b1) $\phi_{\mu_n}(z) \to \phi(z)$*, as* $n \to \infty$*, uniformly on compact subsets of* $\Gamma_{\eta,M}$*,*
  (b2) $\sup_{n\in\mathbb{N}} \left|\frac{\phi_{\mu_n}(z)}{z}\right| \to 0$*, as* $|z| \to \infty$*,* $z \in \Gamma_{\eta,M}$*.*

(c) *There exist positive numbers* η *and* M*, such that all the functions* ϕ_{μ_n} *are defined on* $\Gamma_{\eta,M}$*, and such that*
  (c1) $\lim_{n\to\infty} \phi_{\mu_n}(iy)$ *exists for all* y *in* $[M, \infty[$*.*
  (c2) $\sup_{n\in\mathbb{N}} \left|\frac{\phi_{\mu_n}(iy)}{y}\right| \to 0$*, as* $y \to \infty$*.*

If the conditions (a),(b) and (c) are satisfied, then $\phi = \phi_\mu$ *on* $\Gamma_{\eta,M}$*.*

Remark 4.13 (Cumulants I). Under the assumption of finite moments of all orders, both classical and free convolution can be handled completely by a combinatorial approach based on cumulants. Suppose, for simplicity, that μ is a compactly supported probability measure on $\mathbb{R}$. Then for n in $\mathbb{N}$, the classical cumulant c_n of μ may be defined as the n'th derivative at 0 of the cumulant transform $\log f_\mu$. In other words, we have the Taylor expansion:

$$\log f_\mu(z) = \sum_{n=1}^{\infty} \frac{c_n}{n!} z^n.$$

Consider further the sequence $(m_n)_{n\in\mathbb{N}_0}$ of moments of μ. Then the sequence (m_n) is uniquely determined by the sequence (c_n) (and vice versa). The formulas determining m_n from (c_n) are generally quite complicated. However, by viewing the sequences (m_n) and (c_n) as multiplicative functions M and C on the lattice of all partitions of $\{1, 2, \ldots, n\}$, $n \in \mathbb{N}$ (cf. e.g. [Sp97]), the relationship between (m_n) and (c_n) can be elegantly expressed by the formula:

$$C = M \star \text{Moeb},$$

where Moeb denotes the Möbius transform and where $\star$ denotes *combinatorial convolution* of multiplicative functions on the lattice of all partitions (see [Sp97],[Ro64] or [BaCo89]).

The *free* cumulants (k_n) of μ were introduced by R. Speicher in [Sp94]. They may, similarly, be defined as the coefficients in the Taylor expansion of the free cumulant transform $\mathcal{C}_\mu$:

$$\mathcal{C}_\mu(z) = \sum_{n=1}^{\infty} k_n z^n,$$

(see (4.3)). Viewing then (k_n) and (m_n) as multiplicative functions k and m on the lattice of all *non-crossing* partitions of $\{1, 2, \ldots, n\}$, $n \in \mathbb{N}$, the relationship between (k_n) and (m_n) is expressed by the exact same formula:

$$k = m \star \text{Moeb}, \tag{4.4}$$

where now $\star$ denotes combinatorial convolution of multiplicative functions on the lattice of all *non-crossing* partitions (see [Sp97]).

For a family $a_1, a_2, \dots, a_r$ of selfadjoint operators in a W^*-probability space $(\mathcal{A}, \tau)$ it is also possible to define generalized cumulants, which are related to the family of all mixed moments (w.r.t. τ) of $a_1, a_2, \dots, a_r$ by a formula similar to (4.4) (see e.g. [Sp97]). In terms of these multivariate cumulants, free independence of $a_1, a_2, \dots, a_r$ has a rather simple formulation, and using this formulation, R. Speicher gave a simple and completely combinatorial proof of the fact that the free cumulants (and hence the free cumulant transform) linearize free convolution (see [Sp94]). A treatment of the theory of classical multivariate cumulants can be found in [BaCo89].

4.5 Basic Results in Free Infinite Divisibility

In this section we recall the definition and some basic facts about infinite divisibility w.r.t. free additive convolution. In complete analogy with the classical case, a probability measure μ on $\mathbb{R}$ is $\boxplus$-infinitely divisible, if for any n in $\mathbb{N}$ there exists a probability measure μ_n on $\mathbb{R}$, such that

$$\mu = \underbrace{\mu_n \boxplus \mu_n \boxplus \cdots \boxplus \mu_n}_{n \text{ terms}}.$$

It was proved in [Pa96] that the class $\mathcal{ID}(\boxplus)$ of $\boxplus$-infinitely divisible probability measures on $\mathbb{R}$ is closed w.r.t. weak convergence. For the corresponding classical result, see [GnKo68, §17, Theorem 3]. As in classical probability, $\boxplus$-infinitely divisible probability measures are characterized as those probability measures that have a (free) Lévy-Khintchine representation:

Theorem 4.14 ([Vo86],[Ma92],[BeVo93]).
Let μ be a probability measure on $\mathbb{R}$. Then μ is $\boxplus$-infinitely divisible, if and only if there exist a finite measure σ on $\mathbb{R}$ and a real constant γ, such that

$$\phi_\mu(z) = \gamma + \int_{\mathbb{R}} \frac{1+tz}{z-t}\,\sigma(\mathrm{d}t), \qquad (z \in \mathbb{C}). \tag{4.5}$$

Moreover, for a $\boxplus$-infinitely divisible probability measure μ on $\mathbb{R}$, the real constant γ and the finite measure σ, described above, are uniquely determined.

Proof. The equivalence between $\boxplus$-infinite divisibility and the existence of a representation in the form (4.5) was proved (in the general case) by Voiculescu and Bercovici in [BeVo93, Theorem 5.10]. They proved first that μ is $\boxplus$-infinitely divisible, if and only if ϕ_μ has an extension to a function of the form: $\phi\colon \mathbb{C}^+ \to \mathbb{C}^- \cup \mathbb{R}$, i.e. a Pick function multiplied by -1. Equation (4.5) (and its uniqueness) then follows from the existence (and uniqueness) of the integral representation of Pick functions (cf. [Do74, Chapter 2, Theorem I]). Compared to the general integral representation for Pick functions, just referred to, there is a linear term missing on the right hand side of (4.5), but this corresponds to the fact that $\frac{\phi(iy)}{y} \to 0$ as $y \to \infty$, if ϕ is a Voiculescu transform (cf. Theorem 4.11 above). □

Definition 4.15. *Let μ be a $\boxplus$-infinitely divisible probability measure on $\mathbb{R}$, and let γ and σ be, respectively, the (uniquely determined) real constant and finite measure on $\mathbb{R}$ appearing in* (4.5). *We say then that the pair (γ, σ) is the free generating pair for μ.*

In terms of the free cumulant transform, the free Lévy-Khintchine representation resembles more closely the classical Lévy-Khintchine representation, as the following proposition shows.

Proposition 4.16. *A probability measure ν on $\mathbb{R}$ is $\boxplus$-infinitely divisible if and only if there exist a non-negative number a, a real number η and a Lévy measure ρ, such that the free cumulant transform $\mathcal{C}_\nu$ has the representation:*

$$\mathcal{C}_\nu(z) = \eta z + az^2 + \int_{\mathbb{R}} \Big(\frac{1}{1-tz} - 1 - tz 1_{[-1,1]}(t) \Big) \, \rho(\mathrm{d}t), \quad (z \in \mathbb{C}^-). \quad (4.6)$$

In that case, the triplet (a, ρ, η) is uniquely determined and is called the free characteristic triplet for ν.

Proof. Let ν be a measure in $\mathcal{ID}(\boxplus)$ with free generating pair (γ, σ), and consider its free Lévy-Khintchine representation (in terms of the Voiculescu transform):

$$\phi_\nu(z) = \gamma + \int_{\mathbb{R}} \frac{1+tz}{z-t} \, \sigma(\mathrm{d}t), \quad (z \in \mathbb{C}^+). \quad (4.7)$$

Then define the triplet (a, ρ, η) by (2.3), and note that

$$\sigma(\mathrm{d}t) = a\delta_0(\mathrm{d}t) + \frac{t^2}{1+t^2}\rho(\mathrm{d}t),$$

$$\gamma = \eta - \int_{\mathbb{R}} t\Big(1_{[-1,1]}(t) - \frac{1}{1+t^2}\Big) \, \rho(\mathrm{d}t).$$

Now, for z in $\mathbb{C}^-$, the corresponding free cumulant transform $\mathcal{C}_\nu$ is given by

$$\begin{aligned}
&\mathcal{C}_\nu(z) \\
&= z\phi_\nu(1/z) = z\Big(\gamma + \int_{\mathbb{R}} \frac{1+t(1/z)}{(1/z)-t} \, \sigma(\mathrm{d}t)\Big) \\
&= \gamma z + z\int_{\mathbb{R}} \frac{z+t}{1-tz} \, \sigma(\mathrm{d}t) = \gamma z + \int_{\mathbb{R}} \frac{z^2+tz}{1-tz} \, \sigma(\mathrm{d}t) \\
&= \eta z - \Big[\int_{\mathbb{R}} t\Big(1_{[-1,1]}(t) - \frac{1}{1+t^2}\Big) \, \rho(\mathrm{d}t)\Big] z + az^2 + \int_{\mathbb{R}} \frac{z^2+tz}{1-tz}\frac{t^2}{1+t^2} \, \rho(\mathrm{d}t).
\end{aligned}$$

Note here that

$$1_{[-1,1]}(t) - \frac{1}{1+t^2} = 1 - \frac{1}{1+t^2} - 1_{\mathbb{R}\setminus[-1,1]}(t) = \frac{t^2}{1+t^2} - 1_{\mathbb{R}\setminus[-1,1]}(t),$$

so that

$$\int_{\mathbb{R}} t\Big(1_{[-1,1]}(t) - \frac{1}{1+t^2}\Big)\,\rho(\mathrm{d}t) = \int_{\mathbb{R}} \Big(\frac{t}{1+t^2} - t^{-1}1_{\mathbb{R}\setminus[-1,1]}(t)\Big)\,t^2\rho(\mathrm{d}t).$$

Note also that

$$\frac{z^2+tz}{(1-tz)(1+t^2)} = \frac{z^2}{1-tz} + \frac{tz}{1+t^2}.$$

Therefore,

$$\begin{aligned}
\mathcal{C}_\nu(z) &= \eta z - \Big[\int_{\mathbb{R}} \Big(\frac{t}{1+t^2} - t^{-1}1_{\mathbb{R}\setminus[-1,1]}(t)\Big)\,t^2\rho(\mathrm{d}t)\Big]z + az^2 \\
&\qquad + \int_{\mathbb{R}} \Big(\frac{z^2}{1-tz} + \frac{tz}{1+t^2}\Big)\,t^2\rho(\mathrm{d}t) \\
&= \eta z + az^2 + \int_{\mathbb{R}} \Big(\frac{z^2}{1-tz} + t^{-1}z1_{\mathbb{R}\setminus[-1,1]}(t)\Big)\,t^2\rho(\mathrm{d}t) \\
&= \eta z + az^2 + \int_{\mathbb{R}} \Big(\frac{(tz)^2}{1-tz} + tz1_{\mathbb{R}\setminus[-1,1]}(t)\Big)\,\rho(\mathrm{d}t).
\end{aligned}$$

Further,

$$\begin{aligned}
\frac{(tz)^2}{1-tz} + tz1_{\mathbb{R}\setminus[-1,1]}(t) &= \Big(\frac{(tz)^2}{1-tz} + tz\Big) - tz1_{[-1,1]}(t) \\
&= \frac{tz}{1-tz} - tz1_{[-1,1]}(t) \\
&= \frac{1}{1-tz} - 1 - tz1_{[-1,1]}(t).
\end{aligned}$$

We conclude that

$$\mathcal{C}_\nu(z) = \eta z + az^2 + \int_{\mathbb{R}} \Big(\frac{1}{1-tz} - 1 - tz1_{[-1,1]}(t)\Big)\,\rho(\mathrm{d}t). \tag{4.8}$$

Clearly the above calculations may be reversed, so that (4.7) and (4.8) are equivalent. □

Apart from the striking similarity between (2.2) and (4.6), note that these particular representations clearly exhibit how μ (respectively ν) is always the convolution of a Gaussian distribution (respectively a semi-circle distribution) and a distribution of generalized Poisson (respectively free Poisson) type (cf. also the Lévy-Itô decomposition described in Section 6.5). In particular, the cumulant transform for the Gaussian distribution with mean η and variance a is: $u \mapsto i\eta u - \frac{1}{2}au^2$, and the free cumulant transform for the semi-circle distribution with mean η and variance a is $z \mapsto \eta z + az^2$ (see [VoDyNi92]).

The next result, due to Bercovici and Pata, is the free analogue of Khintchine's characterization of classically infinitely divisible probability measures. It plays an important role in Section 4.6.

Definition 4.17. *Let $(k_n)_{n\in\mathbb{N}}$ be a sequence of positive integers, and let*

$$A = \{\mu_{nj} \mid n \in \mathbb{N},\ j \in \{1, 2, \ldots, k_n\}\},$$

be an array of probability measures on $\mathbb{R}$. We say then that A is a null array, if the following condition is fulfilled:

$$\forall \epsilon > 0\colon \lim_{n\to\infty} \max_{1\le j\le k_n} \mu_{nj}(\mathbb{R} \setminus [-\epsilon, \epsilon]) = 0.$$

Theorem 4.18 ([BePa00]). *Let $\{\mu_{nj} \mid n \in \mathbb{N},\ j \in \{1, 2, \ldots, k_n\}\}$ be a null-array of probability measures on $\mathbb{R}$, and let $(c_n)_{n\in\mathbb{N}}$ be a sequence of real numbers. If the probability measures $\mu_n = \delta_{c_n} \boxplus \mu_{n1} \boxplus \mu_{n2} \boxplus \cdots \boxplus \mu_{nk_n}$ converge weakly, as $n \to \infty$, to a probability measure μ on $\mathbb{R}$, then μ has to be $\boxplus$-infinitely divisible.*

4.6 Classes of Freely Infinitely Divisible Probability Measures

In this section we study the free counterparts $\mathcal{S}(\boxplus)$ and $\mathcal{L}(\boxplus)$ to the classes $\mathcal{S}(*)$ and $\mathcal{L}(*)$ of stable and selfdecomposable distributions. We show in particular that we have the following hierarchy

$$\mathcal{G}(\boxplus) \subset \mathcal{S}(\boxplus) \subset \mathcal{L}(\boxplus) \subset \mathcal{ID}(\boxplus), \tag{4.9}$$

where $\mathcal{G}(\boxplus)$ denotes the class of semi-circle distributions. We start with the formal definitions of and $\mathcal{S}(\boxplus)$ and $\mathcal{L}(\boxplus)$.

Definition 4.19. (i) *A probability measure μ on $\mathbb{R}$ is called stable w.r.t. free convolution (or just $\boxplus$-stable), if the class*

$$\{\psi(\mu) \mid \psi\colon \mathbb{R} \to \mathbb{R} \text{ is an increasing affine transformation}\}$$

is closed under the operation $\boxplus$. By $\mathcal{S}(\boxplus)$ we denote the class of $\boxplus$-stable probability measures on $\mathbb{R}$.

(ii) *A probability measure μ on $\mathbb{R}$ is selfdecomposable w.r.t. free additive convolution (or just $\boxplus$-selfdecomposable), if for any c in $]0,1[$ there exists a probability measure μ_c on $\mathbb{R}$, such that*

$$\mu = D_c\mu \boxplus \mu_c. \tag{4.10}$$

By $\mathcal{L}(\boxplus)$ we denote the class of $\boxplus$-selfdecomposable probability measures on $\mathbb{R}$.

Note that for a probability measure μ on $\mathbb{R}$ and a constant c in $]0,1[$, there can be only one probability measure μ_c, such that $\mu = D_c\mu \boxplus \mu_c$. Indeed, choose positive numbers η and M, such that all three Voiculescu transforms ϕ_μ, $\phi_{D_c\mu}$ and ϕ_{μ_c} are defined on the region $\Gamma_{\eta,M}$. Then by Theorem 4.9, ϕ_{μ_c} is uniquely determined on $\Gamma_{\eta,M}$, and hence, by Remark 4.10, μ_c is uniquely determined too.

In order to prove the inclusions in (4.9), we need the following technical result.

Lemma 4.20. *Let μ be a probability measure on $\mathbb{R}$, and let η and M be positive numbers such that the Voiculescu transform ϕ_μ is defined on $\Gamma_{\eta,M}$ (see Section 4.4). Then for any constant c in $\mathbb{R}\setminus\{0\}$, $\phi_{D_c\mu}$ is defined on $|c|\Gamma_{\eta,M} = \Gamma_{\eta,|c|M}$, and*

(i) *if $c > 0$, then $\phi_{D_c\mu}(z) = c\phi_\mu(c^{-1}z)$ for all z in $c\Gamma_{\eta,M}$,*
(ii) *if $c < 0$, then $\phi_{D_c\mu}(z) = c\overline{\phi_\mu(c^{-1}\overline{z})}$ for all z in $|c|\Gamma_{\eta,M}$.*

In particular, for a constant c in $[-1,1]$, the domain of $\phi_{D_c\mu}$ contains the domain of ϕ_μ.

Proof. (i) This is a special case of [BeVo93, Lemma 7.1].

(ii) Note first that by virtue of (i), it suffices to prove (ii) in the case $c = -1$.

We start by noting that the Cauchy transform G_μ (see Section 4.4) is actually well-defined for all z in $\mathbb{C}\setminus\mathbb{R}$ (even for all z outside $\mathrm{supp}(\mu)$), and that $G_\mu(\overline{z}) = \overline{G_\mu(z)}$, for all such z. Similarly, F_μ is defined for all z in $\mathbb{C}\setminus\mathbb{R}$, and $F_\mu(z) = \overline{F_\mu(\overline{z})}$, for such z.

Note next that for any z in $\mathbb{C}\setminus\mathbb{R}$, $G_{D_{-1}\mu}(z) = -G_\mu(-z)$, and consequently

$$F_{D_{-1}\mu}(z) = -F_\mu(-z) = -\overline{F_\mu(-\overline{z})}.$$

Now, since $-\overline{\Gamma_{\eta,M}} = \Gamma_{\eta,M}$, it follows from the equation above, that $F_{D_{-1}\mu}$ has a right inverse on $\Gamma_{\eta,M}$, given by $F^{-1}_{D_{-1}\mu}(z) = -\overline{F^{-1}_\mu(-\overline{z})}$, for all z in $\Gamma_{\eta,M}$. Consequently, for z in $\Gamma_{\eta,M}$, we have

$$\phi_{D_{-1}\mu}(z) = F^{-1}_{D_{-1}\mu}(z) - z = -\overline{F^{-1}_\mu(-\overline{z})} - z = -\overline{(F^{-1}_\mu(-\overline{z}) - (-\overline{z}))} = -\overline{\phi_\mu(-\overline{z})},$$

as desired. □

Remark 4.21. With respect to dilation the free cumulant transform behaves exactly as the classical cumulant function, i.e.

$$\mathcal{C}_{D_c\mu}(z) = \mathcal{C}_\mu(cz), \tag{4.11}$$

for any probability measure μ on $\mathbb{R}$ and any positive constant c. This follows easily from Lemma 4.20. As a consequence, it follows as in the classical case that a probability measure μ on $\mathbb{R}$ belongs to $\mathcal{S}(\boxplus)$, if and only if the following condition is satisfied (for z^{-1} in a region of the form $\Gamma(\eta, M)$)

$$\forall a, a' > 0\, \forall b, b' \in \mathbb{R}\, \exists a'' > 0\, \exists b'' \in \mathbb{R}\colon \mathcal{C}_\mu(az) + bz + \mathcal{C}_\mu(a'z) + b'z = \mathcal{C}_\mu(a''z) + b''z.$$

It is easy to see that the above condition is equivalent to the following

$$\forall a > 0\ \exists a'' > 0\ \exists b'' \in \mathbb{R}\colon \mathcal{C}_\mu(z) + \mathcal{C}_\mu(az) = \mathcal{C}_\mu(a''z) + b''z. \tag{4.12}$$

Similarly, a probability measure μ on $\mathbb{R}$ is $\boxplus$-selfdecomposable, if and only if there exists, for any c in $]0,1[$, a probability measure μ_c on $\mathbb{R}$, such that

$$\mathcal{C}_\mu(z) = \mathcal{C}_\mu(cz) + \mathcal{C}_{\mu_c}(z), \tag{4.13}$$

for z^{-1} in a region of the form $\Gamma(\eta, M)$. In terms of the Voiculescu transform ϕ_μ, formula (4.13) takes the equivalent form

$$\phi_\mu(z) = c\phi_\mu(c^{-1}z) + \phi_{\mu_c}(z),$$

for all z in a region $\Gamma_{\eta,M}$.

Proposition 4.22. (i) *Any semi-circle law is $\boxplus$-stable.*
(ii) *Let μ be a $\boxplus$-stable probability measure on $\mathbb{R}$. Then μ is necessarily $\boxplus$-selfdecomposable.*

Proof. (i) Let $\gamma_{0,2}$ denote the standard semi-circle distribution, i.e.

$$\gamma_{0,2}(\mathrm{d}x) = 1_{[-2,2]}(x)\sqrt{4 - x^2}\,\mathrm{d}x.$$

Then, by definition,

$$\mathcal{G}(\boxplus) = \{D_a\gamma_{0,2} \boxplus \delta_b \mid a \geq 0,\ b \in \mathbb{R}\}.$$

It is easy to see that $\mathcal{S}(\boxplus)$ is closed under the operations D_a $(a > 0)$, and under (free) convolution with δ_b $(b \in \mathbb{R})$. Therefore, it suffices to show that $\gamma_{0,2} \in \mathcal{S}(\boxplus)$. By [VoDyNi92, Example 3.4.4], the free cumulant transform of $\gamma_{0,2}$ is given by

$$\mathcal{C}_{\gamma_{0,2}}(z) = z^2, \qquad (z \in \mathbb{C}^+),$$

and clearly this function satisfies condition (4.12) above.

(ii) Let μ be a measure in $\mathcal{S}(\boxplus)$. The relationship between the constants a and a'' in (4.12) is of the form $a'' = f(a)$, where $f\colon]0,\infty[\to]1,\infty[$ is a continuous, strictly increasing function, satisfying that $f(t) \to 1$ as $t \to 0^+$ and $f(t) \to \infty$ as $t \to \infty$ (see the proof of [BeVo93, Lemma 7.4]). Now, given c in $]0,1[$, put $a = f^{-1}(1/c) \in]0,\infty[$, so that

$$\mathcal{C}_\mu(z) + \mathcal{C}_\mu(az) = \mathcal{C}_\mu(c^{-1}z) + bz,$$

for suitable b in $\mathbb{R}$. Putting $z = cw$, it follows that

$$\mathcal{C}_\mu(w) - \mathcal{C}_\mu(cw) = \mathcal{C}_\mu(acw) - bcw.$$

Based on Theorem 4.11 is is not hard to see that $z \mapsto \mathcal{C}_\mu(acw) - bcw$ is the free cumulant transform of some measure μ_c in $\mathcal{P}$. With this μ_c, condition (4.13) is satisfied. □

We turn next to the last inclusion in (4.9).

Lemma 4.23. *Let μ be a $\boxplus$-selfdecomposable probability measure on $\mathbb{R}$, let c be a number in $]0,1[$, and let μ_c be the probability measure on $\mathbb{R}$ determined by the equation:*

$$\mu = D_c\mu \boxplus \mu_c.$$

Let η and M be positive numbers, such that ϕ_μ is defined on $\Gamma_{\eta,M}$. Then ϕ_{μ_c} is defined on $\Gamma_{\eta,M}$ as well.

Proof. Choose positive numbers η' and M' such that $\Gamma_{\eta',M'} \subseteq \Gamma_{\eta,M}$ and such that ϕ_μ and ϕ_{μ_c} are both defined on $\Gamma_{\eta',M'}$. For z in $\Gamma_{\eta',M'}$, we then have (cf. Lemma 4.20):

$$\phi_\mu(z) = c\phi_\mu(c^{-1}z) + \phi_{\mu_c}(z).$$

Recalling the definition of the Voiculescu transform, the above equation means that

$$F_\mu^{-1}(z) - z = c\phi_\mu(c^{-1}z) + F_{\mu_c}^{-1}(z) - z, \quad (z \in \Gamma_{\eta',M'}),$$

so that

$$F_{\mu_c}^{-1}(z) = F_\mu^{-1}(z) - c\phi_\mu(c^{-1}z), \quad (z \in \Gamma_{\eta',M'}).$$

Now put $\psi(z) = F_\mu^{-1}(z) - c\phi_\mu(c^{-1}z)$ and note that ψ is defined and holomorphic on all of $\Gamma_{\eta,M}$ (cf. Lemma 4.20), and that

$$F_{\mu_c}(\psi(z)) = z, \quad (z \in \Gamma_{\eta',M'}). \tag{4.14}$$

We note next that ψ takes values in $\mathbb{C}^+$. Indeed, since F_μ is defined on $\mathbb{C}^+$, we have that $\mathrm{Im}(F_\mu^{-1}(z)) > 0$, for any z in $\Gamma_{\eta,M}$ and furthermore, for all such z, $\mathrm{Im}(\phi_\mu(c^{-1}z)) \leq 0$, as noted in Section 4.4.

Now, since F_{μ_c} is defined and holomorphic on all of $\mathbb{C}^+$, both sides of (4.14) are holomorphic on $\Gamma_{\eta,M}$. Since $\Gamma_{\eta',M'}$ has an accumulation point in $\Gamma_{\eta,M}$, it follows, by uniqueness of analytic continuation, that the equality in (4.14) actually holds for all z in $\Gamma_{\eta,M}$. Thus, F_{μ_c} has a right inverse on $\Gamma_{\eta,M}$, which means that ϕ_{μ_c} is defined on $\Gamma_{\eta,M}$, as desired. □

Lemma 4.24. *Let μ be a $\boxplus$-selfdecomposable probability measure on $\mathbb{R}$, and let (c_n) be a sequence of numbers in $]0,1[$. For each n, let μ_{c_n} be the probability measure on $\mathbb{R}$ satisfying*

$$\mu = D_{c_n}\mu \boxplus \mu_{c_n}.$$

Then, if $c_n \to 1$ as $n \to \infty$, we have $\mu_{c_n} \xrightarrow{\text{w}} \delta_0$, as $n \to \infty$.

Proof. Choose positive numbers η and M, such that ϕ_μ is defined on $\Gamma_{\eta,M}$. Note then that, by Lemma 4.23, $\phi_{\mu_{c_n}}$ is also defined on $\Gamma_{\eta,M}$ for each n in $\mathbb{N}$ and, moreover,

$$\phi_{\mu_{c_n}}(z) = \phi_\mu(z) - c_n\phi_\mu(c_n^{-1}z), \quad (z \in \Gamma_{\eta,M},\ n \in \mathbb{N}). \tag{4.15}$$

Assume now that $c_n \to 1$ as $n \to \infty$. From (4.15) and continuity of ϕ_μ it is then straightforward that $\phi_{\mu_{c_n}}(z) \to 0 = \phi_{\delta_0}(z)$, as $n \to \infty$, uniformly on compact subsets of $\Gamma_{\eta,M}$. Note furthermore that

$$\sup_{n\in\mathbb{N}}\left|\frac{\phi_{\mu_{c_n}}(z)}{z}\right| = \sup_{n\in\mathbb{N}}\left|\frac{\phi_\mu(z)}{z} - \frac{\phi_\mu(c_n^{-1}z)}{c_n^{-1}z}\right| \to 0, \quad \text{as } |z| \to \infty,\ z \in \Gamma_{\eta,M},$$

since $\frac{\phi_\mu(z)}{z} \to 0$ as $|z| \to \infty$, $z \in \Gamma_{\eta,M}$, and since $c_n^{-1} \geq 1$ for all n. It follows thus from Proposition 4.12 that $\mu_{c_n} \xrightarrow{w} \delta_0$, for $n \to \infty$, as desired. □

Theorem 4.25. *Let μ be a probability measure on $\mathbb{R}$. If μ is $\boxplus$-selfdecomposable, then μ is $\boxplus$-infinitely divisible.*

Proof. Assume that μ is $\boxplus$-selfdecomposable. Then by successive applications of (4.10), we get for any c in $]0,1[$ and any n in $\mathbb{N}$ that

$$\mu = D_{c^n}\mu \boxplus D_{c^{n-1}}\mu_c \boxplus D_{c^{n-2}}\mu_c \boxplus \cdots \boxplus D_c\mu_c \boxplus \mu_c. \tag{4.16}$$

The idea now is to show that for a suitable choice of $c = c_n$, the probability measures:

$$D_{c_n^n}\mu, D_{c_n^{n-1}}\mu_{c_n}, D_{c_n^{n-2}}\mu_{c_n}, \dots, D_{c_n}\mu_{c_n}, \mu_{c_n}, \quad (n \in \mathbb{N}), \tag{4.17}$$

form a null-array (cf. Theorem 4.18). Note for this, that for any choice of c_n in $]0,1[$, we have that

$$D_{c_n^j}\mu_{c_n}(\mathbb{R} \setminus [-\epsilon,\epsilon]) \leq \mu_{c_n}(\mathbb{R} \setminus [-\epsilon,\epsilon]),$$

for any j in $\mathbb{N}$ and any ϵ in $]0,\infty[$. Therefore, in order that the probability measures in (4.17) form a null-array, it suffices to choose c_n in such a way that

$$D_{c_n^n}\mu \xrightarrow{w} \delta_0 \quad \text{and} \quad \mu_{c_n} \xrightarrow{w} \delta_0, \quad \text{as } n \to \infty.$$

We claim that this will be the case if we put (for example)

$$c_n = \mathrm{e}^{-\frac{1}{\sqrt{n}}}, \quad (n \in \mathbb{N}). \tag{4.18}$$

To see this, note that with the above choice of c_n, we have:

$$c_n \to 1 \quad \text{and} \quad c_n^n \to 0, \quad \text{as } n \to \infty.$$

Thus, it follows immediately from Lemma 4.24, that $\mu_{c_n} \xrightarrow{w} \delta_0$, as $n \to \infty$. Moreover, if we choose a (classical) real valued random variable X with distribution μ, then, for each n, $D_{c_n^n}\mu$ is the distribution of c_n^nX. Now, $c_n^nX \to 0$, *almost surely*, as $n \to \infty$, and this implies that $c_n^n X \to 0$, *in distribution*, as $n \to \infty$.

We have verified, that if we choose c_n according to (4.18), then the probability measures in (4.17) form a null-array. Hence by (4.16) (with $c = c_n$) and Theorem 4.18, μ is $\boxplus$-infinitely divisible. □

Proposition 4.26. *Let μ be a $\boxplus$-selfdecomposable probability measure on $\mathbb{R}$, let c be a number in $]0,1[$ and let μ_c be the probability measure on $\mathbb{R}$ satisfying the condition:*

$$\mu = D_c\mu \boxplus \mu_c.$$

Then μ_c is $\boxplus$-infinitely divisible.

Proof. As noted in the proof of Theorem 4.25, for any d in $]0,1[$ and any n in $\mathbb{N}$ we have

$$\mu = D_{d^n}\mu \boxplus D_{d^{n-1}}\mu_d \boxplus D_{d^{n-2}}\mu_d \boxplus \cdots \boxplus D_d\mu_d \boxplus \mu_d,$$

where μ_d is defined by the case $n = 1$. Using now the above equation with $d = c^{1/n}$, we get for each n in $\mathbb{N}$ that

$$D_c\mu \boxplus \mu_c = \mu = D_c\mu \boxplus D_{c^{(n-1)/n}}\mu_{c^{1/n}} \boxplus D_{c^{(n-2)/n}}\mu_{c^{1/n}} \boxplus \cdots \boxplus D_{c^{1/n}}\mu_{c^{1/n}} \boxplus \mu_{c^{1/n}}. \tag{4.19}$$

From this it follows that

$$\mu_c = D_{c^{(n-1)/n}}\mu_{c^{1/n}} \boxplus D_{c^{(n-2)/n}}\mu_{c^{1/n}} \boxplus \cdots \boxplus D_{c^{1/n}}\mu_{c^{1/n}} \boxplus \mu_{c^{1/n}}, \quad (n \in \mathbb{N}). \tag{4.20}$$

Indeed, by taking Voiculescu transforms in (4.19) and using Theorem 4.9, it follows that the Voiculescu transforms of the right and left hand sides of (4.20) coincide on some region $\Gamma_{\eta,M}$. By Remark 4.10, this implies the validity of (4.20).

By (4.20) and Theorem 4.18, it remains now to show that the probability measures:

$$D_{c^{(n-1)/n}}\mu_{c^{1/n}}, D_{c^{(n-2)/n}}\mu_{c^{1/n}}, \ldots, D_{c^{1/n}}\mu_{c^{1/n}}, \mu_{c^{1/n}},$$

form a null-array. Since $c^{j/n} \in]0,1[$ for any j in $\{1,2,\ldots,n-1\}$, this is the case if and only if $\mu_{c^{1/n}} \xrightarrow{\text{w}} \delta_0$, as $n \to \infty$. But since $c^{1/n} \to 1$, as $n \to \infty$, Lemma 4.24 guarantees the validity of the latter assertion. □

4.7 Free Lévy Processes

Let $(\mathcal{A}, \tau)$ be a W^*-probability space acting on a Hilbert space $\mathcal{H}$ (see Section 4.1 and the Appendix). By a (stochastic) process affiliated with $\mathcal{A}$, we shall simply mean a family $(Z_t)_{t\in[0,\infty[}$ of *selfadjoint* operators in $\overline{\mathcal{A}}$, which is indexed by the non-negative reals. For such a process (Z_t), we let μ_t denote the (spectral) distribution of Z_t, i.e. $\mu_t = L\{Z_t\}$. We refer to the family

(μ_t) of probability measures on $\mathbb{R}$ as the family of *marginal distributions* of (Z_t). Moreover, if $s, t \in [0, \infty[$, such that $s < t$, then $Z_t - Z_s$ is again a selfadjoint operator in $\overline{\mathcal{A}}$ (see the Appendix), and we may consider its distribution $\mu_{s,t} = L\{Z_t - Z_s\}$. We refer to the family $(\mu_{s,t})_{0 \le s < t}$ as the family of *increment distributions* of (Z_t).

Definition 4.27. *A free Lévy process (in law), affiliated with a W^*-probability space $(\mathcal{A}, \tau)$, is a process $(Z_t)_{t \ge 0}$ of selfadjoint operators in $\overline{\mathcal{A}}$, which satisfies the following conditions:*

(i) *whenever $n \in \mathbb{N}$ and $0 \le t_0 < t_1 < \cdots < t_n$, the increments*

$$Z_{t_0}, Z_{t_1} - Z_{t_0}, Z_{t_2} - Z_{t_1}, \ldots, Z_{t_n} - Z_{t_{n-1}},$$

are freely independent random variables.

(ii) $Z_0 = 0$.

(iii) *for any s, t in $[0, \infty[$, the (spectral) distribution of $Z_{s+t} - Z_s$ does not depend on s.*

(iv) *for any s in $[0, \infty[$, $Z_{s+t} - Z_s \to 0$ in distribution, as $t \to 0$, i.e. the spectral distributions $L\{Z_{s+t} - Z_s\}$ converge weakly to δ_0, as $t \to 0$.*

Note that under the assumption of (ii) and (iii) in the definition above, condition (iv) is equivalent to saying that $Z_t \to 0$ in distribution, as $t \searrow 0$.

Remark 4.28. **(Free additive processes I)** A process (Z_t) of selfadjoint operators in $\overline{\mathcal{A}}$, which satisfies conditions (i), (ii) and (iv) of Definition 4.27, is called a *free additive process (in law)*. Given such a process (Z_t), let, as above, $\mu_s = L\{Z_s\}$ and $\mu_{s,t} = L\{Z_t - Z_s\}$, whenever $0 \le s < t$. It follows then that whenever $0 \le r < s < t$, we have

$$\mu_s = \mu_r \boxplus \mu_{r,s} \quad \text{and} \quad \mu_{r,t} = \mu_{r,s} \boxplus \mu_{s,t}, \tag{4.21}$$

and furthermore

$$\mu_{s+t,s} \xrightarrow{\text{w}} \delta_0, \quad \text{as} \quad t \to 0, \tag{4.22}$$

for any s in $[0, \infty[$.

Conversely, given any family $\{\mu_t \mid t \ge 0\} \cup \{\mu_{s,t} \mid 0 \le s < t\}$ of probability measures on $\mathbb{R}$, such that (4.21) and (4.22) are satisfied, there exists a free additive process (in law) (Z_t) affiliated with a W^*-probability space $(\mathcal{A}, \tau)$, such that $\mu_s = L\{Z_s\}$ and $\mu_{s,t} = L\{Z_t - Z_s\}$, whenever $0 \le s < t$. In fact, for any families (μ_t) and $(\mu_{s,t})$ satisfying condition (4.21), there exists a process (Z_t) affiliated with some W^*-probability space $(\mathcal{A}, \tau)$, such that conditions (i) and (ii) in Definition 4.27 are satisfied, and such that $\mu_s = L\{Z_s\}$ and $\mu_{s,t} = L\{Z_t - Z_s\}$. This was noted in [Bi98] and [Vo98] (see also Remark 6.29 below). Note that with the notation introduced above, the free Lévy processes (in law) are exactly those free additive processes (in law), for which $\mu_{s,t} = \mu_{t-s}$ for all s, t such that $0 \le s < t$. In this case the condition (4.21) simplifies to

$$\mu_t = \mu_s \boxplus \mu_{t-s}, \quad (0 \le s < t). \tag{4.23}$$

In particular, for any family (μ_t) of probability measures on $\mathbb{R}$, such that (4.23) is satisfied, and such that $\mu_t \xrightarrow{w} \delta_0$ as $t \searrow 0$, there exists a free Lévy process (in law) (Z_t), such that $\mu_t = L\{Z_t\}$ for all t.

Consider now a free Lévy process $(Z_t)_{t\ge 0}$, with marginal distributions (μ_t). As for (classical) Lévy processes, it follows then, that each μ_t is necessarily $\boxplus$-infinitely divisible. Indeed, for any n in $\mathbb{N}$ we have:

$$Z_t = \sum_{j=1}^{n} (Z_{jt/n} - Z_{(j-1)t/n}),$$

and thus, in view of conditions (i) and (iii) in Definition 4.27,

$$\mu_t = \mu_{t/n} \boxplus \cdots \boxplus \mu_{t/n} \quad (n \text{ terms}).$$

5 Connections between Free and Classical Infinite Divisibility

An important connection between free and classical infinite divisibility was established by Bercovici and Pata, in the form of a bijection Λ from the class of classical infinitely divisible laws to the class of free infinitely divisible laws. The mapping Υ of Section 3.2 embodies a direct version of the Bercovici-Pata bijection and shows rather surprisingly that, in a sense, the class of free infinitely divisible laws corresponds to a regular subset of the class of all classical infinitely divisible laws. The mapping Λ also give rise to a direct connection between the classical and the free Lévy processes, as discussed at the end of the section.

5.1 The Bercovici-Pata Bijection Λ

The bijection to be defined next was introduced by Bercovici and Pata in [BePa99].

Definition 5.1. *By the Bercovici-Pata bijection $\Lambda\colon \mathcal{ID}(*) \to \mathcal{ID}(\boxplus)$ we denote the mapping defined as follows: Let μ be a measure in $\mathcal{ID}(*)$, and consider its generating pair (γ, σ) (see formula (2.1)). Then $\Lambda(\mu)$ is the measure in $\mathcal{ID}(\boxplus)$ that has (γ, σ) as free generating pair (see Definition 4.15).*

Since the $*$-infinitely divisible (respectively $\boxplus$-infinitely divisible) probability measures on $\mathbb{R}$ are exactly those measures that have a (unique) Lévy-Khintchine representation (respectively free Lévy-Khintchine representation), it follows immediately that Λ is a (well-defined) bijection between $\mathcal{ID}(*)$ and $\mathcal{ID}(\boxplus)$. In terms of characteristic triplets, the Bercovici-Pata bijection may be characterized as follows.

Proposition 5.2. *If μ is a measure in $\mathcal{ID}(*)$ with (classical) characteristic triplet (a, ρ, η), then $\Lambda(\mu)$ has free characteristic triplet (a, ρ, η) (cf. Proposition 4.16).*

Proof. Suppose $\mu \in \mathcal{ID}(*)$ with generating pair (γ, σ) and characteristic triplet (a, ρ, η), the relationship between which is given by (2.3). Then, by definition of Λ, $\Lambda(\mu)$ has free generating pair (γ, σ), and the calculations in the proof of Proposition 4.16 (with ν replaced by $\Lambda(\mu)$) show that $\Lambda(\mu)$ has free characteristic triplet (a, ρ, η). □

Example 5.3. (a) Let μ be the standard Gaussian distribution, i.e.

$$\mu(\mathrm{d}x) = \frac{1}{\sqrt{2\pi}} \exp(-\tfrac{1}{2}x^2)\,\mathrm{d}x.$$

Then $\Lambda(\mu)$ is the semi-circle distribution, i.e.

$$\Lambda(\mu)(\mathrm{d}x) = \frac{1}{2\pi}\sqrt{4 - x^2} \cdot 1_{[-2,2]}(x)\,\mathrm{d}x.$$

(b) Let μ be the classical Poisson distribution $\mathrm{Poiss}^*(\lambda)$ with mean $\lambda > 0$, i.e.

$$\mu(\{n\}) = \mathrm{e}^{-\lambda}\frac{\lambda^n}{n!}, \qquad (n \in \mathbb{N}_0).$$

Then $\Lambda(\mu)$ is the free Poisson distribution $\mathrm{Poiss}^{\boxplus}(\lambda)$ with mean λ, i.e.

$$\Lambda(\mu)(\mathrm{d}x) = \begin{cases} (1-\lambda)\delta_0 + \frac{1}{2\pi x}\sqrt{(x-a)(b-x)} \cdot 1_{[a,b]}(x)\,\mathrm{d}x, & \text{if } 0 \le \lambda \le 1, \\ \frac{1}{2\pi x}\sqrt{(x-a)(b-x)} \cdot 1_{[a,b]}(x)\,\mathrm{d}x, & \text{if } \lambda > 1, \end{cases}$$

where $a = (1-\sqrt{\lambda})^2$ and $b = (1+\sqrt{\lambda})^2$.

Remark 5.4 (Cumulants II). Let μ be a compactly supported probability measure in $\mathcal{ID}(*)$, and consider its sequence (c_n) of classical cumulants (cf. Remark 4.13). Then the Bercovici-Pata bijection Λ may also be defined as the mapping that sends μ to the probability measure on $\mathbb{R}$ with free cumulants (c_n). In other words, the free cumulants for $\Lambda(\mu)$ are the classical cumulants for μ. This fact was noted by M. Anshelevich in [An01, Lemma 6.5]. In view of the theory of free cumulants for several variables (cf. Remark 4.13), this point of view might be used to generalize the Bercovici-Pata bijection to multidimensional probability measures.

5.2 Connection between Υ and Λ

The starting point of this section is the following observation that links the Bercovici-Pata bijection Λ to the Υ-transformation of Section 3.

Theorem 5.5. *For any $\mu \in \mathcal{ID}(*)$ we have*

$$\mathcal{C}_{\Upsilon(\mu)}(\zeta) = \mathcal{C}_{\Lambda(\mu)}(\mathrm{i}\zeta) = \int_0^\infty C_\mu(\zeta x)\mathrm{e}^{-x}\,\mathrm{d}x, \qquad (\zeta \in]-\infty, 0[). \tag{5.1}$$

Proof. These identities follow immediately by combining Proposition 5.2, Proposition 4.16, Theorem 3.16 and Theorem 3.17. □

Remark 5.6. Theorem 5.5 shows, in particular, that any free cumulant function of an element in $\mathcal{ID}(\boxplus)$ is, in fact, identical to a classical cumulant function of an element of $\mathcal{ID}(*)$. The second equality in (5.1) provides an alternative, more direct, way of passing from the measure μ to its free counterpart, $\Lambda(\mu)$, without passing through the Lévy-Khintchine representations. This way is often quite effective, when it comes to calculating $\Lambda(\mu)$ for specific examples of μ. Taking Theorem 3.43 into account, we note that for any measure μ in $\mathcal{ID}(*)$, the free cumulant transform of the measure $\Lambda(\mu)$ is equal to the classical cumulant transform of the stochastic integral $\int_0^1 -\log(1-t)\,\mathrm{d}X_t$, where (X_t) is a classical Lévy process (in law), such that $L\{X_1\} = \mu$.

In analogy with the proof of Proposition 3.38, The second equality in (5.1) provides an easy proof of the following algebraic properties of Λ:

Theorem 5.7. *The Bercovici-Pata bijection $\Lambda\colon \mathcal{ID}(*) \to \mathcal{ID}(\boxplus)$, has the following (algebraic) properties:*

(i) *If $\mu_1, \mu_2 \in \mathcal{ID}(*)$, then $\Lambda(\mu_1 * \mu_2) = \Lambda(\mu_1) \boxplus \Lambda(\mu_2)$.*
(ii) *If $\mu \in \mathcal{ID}(*)$ and $c \in \mathbb{R}$, then $\Lambda(D_c\mu) = D_c\Lambda(\mu)$.*
(iii) *For any constant c in $\mathbb{R}$, we have $\Lambda(\delta_c) = \delta_c$.*

Proof. The proof is similar to that of Proposition 3.38. Indeed, property (ii), say, may be proved as follows: For μ in $\mathcal{ID}(*)$ and ζ in $]-\infty, 0[$, we have

$$\begin{aligned}\mathcal{C}_{\Lambda(D_c\mu)}(\mathrm{i}\zeta) &= \int_{\mathbb{R}} C_{D_c\mu}(\zeta x)\mathrm{e}^{-x}\,\mathrm{d}x = \int_{\mathbb{R}} C_\mu(c\zeta x)\mathrm{e}^{-x}\,\mathrm{d}x \\ &= \mathcal{C}_{\Lambda(\mu)}(\mathrm{i}c\zeta) = \mathcal{C}_{D_c\Lambda(\mu)}(\mathrm{i}\zeta),\end{aligned}$$

and the result then follows from uniqueness of analytic continuation. □

Corollary 5.8. *The bijection $\Lambda\colon \mathcal{ID}(*) \to \mathcal{ID}(\boxplus)$ is invariant under affine transformations, i.e. if $\mu \in \mathcal{ID}(*)$ and $\psi\colon \mathbb{R} \to \mathbb{R}$ is an affine transformation, then*

$$\Lambda(\psi(\mu)) = \psi(\Lambda(\mu)).$$

Proof. Let $\psi\colon \mathbb{R} \to \mathbb{R}$ be an affine transformation, i.e. $\psi(t) = ct + d$, $(t \in \mathbb{R})$, for some constants c, d in $\mathbb{R}$. Then for a probability measure μ on $\mathbb{R}$, $\psi(\mu) = D_c\mu * \delta_d$, and also $\psi(\mu) = D_c\mu \boxplus \delta_d$. Assume now that $\mu \in \mathcal{ID}(*)$. Then by Theorem 5.7,

$$\Lambda(\psi(\mu)) = \Lambda(D_c\mu * \delta_d) = D_c\Lambda(\mu) \boxplus \Lambda(\delta_d) = D_c\Lambda(\mu) \boxplus \delta_d = \psi(\Lambda(\mu)),$$

as desired. □

As a consequence of the corollary above, we get a short proof of the following result, which was proved by Bercovici and Pata in [BePa99].

Corollary 5.9 ([BePa99]). *The bijection $\Lambda\colon \mathcal{ID}(*) \to \mathcal{ID}(\boxplus)$ maps the $*$-stable probability measures on $\mathbb{R}$ onto the $\boxplus$-stable probability measures on $\mathbb{R}$.*

Proof. Assume that μ is a $*$-stable probability measure on $\mathbb{R}$, and let $\psi_1, \psi_2\colon \mathbb{R} \to \mathbb{R}$ be increasing affine transformations on $\mathbb{R}$. Then $\psi_1(\mu) * \psi_2(\mu) = \psi_3(\mu)$, for yet another increasing affine transformation $\psi_3\colon \mathbb{R} \to \mathbb{R}$. Now by Corollary 5.8 and Theorem 5.7(i),

$$\begin{aligned}\psi_1(\Lambda(\mu)) \boxplus \psi_2(\Lambda(\mu)) &= \Lambda(\psi_1(\mu)) \boxplus \Lambda(\psi_2(\mu)) = \Lambda(\psi_1(\mu) * \psi_2(\mu)) \\ &= \Lambda(\psi_3(\mu)) = \psi_3(\Lambda(\mu)),\end{aligned}$$

which shows that $\Lambda(\mu)$ is $\boxplus$-stable.

The same line of argument shows that μ is $*$-stable, if $\Lambda(\mu)$ is $\boxplus$-stable. □

Corollary 5.10. *Let μ be a $*$-selfdecomposable probability measure on $\mathbb{R}$ and let $(\mu_c)_{c\in]0,1[}$ be the family of probability measures on $\mathbb{R}$ defined by the equation:*

$$\mu = D_c\mu * \mu_c.$$

Then, for any c in $]0,1[$, we have the decomposition:

$$\Lambda(\mu) = D_c\Lambda(\mu) \boxplus \Lambda(\mu_c). \tag{5.2}$$

Consequently, a probability measure μ on $\mathbb{R}$ is $*$-selfdecomposable, if and only if $\Lambda(\mu)$ is $\boxplus$-selfdecomposable, and thus the bijection $\Lambda\colon \mathcal{ID}(*) \to \mathcal{ID}(\boxplus)$ maps the class $\mathcal{L}(*)$ of $*$-selfdecomposable probability measures onto the class $\mathcal{L}(\boxplus)$ of $\boxplus$-selfdecomposable probability measures.

Proof. For any c in $]0,1[$, the measures $D_c\mu$ and μ_c are both $*$-infinitely divisible (see Section 2.5), and hence, by (i) and (ii) of Theorem 5.7,

$$\Lambda(\mu) = \Lambda(D_c\mu * \mu_c) = D_c\Lambda(\mu) \boxplus \Lambda(\mu_c).$$

Since this holds for all c in $]0,1[$, it follows that $\Lambda(\mu)$ is $\boxplus$-selfdecomposable.

Assume conversely that μ' is a $\boxplus$-selfdecomposable probability measure on $\mathbb{R}$, and let $(\mu'_c)_{c\in]0,1[}$ be the family of probability measures on $\mathbb{R}$ defined by:

$$\mu' = D_c\mu' \boxplus \mu'_c.$$

By Theorem 4.25 and Proposition 4.26, $\mu', \mu'_c \in \mathcal{ID}(\boxplus)$, so we may consider the $*$-infinitely divisible probability measures $\mu := \Lambda^{-1}(\mu')$ and $\mu_c := \Lambda^{-1}(\mu'_c)$. Then by (i) and (ii) of Theorem 5.7,

$$\begin{aligned}\mu = \Lambda^{-1}(\mu') &= \Lambda^{-1}(D_c(\mu') \boxplus \mu'_c) = \Lambda^{-1}(D_c\Lambda(\mu) \boxplus \Lambda(\mu_c)) \\ &= \Lambda^{-1}(\Lambda(D_c\mu * \mu_c)) = D_c\mu * \mu_c.\end{aligned}$$

Since this holds for any c in $]0,1[$, μ is $*$-selfdecomposable. □

To summarize, we note that the Bercovici-Pata bijection Λ maps each of the classes $\mathcal{G}(*), \mathcal{S}(*), \mathcal{L}(*), \mathcal{ID}(*)$ in the hierarchy (2.13) onto the corresponding free class in (4.9).

Remark 5.11. Above we have discussed the free analogues of the classical stable and selfdecomposable laws, defining the free versions via free convolution properties. Alternatively, one may define the classes of free stable and free selfdecomposable laws in terms of monotonicity properties of the associated Lévy measures, simply using the same characterizations as those holding in the classical case, see Section 2.5. The same approach leads to free analogues $\mathcal{R}(\boxplus)$, $\mathcal{T}(\boxplus)$ and $\mathcal{B}(\boxplus)$ of the classes $\mathcal{R}(*)$, $\mathcal{T}(*)$ and $\mathcal{B}(*)$. We shall however not study these latter analogues here.

Remark 5.12. We end this section by mentioning the possible connection between the mapping Υ^α, introduced in Section 3.4, and the notion of α-probability theory (usually denoted q-deformed probability). For each q in $[-1, 1]$, the so called q-deformed probability theory has been developed by a number of authors (see e.g. [BoSp91] and [Ni95]). For $q = 0$, this corresponds to Voiculescu's free probability and for $q = 1$ to classical probability. Since the right hand side of (3.60) interpolates correspondingly between the free and classical Lévy-Khintchine representations, one may speculate whether the right hand side of (3.60) (for $\alpha = q$) might be interpreted as a kind of Lévy-Khintchine representation for the q-analogue of the cumulant transform (see [Ni95]).

5.3 Topological Properties of Λ

In this section, we study some topological properties of Λ. The key result is the following theorem, which is the free analogue of a result due to B.V. Gnedenko (cf. [GnKo68, §19, Theorem 1]).

Theorem 5.13. *Let μ be a measure in $\mathcal{ID}(\boxplus)$, and let (μ_n) be a sequence of measures in $\mathcal{ID}(\boxplus)$. For each n, let (γ_n, σ_n) be the free generating pair for μ_n, and let (γ, σ) be the free generating pair for μ. Then the following two conditions are equivalent:*

(i) $\mu_n \xrightarrow{\text{w}} \mu$, *as* $n \to \infty$.
(ii) $\gamma_n \to \gamma$ *and* $\sigma_n \xrightarrow{\text{w}} \sigma$, *as* $n \to \infty$.

Proof. (ii) $\Rightarrow$ (i): Assume that (ii) holds. By Theorem 4.12 it is sufficient to show that

(a) $\phi_{\mu_n}(\mathrm{i}y) \to \phi(\mathrm{i}y)$, as $n \to \infty$, for all y in $]0, \infty[$.

(b) $\sup_{n \in \mathbb{N}} \left| \frac{\phi_{\mu_n}(\mathrm{i}y)}{y} \right| \to 0$, as $y \to \infty$.

Regarding (a), note that for any y in $]0,\infty[$, the function $t \mapsto \frac{1+tiy}{iy-t}$, $t \in \mathbb{R}$, is continuous and bounded. Therefore, by the assumptions in (ii),

$$\phi_{\mu_n}(iy) = \gamma_n + \int_{\mathbb{R}} \frac{1+tiy}{iy-t}\,\sigma_n(dt) \underset{n\to\infty}{\longrightarrow} \gamma + \int_{\mathbb{R}} \frac{1+tiy}{iy-t}\,\sigma(dt) = \phi_\mu(iy).$$

Turning then to (b), note that for n in $\mathbb{N}$ and y in $]0,\infty[$,

$$\frac{\phi_{\mu_n}(iy)}{y} = \frac{\gamma_n}{y} + \int_{\mathbb{R}} \frac{1+tiy}{y(iy-t)}\,\sigma_n(dt).$$

Since the sequence (γ_n) is, in particular, bounded, it suffices thus to show that

$$\sup_{n\in\mathbb{N}} \Big| \int_{\mathbb{R}} \frac{1+tiy}{y(iy-t)}\,\sigma_n(dt) \Big| \to 0, \quad \text{as } y \to \infty. \tag{5.3}$$

For this, note first that since $\sigma_n \xrightarrow{w} \sigma$, as $n \to \infty$, and since $\sigma(\mathbb{R}) < \infty$, it follows by standard techniques that the family $\{\sigma_n \mid n \in \mathbb{N}\}$ is tight (cf. [Br92, Corollary 8.11]).

Note next, that for any t in $\mathbb{R}$ and any y in $]0,\infty[$,

$$\Big| \frac{1+tiy}{y(iy-t)} \Big| \le \frac{1}{y(y^2+t^2)^{1/2}} + \frac{|t|}{(y^2+t^2)^{1/2}}.$$

From this estimate it follows that

$$\sup_{y\in[1,\infty[,\,t\in\mathbb{R}} \Big| \frac{1+tiy}{y(iy-t)} \Big| \le 2,$$

and that for any N in $\mathbb{N}$ and y in $[1,\infty[$,

$$\sup_{t\in[-N,N]} \Big| \frac{1+tiy}{y(iy-t)} \Big| \le \frac{N+1}{y}.$$

From the two estimates above, it follows that for any N in $\mathbb{N}$, and any y in $[1,\infty[$, we have

$$\begin{aligned} \sup_{n\in\mathbb{N}} \Big| \int_{\mathbb{R}} \frac{1+tiy}{y(iy-t)}\,\sigma_n(dt) \Big| &\le \frac{N+1}{y} \sup_{n\in\mathbb{N}} \sigma_n([-N,N]) + 2 \cdot \sup_{n\in\mathbb{N}} \sigma_n([-N,N]^c) \\ &\le \frac{N+1}{y} \sup_{n\in\mathbb{N}} \sigma_n(\mathbb{R}) + 2 \cdot \sup_{n\in\mathbb{N}} \sigma_n([-N,N]^c). \end{aligned} \tag{5.4}$$

Now, given ϵ in $]0,\infty[$ we may, since $\{\sigma_n \mid n \in \mathbb{N}\}$ is tight, choose N in $\mathbb{N}$, such that $\sup_{n\in\mathbb{N}} \sigma_n([-N,N]^c) \le \frac{\epsilon}{4}$. Moreover, since $\sigma_n \xrightarrow{w} \sigma$ and $\sigma(\mathbb{R}) < \infty$, the sequence $\{\sigma_n(\mathbb{R}) \mid n \in \mathbb{N}\}$ is, in particular, bounded, and hence, for the chosen

N, we may subsequently choose y_0 in $[1,\infty[$, such that $\frac{N+1}{y_0}\sup_{n\in\mathbb{N}}\sigma_n(\mathbb{R})\le\frac{\epsilon}{2}$. Using then the estimate in (5.4), it follows that

$$\sup_{n\in\mathbb{N}}\Big|\int_{\mathbb{R}}\frac{1+t\mathrm{i}y}{y(\mathrm{i}y-t)}\,\sigma_n(\mathrm{d}t)\Big|\le\epsilon,$$

whenever $y\ge y_0$. This verifies (5.3).

(i) $\Rightarrow$ (ii): Suppose that $\mu_n\xrightarrow{\mathrm{w}}\mu$, as $n\to\infty$. Then by Theorem 4.12, there exists a number M in $]0,\infty[$, such that

(c) $\forall y\in[M,\infty[\colon \phi_{\mu_n}(\mathrm{i}y)\to\phi_\mu(\mathrm{i}y)$, as $n\to\infty$.

(d) $\displaystyle\sup_{n\in\mathbb{N}}\Big|\frac{\phi_{\mu_n}(\mathrm{i}y)}{y}\Big|\to 0$, as $y\to\infty$.

We show first that the family $\{\sigma_n\mid n\in\mathbb{N}\}$ is conditionally compact w.r.t. weak convergence, i.e. that any subsequence $(\sigma_{n'})$ has a subsequence $(\sigma_{n''})$, which converges weakly to some finite measure σ^* on $\mathbb{R}$. By [GnKo68, §9, Theorem 3 bis], it suffices, for this, to show that $\{\sigma_n\mid n\in\mathbb{N}\}$ is tight, and that $\{\sigma_n(\mathbb{R})\mid n\in\mathbb{N}\}$ is bounded. The key step in the argument is the following observation: For any n in $\mathbb{N}$ and any y in $]0,\infty[$, we have,

$$\begin{aligned}-\mathrm{Im}\phi_{\mu_n}(\mathrm{i}y)&=-\mathrm{Im}\Big(\gamma_n+\int_{\mathbb{R}}\frac{1+t\mathrm{i}y}{\mathrm{i}y-t}\,\sigma_n(\mathrm{d}t)\Big)\\&=-\mathrm{Im}\Big(\int_{\mathbb{R}}\frac{1+t\mathrm{i}y}{\mathrm{i}y-t}\,\sigma_n(\mathrm{d}t)\Big)=y\int_{\mathbb{R}}\frac{1+t^2}{y^2+t^2}\,\sigma_n(\mathrm{d}t).\end{aligned}\tag{5.5}$$

We show now that $\{\sigma_n\mid n\in\mathbb{N}\}$ is tight. For fixed y in $]0,\infty[$, note that

$$\{t\in\mathbb{R}\mid |t|\ge y\}\subseteq\{t\in\mathbb{R}\mid \tfrac{1+t^2}{y^2+t^2}\ge\tfrac12\},$$

so that, for any n in $\mathbb{N}$,

$$\sigma_n(\{t\in\mathbb{R}\mid |t|\ge y\})\le 2\int_{\mathbb{R}}\frac{1+t^2}{y^2+t^2}\,\sigma_n(\mathrm{d}t)=-2\mathrm{Im}\Big(\frac{\phi_{\mu_n}(\mathrm{i}y)}{y}\Big)\le 2\Big|\frac{\phi_{\mu_n}(\mathrm{i}y)}{y}\Big|.$$

Combining this estimate with (d), it follows immediately that $\{\sigma_n\mid n\in\mathbb{N}\}$ is tight.

We show next that the sequence $\{\sigma_n(\mathbb{R})\mid n\in\mathbb{N}\}$ is bounded. For this, note first that with M as in (c), there exists a constant c in $]0,\infty[$, such that

$$c\le\frac{M(1+t^2)}{M^2+t^2},\quad\text{for all } t \text{ in } \mathbb{R}.$$

It follows then, by (5.5), that for any n in $\mathbb{N}$,

$$c\sigma_n(\mathbb{R})\le\int_{\mathbb{R}}\frac{M(1+t^2)}{M^2+t^2}\,\sigma_n(\mathrm{d}t)=-\mathrm{Im}\phi_{\mu_n}(\mathrm{i}M),$$

and therefore by (c),

$$\limsup_{n\to\infty} \sigma_n(\mathbb{R}) \le \limsup_{n\to\infty} \big\{ -c^{-1}\cdot \mathrm{Im}\phi_{\mu_n}(iM)\big\} = -c^{-1}\cdot \mathrm{Im}\phi_{\mu}(iM) < \infty,$$

which shows that $\{\sigma_n(\mathbb{R}) \mid n\in\mathbb{N}\}$ is bounded.

Having established that the family $\{\sigma_n \mid n\in\mathbb{N}\}$ is conditionally compact, recall next from Remark 2.3, that in order to show that $\sigma_n \xrightarrow{\mathrm{w}} \sigma$, it suffices to show that any subsequence $(\sigma_{n'})$ has a subsequence, which converges weakly to σ. A similar argument works, of course, to show that $\gamma_n \to \gamma$. So consider any subsequence $(\gamma_{n'},\sigma_{n'})$ of the sequence of generating pairs. Since $\{\sigma_n \mid n\in\mathbb{N}\}$ is conditionally compact, there is a subsequence (n'') of (n'), such that the sequence $(\sigma_{n''})$ is weakly convergent to some finite measure σ^* on $\mathbb{R}$. Since the function $t \mapsto \frac{1+t\mathrm{i}y}{\mathrm{i}y-t}$ is continuous and bounded for any y in $]0,\infty[$, we know then that

$$\int_{\mathbb{R}} \frac{1+t\mathrm{i}y}{\mathrm{i}y-t}\,\sigma_{n''}(\mathrm{d}t) \underset{n\to\infty}{\longrightarrow} \int_{\mathbb{R}} \frac{1+t\mathrm{i}y}{\mathrm{i}y-t}\,\sigma^*(\mathrm{d}t),$$

for any y in $]0,\infty[$. At the same time, we know from (c) that

$$\gamma_{n''} + \int_{\mathbb{R}} \frac{1+t\mathrm{i}y}{\mathrm{i}y-t}\,\sigma_{n''}(\mathrm{d}t) = \phi_{\mu_{n''}}(\mathrm{i}y) \underset{n\to\infty}{\longrightarrow} \phi_\mu(\mathrm{i}y) = \gamma + \int_{\mathbb{R}} \frac{1+t\mathrm{i}y}{\mathrm{i}y-t}\,\sigma(\mathrm{d}t),$$

for any y in $[M,\infty[$. From these observations, it follows that the sequence $(\gamma_{n''})$ must converge to some real number γ^*, which then has to satisfy the identity:

$$\gamma^* + \int_{\mathbb{R}} \frac{1+t\mathrm{i}y}{\mathrm{i}y-t}\,\sigma^*(\mathrm{d}t) = \phi_\mu(\mathrm{i}y) = \gamma + \int_{\mathbb{R}} \frac{1+t\mathrm{i}y}{\mathrm{i}y-t}\,\sigma(\mathrm{d}t),$$

for all y in $[M,\infty[$. By uniqueness of the free Lévy-Khintchine representation (cf. Theorem 4.14) and uniqueness of analytic continuation, it follows that we must have $\sigma^* = \sigma$ and $\gamma^* = \gamma$. We have thus verified the existence of a subsequence $(\gamma_{n''},\sigma_{n''})$ which converges (coordinate-wise) to (γ,σ), and that was our objective. □

As an immediate consequence of Theorem 5.13 and the corresponding result in classical probability, we get the following

Corollary 5.14. *The Bercovici-Pata bijection $\Lambda\colon \mathcal{ID}(*) \to \mathcal{ID}(\boxplus)$ is a homeomorphism w.r.t. weak convergence. In other words, if μ is a measure in $\mathcal{ID}(*)$ and (μ_n) is a sequence of measures in $\mathcal{ID}(*)$, then $\mu_n \xrightarrow{\mathrm{w}} \mu$, as $n\to\infty$, if and only if $\Lambda(\mu_n) \xrightarrow{\mathrm{w}} \Lambda(\mu)$, as $n\to\infty$.*

Proof. Let (γ,σ) be the generating pair for μ and, for each n, let (γ_n,σ_n) be the generating pair for μ_n.

Assume first that $\mu_n \xrightarrow{\mathrm{w}} \mu$. Then by [GnKo68, §19, Theorem 1], $\gamma_n \to \gamma$ and $\sigma_n \xrightarrow{\mathrm{w}} \sigma$. Since (γ_n,σ_n) (respectively (γ,σ)) is the free generating pair for $\Lambda(\mu_n)$ (respectively $\Lambda(\mu)$), it follows then from Theorem 5.13 that $\Lambda(\mu_n) \xrightarrow{\mathrm{w}} \Lambda(\mu)$.

The same argument applies to the converse implication. □

We end this section by presenting the announced proof of property (v) in Theorem 3.18. The proof follows easily by combining Theorem 5.5 and Theorem 5.13.

Proof of Theorem 3.18(v).

Let $\mu, \mu_1, \mu_2, \mu_3, \ldots$, be probability measures in $\mathcal{ID}(*)$, such that $\mu_n \xrightarrow{w} \mu$, as $n \to \infty$. We need to show that $\Upsilon(\mu_n) \xrightarrow{w} \Upsilon(\mu)$ as $n \to \infty$. Since Λ is continuous w.r.t. weak convergence, $\Lambda(\mu_n) \xrightarrow{w} \Lambda(\mu)$, as $n \to \infty$, and this implies that $\mathcal{C}_{\Lambda(\mu_n)}(\mathrm{i}\zeta) \to \mathcal{C}_{\Lambda(\mu)}(\mathrm{i}\zeta)$, as $n \to \infty$, for any ζ in $]-\infty, 0[$ (use e.g. Theorem 5.13). Thus,

$$C_{\Upsilon(\mu_n)}(\zeta) = \mathcal{C}_{\Lambda(\mu_n)}(\mathrm{i}\zeta) \underset{n\to\infty}{\longrightarrow} \mathcal{C}_{\Lambda(\mu)}(\mathrm{i}\zeta) = C_{\Upsilon(\mu)}(\zeta),$$

for any negative number ζ, and hence also $f_{\Upsilon(\mu_n)}(\zeta) = \exp(C_{\Upsilon(\mu_n)}(\zeta)) \to \exp(C_{\Upsilon(\mu)}(\zeta)) = f_{\Upsilon(\mu)}(\zeta)$, as $n \to \infty$, for such ζ. Applying now complex conjugation, it follows that $f_{\Upsilon(\mu_n)}(\zeta) \to f_{\Upsilon(\mu)}(\zeta)$, as $n \to \infty$, for any (non-zero) ζ, and this means that $\Upsilon(\mu_n) \xrightarrow{w} \Upsilon(\mu)$, as $n \to \infty$. □

5.4 Classical vs. Free Lévy Processes

Consider now a free Lévy process $(Z_t)_{t\geq 0}$, with marginal distributions (μ_t). As for (classical) Lévy processes, it follows then, that each μ_t is necessarily $\boxplus$-infinitely divisible. Indeed, for any n in $\mathbb{N}$ we have: $Z_t = \sum_{j=1}^{n}(Z_{jt/n} - Z_{(j-1)t/n})$, and thus, in view of conditions (i) and (iii) in Definition 4.27, $\mu_t = \mu_{t/n} \boxplus \cdots \boxplus \mu_{t/n}$ (n terms). From the observation just made, it follows that the Bercovici-Pata bijection $\Lambda\colon \mathcal{ID}(*) \to \mathcal{ID}(\boxplus)$ gives rise to a correspondence between classical and free Lévy processes:

Proposition 5.15. *Let $(Z_t)_{t\geq 0}$ be a free Lévy process (in law) affiliated with a W^*-probability space $(\mathcal{A}, \tau)$, and with marginal distributions (μ_t). Then there exists a (classical) Lévy process $(X_t)_{t\geq 0}$, with marginal distributions $(\Lambda^{-1}(\mu_t))$.*

Conversely, for any (classical) Lévy process (X_t) with marginal distributions (μ_t), there exists a free Lévy process (in law) (Z_t) with marginal distributions $(\Lambda(\mu_t))$.

Proof. Consider a free Lévy process (in law) (Z_t) with marginal distributions (μ_t). Then, as noted above, $\mu_t \in \mathcal{ID}(\boxplus)$ for all t, and hence we may define $\mu'_t = \Lambda^{-1}(\mu_t)$, $t \geq 0$. Then, whenever $0 \leq s < t$,

$$\mu'_t = \Lambda^{-1}(\mu_s \boxplus \mu_{t-s}) = \Lambda^{-1}(\mu_s) * \Lambda^{-1}(\mu_{t-s}) = \mu'_s * \mu'_{t-s}.$$

Hence, by the Kolmogorov Extension Theorem (cf. [Sa99, Theorem 1.8]), there exists a (classical) stochastic process (X_t) (defined on some probability space $(\Omega, \mathcal{F}, P)$), with marginal distributions (μ'_t), and which satisfies conditions (i)-(iii) of Definition 2.2. Regarding condition (iv), note that since (Z_t) is a free Lévy process, $\mu_t \xrightarrow{w} \delta_0$ as $t \searrow 0$, and hence, by continuity of Λ^{-1} (cf. Corollary 5.14),

$$\mu_t' = \Lambda^{-1}(\mu_t) \xrightarrow{w} \Lambda^{-1}(\delta_0) = \delta_0, \quad \text{as } t \searrow 0.$$

Thus, (X_t) is a (classical) Lévy process in law, and hence we can find a modification of (X_t) which is a genuine Lévy process.

The second statement of the proposition follows by a similar argument, using Λ rather than Λ^{-1}, and that the marginal distributions of a classical Lévy process are necessarily $*$-infinitely divisible. Furthermore, we have to call upon the existence statement for free Lévy processes (in law) in Remark 4.28. □

Example 5.16. The *free Brownian motion* is the free Lévy process (in law), $(W_t)_{t\geq 0}$, which corresponds to the classical Brownian motion, $(B_t)_{t\geq 0}$, via the correspondence described in Proposition 5.15. In particular (cf. Example 5.3),

$$L\{W_t\}(\mathrm{d}s) = \frac{1}{2\pi t}\sqrt{4t - s^2} \cdot 1_{[-\sqrt{4t},\sqrt{4t}]}(s)\,\mathrm{d}s, \qquad (t > 0).$$

Remark 5.17. **(Free additive processes II)** Though our main objectives in this section are free Lévy processes, we mention, for completeness, that the Bercovici-Pata bijection Λ also gives rise to a correspondence between classical and free additive processes (in law). Thus, to any classical additive process (in law), with corresponding marginal distributions (μ_t) and increment distributions $(\mu_{s,t})_{0\leq s<t}$, there corresponds a free additive process (in law), with marginal distributions $(\Lambda(\mu_t))$ and increment distributions $(\Lambda(\mu_{s,t}))_{0\leq s<t}$. And vice versa.

This follows by the same method as used in the proof of Proposition 5.15 above, once it has been established that for a free additive process (in law) (Z_t), the distributions $\mu_t = L\{Z_t\}$ and $\mu_{s,t} = L\{Z_t - Z_s\}$, $0 \leq s < t$, are necessarily $\boxplus$-infinitely divisible (for the corresponding classical result, see [Sa99, Theorem 9.1]). The key to this result is Theorem 4.18, together with the fact that (Z_t) is actually uniformly stochastically continuous on compact intervals, in the following sense: For any compact interval $[0, b]$ in $[0, \infty[$, and for any positive numbers ϵ, ρ, there exists a positive number δ such that $\mu_{s,t}(\mathbb{R} \setminus [-\epsilon, \epsilon]) < \rho$, for any s, t in $[0, b]$, for which $s < t < s + \delta$. As in the classical case, this follows from condition (iv) in Definition 4.27, by a standard compactness argument (see [Sa99, Lemma 9.6]). Now for any t in $[0, \infty[$ and any n in $\mathbb{N}$, we have (cf. (4.21)),

$$\mu_t = \mu_{0,t/n} \boxplus \mu_{t/n,2t/n} \boxplus \mu_{2t/n,3t/n} \boxplus \cdots \boxplus \mu_{(n-1)t/n,t}. \tag{5.6}$$

Since (Z_t) is uniformly stochastically continuous on $[0, t]$, it follows that the family $\{\mu_{(j-1)t/n,jt/n} \mid n \in \mathbb{N},\ 1 \leq j \leq n\}$ is a null-array, and hence, by Theorem 4.18, (5.6) implies that μ_t is $\boxplus$-infinitely divisible. Applying then this fact to the free additive process (in law) $(Z_t - Z_s)_{t \geq s}$, it follows that also $\mu_{s,t}$ is $\boxplus$-infinitely divisible whenever $0 \leq s < t$.

Remark 5.18. **(An alternative concept of free Lévy processes)** For a classical Lévy process (X_t), condition (iii) in Definition 2.2 is equivalent to the condition that whenever $0 \leq s < t$, the conditional distribution $\text{Prob}(X_t \mid X_s)$ depends only on $t - s$. Conditional probabilities in free probability were studied by Biane in [Bi98], and he noted, in particular, that in the free case, the condition just stated *is not* equivalent to condition (iii) in Definition 4.27. Consequently, in free probability there are two classes of stochastic processes, that may naturally be called Lévy processes: The ones we defined in Definition 4.27 and the ones for which condition (iii) in Definition 4.27 is replaced by the condition on the conditional distributions, mentioned above. In [Bi98] these two types of processes were denoted FAL1 respectively FAL2. We should mention here that in [Bi98], the assumption of stochastic continuity (condition (iv) in Definition 4.27) was not included in the definitions of neither FAL1 nor FAL2. We have included that condition, primarily because it is crucial for the definition of the stochastic integral to be constructed in the next section.

6 Free Stochastic Integration

In the classical setting, stochastic integration with respect to Lévy processes and to Poisson random measures is of key importance. This Section establishes base elements of a similar theory of free stochastic integration. As applications, a representation of free selfdecomposable variates as stochastic integrals is given and free OU processes are introduced. Furthermore, the free Lévy-Itô decomposition is derived.

6.1 Stochastic Integrals w.r.t. free Lévy Processes

As mentioned in Section 2.3, if (X_t) is a classical Lévy process and $f\colon [A, B] \to \mathbb{R}$ is a continuous function defined on an interval $[A, B]$ in $[0, \infty[$, then the stochastic integral $\int_A^B f(t)\, \mathrm{d}X_t$ may be defined as the limit in probability of approximating Riemann sums. More precisely, for each n in $\mathbb{N}$, let $\mathcal{D}_n = \{t_{n,0}, t_{n,1}, \ldots, t_{n,n}\}$ be a subdivision of $[A, B]$, i.e.

$$A = t_{n,0} < t_{n,1} < \cdots < t_{n,n} = B.$$

Assume that

$$\lim_{n\to\infty} \max_{j=1,2,\ldots,n} (t_{n,j} - t_{n,j-1}) = 0. \tag{6.1}$$

Moreover, for each n, choose intermediate points:

$$t^{\#}_{n,j} \in [t_{n,j-1}, t_{n,j}], \quad j = 1, 2, \ldots, n. \tag{6.2}$$

Then the Riemann sums

$$S_n = \sum_{j=1}^{n} f(t^{\#}_{n,j}) \cdot (X_{t_{n,j}} - X_{t_{n,j-1}}),$$

converge *in probability*, as $n \to \infty$, to a random variable S. Moreover, this random variable S does not depend on the choice of subdivisions $\mathcal{D}_n$ (satisfying (6.1)), nor on the choice of intermediate points $t^{\#}_{n,j}$. Hence, it makes sense to call S the stochastic integral of f over $[A, B]$ w.r.t. (X_t), and we denote S by $\int_A^B f(t)\, \mathrm{d}X_t$.

The construction just sketched depends, of course, heavily on the stochastic continuity of the Lévy process in law (X_t) (condition (iv) in Definition 2.2). A proof of the assertions made above can be found in [Lu75, Theorem 6.2.3]. We show next how the above construction carries over, via the Bercovici-Pata bijection, to a corresponding stochastic integral w.r.t. free Lévy processes (in law).

Theorem 6.1. *Let (Z_t) be a free Lévy process (in law), affiliated with a W^*-probability space $(\mathcal{A}, \tau)$. Then for any compact interval $[A, B]$ in $[0, \infty[$ and any continuous function $f\colon [A, B] \to \mathbb{R}$, the stochastic integral $\int_A^B f(t)\, \mathrm{d}Z_t$ exists as the limit in probability (see Definition 4.3) of approximating Riemann sums. More precisely, there exists a (unique) selfadjoint operator T affiliated with $(\mathcal{A}, \tau)$, such that for any sequence $(\mathcal{D}_n)_{n\in\mathbb{N}}$ of subdivisions of $[A, B]$, satisfying* (6.1)*, and for any choice of intermediate points $t^{\#}_{n,j}$, as in* (6.2)*, the corresponding Riemann sums*

$$T_n = \sum_{j=1}^{n} f(t^{\#}_{n,j}) \cdot (Z_{t_{n,j}} - Z_{t_{n,j-1}}),$$

converge in probability to T as $n \to \infty$. We call T the stochastic integral of f over $[A, B]$ w.r.t. (Z_t), and denote it by $\int_A^B f(t)\, \mathrm{d}Z_t$.

In the proof below, we shall use the notation:

$$*_{j=1}^{r} \mu_j := \mu_1 * \cdots * \mu_r \quad \text{and} \quad \boxplus_{j=1}^{r} \mu_j := \mu_1 \boxplus \cdots \boxplus \mu_r,$$

for probability measures $\mu_1, \ldots, \mu_r$ on $\mathbb{R}$.

Proof of Theorem 6.1. Let $(\mathcal{D}_n)_{n\in\mathbb{N}}$ be a sequence of subdivisions of $[A, B]$ satisfying (6.1), let $t^{\#}_{n,j}$ be a family of intermediate points as in (6.2), and consider, for each n, the corresponding Riemann sum:

$$T_n = \sum_{j=1}^{n} f(t^{\#}_{n,j}) \cdot (Z_{t_{n,j}} - Z_{t_{n,j-1}}) \in \overline{\mathcal{A}}.$$

We show that (T_n) is a Cauchy sequence w.r.t. convergence in probability or, equivalently, w.r.t. the measure topology (see the Appendix). Given any n, m in $\mathbb{N}$, we form the subdivision

$$A = s_0 < s_1 < \cdots < s_{p(n,m)} = B,$$

which consists of the points in $\mathcal{D}_n \cup \mathcal{D}_m$ (so that $p(n,m) \le n+m$). Then, for each j in $\{1, 2, \dots, p(n,m)\}$, we choose (in the obvious way) $s^{\#}_{n,j}$ in $\{t^{\#}_{n,k} \mid k = 1, 2, \dots, n\}$ and $s^{\#}_{m,j}$ in $\{t^{\#}_{m,k} \mid k = 1, 2, \dots, m\}$ such that

$$T_n = \sum_{j=1}^{p(n,m)} f(s^{\#}_{n,j}) \cdot (Z_{s_j} - Z_{s_{j-1}}) \quad \text{and} \quad T_m = \sum_{j=1}^{p(n,m)} f(s^{\#}_{m,j}) \cdot (Z_{s_j} - Z_{s_{j-1}}).$$

It follows then that

$$T_n - T_m = \sum_{j=1}^{p(n,m)} \big(f(s^{\#}_{n,j}) - f(s^{\#}_{m,j})\big) \cdot (Z_{s_j} - Z_{s_{j-1}}).$$

Let (μ_t) denote the family of marginal distributions of (Z_t), and then consider a classical Lévy process (X_t) with marginal distributions $(\Lambda^{-1}(\mu_t))$ (cf. Proposition 5.15). For each n, form the Riemann sum

$$S_n = \sum_{j=1}^{n} f(t^{\#}_{n,j}) \cdot (X_{t_{n,j}} - X_{t_{n,j-1}}),$$

corresponding to the same $\mathcal{D}_n$ and $t^{\#}_{n,j}$ as above. Then for any n, m in $\mathbb{N}$, we have also that

$$S_n - S_m = \sum_{j=1}^{p(n,m)} \big(f(s^{\#}_{n,j}) - f(s^{\#}_{m,j})\big) \cdot (X_{s_j} - X_{s_{j-1}}).$$

From this expression, it follows that

$$\begin{aligned} L\{S_n - S_m\} &= *_{j=1}^{p(n,m)} D_{f(s^{\#}_{n,j}) - f(s^{\#}_{m,j})} L\{X_{s_j} - X_{s_{j-1}}\} \\ &= *_{j=1}^{p(n,m)} D_{f(s^{\#}_{n,j}) - f(s^{\#}_{m,j})} \Lambda^{-1}(\mu_{s_j - s_{j-1}}), \end{aligned}$$

so that (by Theorem 5.7),

$$\begin{aligned} \Lambda(L\{S_n - S_m\}) &= \boxplus_{j=1}^{p(n,m)} D_{f(s^{\#}_{n,j}) - f(s^{\#}_{m,j})} \mu_{s_j - s_{j-1}} \\ &= L\Big\{ \sum_{j=1}^{p(n,m)} \big(f(s^{\#}_{n,j}) - f(s^{\#}_{m,j})\big) \cdot (Z_{s_j} - Z_{s_{j-1}}) \Big\} \\ &= L\{T_n - T_m\}. \end{aligned}$$

We know from the classical theory (cf. [Lu75, Theorem 6.2.3]), that (S_n) is a Cauchy sequence w.r.t. convergence in probability, i.e. that $L\{S_n - S_m\} \xrightarrow{\text{w}} \delta_0$, as $n, m \to \infty$. By continuity of Λ, it follows thus that also

$$L\{T_n - T_m\} = \Lambda(L\{S_n - S_m\}) \xrightarrow{\text{w}} \Lambda(\delta_0) = \delta_0, \quad \text{as } n, m \to \infty.$$

By Proposition A.8, this means that (T_n) is a Cauchy sequence w.r.t. the measure topology, and since $\overline{\mathcal{A}}$ is complete in the measure topology (Proposition A.5), there exists an operator T in $\overline{\mathcal{A}}$, such that $T_n \to T$ in the measure topology, i.e. in probability. Since T_n is selfadjoint for each n (see the Appendix) and since the adjoint operation is continuous w.r.t. the measure topology (Proposition A.5), T is necessarily a selfadjoint operator.

It remains to show that the operator T, found above, does not depend on the choice of subdivisions $(\mathcal{D}_n)$ or intermediate points $t^{\#}_{n,j}$. Suppose thus that (T_n) and (T'_n) are two sequences of Riemann sums of the kind considered above. Then by the argument given above, there exist operators T and T' in $\overline{\mathcal{A}}$, such that $T_n \to T$ and $T'_n \to T'$ in probability. Furthermore, if we consider the "mixed sequence" $T_1, T'_2, T_3, T'_4, \ldots$, then the corresponding sequence of subdivisions also satisfies (6.1), and hence this mixed sequence also converges in probability to an operator T'' in $\overline{\mathcal{A}}$. Since the mixed sequence has subsequences converging, in probability, to T and T' respectively, and since the measure topology is a Hausdorff topology (cf. Proposition A.5), we may thus conclude that $T = T'' = T'$, as desired. □

The stochastic integral $\int_A^B f(t)\, \mathrm{d}Z_t$, introduced above, extends to continuous functions $f\colon [A,B] \to \mathbb{C}$ in the usual way (the result being non-selfadjoint in general). From the construction of $\int_A^B f(t)\, \mathrm{d}Z_t$ as the limit of approximating Riemann sums, it follows immediately that whenever $0 \le A < B < C$, we have

$$\int_A^C f(t)\, \mathrm{d}Z_t = \int_A^B f(t)\, \mathrm{d}Z_t + \int_B^C f(t)\, \mathrm{d}Z_t,$$

for any continuous function $f\colon [A,C] \to \mathbb{C}$. Another consequence of the construction, given in the proof above, is the following correspondence between stochastic integrals w.r.t. classical and free Lévy processes (in law).

Corollary 6.2. *Let (X_t) be a classical Lévy process with marginal distributions (μ_t), and let (Z_t) be a corresponding free Lévy process (in law) with marginal distributions $(\Lambda(\mu_t))$ (cf. Proposition 5.15). Then for any compact interval $[A,B]$ in $[0,\infty[$ and any continuous function $f\colon [A,B] \to \mathbb{R}$, the distributions $L\{\int_A^B f(t)\, \mathrm{d}X_t\}$ and $L\{\int_A^B f(t)\, \mathrm{d}Z_t\}$ are $*$-infinitely divisible respectively $\boxplus$-infinitely divisible and, moreover*

$$L\{\textstyle\int_A^B f(t)\, \mathrm{d}Z_t\} = \Lambda\big[L\{\textstyle\int_A^B f(t)\, \mathrm{d}X_t\}\big].$$

Proof. Let $(\mathcal{D}_n)_{n\in\mathbb{N}}$ be a sequence of subdivisions of $[A,B]$ satisfying (6.1), let $t^{\#}_{n,j}$ be a family of intermediate points as in (6.2), and consider, for each n, the corresponding Riemann sums:

$$S_n = \sum_{j=1}^{n} f(t^{\#}_{n,j}) \cdot (X_{t_{n,j}} - X_{t_{n,j-1}}) \quad \text{and} \quad T_n = \sum_{j=1}^{n} f(t^{\#}_{n,j}) \cdot (Z_{t_{n,j}} - Z_{t_{n,j-1}}).$$

Since convergence in probability implies convergence in distribution (Proposition A.9), it follows from [Lu75, Theorem 6.2.3] and Theorem 6.1 above, that $L\{S_n\} \xrightarrow{w} L\{\int_A^B f(t)\, \mathrm{d}X_t\}$ and $L\{T_n\} \xrightarrow{w} L\{\int_A^B f(t)\, \mathrm{d}Z_t\}$. Since $\mathcal{ID}(*)$ and $\mathcal{ID}(\boxplus)$ are closed w.r.t. weak convergence (as noted in Section 4.5), it follows thus that $L\{\int_A^B f(t)\, \mathrm{d}X_t\} \in \mathcal{ID}(*)$ and $L\{\int_A^B f(t)\, \mathrm{d}Z_t\} \in \mathcal{ID}(\boxplus)$. Moreover, by Theorem 5.7, $L\{T_n\} = \Lambda(L\{S_n\})$, for each n in $\mathbb{N}$, and hence the last assertion follows by continuity of Λ. □

6.2 Integral Representation of Freely Selfdecomposable Variates

As mentioned in Section 2.5, a (classical) random variable Y has distribution in $\mathcal{L}(*)$ if and only if it has a representation in law of the form

$$Y \stackrel{\mathrm{d}}{=} \int_0^\infty \mathrm{e}^{-t}\, \mathrm{d}X_t, \tag{6.3}$$

where $(X_t)_{t\geq 0}$ is a (classical) Lévy process, satisfying the condition $\mathbb{E}[\log(1 + |X_1|)] < \infty$. The aim of this section is to establish a similar correspondence between selfadjoint operators with (spectral) distribution in $\mathcal{L}(\boxplus)$ and free Lévy processes (in law).

The stochastic integral appearing in (6.3) is the limit in probability, as $R \to \infty$, of the stochastic integrals $\int_0^R \mathrm{e}^{-t}\, \mathrm{d}X_t$, i.e. we have

$$\int_0^R \mathrm{e}^{-t}\, \mathrm{d}X_t \xrightarrow{\mathrm{P}} \int_0^\infty \mathrm{e}^{-t}\, \mathrm{d}X_t, \quad \text{as } R \to \infty,$$

(the convergence actually holds almost surely; see Proposition 6.3 below). The stochastic integral $\int_0^R \mathrm{e}^{-t}\, \mathrm{d}X_t$ is, in turn, defined as the limit of approximating Riemann sums as described in Section 6.1

For a free Lévy process (Z_t), we determine next under which conditions the stochastic integral $\int_0^\infty \mathrm{e}^{-t}\, \mathrm{d}Z_t$ makes sense as the limit, for $R \to \infty$, of the stochastic integrals $\int_0^R \mathrm{e}^{-t}\, \mathrm{d}Z_t$, which are defined by virtue of Theorem 6.1. Again, the result we obtain is derived by applications of the mapping Λ and the following corresponding classical result:

Proposition 6.3 ([JuVe83]). *Let (X_t) be a classical Lévy process defined on some probability space $(\Omega, \mathcal{F}, P)$, and let (γ, σ) be the generating pair for the $*$-infinitely divisible probability measure $L\{X_1\}$. Then the following conditions are equivalent:*

(i) $\int_{\mathbb{R}\setminus]-1,1[} \log(1 + |t|)\, \sigma(\mathrm{d}t) < \infty$.
(ii) $\int_0^R \mathrm{e}^{-t}\, \mathrm{d}X_t$ *converges almost surely, as* $R \to \infty$.
(iii) $\int_0^R \mathrm{e}^{-t}\, \mathrm{d}X_t$ *converges in distribution, as* $R \to \infty$.
(iv) $\mathbb{E}[\log(1 + |X_1|)] < \infty$.

Proof. This was proved in [JuVe83, Theorem 3.6.6]. We note, though, that in [JuVe83], the measure σ in condition (i) is replaced by the Lévy measure ρ appearing in the alternative Lévy-Khintchine representation (2.2) for $L\{X_1\}$. However, since $\rho(\mathrm{d}t) = \frac{1+t^2}{t^2} \cdot 1_{\mathbb{R}\setminus\{0\}}(t)\ \sigma(\mathrm{d}t)$, it is clear that the integrals $\int_{\mathbb{R}\setminus]-1,1[} \log(1+|t|)\ \rho(\mathrm{d}t)$ and $\int_{\mathbb{R}\setminus]-1,1[} \log(1+|t|)\ \sigma(\mathrm{d}t)$ are finite simultaneously. □

Proposition 6.4. *Let (Z_t) be a free Lévy process (in law) affiliated with a W^*-probability space $(\mathcal{A},\tau)$, and let (γ,σ) be the free generating pair for the $\boxplus$-infinitely divisible probability measure $L\{Z_1\}$. Then the following statements are equivalent:*

(i) $\int_{\mathbb{R}\setminus]-1,1[} \log(1+|t|)\ \sigma(\mathrm{d}t) < \infty$.
(ii) $\int_0^R \mathrm{e}^{-t}\ \mathrm{d}Z_t$ *converges in probability, as* $R \to \infty$.
(iii) $\int_0^R \mathrm{e}^{-t}\ \mathrm{d}Z_t$ *converges in distribution, as* $R \to \infty$.

Proof. Let (μ_t) be the family of marginal distributions of (Z_t) and consider then a classical Lévy process (X_t) with marginal distributions $(\Lambda^{-1}(\mu_t))$ (cf. Proposition 5.15). By the definition of Λ, it follows then that (γ,σ) is the generating pair for the $*$-infinitely divisible probability measure $L\{X_1\}$.

(i) $\Rightarrow$ (ii): Assume that (i) holds. Then condition (i) in Proposition 6.3 is satisfied for the classical Lévy process (X_t). Hence by (ii) of that proposition, $\int_0^R \mathrm{e}^{-t}\ \mathrm{d}X_t$ converges almost surely, and hence in probability, as $R \to \infty$. Consider now any increasing sequence (R_n) of positive numbers, such that $R_n \nearrow \infty$, as $n \to \infty$. Then for any m,n in $\mathbb{N}$ such that $m > n$, we have by Corollary 6.2

$$\begin{aligned} L\big\{ \textstyle\int_0^{R_m} \mathrm{e}^{-t}\ \mathrm{d}Z_t - \int_0^{R_n} \mathrm{e}^{-t}\ \mathrm{d}Z_t\big\} &= L\big\{ \textstyle\int_{R_n}^{R_m} \mathrm{e}^{-t}\ \mathrm{d}Z_t\big\} = \Lambda\big[L\big\{ \textstyle\int_{R_n}^{R_m} \mathrm{e}^{-t}\ \mathrm{d}X_t\big\}\big] \\ &= \Lambda\big[L\big\{ \textstyle\int_0^{R_m} \mathrm{e}^{-t}\ \mathrm{d}X_t - \int_0^{R_n} \mathrm{e}^{-t}\ \mathrm{d}X_t\big\}\big]. \end{aligned} \tag{6.4}$$

Since the sequence $(\int_0^{R_n} \mathrm{e}^{-t}\ \mathrm{d}X_t)_{n\in\mathbb{N}}$ is a Cauchy sequence with respect to convergence in probability, it follows thus, by continuity of Λ, that so is the sequence $(\int_0^{R_n} \mathrm{e}^{-t}\ \mathrm{d}Z_t)_{n\in\mathbb{N}}$. Hence, by Proposition A.5, there exists a selfadjoint operator W affiliated with $(\mathcal{A},\tau)$, such that $\int_0^{R_n} \mathrm{e}^{-t}\ \mathrm{d}Z_t \to W$ in probability. It remains to argue that W does not depend on the sequence (R_n). This follows, for example, as in the proof of Theorem 6.1, by considering, for two given sequences (R_n) and (R_n'), a third increasing sequence (R_n''), containing infinitely many elements from both of the original sequences.

(ii) $\Rightarrow$ (i): Assume that (ii) holds. It follows then by (6.4) and continuity of Λ^{-1} that for any increasing sequence (R_n), as above, $(\int_0^{R_n} \mathrm{e}^{-t}\ \mathrm{d}X_t)$ is a Cauchy sequence w.r.t. convergence in probability. We deduce that (iii) of Proposition 6.3 is satisfied for (X_t), and hence so is (i) of that proposition. By definition of (X_t), this means exactly that (i) of Proposition 6.4 is satisfied for (Z_t).

(ii) $\Rightarrow$ (iii): This follows from Proposition A.9.

(iii)$\Rightarrow$(i): Suppose (iii) holds, and note that the limit distribution is necessarily $\boxplus$-infinitely divisible. Now by Corollary 6.2 and continuity of Λ^{-1}, condition (iii) of Proposition 6.3 is satisfied for (X_t), and hence so is (i) of that proposition. This means, again, that (i) in Proposition 6.4 is satisfied for (Z_t). □

If (Z_t) is a free Lévy process (in law) affiliated with $(\mathcal{A}, \tau)$, such that (i) of Proposition 6.4 is satisfied, then we denote by $\int_0^\infty e^{-t}\, dZ_t$ the selfadjoint operator affiliated with $(\mathcal{A}, \tau)$, to which $\int_0^R e^{-t}\, dZ_t$ converges, in probability, as $R \to \infty$. We note that $L\{\int_0^\infty e^{-t}\, dZ_t\}$ is $\boxplus$-infinitely divisible, and that Corollary 6.2 and Proposition A.9 yield the following relation:

$$L\{\textstyle\int_0^\infty e^{-t}\, dZ_t\} = \Lambda\big[L\{\textstyle\int_0^\infty e^{-t}\, dX_t\}\big], \tag{6.5}$$

where (X_t) is a classical Lévy process corresponding to (Z_t) as in Proposition 5.15.

Theorem 6.5. *Let y be a selfadjoint operator affiliated with a W^*-probability space $(\mathcal{A}, \tau)$. Then the distribution of y is $\boxplus$-selfdecomposable if and only if y has a representation in law in the form:*

$$y \stackrel{d}{=} \int_0^\infty e^{-t}\, dZ_t, \tag{6.6}$$

for some free Lévy process (in law) (Z_t) affiliated with some W^*-probability space $(\mathcal{B}, \psi)$, and satisfying condition (i) of Proposition 6.4.

Proof. Put $\mu = L\{y\}$. Suppose first that μ is $\boxplus$-selfdecomposable and put $\mu' = \Lambda^{-1}(\mu)$. Then, by Corollary 5.10, μ' is $*$-selfdecomposable, and hence by the classical version of this theorem (cf. [JuVe83, Theorem 3.2]), there exists a classical Lévy process (X_t) defined on some probability space $(\Omega, \mathcal{F}, P)$, such that condition (i) in Proposition 6.3 is satisfied, and such that $\Lambda^{-1}(\mu) = L\{\int_0^\infty e^{-t}\, dX_t\}$. Let (Z_t) be a free Lévy process (in law) affiliated with some W^*-probability space $(\mathcal{B}, \psi)$, and corresponding to (X_t) as in Proposition 5.15. Then, by definition of Λ, condition (i) in Proposition 6.4 is satisfied for (Z_t) and, by formula (6.5), $L\{\int_0^\infty e^{-t}\, dZ_t\} = \mu$.

Assume, conversely, that there exists a free Lévy process (in law) (Z_t) affiliated with some W^*-probability space $(\mathcal{B}, \psi)$, such that condition (i) of Proposition 6.4 is satisfied, and such that $\mu = L\{\int_0^\infty e^{-t}\, dZ_t\}$. Then consider a classical Lévy process (X_t) defined on some probability space $(\Omega, \mathcal{F}, P)$, and corresponding to (Z_t) as in Proposition 5.15. Condition (i) in Proposition 6.3 is then satisfied for (X_t) and, by (6.5), $\Lambda^{-1}(\mu) = L\{\int_0^\infty e^{-t}\, dX_t\}$. Thus, by the classical version of this theorem, $\Lambda^{-1}(\mu)$ is $*$-selfdecomposable, and hence μ is $\boxplus$-selfdecomposable. □

Remark 6.6 (Free OU processes). Let y be a selfadjoint operator affiliated with some W^*-probability space $(\mathcal{A},\tau)$, and assume that there exists a free Lévy process (in law) (Z_t) affiliated with some W^*-probability space $(\mathcal{B},\psi)$, such that condition (i) of Proposition 6.4 is satisfied, and such that $y \stackrel{\mathrm{d}}{=} \int_0^\infty \mathrm{e}^{-t}\,\mathrm{d}Z_t$. Note then, that for any positive numbers s,λ, we have

$$\begin{aligned}\int_0^\infty \mathrm{e}^{-t}\,\mathrm{d}Z_t &= \int_0^\infty \mathrm{e}^{-\lambda t}\,\mathrm{d}Z_{\lambda t} = \int_s^\infty \mathrm{e}^{-\lambda t}\,\mathrm{d}Z_{\lambda t} + \int_0^s \mathrm{e}^{-\lambda t}\,\mathrm{d}Z_{\lambda t} \\ &= \mathrm{e}^{-\lambda s}\int_0^\infty \mathrm{e}^{-\lambda t}\,\mathrm{d}Z_{\lambda(s+t)} + \int_0^{\lambda s} \mathrm{e}^{-t}\,\mathrm{d}Z_t,\end{aligned} \tag{6.7}$$

where we have introduced integration w.r.t. the processes $V_t = Z_{\lambda t}$ and $W_t = Z_{\lambda(s+t)}$, $t \geq 0$. The rules of transformation for stochastic integrals, used above, are easily verified by considering the integrals as limits of Riemann sums. That same point of view, together with the fact that (Z_t) has freely independent stationary increments (conditions (i) and (iii) in Definition 4.27), implies, furthermore, that $\int_0^\infty \mathrm{e}^{-\lambda t}\,\mathrm{d}Z_{\lambda(s+t)} \stackrel{\mathrm{d}}{=} \int_0^\infty \mathrm{e}^{-\lambda t}\,\mathrm{d}Z_{\lambda t} \stackrel{\mathrm{d}}{=} y$. Note also that the two terms in the last expression of (6.7) are freely independent. Thus, (6.7) shows, that for any positive numbers s,λ, we have a decomposition in the form: $y \stackrel{\mathrm{d}}{=} \mathrm{e}^{-\lambda s}y(\lambda,s) + u(\lambda,s)$, where $y(\lambda,s)$ and $u(\lambda,s)$ are freely independent, and where $y(\lambda,s) \stackrel{\mathrm{d}}{=} y$. In particular, we have verified, directly, that $L\{y\}$ is $\boxplus$-selfdecomposable. Moreover, if we choose a selfadjoint operator Y_0 affiliated with $(\mathcal{B},\psi)$, which is freely independent of (Z_t), and such that $L\{Y_0\} = L\{y\}$ (extend $(\mathcal{B},\psi)$ if necessary), then the expression:

$$Y_s = \mathrm{e}^{-\lambda s}Y_0 + \int_0^{\lambda s} \mathrm{e}^{-t}\,\mathrm{d}Z_t, \quad (s \geq 0),$$

defines an operator valued stochastic process (Y_s) affiliated with $(\mathcal{B},\psi)$, satisfying that $Y_s \stackrel{\mathrm{d}}{=} y$ for all s. If we replace (Z_t) above by a classical Lévy process (X_t), satisfying condition (i) in Proposition 6.3, and let Y_0 be a (classical) random variable, which is independent of (X_t), then the corresponding process (Y_s) is a solution to the stochastic differential equation:

$$\mathrm{d}Y_s = -\lambda Y_s\,\mathrm{d}s + \mathrm{d}X_{\lambda s},$$

and (Y_s) is said to be a process of *Ornstein-Uhlenbeck type* or an *OU process*, for short (cf. [BaSh01a],[BaSh01b] and references given there).

6.3 Free Poisson Random Measures

In this section, we introduce free Poisson random measures and prove their existence. We mention in passing the related notions of free stochastic measures (cf. [An00]) and free white noise (cf. [Sp90]). We mention also that the existence of free Poisson random measures was established by Voiculescu in [Vo98] in a different way than the one presented below. Recall, that for any number λ in $[0,\infty[$, we denote by $\mathrm{Poiss}^{\boxplus}(\lambda)$ the free Poisson distribution with mean λ (cf. Example 5.3).

Definition 6.7. *Let $(\Theta,\mathcal{E},\nu)$ be a measure space, and put*

$$\mathcal{E}_0 = \{E \in \mathcal{E} \mid \nu(E) < \infty\}.$$

Let further $(\mathcal{A},\tau)$ be a W^*-probability space, and let $\mathcal{A}_+$ denote the cone of positive operators in $\mathcal{A}$. Then a free Poisson random measure on $(\Theta,\mathcal{E},\nu)$ with values in $(\mathcal{A},\tau)$, is a mapping $M\colon \mathcal{E}_0 \to \mathcal{A}_+$, with the following properties:

(i) *For any set E in $\mathcal{E}_0$, $L\{M(E)\} = \mathrm{Poiss}^{\boxplus}(\nu(E))$.*
(ii) *If $r \in \mathbb{N}$ and $E_1,\dots,E_r$ are disjoint sets from $\mathcal{E}_0$, then $M(E_1),\dots,M(E_r)$ are freely independent operators.*
(iii) *If $r \in \mathbb{N}$ and $E_1,\dots,E_r$ are disjoint sets from $\mathcal{E}_0$, then $M(\cup_{j=1}^r E_j) = \sum_{j=1}^r M(E_j)$.*

In the setting of Definition 6.7, the measure ν is called the *intensity measure* for the free Poisson random measure M. Note, in particular, that $M(E)$ is a *bounded* positive operator for all E in $\mathcal{E}_0$. The definition above might seem a little "poor"compared to that of a classical Poisson random measure. The following remark might offer a bit of consolation.

Remark 6.8. Suppose M is a free Poisson random measure on the measure space $(\Theta,\mathcal{E},\nu)$ with values in the W^*-probability space $(\mathcal{A},\tau)$. Let further (E_n) be a sequence of disjoint sets from $\mathcal{E}_0$. If we assume, in addition, that $\cup_{j\in\mathbb{N}} E_j \in \mathcal{E}_0$, then we also have that

$$M\Big(\bigcup_{j\in\mathbb{N}} E_j\Big) = \sum_{j=1}^{\infty} M(E_j),$$

where the right hand side should be understood as the limit *in probability* (see Definition 4.3) of $\sum_{j=1}^n M(E_j)$ as $n \to \infty$.

Indeed, put $E = \cup_{j\in\mathbb{N}} E_j$, and assume that $E \in \mathcal{E}_0$. Then for any n in $\mathbb{N}$,

$$M(E) - \sum_{j=1}^{n} M(E_j) = M(E) - M(\cup_{j=1}^n E_j) = M(\cup_{j=n+1}^{\infty} E_j),$$

so that

$$\begin{aligned} L\Big\{M(E) - \sum_{j=1}^{n} M(E_j)\Big\} &= \mathrm{Poiss}^{\boxplus}\big(\nu(\cup_{j=n+1}^{\infty} E_j)\big) \\ &= \mathrm{Poiss}^{\boxplus}\big(\textstyle\sum_{j=n+1}^{\infty} \nu(E_j)\big) \xrightarrow{\text{w}} \delta_0, \end{aligned}$$

as $n \to \infty$, since $\sum_{j=n+1}^{\infty} \nu(E_j) \to 0$ as $n \to \infty$, because $\sum_{j=1}^{\infty} \nu(E_j) = \nu(E) < \infty$.

The main purpose of the section is to prove the general existence of free Poisson random measures.

Theorem 6.9. *Let $(\Theta, \mathcal{E}, \nu)$ be a measure space. Then there exists a W^*-probability space $(\mathcal{A}, \tau)$ and a free Poisson random measure M on $(\Theta, \mathcal{E}, \nu)$ with values in $(\mathcal{A}, \tau)$.*

The proof of Theorem 6.9 is given in a series of lemmas. First of all, though, we introduce some notation:

If $\mu_1, \mu_2, \dots, \mu_r$ are probability measures on $\mathbb{R}$, we put (as in Section 6.1)

$$\mathop{*}_{h=1}^{r} \mu_h = \mu_1 * \mu_2 * \cdots * \mu_r \quad \text{and} \quad \mathop{\boxplus}_{h=1}^{r} \mu_h = \mu_1 \boxplus \mu_2 \boxplus \cdots \boxplus \mu_r.$$

In the remaining part of this section, we consider the measure space $(\Theta, \mathcal{E}, \nu)$ appearing in Theorem 6.9. Consider then the set

$$\mathcal{I} = \bigcup_{k \in \mathbb{N}} \{(E_1, \dots, E_k) \mid E_1, \dots, E_k \in \mathcal{E}_0 \setminus \{\emptyset\} \text{ and } E_1, \dots, E_k \text{ are disjoint}\},$$

where we think of $(E_1, \dots, E_k)$ merely as a collection of sets from $\mathcal{E}_0$. In particular, we identify $(E_1, \dots, E_k)$ with $(E_{\pi(1)}, \dots, E_{\pi(k)})$ for any permutation π of $\{1, 2, \dots, k\}$. We introduce, furthermore, a partial order $\leq$ on $\mathcal{I}$ by the convention:

$$(E_1, \dots, E_k) \leq (F_1, \dots, F_l) \iff \text{each } E_i \text{ is a union of some of the } F_j\text{'s}.$$

Lemma 6.10. *Given a tuple $S = (E_1, \dots, E_k)$ from $\mathcal{I}$, there exists a W^*-probability space $(\mathcal{A}_S, \tau_S)$, which is generated by freely independent positive operators $M_S(E_1), \dots, M_S(E_k)$ from $\mathcal{A}_S$, satisfying that*

$$L\{M_S(E_i)\} = \text{Poiss}^{\boxplus}(\nu(E_i)), \qquad (i = 1, \dots, k).$$

Proof. This is an immediate consequence of Voiculescu's theory of (reduced) free products of von Neumann algebras (cf. [VoDyNi92]). Indeed, we may take $(\mathcal{A}_S, \tau_S)$ to be the (reduced) von Neumann algebra free product of the Abelian W^*-probability spaces $(L^\infty(\mathbb{R}, \mu_i), \mathbb{E}_{\mu_i})$, $i = 1, \dots, k$, where $\mu_i = \text{Poiss}^{\boxplus}(\nu(E_i))$ and $\mathbb{E}_{\mu_i}$ denotes expectation with respect to μ_i. □

Lemma 6.11. *Consider two elements $S = (E_1, \dots, E_k)$ and $T = (F_1, \dots, F_l)$ of $\mathcal{I}$, and suppose that $S \leq T$. Consider the W^*-probability spaces $(\mathcal{A}_S, \tau_S)$ and $(\mathcal{A}_T, \tau_T)$ given by Lemma 6.10. Then there exists an injective, unital, normal $*$-homomorphism $\iota_{S,T} \colon \mathcal{A}_S \to \mathcal{A}_T$, such that $\tau_S = \tau_T \circ \iota_{S,T}$.*

Proof. We adapt the notation from Lemma 6.10. For any fixed i in $\{1, \dots, k\}$, we have that $E_i = F_{j(i,1)} \cup \cdots \cup F_{j(i,l_i)}$, for suitable (distinct) $j(i,1), \dots, j(i,l_i)$ from $\{1, 2, \dots, l\}$. Note then that

$$\begin{aligned} L\{M_T(F_{j(i,1)}) + \cdots + M_T(F_{j(i,l_i)})\} &= \boxplus_{h=1}^{l_i} \mathrm{Poiss}^{\boxplus}(\nu(F_{j(i,h)})) \\ &= \mathrm{Poiss}^{\boxplus}\big(\nu(F_{j(i,1)}) + \cdots + \nu(F_{j(i,l_i)})\big) \\ &= \mathrm{Poiss}^{\boxplus}\big(\nu(F_{j(i,1)} \cup \cdots \cup F_{j(i,l_i)})\big) \\ &= \mathrm{Poiss}^{\boxplus}(\nu(E_i)) = L\{M_S(E_i)\}. \end{aligned}$$

In addition, $M_S(E_1), \dots, M_S(E_k)$ are freely independent selfadjoint operators, and, similarly, the operators $\sum_{h=1}^{l_i} M_T(F_{j(i,h)})$, $i = 1, \dots, k$ are freely independent and selfadjoint. Combining these observations with [Vo90, Remark 1.8], it follows that there exists an injective, unital, normal $*$-homomorphism $\iota_{S,T} \colon \mathcal{A}_S \to \mathcal{A}_T$, such that

$$\iota_{S,T}(M_S(E_i)) = M_T(F_{j(i,1)}) + \cdots + M_T(F_{j(i,l_i)}), \qquad (i = 1, 2, \dots, r), \quad (6.8)$$

and such that $\tau_S = \tau_T \circ \iota_{S,T}$. □

Lemma 6.12. *Adapting the notation from Lemmas 6.10-6.11, the system*

$$(\mathcal{A}_S, \tau_S)_{S \in \mathcal{I}}, \quad \{\iota_{S,T} \mid S, T \in \mathcal{I},\ S \leq T\}, \tag{6.9}$$

is a directed system of W^*-algebras and injective, unital, normal $*$-homomorphisms (cf. [KaRi83, Section 11.4]).

Proof. Suppose that $R = (D_1, \dots, D_m)$, $S = (E_1, \dots, E_k)$ and $T = (F_1, \dots, F_l)$ are elements of $\mathcal{I}$, such that $R \leq S \leq T$. We have to show that $\iota_{R,T} = \iota_{S,T} \circ \iota_{R,S}$. We may write (unambiguously),

$$D_h = E_{i(h,1)} \cup \cdots \cup E_{i(h,k_h)}, \qquad (h = 1, \dots, m),$$

$$E_i = F_{j(i,1)} \cup \cdots \cup E_{j(i,l_i)}, \qquad (i = 1, \dots, k),$$

for suitable $i(h,1), \dots, i(h,k_h)$ in $\{1, 2, \dots, k\}$ and $j(i,1), \dots, j(i,l_i)$ in $\{1, 2, \dots, l\}$. Then for any h in $\{1, \dots, m\}$, we have

$$D_h = E_{i(h,1)} \cup \cdots \cup E_{i(h,k_h)} = \Big(\bigcup_{r=1}^{l_{i(h,1)}} F_{j(i(h,1),r)}\Big) \cup \cdots \cup \Big(\bigcup_{r=1}^{l_{i(h,k_h)}} F_{j(i(h,k_h),r)}\Big)$$

so that, by definition of $\iota_{R,T}$, $\iota_{R,S}$ and $\iota_{S,T}$ (cf. (6.8)),

$$\begin{aligned} \iota_{R,T}(D_h) &= \sum_{r=1}^{l_{i(h,1)}} M_T(F_{j(i(h,1),r)}) + \cdots + \sum_{r=1}^{l_{i(h,k_h)}} M_T(F_{j(i(h,k_h),r)}) \\ &= \iota_{S,T}\big[M_S(E_{i(h,1)})\big] + \cdots + \iota_{S,T}\big[M_S(E_{i(h,k_h)})\big] \\ &= \iota_{S,T}\big[M_S(E_{i(h,1)}) + \cdots + M_S(E_{i(h,k_h)})\big] \\ &= \iota_{S,T}\big[\iota_{R,S}(D_h)\big]. \end{aligned}$$

Since $\mathcal{A}_R$ is generated, as a von Neumann algebra, by the operators

$$M_R(D_1), \dots, M_R(D_m),$$

and since $\iota_{R,T}$ and $\iota_{S,T} \circ \iota_{R,S}$ are both normal $*$-homomorphisms, it follows by Kaplansky's density theorem (cf. [KaRi83, Theorem 5.3.5]) and the calculation above that $\iota_{R,T} = \iota_{S,T} \circ \iota_{R,S}$, as desired. □

Lemma 6.13. *Let $\mathcal{A}^0$ denote the C^*-inductive limit of the directed system (6.9) and let $\iota_S \colon \mathcal{A}_S \to \mathcal{A}^0$ denote the canonical embedding of $\mathcal{A}_S$ into $\mathcal{A}^0$ (cf. [KaRi83, Proposition 11.4.1]). Then there is a unique tracial state τ^0 on $\mathcal{A}^0$, satisfying that*

$$\tau_S = \tau^0 \circ \iota_S, \qquad \text{for all } S \text{ in } \mathcal{I}. \tag{6.10}$$

Proof. Recall that the canonical embeddings $\iota_S \colon \mathcal{A}_S \to \mathcal{A}^0$ $(S \in \mathcal{I})$ satisfy the condition:

$$\iota_R = \iota_S \circ \iota_{R,S}, \qquad \text{whenever } R, S \in \mathcal{I} \text{ and } R \le S.$$

We note first that (6.10) gives rise to a well-defined mapping τ^0 on the set $\mathcal{A}^{00} = \cup_{S \in \mathcal{I}} \iota_S(\mathcal{A}_S)$. Indeed, suppose that $\iota_S(a') = \iota_T(a'')$ for some S, T in $\mathcal{I}$ and $a' \in \mathcal{A}_S$, $a'' \in \mathcal{A}_T$. We need to show that $\tau_S(a') = \tau_T(a'')$. Let $S \vee T$ denote the tuple in $\mathcal{I}$ consisting of all non-empty sets of the form $E \cap F$, where $E \in S$ and $F \in T$. Note that $S, T \le S \vee T$. Since $\iota_S = \iota_{S\vee T} \circ \iota_{S,S\vee T}$ and $\iota_T = \iota_{S\vee T} \circ \iota_{T,S\vee T}$, it follows, by injectivity of $\iota_{S\vee T}$, that $\iota_{S,S\vee T}(a') = \iota_{T,S\vee T}(a'')$. Hence, by Lemma 6.11,

$$\tau_S(a') = \tau_{S\vee T} \circ \iota_{S,S\vee T}(a') = \tau_{S\vee T} \circ \iota_{T,S\vee T}(a'') = \tau_T(a''),$$

as desired. Now, given a, b in $\mathcal{A}^{00}$, we can find S from $\mathcal{I}$, such that a, b are both in $\iota_S(\mathcal{A}_S)$, and hence it follows immediately that τ^0 is a linear tracial functional on the vector space $\mathcal{A}^{00}$. Furthermore, if $a = \iota_S(a')$ for some a' in $\mathcal{A}_S$, then

$$|\tau^0(a)| = |\tau_S(a')| \le \|a'\| = \|\iota_S(a')\| = \|a\|,$$

so that τ^0 is norm decreasing. Since $\mathcal{A}^{00}$ is norm dense in $\mathcal{A}^0$ (cf. [KaRi83, Proposition 11.4.1]), if follows then that τ^0 has a unique extension to a mapping $\tau^0 \colon \mathcal{A}^0 \to \mathbb{C}$, which is automatically linear, tracial and norm-decreasing. In addition, $\tau^0(\mathbf{1}_{\mathcal{A}^0}) = 1 = \|\tau^0\|$, so, altogether, it follows that τ^0 is a tracial state on $\mathcal{A}^0$, satisfying (6.10). □

Lemma 6.14. *Let $(\mathcal{A}^0, \tau^0)$ be as in Lemma 6.13. There exists a mapping $M^0 \colon \mathcal{E}_0 \to \mathcal{A}^0_+$, which satisfies conditions (i)-(iii) of Definition 6.7.*

Proof. We define M^0 by the equation:

$$M^0(E) = \iota_{\{E\}}(M_{\{E\}}(E)), \qquad (E \in \mathcal{E}_0).$$

Then $M^0(E)$ is positive for each E in $\mathcal{E}_0$, since $\iota_{\{E\}}$ is a $*$-homomorphism. Note also that if $E \in \mathcal{E}_0$ and $S \in \mathcal{I}$ such that $E \in S$, then $\{E\} \leq S$ and

$$M^0(E) = \iota_{\{E\}}(M_{\{E\}}(E)) = \iota_S \circ \iota_{\{E\},S}(M_{\{E\}}(E)) = \iota_S(M_S(E)). \quad (6.11)$$

We now have

(i) For each E in $\mathcal{E}_0$, we have that $\tau_{\{E\}} = \tau^0 \circ \iota_{\{E\}}$, and hence, since $\iota_{\{E\}}$ is a $*$-homomorphism, $M_{\{E\}}(E)$ and $M^0(E)$ have the same moments with respect to $\tau_{\{E\}}$ and τ^0, respectively. Since both operators are bounded, this implies that $L\{M^0(E)\} = L\{M_{\{E\}}(E)\} = \text{Poiss}^{\boxplus}(\nu(E))$.
(ii) Let $E_1, \dots, E_k$ be disjoint sets from $\mathcal{E}_0$ and consider the tuple $S = (E_1, \dots, E_k) \in \mathcal{I}$. Then, since $\tau_S = \tau^0 \circ \iota_S$ and ι_S is a $*$-homomorphism, we find, using (6.11),

$$\tau^0\big(M^0(E_{i_1})M^0(E_{i_2}) \cdots M^0(E_{i_p})\big) = \tau_S\big(M_S(E_{i_1})M_S(E_{i_2}) \cdots M_S(E_{i_p})\big),$$

for any $i_1, \dots, i_p$ in $\{1, 2, \dots, k\}$. Since $M_S(E_1), \dots, M_S(E_k)$ are freely independent, this implies that so are $M^0(E_1), \dots, M^0(E_k)$.
(iii) Let $E_1, \dots, E_k$ be disjoint sets from $\mathcal{E}_0$, put $E = \cup_{i=1}^k E_i$ and consider the tuple $S = (E_1, \dots, E_k) \in \mathcal{I}$. Then, by definition of $\iota_{\{E\},S}$, we have

$$\begin{aligned} M^0(E) &= \iota_{\{E\}}(M_{\{E\}}(E)) = \iota_S \circ \iota_{\{E\},S}(M_{\{E\}}(E)) \\ &= \iota_S\big(M_S(E_1) + \cdots + M_S(E_k)\big) \\ &= \iota_S(M_S(E_1)) + \cdots + \iota_S(M_S(E_k)) \\ &= M^0(E_1) + \cdots + M^0(E_k). \end{aligned}$$

This concludes the proof. □

Lemma 6.15. *Let $(\mathcal{A}^0, \tau^0)$ be as in Lemma 6.13, let $\Phi^0 : \mathcal{A}^0 \to \mathcal{B}(\mathcal{H}^0)$ denote the GNS representation*[9] *of $\mathcal{A}^0$ associated to τ^0, and let $\mathcal{A}$ be the closure of $\Phi^0(\mathcal{A}^0)$ in $\mathcal{B}(\mathcal{H}^0)$ with respect to the weak operator topology. Let, further, ξ^0 denote the unit vector in $\mathcal{H}^0$, which corresponds to the unit $\mathbf{1}_{\mathcal{A}^0}$ via the GNS-construction, and let τ denote the vector state on $\mathcal{A}$ given by ξ^0. Then $(\mathcal{A}, \tau)$ is a W^*-probability space, and $\tau^0 = \tau \circ \Phi^0$.*

Proof. It follows immediately from the GNS-construction that

$$\tau^0 = \tau \circ \Phi^0, \quad (6.12)$$

so we only have to prove that τ is a faithful trace on $\mathcal{A}$. To see that τ is a trace, note that since τ^0 is a trace, it follows from (6.12) that τ is a trace on the weakly dense C^*-subalgebra $\Phi^0(\mathcal{A}^0)$ of $\mathcal{A}$. Since the multiplication of operators

[9] GNS stands for Gelfand-Naimark-Segal; see [KaRi83, Theorem 4.5.2].

is separately continuous in each variable in the weak operator topology, and since τ is a vector state, we may subsequently conclude that $\tau(ab) = \tau(ba)$ whenever, say, $a \in \mathcal{A}$ and $b \in \Phi^0(\mathcal{A}^0)$. Repeating the argument just given, it follows that τ is a trace on all of $\mathcal{A}$. This means, furthermore, that ξ^0 is a generating trace vector for $\mathcal{A}$, and hence, by [KaRi83, Lemma 7.2.14], it is also a generating trace vector for the commutant $\mathcal{A}' \subseteq \mathcal{B}(\mathcal{H}^0)$. This implies, in particular, that ξ^0 is separating for $\mathcal{A}$ (cf. [KaRi83, Corollary 5.5.12]), which, in turn, implies that τ is faithful on $\mathcal{A}$. □

Proof of Theorem 6.9. Let Φ^0 and $(\mathcal{A}, \tau)$ be as in Lemma 6.15. We then define the mapping $M\colon \mathcal{E}_0 \to \mathcal{A}_+$ by setting

$$M(E) = \Phi^0(M^0(E)), \qquad (E \in \mathcal{E}_0).$$

Now, Φ^0 is a $*$-homomorphism and $\tau^0 = \tau \circ \Phi^0$, so Φ^0 preserves all (mixed) moments of the elements $M^0(E)$, $E \in \mathcal{E}_0$. Since M^0 satisfies conditions (i)-(iii) of Definition 6.7, it follows thus, using the same line of argumentation as in the proof of Lemma 6.14, that M satisfies conditions (i)-(iii) too. Consequently, M is a free Poisson random measure on $(\Theta, \mathcal{E}, \nu)$ with values in $(\mathcal{A}, \tau)$. □

6.4 Integration with Respect to Free Poisson Random Measures

Throughout this section, we consider a free Poisson random measure M on the σ-finite measure space $(\Theta, \mathcal{E}, \nu)$ and with values in the W^*-probability space $(\mathcal{A}, \tau)$. We consider also a classical Poisson random measure N on $(\Theta, \mathcal{E}, \nu)$ defined on a classical probability space $(\Omega, \mathcal{F}, P)$. The aim of this section is to establish a theory of integration with respect to M, making sense, thus, to the integral $\int_\Theta f \,\mathrm{d}M$ for any function f in $\mathcal{L}^1(\Theta, \mathcal{E}, \nu)$. As in most theories of integration, we start by defining integration for simple ν-integrable functions.

Definition 6.16. *Let s be a real-valued simple function in $\mathcal{L}^1(\Theta, \mathcal{E}, \nu)$, i.e. s can be written, unambiguously, in the form*

$$s = \sum_{j=1}^{r} a_j 1_{E_j},$$

where $r \in \mathbb{N}$, $a_1, \ldots, a_r$ are distinct numbers in $\mathbb{R} \setminus \{0\}$ and $E_1, \ldots, E_r$ are disjoint sets from $\mathcal{E}_0$ (since s is ν-integrable). We then define the integral $\int_\Theta s \,\mathrm{d}M$ of s with respect to M as follows:

$$\int_\Theta s \,\mathrm{d}M = \sum_{j=1}^{r} a_j M(E_j) \in \mathcal{A}.$$

Remark 6.17. (a) Since $M(E) \in \mathcal{A}_+$ for any E in $\mathcal{E}_0$, it follows immediately from Definition 6.16 that $\int_\Theta s \,\mathrm{d}M$ is a selfadjoint operator in $\mathcal{A}$ for any real-valued simple function s in $\mathcal{L}^1(\Theta, \mathcal{E}, \mu)$.

(b) Suppose s and t are real-valued simple functions in $\mathcal{L}^1(\Theta,\mathcal{E},\nu)$ and that $c \in \mathbb{R}$. Then $s+t$ and $c\cdot s$ are clearly simple functions too, and, using standard arguments, it is not hard to see that

$$\int_\Theta (s+t)\,\mathrm{d}M = \int_\Theta s\,\mathrm{d}M + \int_\Theta t\,\mathrm{d}M, \quad \text{and} \quad \int_\Theta c\cdot s\,\mathrm{d}M = c\int_\Theta s\,\mathrm{d}M.$$

(c) Consider now, in addition, the classical Poisson random measure N on $(\Theta,\mathcal{E},\nu)$, defined on $(\Omega,\mathcal{F},P)$. Let, further, s be a real-valued simple function in $\mathcal{L}^1(\Theta,\mathcal{E},\nu)$. Then $L\{\int_\Theta s\,\mathrm{d}N\} \in \mathcal{ID}(*)$, $L\{\int_\Theta s\,\mathrm{d}M\} \in \mathcal{ID}(\boxplus)$, and

$$\Lambda\Big(L\Big\{\int_\Theta s\,\mathrm{d}N\Big\}\Big) = L\Big\{\int_\Theta s\,\mathrm{d}M\Big\},$$

where Λ is the Bercovici-Pata bijection. Indeed, we may write s in the form $s = \sum_{j=1}^r a_j 1_{E_j}$, where $r \in \mathbb{N}$, $a_1,\dots,a_r$ are distinct numbers in $\mathbb{R}\setminus\{0\}$ and $E_1,\dots,E_r$ are disjoint sets from $\mathcal{E}_0$. Then, using the properties of Λ, we find that

$$\begin{aligned} L\Big\{\int_\Theta s\,\mathrm{d}M\Big\} &= L\Big\{\sum_{j=1}^r a_j M(E_j)\Big\} = \boxplus_{j=1}^r D_{a_j}\mathrm{Poiss}^{\boxplus}(\nu(E_j)) \\ &= \boxplus_{j=1}^r D_{a_j}\Lambda\big[\mathrm{Poiss}^*(\nu(E_j))\big] = \Lambda\Big[\ast_{j=1}^r D_{a_j}\mathrm{Poiss}^*(\nu(E_j))\Big] \\ &= \Lambda\Big[L\Big\{\sum_{j=1}^r a_j N(E_j)\Big\}\Big] = \Lambda\Big[L\Big\{\int_\Theta s\,\mathrm{d}N\Big\}\Big]. \end{aligned}$$

By $\mathcal{L}^1(\Theta,\mathcal{E},\nu)_+$, we denote the set of positive functions from $\mathcal{L}^1(\Theta,\mathcal{E},\nu)$.

Proposition 6.18. *Let f be a real-valued function in $\mathcal{L}^1(\Theta,\mathcal{E},\nu)$, and choose a sequence (s_n) of real-valued simple $\mathcal{E}$-measurable functions, satisfying the conditions:*

$$\exists h \in \mathcal{L}^1(\Theta,\mathcal{E},\nu)_+ \ \forall \theta \in \Theta \ \forall n \in \mathbb{N}\colon |s_n(\theta)| \le h(\theta), \tag{6.13}$$

and

$$\lim_{n\to\infty} s_n(\theta) = f(\theta), \qquad (\theta \in \Theta). \tag{6.14}$$

Then $s_n \in \mathcal{L}^1(\Theta,\mathcal{E},\nu)$ for all n, and the integrals $\int_\Theta s_n\,\mathrm{d}M$ converge in probability to a selfadjoint (possibly unbounded) operator $I(f)$ affiliated with $\mathcal{A}$.

Furthermore, the limit $I(f)$ is independent of the choice of approximating sequence (s_n) of simple functions (subject to conditions (6.13) *and* (6.14)*).*

In condition (6.13), we might have taken $h = |f|$, but it is convenient to allow for more general dominators.

Proof of Proposition 6.18. Let f, (s_n) and h be as set out in the proposition. Then, for any n in $\mathbb{N}$, $\int_\Theta |s_n|\,\mathrm{d}\nu \le \int_\Theta h\,\mathrm{d}\nu < \infty$, so that $s_n \in \mathcal{L}^1(\Theta,\mathcal{E},\nu)$ and

$\int_\Theta s_n \, \mathrm{d}M$ is well-defined. Note further that for any n, m in $\mathbb{N}$, $s_n - s_m$ is again a simple function in $\mathcal{L}^1(\Theta, \mathcal{E}, \nu)$, and, using Remark 6.17(c),(d), it follows that

$$\begin{aligned} L\Big\{\int_\Theta s_n \, \mathrm{d}M - \int_\Theta s_m \, \mathrm{d}M\Big\} &= L\Big\{\int_\Theta (s_n - s_m) \, \mathrm{d}M\Big\} \\ &= \Lambda\Big[L\Big\{\int_\Theta (s_n - s_m) \, \mathrm{d}N\Big\}\Big], \end{aligned} \tag{6.15}$$

with N the classical Poisson random measure introduced before. Since $h \in \mathcal{L}^1(\Theta, \mathcal{E}, \nu)$, it follows from Proposition 2.8 that $h \in \mathcal{L}^1(\Theta, \mathcal{E}, N(\cdot, \omega))$ for almost all ω in Ω. Hence, by Lebesgue's theorem on dominated convergence, we have that

$$\int_\Theta s_n(\theta) \, N(\mathrm{d}\theta, \omega) \longrightarrow \int_\Theta f(\theta) \, N(\mathrm{d}\theta, \omega), \quad \text{as } n \to \infty,$$

for almost all ω in Ω. In other words, $\int_\Theta s_n \, \mathrm{d}N \to \int_\Theta f \, \mathrm{d}N$, almost surely, as $n \to \infty$. In particular $\int_\Theta s_n \, \mathrm{d}N \to \int_\Theta f \, \mathrm{d}N$, in probability as $n \to \infty$, so the sequence $(\int_\Theta s_n \, \mathrm{d}N)_{n \in \mathbb{N}}$ is a Cauchy sequence in probability, i.e.

$$L\Big\{\int_\Theta (s_n - s_m) \, \mathrm{d}N\Big\} \xrightarrow{\text{w}} \delta_0, \quad \text{as } n, m \to \infty.$$

Combining this with (6.15) and the continuity of Λ (cf. Corollary 5.14), it follows that $(\int_\Theta s_n \, \mathrm{d}M)_{n \in \mathbb{N}}$ is also a Cauchy sequence in probability, i.e. with respect to the measure topology. Since $\overline{\mathcal{A}}$ is complete in the measure topology (cf. Proposition A.5), there exists, thus, an operator $I(f)$ in $\overline{\mathcal{A}}$, such that $\int_\Theta s_n \, \mathrm{d}M \to I(f)$, in probability as $n \to \infty$. Since $\int_\Theta s_n \, \mathrm{d}M$ is selfadjoint for each n, and since the adjoint operation is continuous in the measure topology, $I(f)$ is a selfadjoint operator in $\overline{\mathcal{A}}$.

Suppose, finally, that (t_n) is another sequence of simple real-valued $\mathcal{E}$-measurable functions satisfying conditions (6.13) and (6.14) (with s_n replaced by t_n). Then, by the argument given above, $\int_\Theta t_n \, \mathrm{d}M \to I'(f)$, in probability as $n \to \infty$, for some selfadjoint operator $I'(f)$ in $\overline{\mathcal{A}}$. Consider now the mixed sequence (u_n) of simple real-valued $\mathcal{E}$-measurable functions given by:

$$u_1 = s_1, u_2 = t_1, u_3 = s_2, u_4 = t_2, \ldots,$$

and note that this sequence satisfies (6.13) and (6.14) too, so that $\int_\Theta u_n \, \mathrm{d}M \to I''(f)$, in probability as $n \to \infty$, for some selfadjoint operator $I''(f)$ in $\overline{\mathcal{A}}$. Now the subsequence (u_{2n-1}) converges in probability to both $I''(f)$ and $I(f)$ as $n \to \infty$, and the subsequence (u_{2n}) converges in probability to both $I''(f)$ and $I'(f)$ as $n \to \infty$. Since the measure topology is a Hausdorff topology, we may conclude, thus, that $I(f) = I''(f) = I'(f)$. This completes the proof. □

Definition 6.19. *Let f be a real-valued function in $\mathcal{L}^1(\Theta, \mathcal{E}, \nu)$, and let $I(f)$ be the selfadjoint operator in $\overline{\mathcal{A}}$ described in Proposition 6.18. We call $I(f)$ the integral of f with respect to M and denote it by $\int_\Theta f \, \mathrm{d}M$.*

Corollary 6.20. *Let M and N be the free and classical Poisson random measures on $(\Theta, \mathcal{E}, \nu)$ introduced above. Then for any f in $\mathcal{L}^1(\Theta, \mathcal{E}, \nu)$, we have $L\{\int_\Theta f \, \mathrm{d}N\} \in \mathcal{ID}(*)$, $L\{\int_\Theta f \, \mathrm{d}M\} \in \mathcal{ID}(\boxplus)$ and*

$$\Lambda\Big(L\Big\{\int_\Theta f \, \mathrm{d}N\Big\}\Big) = L\Big\{\int_\Theta f \, \mathrm{d}M\Big\}.$$

Proof. Choose a sequence (s_n) of real-valued simple $\mathcal{E}$-measurable functions satisfying conditions (6.13) and (6.14) of Proposition 6.18. Then, by Remark 6.17, $L\{\int_\Theta s_n \, \mathrm{d}N\} \in \mathcal{ID}(*)$, $L\{\int_\Theta s_n \, \mathrm{d}M\} \in \mathcal{ID}(\boxplus)$ and $\Lambda(L\{\int_\Theta s_n \mathrm{d}N\}) = L\{\int_\Theta s_n \, \mathrm{d}M\}$ for all n in $\mathbb{N}$. Furthermore,

$$\int_\Theta s_n \, \mathrm{d}N \xrightarrow{\text{a.s.}} \int_\Theta f \, \mathrm{d}N \quad \text{and} \quad \int_\Theta s_n \, \mathrm{d}M \xrightarrow{\text{p}} \int_\Theta f \, \mathrm{d}M, \quad \text{as } n \to \infty.$$

In particular (cf. Proposition A.9),

$$L\Big\{\int_\Theta s_n \, \mathrm{d}N\Big\} \xrightarrow{\text{w}} L\Big\{\int_\Theta f \, \mathrm{d}N\Big\} \quad \text{and} \quad L\Big\{\int_\Theta s_n \, \mathrm{d}M\Big\} \xrightarrow{\text{w}} L\Big\{\int_\Theta f \, \mathrm{d}M\Big\},$$

as $n \to \infty$. Since $\mathcal{ID}(*)$ and $\mathcal{ID}(\boxplus)$ are both closed with respect to weak convergence (see Section 4.5), this implies that $L\{\int_\Theta f \, \mathrm{d}N\} \in \mathcal{ID}(*)$ and $L\{\int_\Theta f \, \mathrm{d}M\} \in \mathcal{ID}(\boxplus)$. Furthermore, by continuity of Λ, $\Lambda(L\{\int_\Theta f \, \mathrm{d}N\}) = L\{\int_\Theta f \, \mathrm{d}M\}$. □

Proposition 6.21. *For any real-valued functions f, g in $\mathcal{L}^1(\Theta, \mathcal{E}, \nu)$ and any real number c, we have that*

$$\int_\Theta (f+g) \, \mathrm{d}M = \int_\Theta f \, \mathrm{d}M + \int_\Theta g \, \mathrm{d}M \quad \textit{and} \quad \int_\Theta c \cdot f \, \mathrm{d}M = c \int_\Theta f \, \mathrm{d}M.$$

Proof. If f and g are simple functions, this was noted in Remark 6.17. The general case follows by approximating f and g by simple functions as in Proposition 6.18 and using that addition and scalar-multiplication are continuous operations in the measure topology (cf. Proposition A.5). □

Proposition 6.22. *Let M be a free Poisson random measure on the σ-finite measure space $(\Theta, \mathcal{E}, \nu)$ with values in the W^*-probability space $(\mathcal{A}, \tau)$. Let, further, $f_1, f_2, \ldots, f_r$ be real-valued functions in $\mathcal{L}^1(\Theta, \mathcal{E}, \nu)$ and let $\Theta_1, \Theta_2, \ldots, \Theta_r$ be disjoint $\mathcal{E}$-measurable subsets of Θ. Then the integrals*

$$\int_{\Theta_1} f_1 \, \mathrm{d}M, \int_{\Theta_2} f_2 \, \mathrm{d}M, \ldots, \int_{\Theta_r} f_r \, \mathrm{d}M,$$

are freely independent selfadjoint operators affiliated with $(\mathcal{A}, \tau)$.

Proof. For each j in $\{1,2,\dots,r\}$, let $(s_{j,n})_{n\in\mathbb{N}}$ be a sequence of real valued simple $\mathcal{E}$-measurable functions, such that

$$|s_{j,n}(\theta)| \le |f_j(\theta)|, \qquad (\theta\in\Theta,\ n\in\mathbb{N}),$$

and

$$\lim_{n\to\infty} s_{j,n}(\theta) = f_j(\theta), \qquad (\theta\in\Theta).$$

Then, for each j in $\{1,2,\dots,r\}$ and each n in $\mathbb{N}$, we may write $s_{j,n}\cdot 1_{\Theta_j}$ in the form:

$$s_{j,n}\cdot 1_{\Theta_j} = \sum_{l=1}^{k_{j,n}} \alpha(l,j,n) 1_{A(l,j,n)},$$

where $\alpha(1,j,n),\dots,\alpha(k_{j,n},j,n)\in\mathbb{R}\setminus\{0\}$ and $A(1,j,n),\dots,A(k_{j,n},j,n)$ are disjoint sets from $\mathcal{E}_0$, such that $A(l,j,n)\subseteq\Theta_j$ for all l. Now,

$$\int_\Theta s_{j,n}\cdot 1_{\Theta_j}\,\mathrm{d}M = \sum_{l=1}^{k_{j,n}} \alpha(l,j,n)M((A(l,j,n)), \qquad (j=1,2,\dots,r,\ n\in\mathbb{N}),$$

so by the properties of free Poisson random measures, the integrals

$$\int_\Theta s_{1,n}\cdot 1_{\Theta_1}\,\mathrm{d}M,\dots,\int_\Theta s_{r,n}\cdot 1_{\Theta_r}\,\mathrm{d}M,$$

are freely independent for each n in $\mathbb{N}$. Finally, for each j in $\{1,2,\dots,r\}$ we have (cf. Proposition 6.18)

$$\int_{\Theta_j} f_j\,\mathrm{d}M = \int_\Theta f_j\cdot 1_{\Theta_j}\,\mathrm{d}M = \lim_{n\to\infty}\int_\Theta s_{j,n}\cdot 1_{\Theta_j}\,\mathrm{d}M,$$

where the limit is taken in probability. Taking now Proposition 4.7 into account, we obtain the desired conclusion. □

6.5 The Free Lévy-Itô Decomposition

In this section we derive the free version of the Lévy-Itô decomposition. We mention in passing the related decomposition of free white noises, which was established in [GlScSp92].

Throughout this section we put

$$\mathsf{H} =]0,\infty[\times\mathbb{R}\subseteq\mathbb{R}^2,$$

and we denote by $\mathsf{B}(\mathsf{H})$ the set of all Borel subsets of H. Furthermore, for any ϵ, t in $]0,\infty[$, such that $\epsilon < t$, we put

$$D(\epsilon,\infty) = \{s\in\mathbb{R} \mid \epsilon < |s| < \infty\} = \mathbb{R}\setminus[-\epsilon,\epsilon],$$

$$D(\epsilon,t) = \{s\in\mathbb{R} \mid \epsilon < |s| \le t\} = [-t,t]\setminus[-\epsilon,\epsilon].$$

We shall need the following well-known result about classical Poisson random measures.

Lemma 6.23. *Let ν be a Lévy measure on $\mathbb{R}$ and consider the σ-finite measure $\mathrm{Leb}\otimes\nu$ on H. Consider further a (classical) Poisson random measure N on $(\mathsf{H},\mathsf{B}(\mathsf{H}),\mathrm{Leb}\otimes\nu)$, defined on some probability space $(\Omega,\mathcal{F},P)$.*

Then there is a subset Ω_0 of Ω, such that $\Omega_0\in\mathcal{F}$, $P(\Omega_0)=1$ and such that the following holds for any ω in Ω_0: For any ϵ,t in $]0,\infty[$, the restriction $[N(\cdot,\omega)]_{]0,t]\times D(\epsilon,\infty)}$ of the measure $N(\cdot,\omega)$ to the set $]0,t]\times D(\epsilon,\infty)$ is supported on a finite number of points, each of which has mass 1.

Proof. See [Sa99, Lemma 20.1] □

Lemma 6.24. *Let ν and N be as in Lemma 6.23, and consider a positive Borel function $\varphi\colon\mathbb{R}\to[0,\infty[$.*

(i) *For almost all ω in Ω, the following holds:*

$$\forall\epsilon>0\ \forall 0\le s<t\colon\int_{]s,t]\times D(\epsilon,\infty)}\varphi(x)\,N(\mathrm{d}u,\mathrm{d}x,\omega)<\infty.$$

(ii) *If $\int_{[-1,1]}\varphi(x)\,\nu(\mathrm{d}x)<\infty$, then for almost all ω in Ω, the following holds:*

$$\forall 0\le s<t\colon\int_{]s,t]\times\mathbb{R}}\varphi(x)\,N(\mathrm{d}u,\mathrm{d}x,\omega)<\infty.$$

Proof. Since φ is positive, it suffices to consider the case $s=0$ in (i) and (ii). Moreover, since φ only takes finite values, statement (i) follows immediately from Lemma 6.23.

To prove (ii), assume that $\int_{[-1,1]}\varphi(x)\,\nu(\mathrm{d}x)<\infty$. By virtue of (i), it suffices then to prove, for instance, that for almost all ω in Ω, the following holds:

$$\forall t>0\colon\int_{]0,t]\times[-1,1]}\varphi(x)\,N(\mathrm{d}u,\mathrm{d}x,\omega)<\infty.\tag{6.16}$$

Since the integrals in (6.16) increase with t, it suffices to prove that for any fixed t in $]0,\infty[$,

$$\int_{]0,t]\times[-1,1]}\varphi(x)\,N(\mathrm{d}u,\mathrm{d}x,\omega)<\infty,\quad\text{for almost all }\omega.$$

This, in turn, follows immediately from the following calculation:

$$\begin{aligned}\mathbb{E}\Big\{\int_{]0,t]\times[-1,1]}\varphi(x)\,N(\mathrm{d}u,\mathrm{d}x)\Big\}&=\int_{]0,t]\times[-1,1]}\varphi(x)\,\mathrm{Leb}\otimes\nu(\mathrm{d}u,\mathrm{d}x)\\&=t\int_{[-1,1]}\varphi(x)\,\nu(\mathrm{d}x)<\infty,\end{aligned}$$

where we have used Proposition 2.8. □

Lemma 6.25. *Let ν be a Lévy measure on $\mathbb{R}$, and let M be a Free Poisson random measure on $(\mathsf{H}, \mathsf{B}(\mathsf{H}), \mathrm{Leb} \otimes \nu)$ with values in the W^*-probability space $(\mathcal{A}, \tau)$. Let, further, N be a (classical) Poisson random measure on $(\mathsf{H}, \mathsf{B}(\mathsf{H}), \mathrm{Leb} \otimes \nu)$, defined on a classical probability space $(\Omega, \mathcal{F}, P)$.*

(i) *For any ϵ, s, t in $[0, \infty[$, such that $s < t$ and $\epsilon > 0$, the integrals*

$$\int_{]s,t] \times D(\epsilon,n)} x\, M(\mathrm{d}u, \mathrm{d}x), \qquad (n \in \mathbb{N}),$$

converge in probability, as $n \to \infty$, to some (possibly unbounded) selfadjoint operator affiliated with $\mathcal{A}$, which we denote by $\int_{]s,t] \times D(\epsilon,\infty)} x\, M(\mathrm{d}u, \mathrm{d}x)$. Furthermore (cf. Lemma 6.24), $L\{\int_{]s,t] \times D(\epsilon,\infty)} x\, N(\mathrm{d}u, \mathrm{d}x)\} \in \mathcal{ID}(*)$, $L\{\int_{]s,t] \times D(\epsilon,\infty)} x\, M(\mathrm{d}u, \mathrm{d}x)\} \in \mathcal{ID}(\boxplus)$ and

$$L\Big\{\int_{]s,t] \times D(\epsilon,\infty)} x\, M(\mathrm{d}u, \mathrm{d}x)\Big\} = \Lambda\Big(L\Big\{\int_{]s,t] \times D(\epsilon,\infty)} x\, N(\mathrm{d}s, \mathrm{d}x)\Big\}\Big). \tag{6.17}$$

(ii) *If $\int_{[-1,1]} |x|\, \nu(\mathrm{d}x) < \infty$, then for any s, t in $[0, \infty[$, such that $s < t$, the integrals*

$$\int_{]s,t] \times [-n,n]} x\, M(\mathrm{d}u, \mathrm{d}x), \qquad (n \in \mathbb{N}),$$

converge in probability, as $n \to \infty$, to some (possibly unbounded) selfadjoint operator affiliated with $\mathcal{A}$, which we denote by $\int_{]s,t] \times \mathbb{R}} x\, M(\mathrm{d}u, \mathrm{d}x)$. Furthermore (cf. Lemma 6.24),

$$L\Big\{\int_{]s,t] \times \mathbb{R}} x\, N(\mathrm{d}u, \mathrm{d}x)\Big\} \in \mathcal{ID}(*), \quad L\Big\{\int_{]s,t] \times \mathbb{R}} x\, M(\mathrm{d}u, \mathrm{d}x)\Big\} \in \mathcal{ID}(\boxplus)$$

and

$$L\Big\{\int_{]s,t] \times \mathbb{R}} x\, M(\mathrm{d}u, \mathrm{d}x)\Big\} = \Lambda\Big(L\Big\{\int_{]s,t] \times \mathbb{R}} x\, N(\mathrm{d}s, \mathrm{d}x)\Big\}\Big).$$

Proof. (i) Note first that for any n in $\mathbb{N}$ and any ϵ, s, t in $[0, \infty[$, such that $s < t$ and $\epsilon > 0$, we have that

$$\int_{]s,t] \times D(\epsilon,n)} |x|\, \mathrm{Leb} \otimes \nu(\mathrm{d}u, \mathrm{d}x) = (t - s) \int_{D(\epsilon,n)} |x|\, \nu(\mathrm{d}x) < \infty,$$

since ν is a Lévy measure. Hence, by application of Proposition 6.18, the integral $\int_{]s,t] \times D(\epsilon,n)} x\, M(\mathrm{d}u, \mathrm{d}x)$ is well-defined and furthermore, by Corollary 6.20,

$$L\Big\{\int_{]s,t] \times D(\epsilon,n)} x\, M(\mathrm{d}u, \mathrm{d}x)\Big\} = \Lambda\Big(L\Big\{\int_{]s,t] \times D(\epsilon,n)} x\, N(\mathrm{d}u, \mathrm{d}x)\Big\}\Big). \tag{6.18}$$

Note now that by Lemma 6.24(i) there is a subset Ω_0 of Ω, such that $\Omega_0 \in \mathcal{F}$, $P(\Omega_0) = 1$ and

$$\int_{]s,t]\times D(\epsilon,\infty)} |x|\, N(\mathrm{d}u, \mathrm{d}x, \omega) < \infty, \quad \text{for all } \omega \text{ in } \Omega_0.$$

Then $\int_{]s,t]\times D(\epsilon,\infty)} x\, N(\mathrm{d}u, \mathrm{d}x, \omega)$ is well-definedforall ω in Ω_0 andbyLebesgue's theorem on dominated convergence,

$$\int_{]s,t]\times D(\epsilon,n)} x\, N(\mathrm{d}u, \mathrm{d}x, \omega) \underset{n\to\infty}{\longrightarrow} \int_{]s,t]\times D(\epsilon,\infty)} x\, N(\mathrm{d}u, \mathrm{d}x, \omega),$$

for all ω in Ω_0, i.e. almost surely. In particular

$$\int_{]s,t]\times D(\epsilon,n)} x\, N(\mathrm{d}u, \mathrm{d}x) \underset{n\to\infty}{\longrightarrow} \int_{]s,t]\times D(\epsilon,\infty)} x\, N(\mathrm{d}u, \mathrm{d}x), \quad \text{in probability},$$

and hence $(\int_{]s,t]\times D(\epsilon,n)} x\, N(\mathrm{d}u, \mathrm{d}x))_{n\in\mathbb{N}}$ is a Cauchy sequence in probability. Now, for any n, m in $\mathbb{N}$, such that $n \le m$, we have, by Proposition 6.21 and Corollary 6.20,

$$\begin{aligned}
&L\Big\{\int_{]s,t]\times D(\epsilon,m)} x\, M(\mathrm{d}u, \mathrm{d}x) - \int_{]s,t]\times D(\epsilon,n)} x\, M(\mathrm{d}u, \mathrm{d}x)\Big\} \\
&\quad = L\Big\{\int_{]s,t]\times D(n,m)} x\, M(\mathrm{d}u, \mathrm{d}x)\Big\} \\
&\quad = \Lambda\Big(L\Big\{\int_{]s,t]\times D(n,m)} x\, N(\mathrm{d}u, \mathrm{d}x)\Big\}\Big) \\
&\quad = \Lambda\Big(L\Big\{\int_{]s,t]\times D(\epsilon,m)} x\, N(\mathrm{d}u, \mathrm{d}x) - \int_{]s,t]\times D(\epsilon,n)} x\, N(\mathrm{d}u, \mathrm{d}x)\Big\}\Big).
\end{aligned}$$

By continuity of Λ, this shows that $(\int_{]s,t]\times D(\epsilon,n)} x\, M(\mathrm{d}u, \mathrm{d}x))_{n\in\mathbb{N}}$ is a Cauchy sequence in probability, and hence, by completeness of $\overline{\mathcal{A}}$ in the measure topology,

$$\int_{]s,t]\times D(\epsilon,\infty)} x\, M(\mathrm{d}u, \mathrm{d}x) := \lim_{n\to\infty} \int_{]s,t]\times D(\epsilon,n)} x\, M(\mathrm{d}u, \mathrm{d}x),$$

exists in $\overline{\mathcal{A}}$ as the limit in probability.

Finally, since $\mathcal{ID}(*)$ and $\mathcal{ID}(\boxplus)$ are closed with respect to weak convergence, we have that

$$L\Big\{\int_{]s,t]\times D(\epsilon,\infty)} x\, N(\mathrm{d}u, \mathrm{d}x)\Big\} \in \mathcal{ID}(*)$$

and

$$L\Big\{\int_{]s,t]\times D(\epsilon,\infty)} x\, M(\mathrm{d}u,\mathrm{d}x)\Big\} \in \mathcal{ID}(\boxplus).$$

Moreover, since convergence in probability implies convergence in distribution (cf. Proposition A.9), it follows from (6.18) and continuity of Λ that (6.17) holds.

(ii) Suppose $\int_{[-1,1]} |x|\,\nu(\mathrm{d}x) < \infty$. Then for any n in $\mathbb{N}$ and any s,t in $[0,\infty[$, such that $s<t$, we have that

$$\begin{aligned}\int_{]s,t]\times[-n,n]} |x|\,\mathrm{Leb}\otimes\nu(\mathrm{d}u,\mathrm{d}x) &= (t-s)\int_{[-n,n]} |x|\,\nu(\mathrm{d}x)\\ &= (t-s)\Big(\int_{[-1,1]} |x|\,\nu(\mathrm{d}x) + \int_{D(1,n)} |x|\,\nu(\mathrm{d}x)\Big)\\ &< \infty,\end{aligned}$$

since ν is a Lévy measure. Hence, by application of Proposition 6.18, the integral $\int_{]s,t]\times[-n,n]} x\, M(\mathrm{d}u,\mathrm{d}x)$ is well-defined and, by Corollary 6.20,

$$L\Big\{\int_{]s,t]\times[-n,n]} x\, M(\mathrm{d}u,\mathrm{d}x)\Big\} = \Lambda\Big(L\Big\{\int_{]s,t]\times[-n,n]} x\, N(\mathrm{d}u,\mathrm{d}x)\Big\}\Big).$$

From this point on, the proof is exactly the same as that of (i) given above; the only difference being that the application of Lemma 6.24(i) above must be replaced by an application of Lemma 6.24(ii). □

We are now ready to give a proof of the Lévy-Itô decomposition for free Lévy processes (in law). As is customary in the classical case (cf. [Sa99]), we divide the general formulation into two parts.

Theorem 6.26 (Free Lévy-Itô Decomposition I). *Let (Z_t) be a free Lévy process (in law) affiliated with a W^*-probability space $(\mathcal{A},\tau)$, let ν be the Lévy measure appearing in the free generating triplet for $L\{Z_1\}$ and assume that $\int_{-1}^{1} |x|\ \nu(\mathrm{d}x) < \infty$. Then (Z_t) has a representation in the form:*

$$Z_t \stackrel{\mathrm{d}}{=} \gamma t \mathbf{1}_{\mathcal{A}^0} + \sqrt{a} W_t + \int_{]0,t]\times\mathbb{R}} x\ M(\mathrm{d}u,\mathrm{d}x), \qquad (t\ge 0), \tag{6.19}$$

where $\gamma\in\mathbb{R}$, $a\ge 0$, (W_t) is a free Brownian motion in some W^*-probability space $(\mathcal{A}^0,\tau^0)$ (see Example 5.16) and M is a free Poisson random measure on $(\mathsf{H},\mathsf{B}(\mathsf{H}),\mathrm{Leb}\otimes\nu)$ with values in $(\mathcal{A}^0,\tau^0)$. Furthermore, the process

$$U_t := \int_{]0,t]\times\mathbb{R}} x\ M(\mathrm{d}u,\mathrm{d}x), \qquad (t\ge 0),$$

is a free Lévy process (in law), which is freely independent of (W_t), and the right hand side of (6.19), as a whole, is a free Lévy process (in law).

As the symbol $\stackrel{\mathrm{d}}{=}$ appearing in (6.19) just means that the two operators have the same (spectral) distribution, it does not follow directly from (6.19) that the right hand side is a free Lévy process (in law) (contrary to the situation in the classical Lévy-Itô decomposition).

Proof of Theorem 6.26. By Proposition 5.15, we may choose a classical Lévy process (X_t), defined on some probability space $(\Omega, \mathcal{F}, P)$, such that $\Lambda(L\{X_t\}) = L\{Z_t\}$ for all t in $[0,\infty[$. Then ν is the Lévy measure for $L\{X_1\}$, so by the classical Lévy-Itô Theorem (cf. Theorem 2.9), (X_t) has a representation in the form:

$$X_t \stackrel{\text{a.s.}}{=} \gamma t + \sqrt{a} B_t + \int_{]0,t]\times\mathbb{R}} x N(\mathrm{d}u, \mathrm{d}x), \qquad (t \geq 0),$$

where (B_t) is a (classical) Brownian motion on $(\Omega, \mathcal{F}, P)$, N is a (classical) Poisson random measure on $(\mathsf{H}, \mathsf{B}(\mathsf{H}), \mathrm{Leb} \otimes \nu)$, defined on $(\Omega, \mathcal{F}, P)$ and (B_t) and N are independent. Put

$$Y_t := \int_{]0,t]\times\mathbb{R}} x\, N(\mathrm{d}u, \mathrm{d}x), \qquad (t \geq 0).$$

Now choose a free Brownian motion (W_t) in some W^*-probability space $(\mathcal{A}^1, \tau^1)$, and recall that $L\{W_t\} = \Lambda(L\{B_t\})$ for all t. Choose, further, a free Poisson random measure M on $(\mathsf{H}, \mathsf{B}(\mathsf{H}), \mathrm{Leb} \otimes \nu)$ with values in some W^*-probability space $(\mathcal{A}^2, \tau^2)$. Next, let $(\mathcal{A}^0, \tau^0)$ be the (reduced) free product of the two W^*-probability spaces $(\mathcal{A}^1, \tau^1)$ and $(\mathcal{A}^2, \tau^2)$ (cf. [VoDyNi92, Definition 1.6.1]). We may then consider $\mathcal{A}^1$ and $\mathcal{A}^2$ as two freely independent unital W^*-subalgebras of $\mathcal{A}^0$, such that $\tau^0_{|\mathcal{A}^1} = \tau^1$ and $\tau^0_{|\mathcal{A}^2} = \tau^2$. In particular, (W_t) and M are freely independent in $(\mathcal{A}^0, \tau^0)$.

Since $\int_{[-1,1]} |x|\, \nu(\mathrm{d}x) < \infty$, it follows from Lemma 6.25(ii) that for any t in $]0,\infty[$, the integral $U_t = \int_{]0,t]\times\mathbb{R}} x\, M(\mathrm{d}u, \mathrm{d}x)$ is well-defined, and $L\{U_t\} = \Lambda(L\{Y_t\})$. Furthermore, it follows immediately from Definition 6.16, Proposition 6.18 and Lemma 6.25 that for any t in $[0,t[$, $U_t = \int_{]0,t]\times\mathbb{R}} x\, M(\mathrm{d}u, \mathrm{d}x)$ is in the closure of $\mathcal{A}^2$ with respect to the measure topology. As noted in Remark 4.8, the set $\overline{\mathcal{A}^2}$ of closed, densely defined operators affiliated with $\mathcal{A}^2$ is complete (and hence closed) in the measure topology, and therefore U_t is affiliated with $\mathcal{A}^2$ for all t. This implies, in particular, that the two processes (W_t) and (U_t) are freely independent.

Now, for any t in $]0,\infty[$, we have

$$\begin{aligned} L\{\gamma t\mathbf{1}_{\mathcal{A}^0} + \sqrt{a}W_t + U_t\} &= \delta_{\gamma t} \boxplus D_{\sqrt{a}} L\{W_t\} \boxplus L\{U_t\} \\ &= \Lambda(\delta_{\gamma t}) \boxplus D_{\sqrt{a}} \Lambda(L\{B_t\}) \boxplus \Lambda(L\{Y_t\}) \\ &= \Lambda\big(\delta_{\gamma t} * D_{\sqrt{a}} L\{B_t\} * L\{Y_t\}\big) \\ &= \Lambda\big(L\{\gamma t + \sqrt{a}B_t + Y_t\}\big) \\ &= \Lambda(L\{X_t\}) \\ &= L\{Z_t\}, \end{aligned}$$

and this proves (6.19). We prove next that the process (U_t) is a free Lévy process (in law). For this, recall that (Y_t) is a (classical) Lévy process defined on $(\Omega, \mathcal{F}, P)$ (cf. [Sa99, Theorem 19.3]), and such that $L\{U_t\} = \Lambda(L\{Y_t\})$ for all t. Since (Y_t) has stationary increments, we find for any s, t in $[0, \infty[$ that

$$\begin{aligned} L\{U_{s+t} - U_s\} &= L\Big\{\int_{]s,s+t]\times\mathbb{R}} x\, M(\mathrm{d}u, \mathrm{d}x)\Big\} = \Lambda\Big(L\Big\{\int_{]s,s+t]\times\mathbb{R}} x\, N(\mathrm{d}u, \mathrm{d}x)\Big\}\Big) \\ &= \Lambda\big(L\{Y_{s+t} - Y_s\}\big) = \Lambda(L\{Y_t\}) = L\{U_t\}, \end{aligned}$$

where we have used Lemma 6.25(ii). Thus, (U_t) has stationary increments too. Furthermore, by continuity of Λ,

$$L\{U_t\} = \Lambda\big(L\{Y_t\}\big) \xrightarrow{\mathrm{w}} \Lambda(\delta_0) = \delta_0, \quad \text{as } t \searrow 0,$$

so that (U_t) is stochastically continuous. Finally, to prove that (U_t) has freely independent increments, consider r in $\mathbb{N}$ and $t_0, t_1, \ldots, t_r$ in $[0, \infty[$, such that $0 = t_0 < t_1 < \cdots < t_r$. Then for any j in $\{1, 2, \ldots, r\}$ we have (cf. Lemma 6.25) that

$$U_{t_j} - U_{t_{j-1}} = \int_{]t_{j-1},t_j]\times\mathbb{R}} x\, M(\mathrm{d}u, \mathrm{d}x) = \lim_{n\to\infty} \int_{]t_{j-1},t_j]\times[-n,n]} x\, M(\mathrm{d}u, \mathrm{d}x),$$

where the limit is taken in probability. Since

$$\int_{]t_{j-1},t_j]\times[-n,n]} |x| \,\mathrm{Leb} \otimes \nu(\mathrm{d}u, \mathrm{d}x) < \infty$$

for any n in $\mathbb{N}$ and any j in $\{1, 2, \ldots, r\}$, it follows from Proposition 6.22 that for any n in $\mathbb{N}$, the integrals

$$\int_{]t_{j-1},t_j]\times[-n,n]} x\, M(\mathrm{d}u, \mathrm{d}x), \qquad j = 1, 2, \ldots, r,$$

are freely independent operators. Hence, by Proposition 4.7, the increments

$$U_{t_1}, U_{t_2} - U_{t_1}, \ldots, U_{t_r} - U_{t_{r-1}}$$

are also freely independent.

It remains to note that the right hand side of (6.19) is a free Lévy process (in law). This follows immediately from the fact that the sum of two freely independent free Lévy processes (in law) is again a free Lévy process (in law). Indeed, the stochastic continuity condition follows from the fact that addition is a continuous operation in the measure topology, and the remaining conditions are immediate consequences of basic properties of free independence. This concludes the proof. □

Theorem 6.27 (Free Lévy-Itô Decomposition II). *Let (Z_t) be a free Lévy process (in law) affiliated with a W^*-probability space $(\mathcal{A},\tau)$ and let ν be the Lévy measure appearing in the free characteristic triplet for $L\{Z_1\}$. Then (Z_t) has a representation in the form:*

$$Z_t \overset{\mathrm{d}}{=} \eta t \mathbf{1}_{\mathcal{A}^0} + \sqrt{a} W_t + V_t, \qquad (t \geq 0), \tag{6.20}$$

where

- *$\eta \in \mathbb{R}$, $a \geq 0$ and (W_t) is a free Brownian motion in a W^*-probability space $(\mathcal{A}^0,\tau^0)$.*
- *(V_t) is a free Lévy process (in law) given by*

$$V_t := \lim_{\epsilon \searrow 0} \Big[\int_{]0,t]\times D(\epsilon,\infty)} x\, M(\mathrm{d}u,\mathrm{d}x) - \Big(\int_{]0,t]\times D(\epsilon,1)} x\, \mathrm{Leb}\otimes\nu(\mathrm{d}u,\mathrm{d}x) \Big) \mathbf{1}_{\mathcal{A}^0} \Big],$$

  where M is a free Poisson random measure on $(\mathsf{H},\mathsf{B}(\mathsf{H}),\mathrm{Leb}\otimes\nu)$ with values in $(\mathcal{A}^0,\tau^0)$, and the limit is taken in probability.
- *(W_t) and (V_t) are freely independent processes.*

Furthermore, the right hand side of (6.20), as a whole, is a free Lévy process (in law).

Proof. The proof proceeds along the same lines as that of Theorem 6.26, and we shall not repeat all the arguments. Let (X_t) be a classical Lévy process defined on a probability space $(\Omega,\mathcal{F},P)$ such that $L\{Z_t\} = \Lambda(L\{X_t\})$ for all t. In particular, the Lévy measure for $L\{X_1\}$ is ν. Hence, by Theorem 2.9(ii), (X_t) has a representation in the form

$$X_t \overset{\mathrm{a.s.}}{=} \eta t + \sqrt{a} B_t + Y_t, \qquad (t \geq 0),$$

where

- $\eta \in \mathbb{R}$, $a \geq 0$ and (B_t) is a (classical) Brownian motion on $(\Omega,\mathcal{F},P)$.
- (Y_t) is a classical Lévy process given by

$$Y_t := \lim_{\epsilon \searrow 0} \Big[\int_{]0,t]\times D(\epsilon,\infty)} x\, N(\mathrm{d}u,\mathrm{d}x) - \int_{]0,t]\times D(\epsilon,1)} x\, \mathrm{Leb}\otimes\nu(\mathrm{d}u,\mathrm{d}x) \Big],$$

  where N is a (classical) Poisson random measure on $(\mathsf{H},\mathsf{B}(\mathsf{H}),\mathrm{Leb}\otimes\nu)$, defined on $(\Omega,\mathcal{F},P)$, and the limit is almost surely.

(B_t) and (Y_t) are independent processes.

For all ϵ, t in $]0,\infty[$, we put:

$$Y_{\epsilon,t} = \int_{]0,t]\times D(\epsilon,\infty)} x\ N(\mathrm{d}u,\mathrm{d}x) - \int_{]0,t]\times D(\epsilon,1)} x \,\mathrm{Leb}\otimes\nu(\mathrm{d}u,\mathrm{d}x),$$

so that $Y_t = \lim_{\epsilon\searrow 0} Y_{t,\epsilon}$ almost surely, for each t.

As in the proof of Theorem 6.26 above, we choose, next, a W^*-probability space $(\mathcal{A}^0,\tau^0)$, which contains a free Brownian motion (W_t) and a free Poisson random measure M on $(\mathsf{H},\mathsf{B}(\mathsf{H}),\mathrm{Leb}\otimes\nu)$, which generate freely independent W^*-subalgebras. For any ϵ in $]0,\infty[$, we put (cf. Lemma 6.25(i)),

$$V_{\epsilon,t} = \int_{]0,t]\times D(\epsilon,\infty)} x\ M(\mathrm{d}u,\mathrm{d}x) - \Big(\int_{]0,t]\times D(\epsilon,1)} x\,\mathrm{Leb}\otimes\nu(\mathrm{d}u,\mathrm{d}x)\Big)\mathbf{1}_{\mathcal{A}^0}.$$

Then for any t in $]0,\infty[$ and any ϵ_1,ϵ_2 in $]0,1[$, such that $\epsilon_1 > \epsilon_2$, we have that

$$V_{\epsilon_2,t}-V_{\epsilon_1,t} = \int_{]0,t]\times D(\epsilon_2,\epsilon_1)} x\ M(\mathrm{d}u,\mathrm{d}x) - \Big(\int_{]0,t]\times D(\epsilon_2,\epsilon_1)} x\,\mathrm{Leb}\otimes\nu(\mathrm{d}u,\mathrm{d}x)\Big)\mathbf{1}_{\mathcal{A}^0}.$$

Making the same calculation for $Y_{\epsilon_2,t} - Y_{\epsilon_1,t}$ and taking Corollary 6.20 into account, it follows that $L\{V_{\epsilon_2,t} - V_{\epsilon_1,t}\} = \Lambda(L\{Y_{\epsilon_2,t} - Y_{\epsilon_1,t}\})$. Hence, by continuity of Λ and completeness of the measure topology, we may conclude that the limit $V_t := \lim_{\epsilon\searrow 0} V_{\epsilon,t}$ exists in probability, and that $L\{V_t\} = \Lambda(L\{Y_t\})$. Moreover, as in the proof of Theorem 6.26, it follows that (W_t) and (V_t) are freely independent processes.

Now for any t in $]0,\infty[$, we have:

$$\begin{aligned} L\{\eta t\mathbf{1}_{\mathcal{A}^0} + \sqrt{a}W_t + V_t\} &= \delta_{\eta t} \boxplus D_{\sqrt{a}}L\{W_t\} \boxplus L\{V_t\} \\ &= \Lambda\big(\delta_{\eta t} * D_{\sqrt{a}}L\{B_t\} * L\{Y_t\}\big) = \Lambda\big(L\{X_t\}\big) = L\{Z_t\}. \end{aligned}$$

It remains to prove that (V_t) is a free Lévy process (in law). For this, note first that whenever $s,t \geq 0$, we have (cf. Lemma 6.25(i)),

$$\begin{aligned} &V_{s+t} - V_s \\ &= \lim_{\epsilon\searrow 0} \big(V_{\epsilon,s+t} - V_{\epsilon,s}\big) \\ &= \lim_{\epsilon\searrow 0} \Big[\int_{]s,s+t]\times D(\epsilon,\infty)} x\ M(\mathrm{d}u,\mathrm{d}x) - \Big(\int_{]s,s+t]\times D(\epsilon,1)} x\,\mathrm{Leb}\otimes\nu(\mathrm{d}u,\mathrm{d}x)\Big)\mathbf{1}_{\mathcal{A}^0}\Big]. \end{aligned}$$

Making the same calculation for $Y_{s+t} - Y_s$, and taking Lemma 6.25(i) as well as the continuity of Λ into account, it follows that

$$L\{V_{s+t} - V_s\} = \Lambda(L\{Y_{s+t} - Y_s\}) = \Lambda(L\{Y_t\}) = L\{V_t\},$$

so that (V_t) has stationary increments. The stochastic continuity of (V_t) follows exactly as in the proof of Theorem 6.26. To see, finally, that (V_t) has freely independent increments, assume that $0 = t_0 < t_1 < t_2 < \cdots < t_r$, and consider ϵ in $]0,\infty[$. Then for any j in $\{1,2,\ldots,r\}$,

$$V_{\epsilon,t_j} - V_{\epsilon,t_{j-1}} = \lim_{n\to\infty}\Big[\int_{]t_{j-1},t_j]\times D(\epsilon,n)} x\, M(\mathrm{d}u,\mathrm{d}x) \\ - \Big(\int_{]t_{j-1},t_j]\times D(\epsilon,1)} x\,\mathrm{Leb}\otimes\nu(\mathrm{d}u,\mathrm{d}x)\Big)\mathbf{1}_{\mathcal{A}^0}\Big].$$

Hence, by Proposition 6.22 and Proposition 4.7, the increments $V_{\epsilon,t_j} - V_{\epsilon,t_{j-1}}$, $j = 1,2,\ldots,r$ are freely independent, for any fixed positive ϵ. Yet another application of Proposition 4.7 then yields that the increments

$$V_{t_j} - V_{t_{j-1}} = \lim_{\epsilon\searrow 0}\big(V_{\epsilon,t_j} - V_{\epsilon,t_{j-1}}\big), \qquad (j = 1,2,\ldots,r),$$

are freely independent too. □

Remark 6.28. Let (Z_t) be a free Lévy process in law, such that $L\{Z_1\}$ has Lévy measure ν. If $\int_{[-1,1]}|x|\,\nu(\mathrm{d}x) < \infty$, then Theorems 6.26 and 6.27 provide two different "Lévy-Itô decompositions" of (Z_t). The relationship between the two representations, however, is simply that

$$\eta = \gamma + \int_{[-1,1]} x\,\nu(\mathrm{d}x) \quad\text{and}\quad V_t = U_t - t\Big(\int_{[-1,1]} x\,\nu(\mathrm{d}x)\Big)\mathbf{1}_{\mathcal{A}^0}, \qquad (t \geq 0).$$

Remark 6.29. The proof of the general free Lévy-Itô decomposition, Theorem 6.27, also provides a proof of the general existence of free Lévy processes (in law). Indeed, the conclusion of the proof of Theorem 6.27 might also be formulated in the following way: For any classical Lévy process (X_t), there exists a W^*-probability space $(\mathcal{A}^0,\tau^0)$ containing a free Brownian motion (W_t) and a free Poisson random measure M on $(\mathsf{H},\mathsf{B}(\mathsf{H}),\mathrm{Leb}\otimes\nu)$, which are freely independent, and such that

$$\Lambda(L\{X_t\}) = \\ L\Big\{\eta t\mathbf{1}_{\mathcal{A}^0} + \sqrt{a}W_t + \\ \lim_{\epsilon\searrow 0}\Big[\int_{]0,t]\times D(\epsilon,\infty)} x\, M(\mathrm{d}u,\mathrm{d}x) - \Big(\int_{]0,t]\times D(\epsilon,1)} x\,\mathrm{Leb}\otimes\nu(\mathrm{d}u,\mathrm{d}x)\Big)\mathbf{1}_{\mathcal{A}^0}\Big]\Big\}, \tag{6.21}$$

for suitable constants η in $\mathbb{R}$ and a in $]0,\infty[$. In addition, the process appearing in the right hand side of (6.21) is a free Lévy process (in law) affiliated with $(\mathcal{A}^0,\tau^0)$.

Assume now that $(\nu_t)_{t\geq 0}$ is a family of distributions in $\mathcal{ID}(\boxplus)$, satisfying the two conditions

$$\nu_t = \nu_s \boxplus \nu_{t-s}, \qquad (0 \leq s < t),$$

and

$$\nu_t \xrightarrow{\text{w}} \delta_0, \quad \text{as } t \searrow 0.$$

Then put $\mu_t = \Lambda^{-1}(\nu_t)$ for all t, and note that the family (μ_t) satisfies the corresponding conditions:

$$\mu_t = \mu_s * \mu_{t-s}, \qquad (0 \leq s < t),$$

and

$$\mu_t \xrightarrow{\text{w}} \delta_0, \quad \text{as } t \searrow 0,$$

by the properties of Λ^{-1}. Hence, by the well-known existence result for classical Lévy processes, there exists a classical Lévy process (X_t), such that $L\{X_t\} = \mu_t$ and hence $\Lambda(L\{X_t\}) = \nu_t$ for all t. Therefore, the right hand side of (6.21) is a free Lévy process (in law), (Z_t), such that $L\{Z_t\} = \nu_t$ for all t.

The above argument for the existence of free Lévy processes (in law) is, of course, based on the existence of free Poisson random measures proved in Theorem 6.9. The existence of free Lévy processes (in law) can also, as noted in [Bi98] and [Vo98], be proved directly by a construction similar to that given in the proof of Theorem 6.9. The latter approach, however, is somewhat more complicated than the construction given in the proof of Theorem 6.9, since, in the general case, one has to deal with unbounded operators throughout the construction, whereas free Poisson random measures only involve bounded operators.

A Unbounded Operators Affiliated with a W^*-Probability Space

In this appendix we give a brief account on the theory of closed, densely defined operators affiliated with a finite von Neumann algebra[10]. We start by introducing von Neumann algebras. For a detailed introduction to von Neumann algebras, we refer to [KaRi83], but also the paper [Ne74], referred to below, has a nice short introduction to that subject. For background material on unbounded operators, see [Ru91].

Let $\mathcal{H}$ be a Hilbert space, and consider the vector space $\mathcal{B}(\mathcal{H})$ of bounded (or continuous) linear mappings (or operators) $a\colon \mathcal{H} \to \mathcal{H}$. Recall that composition of operators constitutes a multiplication on $\mathcal{B}(\mathcal{H})$, and that the adjoint operation $a \mapsto a^*$ is an involution on $\mathcal{B}(\mathcal{H})$ (i.e. $(a^*)^* = a$). Altogether $\mathcal{B}(\mathcal{H})$ is a $*$-algebra[11]. For any subset $\mathcal{S}$ of $\mathcal{B}(\mathcal{H})$, we denote by $\mathcal{S}'$ the *commutant*

[10] To make the appendix appear in self-contained form, some of the definitions that already appeared in Section 4.1 will be repeated below.

[11] Throughout this appendix, the $*$ refers to the adjoint operation and not to classical convolution.

of $\mathcal{S}$, i.e.

$$\mathcal{S}' = \{b \in \mathcal{B}(\mathcal{H}) \mid by = yb \text{ for all } y \text{ in } \mathcal{S}\}.$$

A *von Neumann algebra* acting on $\mathcal{H}$ is a subalgebra of $\mathcal{B}(\mathcal{H})$, which contains the multiplicative unit $\mathbf{1}$ of $\mathcal{B}(\mathcal{H})$, and which is closed under the adjoint operation and closed in the weak operator topology (see [KaRi83, Definition 5.1.1]). By von Neumann's fundamental double commutant theorem, a von Neumann algebra may also be characterized as a subset $\mathcal{A}$ of $\mathcal{B}(\mathcal{H})$, which is closed under the adjoint operation and equals the commutant of its commutant: $\mathcal{A}'' = \mathcal{A}$.

A trace (or tracial state) on a von Neumann algebra $\mathcal{A}$ is a positive linear functional $\tau \colon \mathcal{A} \to \mathbb{C}$, satisfying that $\tau(\mathbf{1}) = 1$ and that $\tau(ab) = \tau(ba)$ for all a, b in $\mathcal{A}$. We say that τ is a normal trace on $\mathcal{A}$, if, in addition, τ is continuous on the unit ball of $\mathcal{A}$ w.r.t. the weak operator topology. We say that τ is faithful, if $\tau(a^*a) > 0$ for any non-zero operator a in $\mathcal{A}$.

We shall use the terminology *W^*-probability space* for a pair $(\mathcal{A}, \tau)$, where $\mathcal{A}$ is a von Neumann algebra acting on a Hilbert space $\mathcal{H}$, and $\tau \colon \mathcal{A} \to \mathbb{C}$ is a faithful, normal tracial state on $\mathcal{A}$. In the remaining part of this appendix, $(\mathcal{A}, \tau)$ denotes a W^*-probability space acting on the Hilbert space $\mathcal{H}$.

By a linear operator *in* $\mathcal{H}$, we shall mean a (not necessarily bounded) linear operator $a \colon \mathcal{D}(a) \to \mathcal{H}$, defined on a subspace $\mathcal{D}(a)$ of $\mathcal{H}$. For an operator a in $\mathcal{H}$, we say that

- a is *densely defined*, if $\mathcal{D}(a)$ is dense in $\mathcal{H}$,
- a is *closed*, if the graph $\mathcal{G}(a) = \{(h, ah) \mid h \in \mathcal{D}(a)\}$ of a is a closed subspace of $\mathcal{H} \oplus \mathcal{H}$,
- a is *preclosed*, if the norm closure $\overline{\mathcal{G}(a)}$ is the graph of a (uniquely determined) operator, denoted $[a]$, in $\mathcal{H}$,
- a is *affiliated with* $\mathcal{A}$, if $au = ua$ for any unitary operator u in the commutant $\mathcal{A}'$.

For a densely defined operator a in $\mathcal{H}$, the *adjoint operator* a^* has domain

$$\mathcal{D}(a^*) = \Big\{\eta \in \mathcal{H} \;\Big|\; \sup\{|\langle a\xi, \eta\rangle| \mid \xi \in \mathcal{D}(a),\ \|\xi\| \le 1\} < \infty\Big\},$$

and is given by

$$\langle a\xi, \eta\rangle = \langle \xi, a^*\eta\rangle, \qquad (\xi \in \mathcal{D}(a),\ \eta \in \mathcal{D}(a^*)).$$

We say that a is *selfadjoint* if $a = a^*$ (in particular this requires that $\mathcal{D}(a^*) = \mathcal{D}(a)$).

If a is bounded, a is affiliated with $\mathcal{A}$ if and only if $a \in \mathcal{A}$. In general, a selfadjoint operator a in $\mathcal{H}$ is affiliated with $\mathcal{A}$, if and only if $f(a) \in \mathcal{A}$ for any *bounded Borel* function $f \colon \mathbb{R} \to \mathbb{C}$ (here $f(a)$ is defined in terms of spectral theory). As in the bounded case, if a is a selfadjoint operator affiliated with $\mathcal{A}$, there exists a unique probability measure μ_a on $\mathbb{R}$, concentrated on the spectrum $\mathrm{sp}(a)$, and satisfying that

$$\int_{\mathbb{R}} f(t)\, \mu_a(\mathrm{d}t) = \tau(f(a)),$$

for any bounded Borel function $f: \mathbb{R} \to \mathbb{C}$. We call μ_a the (spectral) distribution of a, and we shall denote it also by $L\{a\}$. Unless a is bounded, $\mathrm{sp}(a)$ is an unbounded subset of $\mathbb{R}$ and, in general, μ_a is not compactly supported.

By $\overline{\mathcal{A}}$ we denote the set of closed, densely defined operators in $\mathcal{H}$, which are affiliated with $\mathcal{A}$. In general, dealing with unbounded operators is somewhat unpleasant, compared to the bounded case, since one needs constantly to take the domains into account. However, the following two important propositions allow us to deal with operators in $\overline{\mathcal{A}}$ in a quite relaxed manner.

Proposition A.1 (cf. [Ne74]). *Let $(\mathcal{A}, \tau)$ be a W^*-probability space. If $a, b \in \overline{\mathcal{A}}$, then $a+b$ and ab are densely defined, preclosed operators affiliated with $\mathcal{A}$, and their closures $[a+b]$ and $[ab]$ belong to $\overline{\mathcal{A}}$. Furthermore, $a^* \in \overline{\mathcal{A}}$.*

By virtue of the proposition above, the adjoint operation may be restricted to an involution on $\overline{\mathcal{A}}$, and we may define operations, the *strong sum* and the *strong product*, on $\overline{\mathcal{A}}$, as follows:

$$(a,b) \mapsto [a+b], \quad \text{and} \quad (a,b) \mapsto [ab], \quad (a,b \in \overline{\mathcal{A}}).$$

Proposition A.2 (cf. [Ne74]). *Let $(\mathcal{A}, \tau)$ be a W^*-probability space. Equipped with the adjoint operation and the strong sum and product, $\overline{\mathcal{A}}$ is a $*$-algebra.*

The effect of the above proposition is, that w.r.t. the adjoint operation and the strong sum and product, we can manipulate with operators in $\overline{\mathcal{A}}$, without worrying about domains etc. So, for example, we have rules like

$$[[a+b]c] = [[ac] + [bc]], \quad [a+b]^* = [a^* + b^*], \quad [ab]^* = [b^* a^*],$$

for operators a, b, c in $\overline{\mathcal{A}}$. Note, in particular, that the strong sum of two selfadjoint operators in $\overline{\mathcal{A}}$ is again a selfadjoint operator. In the following, we shall omit the brackets in the notation for the strong sum and product, and it will be understood that all sums and products are formed in the strong sense.

Remark A.3. If $a_1, a_2 \dots, a_r$ are selfadjoint operators in $\overline{\mathcal{A}}$, we say that they are *freely independent* if, for any bounded Borel functions $f_1, f_2, \dots, f_r: \mathbb{R} \to \mathbb{R}$, the bounded operators $f_1(a_1), f_2(a_2), \dots, f_r(a_r)$ in $\mathcal{A}$ are freely independent in the sense of Section 4. Given any two probability measures μ_1 and μ_2 on $\mathbb{R}$, it follows from a free product construction (see [VoDyNi92]), that one can always find a W^*-probability space $(\mathcal{A}, \tau)$ and selfadjoint operators a and b affiliated with $\mathcal{A}$, such that $\mu_1 = L\{a\}$ and $\mu_2 = L\{b\}$. As noted above, for such operators $a+b$ is again a selfadjoint operator in $\overline{\mathcal{A}}$, and, as was proved in [BeVo93, Theorem 4.6], the (spectral) distribution $L\{a+b\}$ depends only on μ_1 and μ_2. We may thus define the free additive convolution $\mu_1 \boxplus \mu_2$ of μ_1 and μ_2 to be $L\{a+b\}$.

Next, we shall equip $\overline{\mathcal{A}}$ with a topology; the so called measure topology, which was introduced by Segal in [Se53] and later studied by Nelson in [Ne74]. For any positive numbers ϵ, δ, we denote by $N(\epsilon, \delta)$ the set of operators a in $\overline{\mathcal{A}}$, for which there exists an orthogonal projection p in $\mathcal{A}$, satisfying that

$$p(\mathcal{H}) \subseteq \mathcal{D}(a), \quad \|ap\| \leq \epsilon \quad \text{and} \quad \tau(p) \geq 1 - \delta. \tag{A.1}$$

Definition A.4. *Let $(\mathcal{A}, \tau)$ be a W^*-probability space. The measure topology on $\overline{\mathcal{A}}$ is the vector space topology on $\overline{\mathcal{A}}$ for which the sets $N(\epsilon, \delta)$, $\epsilon, \delta > 0$, form a neighbourhood basis for* 0.

It is clear from the definition of the sets $N(\epsilon, \delta)$ that the measure topology satisfies the first axiom of countability. In particular, all convergence statements can be expressed in terms of sequences rather than nets.

Proposition A.5 (cf. [Ne74]). *Let $(\mathcal{A}, \tau)$ be a W^*-probability space and consider the $*$-algebra $\overline{\mathcal{A}}$. We then have*

(i) *Scalar-multiplication, the adjoint operation and strong sum and product are all continuous operations w.r.t. the measure topology. Thus, $\overline{\mathcal{A}}$ is a topological $*$-algebra w.r.t. the measure topology.*
(ii) *The measure topology on $\overline{\mathcal{A}}$ is a complete Hausdorff topology.*

We shall note, next, that the measure topology on $\overline{\mathcal{A}}$ is, in fact, the topology for convergence in probability. Recall first, that for a closed, densely defined operator a in $\mathcal{H}$, we put $|a| = (a^*a)^{1/2}$. In particular, if $a \in \overline{\mathcal{A}}$, then $|a|$ is a selfadjoint operator in $\overline{\mathcal{A}}$ (see [KaRi83, Theorem 6.1.11]), and we may consider the probability measure $L\{|a|\}$ on $\mathbb{R}$.

Definition A.6. *Let $(\mathcal{A}, \tau)$ be a W^*-probability space and let a and a_n, $n \in \mathbb{N}$, be operators in $\overline{\mathcal{A}}$. We say then that $a_n \to a$ in probability, as $n \to \infty$, if $|a_n - a| \to 0$ in distribution, i.e. if $L\{|a_n - a|\} \to \delta_0$ weakly.*

If a and a_n, $n \in \mathbb{N}$, are *selfadjoint* operators in $\overline{\mathcal{A}}$, then, as noted above, $a_n - a$ is selfadjoint for each n, and $L\{|a_n - a|\}$ is the transformation of $L\{a_n - a\}$ by the mapping $t \mapsto |t|$, $t \in \mathbb{R}$. In this case, it follows thus that $a_n \to a$ in probability, if and only if $a_n - a \to 0$ in distribution, i.e. if and only if $L\{a_n - a\} \to \delta_0$ weakly.

From the definition of $L\{|a_n - a|\}$, it follows immediately that we have the following characterization of convergence in probability:

Lemma A.7. *Let $(\mathcal{A}, \tau)$ be a W^*-probability space and let a and a_n, $n \in \mathbb{N}$, be operators in $\overline{\mathcal{A}}$. Then $a_n \to a$ in probability, if and only if*

$$\forall \epsilon > 0\colon \tau\big[1_{]\epsilon,\infty[}(|a_n - a|)\big] \to 0, \quad \textit{as } n \to \infty.$$

Proposition A.8 (cf. [Te81]). *Let $(\mathcal{A},\tau)$ be a W^*-probability space. Then for any positive numbers ϵ, δ, we have*

$$N(\epsilon,\delta) = \{a \in \overline{\mathcal{A}} \mid \tau[1_{]\epsilon,\infty[}(|a|)] \le \delta\}, \tag{A.2}$$

where $N(\epsilon,\delta)$ is defined via (A.1). *In particular, a sequence a_n in $\overline{\mathcal{A}}$ converges, in the measure topology, to an operator a in $\overline{\mathcal{A}}$, if and only if $a_n \to a$ in probability.*

Proof. The last statement of the proposition follows immediately from formula (A.2) and Lemma A.7. To prove (A.2), note first that by considering the polar decomposition of an operator a in $\overline{\mathcal{A}}$ (cf. [KaRi83, Theorem 6.1.11]), it follows that $N(\epsilon,\delta) = \{a \in \overline{\mathcal{A}} \mid |a| \in N(\epsilon,\delta)\}$. From this, the inclusion $\supseteq$ in (A.2) follows easily. Regarding the reverse inclusion, suppose $a \in N(\epsilon,\delta)$, and let p be a projection in $\mathcal{A}$, such that (A.1) is satisfied with a replaced by $|a|$. Then, using spectral theory, it can be shown that the ranges of the projections p and $1_{]\epsilon,\infty[}(|a|)$ only have 0 in common. This implies that $\tau[1_{]\epsilon,\infty[}(|a|)] \le \tau(\mathbf{1}-p) \le \delta$. We refer to [Te81] for further details. □

Finally, we shall need the fact that convergence in probability implies convergence in distribution, also in the non-commutative setting. The key point in the proof given below is that weak convergence can be expressed in terms of the Cauchy transform (cf. [Ma92, Theorem 2.5]).

Proposition A.9. *Let (a_n) be a sequence of selfadjoint operators affiliated with a W^*-probability space $(\mathcal{A},\tau)$, and assume that a_n converges in probability, as $n \to \infty$, to a selfadjoint operator a affiliated with $(\mathcal{A},\tau)$. Then $a_n \to a$ in distribution too, i.e. $L\{a_n\} \xrightarrow{w} L\{a\}$, as $n \to \infty$.*

Proof. Let x, y be real numbers such that $y > 0$, and put $z = x + \mathrm{i}y$. Then define the function $f_z : \mathbb{R} \to \mathbb{C}$ by

$$f_z(t) = \frac{1}{t-z} = \frac{1}{(t-x)-\mathrm{i}y}, \qquad (t \in \mathbb{R}),$$

and note that f_z is continuous and bounded with $\sup_{t\in\mathbb{R}} |f_z(t)| = y^{-1}$. Thus, we may consider the bounded operators $f_z(a_n), f_z(a) \in \mathcal{A}$. Note then that (using strong products and sums),

$$\begin{aligned} f_z(a_n) - f_z(a) &= (a_n - z\mathbf{1})^{-1} - (a - z\mathbf{1})^{-1} \\ &= (a_n - z\mathbf{1})^{-1}\big((a - z\mathbf{1}) - (a_n - z\mathbf{1})\big)(a - z\mathbf{1})^{-1} \\ &= (a_n - z\mathbf{1})^{-1}(a - a_n)(a - z\mathbf{1})^{-1}. \end{aligned} \tag{A.3}$$

Now, given any positive numbers ϵ, δ, we may choose N in $\mathbb{N}$, such that $a_n - a \in N(\epsilon,\delta)$, whenever $n \ge N$. Moreover, since $\|f_z(a_n)\|, \|f_z(a)\| \le y^{-1}$, we have that $f_z(a_n), f_z(a) \in N(y^{-1}, 0)$. Using then the rule: $N(\epsilon_1,\delta_1)N(\epsilon_2,\delta_2) \subseteq N(\epsilon_1\epsilon_2, \delta_1+\delta_2)$, which holds for all ϵ_1, ϵ_2 in $]0,\infty[$ and δ_1, δ_2 in $[0,\infty[$ (see

[Ne74, Formula 17']), it follows from (A.3) that $f_z(a_n) - f_z(a) \in N(\epsilon y^{-2}, \delta)$, whenever $n \geq N$. We may thus conclude that $f_z(a_n) \to f_z(a)$ in the measure topology, i.e. that $L\{|f_z(a_n) - f_z(a)|\} \xrightarrow{w} \delta_0$, as $n \to \infty$. Using now the Cauchy-Schwarz inequality for τ, it follows that

$$\left|\tau(f_z(a_n) - f_z(a))\right|^2 \leq \tau(|f_z(a_n) - f_z(a)|^2) \cdot \tau(\mathbf{1})$$
$$= \int_0^\infty t^2 \, L\{|f_z(a_n) - f_z(a)|\}(\mathrm{d}t) \longrightarrow 0,$$

as $n \to \infty$, since $\mathrm{supp}(L\{|f_z(a_n) - f_z(a)|\}) \subseteq [0, 2y^{-1}]$ for all n, and since $t \mapsto t^2$ is a continuous bounded function on $[0, 2y^{-1}]$.

Finally, let G_n and G denote the Cauchy transforms for $L\{a_n\}$ and $L\{a\}$ respectively. From what we have established above, it follows then that

$$G_n(z) = -\tau(f_z(a_n)) \longrightarrow -\tau(f_z(a)) = G(z), \quad \text{as } n \to \infty,$$

for any complex number $z = x + \mathrm{i}y$ for which $y > 0$. By [Ma92, Theorem 2.5], this means that $L\{a_n\} \xrightarrow{w} L\{a\}$, as desired. □

References

[An00] M. Anchelevich, *Free stochastic measures via noncrossing partitions*, Adv. Math. **155** (2000), 154-179.

[An01] M. Anshelevich, *Partition-dependent Stochastic Measures and q-deformed Cumulants*, Doc. Math. **6** (2001), 343–384.

[An02] M. Anshelevich, *Itô Formula for Free Stochastic Integrals*, J. Funct. Anal. **188** (2002), 292–315.

[BaCo89] O.E. Barndorff-Nielsen and D.R. Cox, *Asymptotic Techniques for Use in Statistics*, Monographs on Statistics and Applied Probability, Chapman and Hall (1989).

[Ba98] O.E. Barndorff-Nielsen, *Processes of normal inverse Gaussian type*, Finance and Stochastics **2** (1998), 41-68.

[BaMiRe01] O.E. Barndorff-Nielsen, T. Mikosch and S. Resnick (Eds.), *Lévy Processes - Theory and Applications*, Boston: Birkhäuser (2001).

[BaPeSa01] O.E. Barndorff-Nielsen, J. Pedersen and K. Sato, *Multivariate subordination, selfdecomposability and stability*, Adv. Appl. Prob. **33** (2001), 160-187.

[BaSh01a] O.E. Barndorff-Nielsen and N. Shephard, *Non-Gaussian OU based models and some of their uses in financial economics (with Discussion)*, J. R. Statist. Soc. B **63** (2001), 167-241.

[BaSh01b] O.E. Barndorff-Nielsen and N. Shephard, *Modelling by Lévy processes for financial econometrics*, in O.E. Barndorff-Nielsen, T. Mikosch and S. Resnick (Eds.): *Lévy Processes - Theory and Applications*, Boston: Birkhäuser (2001), 283-318.

[BaLi04] O.E. Barndorff-Nielsen and A. Lindner, *Some aspects of Lévy copulas.* (2004) (Submitted.).

[BaMaSa04] O.E. Barndorff-Nielsen, M. Maejima and K. Sato, *Some classes of multivariate infinitely divisible distributions admitting stochastic integral representation.* Bernoulli (To appear).

[BaPA05] O.E. Barndorff-Nielsen and V. Pérez-Abreu, *Matrix subordinators and related Upsilon transformations.* (In preparation).

[BaTh02a] O.E. Barndorff-Nielsen and S. Thorbjørnsen, *Selfdecomposability and Lévy processes in free probability,* Bernoulli **8** (2002), 323-366.

[BaTh02b] O.E. Barndorff-Nielsen and S. Thorbjørnsen, *Lévy laws in free probability,* Proc. Nat. Acad. Sci., vol. 99, no. 26 (2002), 16568-16575.

[BaTh02c] O.E. Barndorff-Nielsen and S. Thorbjørnsen, *Lévy processes in free probability,* Proc. Nat. Acad. Sci., vol. 99, no. 26 (2002), 16576-16580.

[BaTh04a] O.E. Barndorff-Nielsen and S. Thorbjørnsen, *A connection between free and classical infinite divisibility,* Inf. Dim. Anal. Quant. Prob. **7** (2004), 573-590.

[BaTh04b] O.E. Barndorff-Nielsen and S. Thorbjørnsen, *Regularising mappings of Lévy measures,* Stoch. Proc. Appl. (To appear).

[BaTh04c] O.E. Barndorff-Nielsen and S. Thorbjørnsen, *Bicontinuity of the Upsilon transformations,* MaPhysto Research Report 2004-**25**, University of Aarhus (Submitted).

[BaTh05] O.E. Barndorff-Nielsen and S. Thorbjørnsen, *The Lévy-Itô Decomposition in Free Probability,* Prob. Theory and Rel. Fields, **131** (2005), 197-228.

[BeVo93] H. Bercovici and D.V. Voiculescu, *Free Convolution of Measures with Unbounded Support,* Indiana Univ. Math. J. **42** (1993), 733-773.

[BePa96] H. Bercovici and V. Pata, *The Law of Large Numbers for Free Identically Distributed Random Variables,* Ann. Probability **24** (1996), 453-465.

[BePa99] H. Bercovici and V. Pata, *Stable Laws and Domains of Attraction in Free Probability Theory,* Ann. Math. **149** (1999), 1023-1060.

[BePa00] H. Bercovici and V. Pata, *A Free Analogue of Hincin's Characterization of Infinite Divisibility,* Proc. AMS. **128** (2000), 1011-1015.

[Be96] J. Bertoin, *Lévy Processes,* Cambridge University Press (1996).

[Be97] J. Bertoin, *Subordinators: Examples and Applications,* in P. Bernard (Ed.): *Lectures on Probability Theory and Statistics,* Ecole d'Éte de St-Flour XXVII, Berlin: Springer-Verlag (1997), 4-91.

[Be00] J. Bertoin, *Subordinators, Lévy processes with no negative jumps and branching processes,* MaPhySto Lecture Notes Series (2000-**8**), (Århus University).

[Bi98] P. Biane, *Processes with free increments,* Math. Zeitschrift **227** (1998), 143-174.

[Bi03] P. Biane, *Free probability for probabilists,* Quantum probability communications, Vol. XI (Grenoble, 1998), 55–71, QP-PQ, XI, World Sci. Publishing (2003).

[BiSp98] P. Biane and R. Speicher, *Stochastic calculus with respect to free Brownian motion and analysis on Wigner space,* Probab. Theory Related Fields **112** (1998), 373-409.

[Bo92] L. Bondesson, *Generalized Gamma Convolutions and Related Classes of Distributions and Densities,* Lecture Notes in Statistics **76**, Berlin: Springer-Verlag (1992).

[BoSp91] M. Bożejko and R. Speicher, *An example of a generalized Brownian motion,* Comm. Math. Phys. **137** (1991), 519-531.

[Br92] L. BREIMAN, *Probability*, Classics In Applied Mathematics 7, SIAM (1992).

[BrReTw82] P.J. BROCKWELL, S.I. RESNICK AND R.L. TWEEDIE *Storage processes with general release rule and additive inputs.*, Adv. Appl. Prob. **14** (1982), 392-433.

[ChYo03] L. CHAUMONT AND M. YOR, *Exercises in Probability.* Cambridge University Press (2003).

[ChSh02] A.S. CHERNY AND A.N. SHIRAYEV, *On Stochastic Integrals up to infinity and Predictable Criteria for integrability*, Notes from a MaPhySto Summer-school, August 2002.

[Ch78] T.S. CHIHARA, *An Introduction to Orthogonal Polynomials*, Gordon and Breach, Science Publishers (1978).

[Do74] W.F. DONOGHUE, JR., *Monotone Matrix Functions and Analytic Continuation*, Grundlehren der mathematichen Wissenschaften **207**, Springer-Verlag (1974).

[Fe71] W. FELLER, *An Introduction to Probability Theory and its Applications, volume II*, Wiley (1971).

[Ge80] S. GEMAN, *A limit theorem for the norm of random matrices*, Annals of Probability **8** (1980), 252-261.

[GlScSp92] P. GLOCKNER, M. SCHÜRMANN AND R. SPEICHER, *Realization of free white noises*, Arch. Math. **58** (1992), 407-416.

[GnKo68] B.V. GNEDENKO AND A.N. KOLMOGOROV, *Limit Distributions for Sums of Independent Random Variables*, Addison-Wesley Publishing Company, Inc. (1968).

[GrRoVaYo99] M. GRADINARU, B. ROYNETTE, P. VALLOIS AND M. YOR, *Abel transform and integrals of Bessel local time*, Ann. Inst. Henri Poincaré **35**, 531-572.

[HiPe00] F. HIAI AND D. PETZ, *The Semicircle Law, Free Random Variables and Entropy*, Mathematical Surveys and Monographs, Vol. 77. Providence: American Mathematical Society (2000).

[JuVe83] Z.J. JUREK AND W. VERWAAT, *An integral representation for selfdecomposable Banach space valued random variables*, Z. Wahrscheinlichkeitstheorie verw. Geb. **62** (1983), 247-262.

[JuMa93] Z.J. JUREK AND J.D. MASON, *Operator-Limit Distributions in Probability Theory*, New York: Wiley (1993).

[KaRi83] R.V. KADISON AND J.R. RINGROSE, *Fundamentals of the theory of operator algebras, vol. I-II*, Academic Press (1983, 1986).

[LG99] J.-F. LE GALL, *Spatial Branching Processes, Random Snakes and Partial Differential Equations*, Basel: Birkhäuser (1999).

[Lu75] E. LUKACS, *Stochastic Convergence (second edition)*, Academic Press (1975).

[Ma92] H. MAASSEN, *Addition of freely independent random variables*, J. Funct. Anal. **106**, (1992), 409-438.

[Ne74] E. NELSON, *Notes on Non-commutative Integration*, J. Funct. Anal. **15** (1974), 103-116.

[Ni95] A. NICA, *A one-parameter family of transforms, linearizing convolution laws for probability distributions*, Comm. Math. Phys. **168** (1995), 187-207.

[MoOs69] S.A. MOLCHANOV AND E. OSTROVSKII, *Symmetric stable processes as traces of degenerate diffusion processes*, Teor. Verojatnost. Primen. **14** (1969), 127-130.

[Pa96] V. PATA, *Domains of partial attraction in non-commutative probability*, Pacific J. Math. **176** (1996), 235-248.

[Pe89] G.K. PEDERSEN, *Analysis Now*, Graduate Texts in Mathematics **118**, Springer Verlag (1989).

[PeSa04] J. PEDERSEN AND K. SATO, *Semigroups and processes with parameter in a cone*, Abstract and applied analysis, 499-513, World Sci. Publ., River Edge, NJ, (2004).

[Ro64] G.-C. ROTA, *On the foundations of combinatorial theory I: Theory of Möbius functions*, Z. Wahrscheinlichkeitstheorie Verw. Geb. **2** (1964), 340-368.

[Ros02] J. ROSINSKI, *Tempered stable processes*, In O.E. Barndorff-Nielsen (Ed.), *Second MaPhySto Conference on Lévy Processes: Theory and Applications*, Aarhus: MaPhySto (2002), 215-220.

[Ros04] J. ROSINSKI, *Tempering stable processes*, Preprint (2004).

[Ru91] W. RUDIN, *Functional Analysis (second edition)*, McGraw-Hill Inc. (1991).

[Sa99] K. SATO, *Lévy Processes and Infinitely Divisible Distributions*, Cambridge studies in advanced math. **68** (1999).

[Sa00] K. SATO, *Subordination and selfdecomposability*, MaPhySto Research Report (2000-**40**), (Århus University).

[Se53] I.E. SEGAL, *A non-commutative extension of abstract integration*, Ann. Math. **57** (1953), 401-457; correction **58** (1953), 595-596.

[SaTa94] G. SAMORODNITSKY AND M.S. TAQQU, *Stable Non-Gaussian Random Processes*, New York: Chapman and Hall (1994).

[Sk91] A.V. SKOROHOD, *Random Processes with Independent Increments*, Kluwer Academic Publisher (1991), Dordrecht, Netherlands (Russian original 1986).

[Sp90] R. SPEICHER, *A new example of 'independence' and 'white noise'*, Probab. Th. Rel. Fields **84** (1990), 141-159.

[Sp94] R. SPEICHER, *Multiplicative functions on the lattice of non-crossing partitions and free convolution*, Math. Ann. **298** (1994), 611-628.

[Sp97] R. SPEICHER, *Free Probability Theory and Non-crossing Partitions*, Sém. Lothar. Combin. **39** (1997), Article B39c (`electronic`).

[St59] W.F. STINESPRING, *Integration theory for gages and duality for unimodular groups*, Transactions of the AMS. **90** (1959), 15-56.

[Te81] M. TERP, *L^p Spaces associated with von Neumann Algebras*, Lecture notes, University of Copenhagen (1981).

[Th77] O. THORIN, *On the infinite divisibility of the Pareto distribution*, Scand. Actuarial J. (1977), 31-40.

[Th78] O. THORIN, *An extension of the notion of a generalized Γ-convolution*, Scand. Actuarial J. (1978), 141-149.

[Vo85] D.V. VOICULESCU, *Symmetries of some reduced free product C^*-algebras*, Operator Algebras and their Connections with Topology and Ergodic Theory, Lecture Notes in Math. **1132** (1985), Springer Verlag, 556-588.

[Vo86] D.V. VOICULESCU, *Addition of certain non-commuting random variables*, J. Funct. Anal. **66**, (1986), 323-346.

[Vo90] D.V. VOICULESCU, *Circular and semicircular systems and free product factors*, in "Operator Algebras, Unitary Representations, Enveloping Algebras and Invariant Theory", *Progress in Mathematics* **92**, Birkhäuser (1990), 45-60.

[Vo91] D.V. VOICULESCU, *Limit laws for random matrices and free products*, Invent. Math. **104** (1991), 201-220.

[VoDyNi92] D.V. VOICULESCU, K.J. DYKEMA AND A. NICA, *Free Random Variables*, CRM Monographs Series, vol. 1, A.M.S. (1992).

[Vo98] D.V. VOICULESCU, *Lectures on Free Probability*, Lecture notes from the 1998 Saint-Fluor Summer School on Probability Theory.

[Wo82] S.J. WOLFE, *On a Continuous Analogue of the Stochastic Difference Equation* $X_n = \rho X_{n-1} + B_n$, Stochastic Process. Appl. **12** (1982), 301-312.

Lévy Processes on Quantum Groups and Dual Groups

Uwe Franz

GSF - Forschungszentrum für Umwelt und Gesundheit
Institut für Biomathematik und Biometrie
Ingolstädter Landstraße 1
85764 Neuherberg
uwe.franz@gsf.de

U. Franz: *Lévy Processes on Quantum Groups and Dual Groups*,
Lect. Notes Math. **1866**, 161–257 (2006)
www.springerlink.com

Introduction

Lévy processes play a fundamental rôle in probability theory and have many important applications in other areas such as statistics, financial mathematics, functional analysis or mathematical physics, as well, see [App05, BNT05] and the references given there.

In quantum probability they first appeared in a model for the laser in [Wal73, Wal84]. Their algebraic framework was formulated in [ASW88]. This lead to the theory of Lévy processes on involutive bialgebras, cf. [ASW88, Sch93, FS99]. These processes are a generalization of both classical stochastic processes with independent and stationary increments, i.e. classical Lévy processes, and factorizable current representations of groups and Lie algebras. The increments of these Lévy processes are independent in the sense of tensor independence, which is a straightforward generalization of the notion of independence used in classical probability theory. However, in quantum probability there exist also other notions of independence like, e.g., freeness [VDN92], see also Section 3. In order to formulate a general theory of Lévy processes for all "nice" independences, $*$-bialgebras or quantum groups have to be replaced by the dual groups introduced in [Voi87], see [Sch95b, BGS99, Fra01, Fra03b].

Quantum Lévy processes play an important rôle in the theory of continuous measurement, cf. [Hol01], and in the theory of dilations, where they describe the evolution of a big system or heat bath, which is coupled to the small system whose evolution one wants to describe.

This chapter is organized as follows.

In the first two sections we review the theory of Lévy processes on involutive bialgebras. In the remaining two sections we discuss the notion of independence in quantum probability and study Lévy processes on dual groups with respect to the five universal independences.

In Section 1, we present the basic theory of Lévy processes on involutive bialgebras. This is the class of quantum Lévy processes that was studied first and where the theory has been developed most. We introduce Schürmann triples and state Schürmann's representation theorem that says that every Lévy process on an involutive bialgebra can be realized as the solution of a quantum stochastic differential equation on a Boson Fock space. The coefficients of the quantum stochastic differential equation are given by the Schürmann triples of the Lévy process. We furthermore present the recent result by Franz, Schürmann, and Skeide that the vacuum vector is cyclic for the realisation of a Lévy processes obtained by Schürmann's representation theorem.

In Section 2, we study Lévy processes on the non-commutative analogue of the coefficient algebra of the unitary group $U(d)$ and classify their generators and Schürmann triples. These Lévy processes play an important role in the construction of dilations of quantum dynamical semigroups on the matrix algebra $\mathcal{M}_d$.

In Section 3, we introduce the notion of a universal independence and recall their classification by Muraki. We show that this notion has a natural formulation in the language of category theory. We also study a notion of reduction of one independence to another that generalizes the bosonisation of Fermi independence. It turns out that three of the five universal independences can be reduced to tensor independence.

Finally, in Section 4, we study Lévy process on dual groups for all five universal independences. We show that in four of the five cases they can be reduced to Lévy process on involutive bialgebras and use the theory developped in Section 1 to construct them and to study their properties. It is still open, if a similar construction is possible for Lévy processes on dual groups with free increments.

1 Lévy Processes on Quantum Groups

In this section we will give the definition of Lévy processes on involutive bialgebras, cf. Subsection 1.1, and develop their general theory.

In Subsection 1.2 we will begin to develop their basic theory. We will see that the marginal distributions of a Lévy process form a convolution semigroup of states and that we can associate a generator with a Lévy process on an involutive bialgebra, that characterizes uniquely its distribution, like in classical probability. By a GNS-type construction we can get a so-called Schürmann triple from the generator.

This Schürmann triple can be used to obtain a realization of the process on a symmetric Fock space, see Subsection 1.3. This realization can be found as the (unique) solution of a quantum stochastic differential equation. It establishes the one-to-one correspondence between Lévy processes, convolution semigroups of states, generators, and Schürmann triples. We will not present the proof of the representation theorem here, but refer to [Sch93, Chapter 2].

In Subsection 1.4, we present a recent unpublished result by Franz, Schürmann, and Skeide. If the cocycle of the Schürmann triple is surjective, then the vacuum vector is cyclic for the Lévy process constructed on the symmetric Fock space via the representation theorem.

Finally, in Subsection 1.5, we look at several examples.

For more information on Lévy processes on involutive bialgebras, see also [Sch93][Mey95, Chapter VII][FS99].

1.1 Definition of Lévy Processes on Involutive Bialgebras

A *quantum probability space* in the purely algebraic sense is a pair $(\mathcal{A}, \Phi)$ consisting of a unital $*$-algebra $\mathcal{A}$ and a state (i.e. a normalized positive linear functional) Φ on $\mathcal{A}$. Positivity in this purely algebraic context simply means $\Phi(a^*a) \geq 0$ for all $a \in \mathcal{A}$. A *quantum random variable* j over a quantum probability space $(\mathcal{A}, \Phi)$ on a $*$-algebra $\mathcal{B}$ is simply a $*$-algebra homomorphism $j : \mathcal{B} \to \mathcal{A}$. A *quantum stochastic process* is an indexed family of random variables $(j_t)_{t\in I}$. For a quantum random variable $j : \mathcal{B} \to \mathcal{A}$ we will call $\varphi_j = \Phi \circ j$ its *distribution* in the state Φ. For a quantum stochastic process $(j_t)_{t\in I}$ the functionals $\varphi_t = \Phi \circ j_t : \mathcal{B} \to \mathbb{C}$ are called *marginal distributions.* The *joint distribution* $\Phi \circ \left(\coprod_{t\in I} j_t\right)$ of a quantum stochastic process is a functional on the free product $\coprod_{t\in I} \mathcal{B}$, see Section 3.

Two quantum stochastic processes $\left(j_t^{(1)} : \mathcal{B} \to \mathcal{A}_1\right)_{t\in I}$ and $\left(j_t^{(2)} : \mathcal{B} \to \mathcal{A}_2\right)_{t\in I}$ on $\mathcal{B}$ over $(\mathcal{A}_1, \Phi_1)$ and $(\mathcal{A}_2, \Phi_2)$ are called *equivalent*, if there joint distributions coincide. This is the case, if and only if all their moments agree, i.e. if

$$\Phi_1\left(j_{t_1}^{(1)}(b_1)\cdots j_{t_n}^{(1)}(b_n)\right) = \Phi_2\left(j_{t_1}^{(2)}(b_1)\cdots j_{t_n}^{(2)}(b_n)\right)$$

holds for all $n \in \mathbb{N}$, $t_1, \ldots, t_n \in I$ and all $b_1, \ldots, b_n \in \mathcal{B}$.

The term 'quantum stochastic process' is sometimes also used for an indexed family $(X_t)_{t\in I}$ of operators on a Hilbert space or more generally of elements of a quantum probability space. We will reserve the name *operator process* for this. An operator process $(X_t)_{t\in I} \subseteq \mathcal{A}$ (where $\mathcal{A}$ is a $*$-algebra of operators) always defines a quantum stochastic process $(j_t : \mathbb{C}\langle a, a^*\rangle \to \mathcal{A})_{t\in I}$ on the free $*$-algebra with one generator, if we set $j_t(a) = X_t$ and extend j_t as a $*$-algebra homomorphism. On the other hand operator processes can be obtained from quantum stochastic processes $(j_t : \mathcal{B} \to \mathcal{A})_{t\in I}$ by choosing an element x of the algebra $\mathcal{B}$ and setting $X_t = j_t(x)$.

The notion of independence we use for Lévy processes on involutive bialgebras is the so-called tensor or boson independence. In Section 3 we will see that other interesting notions of independence exist.

Definition 1.1. *Let $(\mathcal{A}, \Phi)$ be a quantum probability space and $\mathcal{B}$ a $*$-algebra. The quantum random variables $j_1, \ldots, j_n : \mathcal{B} \to \mathcal{A}$ are called* tensor or Bose independent *(w.r.t. the state Φ), if*

(i) $\Phi\big(j_1(b_1)\cdots j_n(b_n)\big) = \Phi\big(j_1(b_1)\big)\cdots\Phi\big(j_n(b_n)\big)$ for all $b_1, \ldots, b_n \in \mathcal{B}$, and
(ii) $[j_l(b_1), j_k(b_2)] = 0$ for all $k \neq l$ and all $b_1, b_2 \in \mathcal{B}$.

Recall that an *involutive bialgebra* $(\mathcal{B}, \Delta, \varepsilon)$ is a unital $*$-algebra $\mathcal{B}$ with two unital $*$-homomorphisms $\Delta : \mathcal{B} \to \mathcal{B} \otimes \mathcal{B}$, $\varepsilon : \mathcal{B} \to \mathbb{C}$ called *coproduct* or *comultiplication* and *counit*, satisfying

$$(\mathrm{id} \otimes \Delta) \circ \Delta = (\Delta \otimes \mathrm{id}) \circ \Delta \qquad \text{(coassociativity)}$$
$$(\mathrm{id} \otimes \varepsilon) \circ \Delta = \mathrm{id} = (\varepsilon \otimes \mathrm{id}) \circ \Delta \qquad \text{(counit property).}$$

Let $j_1, j_2 : \mathcal{B} \to \mathcal{A}$ be two linear maps with values in some algebra $\mathcal{A}$, then we define their *convolution* $j_1 \star j_2$ by

$$j_1 \star j_2 = m_{\mathcal{A}} \circ (j_1 \otimes j_2) \circ \Delta.$$

Here $m_{\mathcal{A}} : \mathcal{A} \otimes \mathcal{A} \to \mathcal{A}$ denotes the multiplication of $\mathcal{A}$, $m(a \otimes b) = ab$ for $a, b \in \mathcal{A}$.

Using *Sweedler's notation* $\Delta(b) = b_{(1)} \otimes b_{(2)}$, this becomes $(j_1 \star j_2)(b) = j_1(b_{(1)}j_2(b_{(2)})$. If j_1 and j_2 are two independent quantum random variables, then $j_1 \star j_2$ is again a quantum random variable, i.e. a $*$-homomorphism. The fact that we can compose quantum random variables allows us to define Lévy process, i.e. processes with independent and stationary increments.

Definition 1.2. *Let $\mathcal{B}$ be an involutive bialgebra. A quantum stochastic process $(j_{st})_{0 \le s \le t}$ on $\mathcal{B}$ over some quantum probability space $(\mathcal{A}, \Phi)$ is called a* Lévy process, *if the following four conditions are satisfied.*

1. *(Increment property) We have*
$$\begin{aligned} j_{rs} \star j_{st} &= j_{rt} \quad \text{for all } 0 \le r \le s \le t, \\ j_{tt} &= \varepsilon \mathbf{1} \quad \text{for all } 0 \le t, \end{aligned}$$
i.e. $j_{tt}(b) = \varepsilon(b)\mathbf{1}$ for all $b \in \mathcal{B}$, where $\mathbf{1}$ denotes the unit of $\mathcal{A}$.
2. *(Independence of increments) The family $(j_{st})_{0 \le s \le t}$ is independent, i.e. the quantum random variables $j_{s_1,t_1}, \ldots, j_{s_n t_n}$ are independent for all $n \in \mathbb{N}$ and all $0 \le s_1 \le t_1 \le s_2 \le \cdots \le t_n$.*
3. *(Stationarity of increments) The distribution $\varphi_{st} = \Phi \circ j_{st}$ of j_{st} depends only on the difference $t - s$.*
4. *(Weak continuity) The quantum random variables j_{st} converge to j_{ss} in distribution for $t \searrow s$.*

Exercise 1.3. Recall that an *(involutive) Hopf algebra* $(\mathcal{B}, \Delta, \varepsilon, S)$ is an (involutive) bialgebra $(\mathcal{B}, \Delta, \varepsilon)$ equipped with a linear map called *antipode* $S : \mathcal{B} \to \mathcal{B}$ satisfying

$$S \star \mathrm{id} = \mathbf{1} \circ \varepsilon = \mathrm{id} \star S. \tag{1.1}$$

The antipode is unique, if it exists. Furthermore, it is an algebra and coalgebra anti-homomorphism, i.e. it satisfies $S(ab) = S(b)S(a)$ for all $a, b \in \mathcal{B}$ and $(S \otimes S) \circ \Delta = \tau \circ \Delta \circ S$, where $\tau : \mathcal{B} \otimes \mathcal{B} \to \mathcal{B} \otimes \mathcal{B}$ is the *flip* $\tau(a \otimes b) = b \otimes a$. If $(\mathcal{B}, \Delta, \varepsilon)$ is an involutive bialgebra and $S : \mathcal{B} \to \mathcal{B}$ a linear map satisfying (1.1), then S satisfies also the relation

$$S \circ * \circ S \circ * = \mathrm{id}.$$

In particular, it follows that the antipode S of an involutive Hopf algebra is invertible. This is not true for Hopf algebras in general.

Show that if $(k_t)_{t \ge 0}$ is any quantum stochastic process on an involutive Hopf algebra, then the quantum stochastic process defined by

$$j_{st} = m_{\mathcal{A}} \circ \big((k_s \circ S) \otimes k_t\big) \circ \Delta,$$

for $0 \leq s \leq t$, satisfies the increment property (1) in Definition 1.2. A one-parameter stochastic process $(k_t)_{t \geq 0}$ on a Hopf $*$-algebra H is called a *Lévy process on* H, if its increment process $(j_{st})_{0 \leq s \leq t}$ with $j_{st} = \big(k_s \circ S\big) \otimes k_t\big) \circ \Delta$ is a Lévy process on H in the sense of Definition 1.2.

Let $(j_{st})_{0 \leq s \leq t}$ be a Lévy process on some involutive bialgebra. We will denote the marginal distributions of $(j_{st})_{0 \leq s \leq t}$ by $\varphi_{t-s} = \Phi \circ j_{st}$. Due to the stationarity of the increments this is well defined.

Lemma 1.4. *The marginal distributions* $(\varphi_t)_{t \geq 0}$ *of a Lévy process on an involutive bialgebra* $\mathcal{B}$ *form a* convolution semigroup of states *on* $\mathcal{B}$, *i.e. they satisfy*

1. $\varphi_0 = \varepsilon$, $\varphi_s \star \varphi_t = \varphi_{s+t}$ *for all* $s, t \geq 0$, *and* $\lim_{t \searrow 0} \varphi_t(b) = \varepsilon(b)$ *for all* $b \in \mathcal{B}$, *and*
2. $\varphi_t(\mathbf{1}) = 1$, *and* $\varphi_t(b^* b) \geq 0$ *for all* $t \geq 0$ *and all* $b \in \mathcal{B}$.

Proof. $\varphi_t = \Phi \circ j_{0t}$ is clearly a state, since j_{0t} is a $*$-homomorphism and Φ a state.

From the first condition in Definition 1.2 we get

$$\varphi_0 = \Phi \circ j_{00} = \Phi(1)\varepsilon = \varepsilon,$$

and

$$\varphi_{s+t}(b) = \Phi\big(j_{0,s+t}(b)\big) = \Phi\left(\sum j_{0s}(b_{(1)}) j_{s,s+t}(b_{(2)})\right),$$

for $b \in \mathcal{B}$, $\Delta(b) = \sum b_{(1)} \otimes b_{(2)}$. Using the independence of increments, we can factorize this and get

$$\begin{aligned}\varphi_{s+t}(b) &= \sum \Phi\big(j_{0s}(b_{(1)})\big)\Phi\big(j_{s,s+t}(b_{(2)})\big) = \sum \varphi_s(b_{(1)})\varphi_t(b_{(2)}) \\ &= \varphi_s \otimes \varphi_t\big(\Delta(b)\big) = \varphi_s \star \varphi_t(b)\end{aligned}$$

for all $\in \mathcal{B}$.

The continuity is an immediate consequence of the last condition in Definition 1.2. □

Lemma 1.5. *The convolution semigroup of states characterizes a Lévy process on an involutive bialgebra up to equivalence.*

Proof. This follows from the fact that the increment property and the independence of increments allow to express all joint moments in terms of the marginals. E.g., for $0 \leq s \leq t \leq u \leq v$ and $a, b, c \in \mathcal{B}$, the moment $\Phi\big(j_{su}(a) j_{st}(b) j_{sv}(c)\big)$ becomes

$$\begin{aligned}\Phi\big(j_{su}(a)j_{st}(b)j_{sv}(c)\big) &= \Phi\big((j_{st}\star j_{tu})(a)j_{st}(b)(j_{st}\star j_{tu}\star j_{uv})(c)\big)\\ &= \Phi\big(j_{st}(a_{(1)})j_{tu}(a_{(2)})j_{st}(b)j_{st}(c_{(1)})j_{tu}(c_{(2)})j_{uv}(c_{(3)})\big)\\ &= \Phi\big(j_{st}(a_{(1)}bc_{(1)})j_{tu}(a_{(2)}c_{(2)})j_{uv}(c_{(3)})\big)\\ &= \varphi_{t-s}(a_{(1)}bc_{(1)})\varphi_{u-t}(a_{(2)}c_{(2)})\varphi_{v-u}(c_{(3)}).\end{aligned}$$

□

It is possible to reconstruct process $(j_{st})_{0\leq s\leq t}$ from its convolution semigroup, see [Sch93, Section 1.9] or [FS99, Section 4.5]. Therefore, we even have a one-to-one correspondence between equivalence classes of Lévy processes on $\mathcal{B}$ and convolution semigroups of states on $\mathcal{B}$.

1.2 The Generator and the Schürmann Triple of a Lévy Process

In this subsection we will meet two more objects that classify Lévy processes, namely their generator and their triple (called Schürmann triple by P.-A. Meyer, see [Mey95, Section VII.1.6]).

We begin with a technical lemma.

Lemma 1.6. *(a) Let $\psi:\mathcal{C}\to\mathbb{C}$ be a linear functional on some coalgebra $\mathcal{C}$. Then the series*

$$\exp_\star\psi(b)\stackrel{\text{def}}{=}\sum_{n=0}\frac{\psi^{\star n}}{n!}(b)=\varepsilon(b)+\psi(b)+\frac{1}{2}\psi\star\psi(b)+\cdots$$

converges for all $b\in\mathcal{C}$.

(b) Let $(\varphi_t)_{t\geq 0}$ be a convolution semigroup on some coalgebra $\mathcal{C}$. Then the limit

$$L(b)=\lim_{t\searrow 0}\frac{1}{t}\big(\varphi_t(b)-\varepsilon(b)\big)$$

exists for all $b\in\mathcal{C}$. Furthermore we have $\varphi_t=\exp_\star tL$ for all $t\geq 0$.

The proof of this lemma relies on the fundamental theorem of coalgebras, see [ASW88, Sch93].

Proposition 1.7. (Schoenberg correspondence) *Let $\mathcal{B}$ be an involutive bialgebra, $(\varphi_t)_{t\geq 0}$ a convolution semigroup of linear functionals on $\mathcal{B}$ and*

$$L=\lim_{t\searrow 0}\frac{1}{t}\big(\varphi_t-\varepsilon\big).$$

Then the following are equivalent.

(i) $(\varphi_t)_{t\geq 0}$ is a convolution semigroup of states.

(ii) $L : \mathcal{B} \to \mathbb{C}$ *satisfies* $L(\mathbf{1}) = 0$, *and it is hermitian and* conditionally positive, *i.e.*

$$L(b^*) = \overline{L(b)}$$

for all $b \in \mathcal{B}$, *and*

$$L(b^*b) \geq 0$$

for all $b \in \mathcal{B}$ *with* $\varepsilon(b) = 0$.

Proof. We prove only the (easy) direction (i)⇒(ii), the converse will follow from the representation theorem 1.15, whose proof can be found in [Sch93, Chapter 2].

The first property follows by differentiating $\varphi_t(\mathbf{1}) = 1$ w.r.t. t.

Let $b \in \mathcal{B}$, $\varepsilon(b) = 0$. If all φ_t are states, then we have $\varphi_t(b^*b) \geq 0$ for all $t \geq 0$ and therefore

$$L(b^*b) = \lim_{t \searrow 0} \frac{1}{t}\big(\varphi_t(b^*b) - \varepsilon(b^*b)\big) = \lim_{t \searrow 0} \frac{\varphi_t(b^*b)}{t} \geq 0.$$

Similarly, L is hermitian, since all φ_t are hermitian. □

We will call a linear functional satisfying condition (ii) of the preceding Proposition a *generator*. Lemma 1.6 and Proposition 1.7 show that Lévy processes can also be characterized by their generator $L = \frac{\mathrm{d}}{\mathrm{d}t}\big|_{t=0} \varphi_t$.

Let D be a pre-Hilbert space. Then we denote by $\mathcal{L}(D)$ the set of all linear operators on D that have an adjoint defined everywhere on D, i.e.

$$\mathcal{L}(D) = \left\{ X : D \to D \text{ linear} \;\middle|\; \begin{array}{l} \text{there exists } X^* : D \to D \text{ linear s.t.} \\ \langle u, Xv \rangle = \langle X^*u, v \rangle \text{ for all } u, v \in D \end{array} \right\}.$$

$\mathcal{L}(D)$ is clearly a unital $*$-algebra.

Definition 1.8. *Let* $\mathcal{B}$ *be a unital* $*$*-algebra equipped with a unital hermitian character* $\varepsilon : \mathcal{B} \to \mathbb{C}$ *(i.e.* $\varepsilon(\mathbf{1}) = 1$, $\varepsilon(b^*) = \overline{\varepsilon(b)}$, *and* $\varepsilon(ab) = \varepsilon(a)\varepsilon(b)$ *for all* $a, b \in \mathcal{B}$*). A* Schürmann triple on $(\mathcal{B}, \varepsilon)$ *is a triple* (ρ, η, L) *consisting of*

- *a unital* $*$*-representation* $\rho : \mathcal{B} \to \mathcal{L}(D)$ *of* $\mathcal{B}$ *on some pre-Hilbert space* D,
- *a* ρ-ε-*1*-cocycle $\eta : \mathcal{B} \to D$, *i.e. a linear map* $\eta : \mathcal{B} \to D$ *such that*

$$\eta(ab) = \rho(a)\eta(b) + \eta(a)\varepsilon(b) \tag{1.2}$$

  for all $a, b \in \mathcal{B}$, *and*
- *a hermitian linear functional* $L : \mathcal{B} \to \mathbb{C}$ *that has the bilinear map* $\mathcal{B} \times \mathcal{B} \ni (a, b) \mapsto -\langle \eta(a^*), \eta(b) \rangle$ *as a* ε-ε-*2*-coboundary, *i.e. that satisfies*

$$-\langle \eta(a^*), \eta(b) \rangle = \partial L(a, b) = \varepsilon(a)L(b) - L(ab) + L(a)\varepsilon(b) \tag{1.3}$$

  for all $a, b \in \mathcal{B}$.

We will call a Schürmann triple surjective, *if the cocycle* $\eta : \mathcal{B} \to D$ *is surjective.*

Theorem 1.9. *Let* $\mathcal{B}$ *be an involutive bialgebra. We have one-to-one correspondences between Lévy processes on* $\mathcal{B}$ *(modulo equivalence), convolution semigroups of states on* $\mathcal{B}$*, generators on* $\mathcal{B}$*, and surjective Schürmann triples on* $\mathcal{B}$ *(modulo unitary equivalence).*

Proof. It only remains to establish the one-to-one correspondence between generators and Schürmann triples.

Let (ρ, η, L) be a Schürmann triple, then we can show that L is a generator, i.e. a hermitian, conditionally positive linear functional with $L(\mathbf{1}) = 0$.

The cocycle has to vanish on the unit element $\mathbf{1}$, since

$$\eta(\mathbf{1}) = \eta(\mathbf{1} \cdot \mathbf{1}) = \rho(\mathbf{1})\eta(\mathbf{1}) + \eta(\mathbf{1})\varepsilon(\mathbf{1}) = 2\eta(\mathbf{1}).$$

This implies

$$L(\mathbf{1}) = L(\mathbf{1} \cdot \mathbf{1}) = \varepsilon(\mathbf{1})L(\mathbf{1}) + \langle \eta(\mathbf{1}), \eta(\mathbf{1}) \rangle + L(\mathbf{1})\varepsilon(\mathbf{1}) = 2L(\mathbf{1}) = 0.$$

Furthermore, L is hermitian by definition and conditionally positive, since by (1.3) we get

$$L(b^*b) = \langle \eta(b), \eta(b) \rangle = ||\eta(b)||^2 \geq 0$$

for $b \in \ker \varepsilon$.

Let now L be a generator. The sesqui-linear form $\langle \cdot, \cdot \rangle_L : \mathcal{B} \times \mathcal{B} \to \mathbb{C}$ defined by

$$\langle a, b \rangle_L = L\Big(\big(a - \varepsilon(a)\mathbf{1}\big)^* \big(b - \varepsilon(b)\mathbf{1}\big) \Big)$$

for $a, b \in \mathcal{B}$ is positive, since L is conditionally positive. Dividing $\mathcal{B}$ by the null-space

$$\mathcal{N}_L = \{ a \in \mathcal{B} | \langle a, a \rangle_L = 0 \}$$

we obtain a pre-Hilbert space $D = \mathcal{B}/\mathcal{N}_L$ with a positive definite inner product $\langle \cdot, \cdot \rangle$ induced by $\langle \cdot, \cdot \rangle_L$. For the cocycle $\eta : \mathcal{B} \to D$ we take the canonical projection, this is clearly surjective and satisfies Equation (1.3).

The $*$-representation ρ is induced from the left multiplication on $\mathcal{B}$ on $\ker \varepsilon$, i.e.

$$\rho(a)\eta\big(b - \varepsilon(b)\mathbf{1}\big) = \eta\Big(a\big(b - \varepsilon(b)\mathbf{1}\big)\Big) \quad \text{or} \quad \rho(a)\eta(b) = \eta(ab) - \eta(a)\varepsilon(b)$$

for $a, b \in \mathcal{B}$. To show that this is well-defined, we have to verify that left multiplication by elements of $\mathcal{B}$ leaves the null-space invariant. Let therefore $a, b \in \mathcal{B}$, $b \in \mathcal{N}_L$, then we have

$$\begin{aligned} \Big|\Big|\Big(a\big(b - \varepsilon(b)\mathbf{1}\big)\Big)\Big|\Big|^2 &= L\Big(\big(ab - a\varepsilon(b)\mathbf{1}\big)^*\big(ab - a\varepsilon(b)\mathbf{1}\big)\Big) \\ &= L\Big(\big(b - \varepsilon(b)\mathbf{1}\big)^* a^* \big(ab - a\varepsilon(b)\mathbf{1}\big)\Big) \\ &= \big\langle b - \varepsilon(b)\mathbf{1}, a^*a\big(b - \varepsilon(b)\mathbf{1}\big)\big\rangle_L \\ &\leq ||b - \varepsilon(b)\mathbf{1}||^2 \; \big|\big|a^*a\big(b - \varepsilon(b)\mathbf{1}\big|\big|^2 = 0, \end{aligned}$$

with Schwarz' inequality.

That the Schürmann triple (ρ, η, L) obtained in this way is unique up to unitary equivalence follows similarly as for the usual GNS construction. □

Exercise 1.10. Let $(X_t)_{t\geq 0}$ be a classical real-valued Lévy process with all moments finite (on some probability space $(\Omega, \mathcal{F}, P)$). Define a Lévy process on the free unital algebra $\mathbb{C}[x]$ generated by one symmetric element $x = x^*$ with the coproduct and counit determined by $\Delta(x) = x \otimes \mathbf{1} + \mathbf{1} \otimes x$ and $\varepsilon(x) = 0$, whose moments agree with those of $(X_t)_{t\geq 0}$. More precisely, such that

$$\Phi\big(j_{st}(x^k)\big) = \mathbb{E}\left((X_t - X_s)^k\right)$$

holds for all $k \in \mathbb{N}$ and all $0 \leq s \leq t$.

Construct the Schürmann triple for Brownian motion and for a compound Poisson process (with finite moments).

For the classification of Gaussian and drift generators on an involutive bialgebra $\mathcal{B}$ with counit ε, we need the ideals

$$\begin{aligned} K &= \ker \varepsilon, \\ K^2 &= \mathrm{span}\,\{ab | a, b \in K\}, \\ K^3 &= \mathrm{span}\,\{abc | a, b, c \in K\}. \end{aligned}$$

Proposition 1.11. *Let L be a conditionally positive, hermitian linear functional on $\mathcal{B}$. Then the following are equivalent.*

(i) $\eta = 0$,
(ii) $L|_{K^2} = 0$,
(iii) L is an ε-derivation, i.e. $L(ab) = \varepsilon(a)L(b) + L(a)\varepsilon(b)$ for all $a, b \in \mathcal{B}$,
(iv) The states φ_t are homomorphisms, i.e. $\varphi_t(ab) = \varphi_t(a)\varphi_t(b)$ for all $a, b \in \mathcal{B}$ and $t \geq 0$.

If a conditionally positive, hermitian linear functional L satisfies one of these conditions, then we call it and the associated Lévy process a *drift*.

Proposition 1.12. *Let L be a conditionally positive, hermitian linear functional on $\mathcal{B}$.*

Then the following are equivalent.

(i) $L|_{K^3} = 0$,
(ii) $L(b^*b) = 0$ for all $b \in K^2$,
(iii) $L(abc) = L(ab)\varepsilon(c) + L(ac)\varepsilon(b) + L(bc)\varepsilon(a) - \varepsilon(ab)L(c) - \varepsilon(ac)L(b) - \varepsilon(bc)L(a)$ for all $a, b, c \in \mathcal{B}$,
(iv) $\rho|_K = 0$ for the representation ρ in the surjective Schürmann triple (ρ, η, L) associated to L by the GNS-type construction presented in the proof of Theorem 1.9,

(v) $\rho = \varepsilon\mathbf{1}$, *for the representation* ρ *in the surjective Schürmann triple* (ρ, η, L) *associated to* L *by the GNS-type construction presented in the proof of Theorem 1.9,*
(vi) $\eta|_{K^2} = 0$ *for the cocycle* η *in any Schürmann triple* (ρ, η, L) *containing* L,
(vii) $\eta(ab) = \varepsilon(a)\eta(b) + \eta(a)\varepsilon(b)$ *for all* $a, b \in \mathcal{B}$ *and the cocycle* η *in any Schürmann triple* (ρ, η, L) *containing* L.

If a conditionally positive, hermitian linear functional L satisfies one of these conditions, then we call it and also the associated Lévy process *quadratic* or *Gaussian.*

The proofs of the preceding two propositions can be carried out as an exercise or found in [Sch93, Section 5.1].

Proposition 1.13. *Let* L *be a conditionally positive, hermitian linear functional on* $\mathcal{B}$. *Then the following are equivalent.*

(i) *There exists a state* $\varphi : \mathcal{B} \to \mathbb{C}$ *and a real number* $\lambda > 0$ *such that*

$$L(b) = \lambda\big(\varphi(b) - \varepsilon(b)\big)$$

for all $b \in \mathcal{B}$.
(ii) *There exists a Schürmann triple* (ρ, η, L) *containing* L, *in which the cocycle* η *is trivial, i.e. of the form*

$$\eta(b) = \big(\rho(b) - \varepsilon(b)\big)\omega, \qquad \text{for all } b \in \mathcal{B},$$

for some non-zero vector $\omega \in D$. *In this case we will also call* η *the* coboundary *of the vector* ω.

If a conditionally positive, hermitian linear functional L satisfies one of these conditions, then we call it a *Poisson generator* and the associated Lévy process a *compound Poisson process.*

Proof. To show that (ii) implies (i), set $\varphi(b) = \frac{\langle \omega, \rho(b)\omega\rangle}{\langle \omega, \omega\rangle}$ and $\lambda = ||\omega||^2$.

For the converse, let (D, ρ, ω) be the GNS triple for $(\mathcal{B}, \varphi)$ and check that (ρ, η, L) with $\eta(b) = \big(\rho(b) - \varepsilon(b)\big)\omega$, $b \in \mathcal{B}$ defines a Schürmann triple. □

Remark 1.14. The Schürmann triple for a Poisson generator $L = \lambda(\varphi - \varepsilon)$ obtained by the GNS construction for φ is not necessarily surjective. Consider, e.g., a classical additive $\mathbb{R}$-valued compound Poisson process, whose Lévy measure μ is not supported on a finite set. Then the construction of a surjective Schürmann triple in the proof of Theorem 1.9 gives the pre-Hilbert space $D_0 = \text{span}\,\{x^k | k = 1, 2, \ldots\} \subseteq L^2(\mathbb{R}, \mu)$. On the other hand, the GNS-construction for φ leads to the pre-Hilbert space $D = \text{span}\,\{x^k | k = 0, 1, 2, \ldots\} \subseteq L^2(\mathbb{R}, \mu)$. The cocycle η is the coboundary of the constant function, which is not contained in D_0.

1.3 The Representation Theorem

The representation theorem gives a direct way to construct a Lévy process from the Schürmann triple, using quantum stochastic calculus.

Theorem 1.15. (Representation theorem) *Let $\mathcal{B}$ be an involutive bialgebra and (ρ, η, L) a Schürmann triple on $\mathcal{B}$. Then the quantum stochastic differential equations*

$$\mathrm{d}j_{st} = j_{st} \star \big(\mathrm{d}A_t^* \circ \eta + \mathrm{d}\Lambda_t \circ (\rho - \varepsilon) + \mathrm{d}A_t \circ \eta \circ * + L\mathrm{d}t\big) \tag{1.4}$$

with the initial conditions

$$j_{ss} = \varepsilon \mathbf{1}$$

have a solution $(j_{st})_{0 \le s \le t}$. Moreover, in the vacuum state $\Phi(\cdot) = \langle \Omega, \cdot \Omega \rangle$, $(j_{st})_{0 \le s \le t}$ is a Lévy process with generator L.

Conversely, every Lévy process with generator L is equivalent to $(j_{st})_{0 \le s \le t}$.

For the proof of the representation theorem we refer to [Sch93, Chapter 2].

Written in integral form and applied to an element $b \in \mathcal{B}$ with $\Delta(b) = b_{(1)} \otimes b_{(2)}$ (Sweedler's notation), Equation (1.4) takes the form

$$j_{st}(b) = \varepsilon(b)\mathbf{1} +$$
$$\int_s^t j_{s\tau}(b_{(1)})\Big(\mathrm{d}A_\tau^*\big(\eta(b_{(2)})\big) + \mathrm{d}\Lambda_\tau\big(\rho(b_{(2)} - \varepsilon(b_{(2)})\big) + \mathrm{d}A_\tau\big(\eta(b_{(2)}^*)\big) + L(b_{(2)})\mathrm{d}\tau\Big).$$

Exercise 1.16. Show that

$$\mathrm{d}M_t = \mathrm{d}A_t^* \circ \eta + \mathrm{d}\Lambda_t \circ (\rho - \varepsilon) + \mathrm{d}A_t \circ \eta \circ * + L\mathrm{d}t$$

formally defines a $*$-homomorphism on $\ker \varepsilon = \mathcal{B}_0$, if we define the algebra of quantum stochastic differentials (or Itô algebra, cf. [Bel98] and the references therein) over some pre-Hilbert space D as follows.

The algebra of quantum stochastic differentials $\mathcal{I}(D)$ over D is the $*$-algebra generated by

$$\{\mathrm{d}\Lambda(F) | F \in \mathcal{L}(D)\} \cup \{\mathrm{d}A^*(u) | u \in D\} \cup \{\mathrm{d}A(u) | u \in D\} \cup \{\mathrm{d}t\},$$

if we identify

$$\begin{aligned} \mathrm{d}\Lambda(\lambda F + \mu G) &\equiv \lambda \mathrm{d}\Lambda(F) + \mu \mathrm{d}\Lambda(G), \\ \mathrm{d}A^*(\lambda u + \mu v) &\equiv \lambda \mathrm{d}A^*(u) + \mu \mathrm{d}A^*(v), \\ \mathrm{d}A(\lambda u + \mu v) &\equiv \overline{\lambda} \mathrm{d}A(u) + \overline{\mu} \mathrm{d}A(v), \end{aligned}$$

for all $F, G \in \mathcal{L}(D)$, $u, v \in D$, $\lambda, \mu \in \mathbb{C}$. The involution of $\mathcal{I}(D)$ is defined by

$$\begin{aligned} \mathrm{d}\Lambda(F)^* &= \mathrm{d}\Lambda(F^*), \\ \left(\mathrm{d}A^*(u)\right)^* &= \mathrm{d}A(u), \\ \mathrm{d}A(u)^* &= \mathrm{d}A^*(u), \\ \mathrm{d}t^* &= \mathrm{d}t, \end{aligned}$$

for $F \in \mathcal{L}(D)$, $u \in D$, and the multiplication by the Itô table

$\bullet$	$\mathrm{d}A^*(u)$	$\mathrm{d}\Lambda(F)$	$\mathrm{d}A(u)$	$\mathrm{d}t$
$\mathrm{d}A^*(v)$	0	0	0	0
$\mathrm{d}\Lambda(G)$	$\mathrm{d}A^*(Gu)$	$\mathrm{d}\Lambda(GF)$	0	0
$\mathrm{d}A(v)$	$\langle v, u\rangle \mathrm{d}t$	$\mathrm{d}A(F^*v)$	0	0
dt	0	0	0	0

for all $F, G \in \mathcal{L}(D)$, $u, v \in D$, i.e. we have, for example,

$$\mathrm{d}A(v) \bullet \mathrm{d}A^*(u) = \langle v, u\rangle \mathrm{d}t, \quad \text{and} \quad \mathrm{d}A^*(u) \bullet \mathrm{d}A(v) = 0.$$

Proposition 1.17. *Let $(j_{st})_{0\le s\le t}$ be a Lévy process on a $*$-bialgebra $\mathcal{B}$ with Schürmann triple (ρ, η, L), realized on the Fock space $\Gamma\big(L^2(\mathbb{R}_+, D)\big)$ over the pre-Hilbert space D. Let furthermore u be a unitary operator on D and $\omega \in D$. Then the quantum stochastic differential equation*

$$\mathrm{d}U_t = U_t\left(\mathrm{d}A_t(\omega) - \mathrm{d}A_t^*(u\omega) + \mathrm{d}\Lambda_t(u - \mathbf{1}) - \frac{||\omega||^2}{2}\mathrm{d}t\right)$$

with the initial condition $U_0 = \mathbf{1}$ has a unique solution $(U_t)_{t\ge 0}$ with U_t a unitary for all $t \ge 0$.

Furthermore, the quantum stochastic process $(\tilde{\jmath}_{st})_{0\le s\le t}$ defined by

$$\tilde{\jmath}_{st}(b) = U_t^* j_{st}(b) U_t, \qquad \textit{for } b \in \mathcal{B},$$

is again a Lévy process with respect to the vacuum state. The Schürmann triple $(\tilde{\rho}, \tilde{\eta}, \tilde{L})$ of $(\tilde{\jmath}_{st})_{0\le s\le t}$ is given by

$$\begin{aligned} \tilde{\rho}(b) &= u^*\rho(b)u, \\ \tilde{\eta}(b) &= u^*\eta(b) - u^*\big(\rho(b) - \varepsilon(b)\big)u\omega, \\ \tilde{L}(b) &= L(b) - \langle u\omega, \eta(b)\rangle - \langle \eta(b^*), u\omega\rangle + \langle u\omega, \big(\rho(b) - \varepsilon(b)\big)u\omega\rangle \\ &= L(b) - \langle \omega, \tilde{\eta}(b)\rangle - \langle \tilde{\eta}(b^*), \omega\rangle - \langle \omega, \big(\tilde{\rho}(b) - \varepsilon(b)\big)\omega\rangle \end{aligned}$$

The proof of this proposition is part of the following Exercise.

Exercise 1.18. Show that (on exponential vectors) the operator process $(U_t)_{t\ge 0}$ is given by

$$U_t = e^{-A_t^*(u\omega)}\Gamma_t(u)e^{A_t(\omega)}e^{-t||\omega||^2/2},$$

where $\Gamma_t(u)$ denotes the second quantization of u. $(U_t)_{t\geq 0}$ is a unitary local cocycle or *HP-cocycle*, cf. [Lin05, Bha05].

Setting

$$k_t(x) = U_t$$

and extending this as a $*$-homomorphism, we get a Lévy process on the group algebra $\mathcal{A} = \mathbb{C}\mathbb{Z}$. $\mathcal{A}$ can be regarded as the $*$-algebra generated by one unitary generator x, i.e. $\mathbb{C}\mathbb{Z} \cong \mathbb{C}\langle x, x^*\rangle / \langle xx^* - \mathbf{1}, x^*x - \mathbf{1}\rangle$. Its Hopf algebra structure is given by

$$\varepsilon(x) = 1, \qquad \Delta(x) = x \otimes x, \qquad S(x) = x^*.$$

Verify that $(\tilde{\jmath}_{st})_{0\leq s\leq t}$ is a Lévy process, using the information on $(U_t)_{t\geq 0}$ we have due to the fact that it is a local unitary cocycle or a Lévy process.

Using the quantum Itô formula, one can then show that $(\tilde{\jmath}_{st})_{0\leq s\leq t}$ satisfies the quantum stochastic differential equation

$$\mathrm{d}\tilde{\jmath}_{st} = j_{st} \star \left(\mathrm{d}A_t^* \circ \tilde{\eta} + \mathrm{d}\Lambda_t \circ (\tilde{\rho} - \varepsilon) + \mathrm{d}A_t \circ \tilde{\eta} \circ * + \tilde{L}\mathrm{d}t\right)$$

with initial condition $\tilde{\jmath}_{ss} = \varepsilon\mathbf{1}$, and deduce that $(\tilde{\rho}, \tilde{\eta}, \tilde{L})$ is a Schürmann triple for $(\tilde{\jmath}_{st})_{0\leq s\leq t}$.

Corollary 1.19. *If the cocycle η is trivial, then $(j_{st})_{\leq s\leq t}$ is cocycle conjugate to the second quantization $\left(\Gamma_{st}(\rho)\right)_{0\leq s\leq t}$ of ρ.*

1.4 Cyclicity of the Vacuum Vector

Recently, Franz, Schürmann, and Skeide[FS03] have shown that the vacuum vector is cyclic for the realization of a Lévy process over the Fock space given by Theorem 1.15, if the cocycle is surjective.

Theorem 1.20. *Let (ρ, η, L) be a surjective Schürmann triple on an involutive bialgebra $\mathcal{B}$ and let $(j_{st})_{0\leq s\leq t}$ be the solution of Equation (1.4) on the Fock space $\Gamma\left(L^2(\mathbb{R}_+, D)\right)$. Then the vacuum vector Ω is cyclic for $(j_{st})_{0\leq s\leq t}$, i.e. the span of*

$$\{j_{s_1t_1}(b_1)\cdots j_{s_nt_n}(b_n)\Omega | n \in \mathbb{N}, 0 \leq s_1 \leq t_1 \leq s_2 \leq \cdots \leq t_n, b_1, \ldots, b_n \in \mathcal{B}\}$$

is dense in $\Gamma\left(L^2(\mathbb{R}_+, D)\right)$.

The proof which we will present here is due to Skeide. It uses the fact that the exponential vectors of indicator functions form a total subset of the Fock space.

Theorem 1.21. *[PS98, Ske00] Let $\mathfrak{h}$ be a Hilbert space and $B \subseteq \mathfrak{h}$ a total subset of $\mathfrak{h}$. Let furthermore $\mathcal{R}$ denote the ring generated by bounded intervals in $\mathbb{R}_+$. Then*

$$\{\mathcal{E}(v\mathbf{1}_I) | v \in B, I \in \mathcal{R}\}$$

is total in $\Gamma\left(L^2(\mathbb{R}_+, \mathfrak{h})\right)$.

We first show how exponential vectors of indicator functions of intervals can be generated from the vacuum vector.

Lemma 1.22. *Let $0 \le s \le t$ and $b \in \ker\varepsilon$. For $n \in \mathbb{N}$, we define*

$$\Pi^n_{[s,t]}(b) = j_{s,s+\delta}(\mathbf{1}+b) j_{s+\delta,s+2\delta}(\mathbf{1}+b)\cdots j_{t-\delta,t}(\mathbf{1}+b)e^{-(t-s)L(b)},$$

where $\delta = (t-s)/n$. Then $\Pi^n_{[s,t]}(b)\Omega$ converges to the exponential vector $\mathcal{E}\big(\eta(b)\mathbf{1}_{[s,t]}\big)$

Proof. Let $b \in \mathcal{B}$ and $k \in D$. Then the fundamental lemma of quantum stochastic calculus, cf. [Lin05], implies

$$\begin{aligned}
&\langle \mathcal{E}(k\mathbf{1}_{[0,T]}), j_{st}(b)\Omega\rangle \\
&= \varepsilon(b) + \int_s^t \langle \mathcal{E}(k\mathbf{1}_{[0,T]}), j_{s\tau}(b_{(1)})\Omega\rangle\big(\langle k,\eta(b_{(2)})\rangle + L(b_{(2)})\big)\mathrm{d}\tau
\end{aligned}$$

for $0 \le s \le t \le T$. This is an integral equation for a linear functional on $\mathcal{B}$, it has a unique solution given by the convolution exponential

$$\langle \mathcal{E}(k\mathbf{1}_{[0,T]}), j_{st}(b)\Omega\rangle = \exp_\star (t-s)\big(\langle k,\eta(b)\rangle + L(b)\big).$$

(On the right-hand-side compute first the convolution exponential of the functional $b \mapsto (t-s)\big(\langle k,\eta(b)\rangle + L(b)\big)$ and then apply it to b.)

Let $b \in \ker\varepsilon$, then we have

$$\langle \mathcal{E}(k\mathbf{1}_{[0,T]}), j_{st}(\mathbf{1}+b)e^{-(t-s)L(b)}\Omega\rangle = 1 + (t-s)\langle k,\eta(b)\rangle + O\left((t-s)^2\right)$$

for all $0 \le s \le t \le T$.

Furthermore, we have

$$\begin{aligned}
&\langle j_{st}(\mathbf{1}+b)e^{-(t-s)L(b)}\Omega, j_{st}(\mathbf{1}+b)e^{-(t-s)L(b)}\Omega\rangle \\
&\quad = \langle \Omega, j_{st}\big((\mathbf{1}+b)^*(\mathbf{1}+b)\big)e^{-(t-s)(L(b)+L(b^*))}\Omega\rangle \\
&\quad = \big(1 + \varphi_{t-s}(b^*) + \varphi_{t-s}(b) + \varphi_{t-s}(b^*b)\big)e^{-(t-s)(L(b)+L(b^*))}
\end{aligned}$$

for $b \in \ker\varepsilon$, and therefore

$$\begin{aligned}
&\langle j_{st}(\mathbf{1}+b)e^{-(t-s)L(b)}\Omega, j_{st}(\mathbf{1}+b)e^{-(t-s)L(b)}\Omega\rangle \\
&\quad = 1 + (t-s)\langle \eta(b),\eta(b)\rangle + O\left((t-s)^2\right).
\end{aligned}$$

These calculations show that $\Pi^n_{[s,t]}(b)\Omega$ converges in norm to the exponential vector $\mathcal{E}\big(\eta(b)\mathbf{1}_{[s,t]}\big)$, since using the independence of increments of $(j_{st})_{0\le s\le t}$, we get

$$\begin{aligned}
&\left|\left|\Pi^n_{[s,t]}(b)\Omega - \mathcal{E}\big(\eta(b)\mathbf{1}_{[s,t]}\big)\right|\right|^2 \\
&= \langle \Pi^n_{[s,t]}(b)\Omega, \Pi^n_{[s,t]}(b)\Omega\rangle - \langle \Pi^n_{[s,t]}(b)\Omega, \mathcal{E}\big(\eta(b)\mathbf{1}_{[s,t]}\big)\rangle \\
&\quad -\langle \mathcal{E}\big(\eta(b)\mathbf{1}_{[s,t]}\big), \Pi^n_{[s,t]}(b)\Omega\rangle + \langle \mathcal{E}\big(\eta(b)\mathbf{1}_{[s,t]}\big), \mathcal{E}\big(\eta(b)\mathbf{1}_{[s,t]}\big)\rangle \\
&= \big(1+\delta||\eta(b)||^2 + O(\delta^2)\big)^n - e^{(t-s)||\eta(b)||^2} \\
&\overset{n\to\infty}{\longrightarrow} 0.
\end{aligned}$$

□

Proof. (of Theorem 1.20) We can generate exponential vectors of the form $\mathcal{E}(v\mathbf{1}_I)$, with $I = I_1 \cup \cdots \cup I_k \in \mathcal{R}$ a union of disjoint intervals by taking products

$$\Pi^n_I(b) = \Pi^n_{I_1}(b)\cdots\Pi^n_{I_k}(b)$$

with an element $b \in \ker\varepsilon$, $\eta(b) = v$. If η is surjective, then it follows from Theorem 1.21 that we can generate a total subset from the vacuum vector. □

If the Lévy process is defined on a Hopf algebra, then it is sufficient to consider time-ordered products of increments corresponding to intervals starting at 0.

Corollary 1.23. *Let H be a Hopf algebra with antipode S. Let furthermore (ρ,η,L) be a surjective Schürmann triple on H over D and $(j_{st})_{0\le s\le t}$ the solution of Equation* (1.4) *on the Fock space $\Gamma\big(L^2(\mathbb{R}_+,D)\big)$. Then the subspaces*

$$\begin{aligned}
\mathcal{H}_\uparrow &= \operatorname{span}\{j_{0t_1}(b_1)\cdots j_{0t_n}(b_n)\Omega | 0\le t_1\le t_2\le\cdots\le t_n, b_1,\ldots,b_n\in H\},\\
\mathcal{H}_\downarrow &= \operatorname{span}\{j_{0t_n}(b_1)\cdots j_{0t_1}(b_n)\Omega | 0\le t_1\le t_2\le\cdots\le t_n, b_1,\ldots,b_n\in H\},
\end{aligned}$$

are dense in $\Gamma\big(L^2(\mathbb{R}_+,D)\big)$.

Remark 1.24. Let (ρ,η,L) be an arbitrary Schürmann triple on some involutive bialgebra $\mathcal{B}$ and let $(j_{st})_{0\le s\le t}$ be the solution of Equation (1.4) on the Fock space $\Gamma\big(L^2(\mathbb{R}_+,D)\big)$. Denote by $\mathcal{H}_0$ the span of the vectors that can be created from the vacuum using arbitrary increments.

Then we have $\mathcal{H}_\uparrow \subseteq \mathcal{H}_0$ and $\mathcal{H}_\downarrow \subseteq \mathcal{H}_0$ for the subspaces $\mathcal{H}_\uparrow,\mathcal{H}_\downarrow,\mathcal{H}_0 \subseteq \Gamma\big(L^2(\mathbb{R}_+,D)\big)$ defined as in Theorem 1.20 and Corollary 1.23. This follows since any product $j_{s_1t_1}(b_1)\cdots j_{s_nt_n}(b_n)$ with arbitrary bounded intervals $[s_1,t_1],\ldots[s_n,t_n] \subseteq \mathbb{R}_+$ can be decomposed in a linear combination of products with disjoint intervals, see the proof of Lemma 1.5.

E.g., for $j_{0s}(a)j_{0t}(b)$, $a,b\in\mathcal{B}$, $0\le s\le t$, we get

$$j_{0s}(a)j_{0t}(b) = j_{0s}(ab_{(1)})j_{st}(b_{(2)})$$

where $\Delta(b) = b_{(1)}\otimes b_{(2)}$.

Proof. The density of $\mathcal{H}_\uparrow$ follows, if we show $\mathcal{H}_\uparrow = \mathcal{H}_0$. This is clear, if we show that the map $T_1 : H \otimes H \to H \otimes H$, $T_1 = (m \otimes \mathrm{id}) \circ (\mathrm{id} \otimes \Delta)$, i.e., $T_1(a \otimes b) = ab_{(1)} \otimes b_{(2)}$ is a bijection, since

$$\begin{aligned} & j_{0t_1}(b_1) \cdots j_{0t_n}(b_n) \\ &= m_{\mathcal{A}}^{(n-1)} \circ (j_{0t_1} \otimes j_{t_1t_2} \otimes \cdots \otimes j_{t_{n-1}t_n})\Big((b_1 \otimes \mathbf{1})(\Delta(b_2) \otimes \mathbf{1}) \cdots (\Delta^{(n-1)})\Big) \\ &= m_{\mathcal{A}}^{(n-1)} \circ (j_{0t_1} \otimes j_{t_1t_2} \otimes \cdots \otimes j_{t_{n-1}t_n}) \circ T_1^{(n)}(b_1 \otimes \cdots \otimes b_n), \end{aligned}$$

where

$$T_1^{(n)} = (T_1 \otimes \mathrm{id}_{H^{\otimes(n-2)}}) \circ (\mathrm{id}_H \otimes T_1 \otimes \mathrm{id}_{H^{\otimes(n-3)}}) \circ \cdots \circ (\mathrm{id}_{H^{\otimes(n-2)}} \otimes T_1)$$

see also [FS99, Section 4.5]. To prove that T_1 is bijective, we give an explicit formula for its inverse,

$$T_1^{-1} = (m \otimes \mathrm{id}) \circ (\mathrm{id} \otimes S \otimes \mathrm{id}) \circ (\mathrm{id} \otimes \Delta).$$

To show $\mathcal{H}_\downarrow = \mathcal{H}_0$ it is sufficient to show that the map $T_2 : H \otimes H \to H \otimes H$, $T_2 = (m \otimes \mathrm{id}) \circ (\mathrm{id} \otimes \tau) \circ (\Delta \otimes \mathrm{id})$, $T_2(a \otimes b) = a_{(1)}b \otimes a_{(2)}$ is bijective. This follows from the first part of the proof, since $T_1 = (* \otimes *) \circ T_2 \circ (* \otimes *)$. □

Exercise 1.25. (a) Prove $T_1 \circ T_1^{-1} = \mathrm{id}_{H \otimes H} = T_1^{-1} \circ T_1$ using associativity, coassociativity, and the antipode axiom.
(b) Find an explicit formula for the inverse of T_2.

The following simple lemma is useful for checking if a Gaussian Schürmann triple is surjective.

Lemma 1.26. *Let (ρ, η, L) be a Gaussian Schürmann triple on a $*$-bialgebra $\mathcal{B}$ and let $G \subseteq \mathcal{B}$ be a set of algebraic generators, i.e.*

$$\mathrm{span}\{g_1 \cdots g_n | n \in \mathbb{N}, g_1, \ldots, g_n \in G\} = \mathcal{B}.$$

Then we have

$$\mathrm{span}\, \eta(G) = \eta(\mathcal{B}).$$

Proof. For Gaussian Schürmann triples one can show by induction over n,

$$\eta(g_1 \cdots g_n) = \sum_{k=1}^{n} \varepsilon(g_1 \cdots g_{k-1} g_{k+1} \cdots g_n) \eta(g_k).$$

□

1.5 Examples

Additive Lévy Processes

For a vector space V the *tensor algebra* $T(V)$ is the vector space

$$T(V) = \bigoplus_{n \in \mathbb{N}} V^{\otimes n},$$

where $V^{\otimes n}$ denotes the n-fold tensor product of V with itself, $V^{\otimes 0} = \mathbb{C}$, with the multiplication given by

$$(v_1 \otimes \cdots \otimes v_n)(w_1 \otimes \cdots \otimes w_m) = v_1 \otimes \cdots \otimes v_n \otimes w_1 \otimes \cdots \otimes w_m,$$

for $n, m \in \mathbb{N}$, $v_1, \ldots, v_n, w_1, \ldots, w_m \in V$. The elements of $\bigcup_{n \in \mathbb{N}} V^{\otimes n}$ are called homogeneous, and the degree of a homogeneous element $a \neq 0$ is n if $a \in V^{\otimes n}$. If $\{v_i | i \in I\}$ is a basis of V, then the tensor algebra $T(V)$ can be viewed as the free algebra generated by v_i, $i \in I$. The tensor algebra can be characterized by the following universal property.

There exists an embedding $\imath : V \to T(V)$ of V into $T(V)$ such that for any linear mapping $R : V \to \mathcal{A}$ from V into an algebra there exists a unique algebra homomorphism $T(R) : T(V) \to \mathcal{A}$ such that the following diagram commutes,

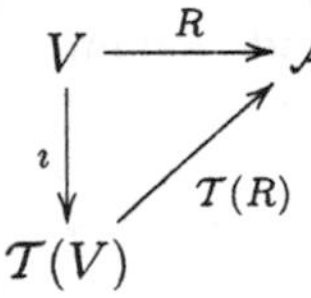

$$\begin{array}{ccc} V & \xrightarrow{\ R\ } & \mathcal{A} \\ {\scriptstyle \imath}\downarrow & \nearrow{\scriptstyle T(R)} & \\ T(V) & & \end{array}$$

i.e. $T(R) \circ \imath = R$.

Conversely, any algebra homomorphism $Q : T(V) \to \mathcal{A}$ is uniquely determined by its restriction to V.

In a similar way, an involution on V gives rise to a unique extension as an involution on $T(V)$. Thus for a $*$-vector space V we can form the tensor $*$-algebra $T(V)$. The tensor $*$-algebra $T(V)$ becomes a $*$-bialgebra, if we extend the linear $*$-maps

$$\varepsilon : V \to \mathbb{C}, \qquad\qquad \varepsilon(v) = 0,$$
$$\Delta : V \to T(V) \otimes T(V), \quad \Delta(v) = v \otimes \mathbf{1} + \mathbf{1} \otimes v,$$

as $*$-homomorphisms to $T(V)$. We will denote the coproduct $T(\Delta)$ and the counit $T(\varepsilon)$ again by Δ and ε. The tensor $*$-algebra is even a Hopf $*$-algebra with the antipode defined by $S(v) = -v$ on the generators and extended as an anti-homomorphism.

We will now study Lévy processes on $T(V)$. Let D be a pre-Hilbert space and suppose we are given

1. a linear $*$-map $R : V \to \mathcal{L}(D)$,

2. a linear map $N : V \to D$, and
3. a linear $*$-map $\psi : V \to \mathbb{C}$ (i.e. a hermitian linear functional),

then

$$J_t(v) = \Lambda_t\big(R(v)\big) + A_t^*\big(N(v)\big) + A_t\big(N(v^*)\big) + t\psi(v) \tag{1.5}$$

for $v \in V$ extends to a Lévy process $(j_t)_{t\geq 0}$, $j_t = \mathcal{T}(J_t)$, on $\mathcal{T}(V)$ (w.r.t. the vacuum state).

In fact, and as we shall prove in the following two exercises, all Lévy processes on $\mathcal{T}(V)$ are of this form, cf. [Sch91b].

Exercise 1.27. Show that (R, N, ψ) can be extended to a Schürmann triple on $\mathcal{T}(V)$ as follows

1. Set $\rho = \mathcal{T}(R)$.
2. Define $\eta : \mathcal{T}(V) \to D$ by $\eta(\mathbf{1}) = 0$, $\eta(v) = N(v)$ for $v \in V$, and
$$\eta(v_1 \otimes \cdots \otimes v_n) = R(v_1)\cdots R(v_{n-1})N(v_n)$$
for homogeneous elements $v_1 \otimes \cdots \otimes v_n \in V^{\otimes n}$, $n \geq 2$.
3. Finally, define $L : \mathcal{T}(V) \to \mathbb{C}$ by $L(\mathbf{1}) = 0$, $L(v) = \psi(v)$ for $v \in V$, and
$$L(v_1 \otimes \cdots \otimes v_n) = \begin{cases} \big\langle N(v_1^*), N(v_2)\big\rangle & \text{if } n = 2, \\ \big\langle N(v_1^*), R(v_2)\cdots R(v_{n-1})N(v_n)\big\rangle & \text{if } n \geq 3, \end{cases}$$
for homogeneous elements $v_1 \otimes \cdots \otimes v_n \in V^{\otimes n}$, $n \geq 2$.

Prove furthermore that all Schürmann triples of $\mathcal{T}(V)$ are of this form.

Exercise 1.28. Let (ρ, η, L) be a Schürmann triple on $\mathcal{T}(V)$. Write down the corresponding quantum stochastic differential equation for homogeneous elements $v \in V$ of degree 1 and show that its solution is given by (1.5).

Lévy Processes on Finite Semigroups

Exercise 1.29. Let $(G, \cdot, e)$ be a finite semigroup with unit element e. Then the complex-valued functions $\mathcal{F}(G)$ on G form an involutive bialgebra. The algebra structure and the involution are given by pointwise multiplication and complex conjugation. The coproduct and counit are defined by

$$\begin{aligned} \Delta(f)(g_1, g_2) &= f(g_1 \cdot g_2) \qquad \text{for } g_1, g_2 \in G, \\ \varepsilon(f) &= f(e), \end{aligned}$$

for $f \in \mathcal{F}(G)$.

Show that the classical Lévy processes in G (in the sense of [App05]) are in one-to-one correspondence to the Lévy processes on the $*$-bialgebra $\mathcal{F}(G)$.

Lévy Processes on Real Lie Algebras

The theory of factorizable representations was developed in the early seventies by Araki, Streater, Parthasarathy, Schmidt, Guichardet, $\cdots$, see, e.g. [Gui72, PS72] and the references therein, or Section 5 of the historical survey by Streater [Str00]. In this Subsection we shall see that in a sense this theory is a special case of the theory of Lévy processes on involutive bialgebras.

Definition 1.30. *A* Lie algebra $\mathfrak{g}$ *over a field* $\mathbb{K}$ *is a* $\mathbb{K}$*-vector space with a linear map* $[\cdot,\cdot]:\mathfrak{g}\times\mathfrak{g}\to\mathfrak{g}$ *called* Lie bracket *that satisfies the following two properties.*

1. *Anti-symmetry: for all* $X,Y\in\mathfrak{g}$*, we have*
$$[X,Y]=-[Y,X].$$
2. *Jacobi identity: for all* $X,Y,Z\in\mathfrak{g}$*, we have*
$$\big[X,[Y,Z]\big]+\big[Y,[Z,X]\big]+\big[Z,[X,Y]\big]=0.$$

For $\mathbb{K}=\mathbb{R}$, we call $\mathfrak{g}$ a *real Lie algebra*, for $\mathbb{K}=\mathbb{C}$ a *complex Lie algebra.*

If $\mathcal{A}$ is an algebra, then $[a,b]=ab-ba$ defines a Lie bracket on $\mathcal{A}$.

We will see below that we can associate a Hopf $*$-algebra to a real Lie algebra, namely its universal enveloping algebra. But it is possible to define Lévy processes on real Lie algebras without explicit reference to any coalgebra structure.

Definition 1.31. *Let* $\mathfrak{g}$ *be a Lie algebra over* $\mathbb{R}$*,* D *be a pre-Hilbert space, and* $\Omega\in D$ *a unit vector. We call a family* $\big(j_{st}:\mathfrak{g}\to\mathcal{L}(D)\big)_{0\le s\le t}$ *of representations of* $\mathfrak{g}$ *by anti-hermitian operators (i.e. satisfying* $j_{st}(X)^*=-j_{st}(X)$ *for all* $X\in\mathfrak{g}$*,* $0\le s\le t$*) a* Lévy process on $\mathfrak{g}$ *over* D *(with respect to* Ω*), if the following conditions are satisfied.*

1. *(Increment property) We have*
$$j_{st}(X)+j_{tu}(X)=j_{su}(X)$$
for all $0\le s\le t\le u$ *and all* $X\in\mathfrak{g}$*.*
2. *(Independence) We have* $[j_{st}(X),j_{s't'}(Y)]=0$ *for all* $X,Y\in\mathfrak{g}$*,* $0\le s\le t\le s'\le t'$ *and*
$$\begin{aligned}&\langle\Omega,j_{s_1t_1}(X_1)^{k_1}\cdots j_{s_nt_n}(X_n)^{k_n}\Omega\rangle\\&=\langle\Omega,j_{s_1t_1}(X_1)^{k_1}\Omega\rangle\cdots\langle\Omega,j_{s_nt_n}(X_n)^{k_n}\Omega\rangle\end{aligned}$$
for all $n,k_1,\ldots,k_n\in\mathbb{N}$*,* $0\le s_1\le t_1\le s_2\le\cdots\le t_n$*,* $X_1,\ldots,X_n\in\mathfrak{g}$*.*
3. *(Stationarity) For all* $n\in\mathbb{N}$ *and all* $X\in\mathfrak{g}$*, the moments*
$$m_n(X;s,t)=\langle\Omega,j_{st}(X)^n\Omega\rangle$$
depend only on the difference $t-s$*.*

4. (Weak continuity) We have $\lim_{t\searrow s}\langle\Omega, j_{st}(X)^n\Omega\rangle = 0$ *for all* $n \in \mathbb{N}$ *and all* $X \in \mathfrak{g}$.

For a classification of several processes on several Lie algebras of interest of physics and for several examples see also [AFS02, Fra03a].

Exercise 1.32. Let $\mathfrak{g}$ be a real Lie algebra. Then the complex vector space $\mathfrak{g}_{\mathbb{C}} = \mathbb{C}\otimes_{\mathbb{R}}\mathfrak{g} = \mathfrak{g}\oplus i\mathfrak{g}$ is a complex Lie algebra with the Lie bracket

$$[X + iY, X' + iY'] = [X, X'] - [Y, Y'] + i([X, Y'] + [Y, X'])$$

for $X, X', Y, Y' \in \mathfrak{g}$.

1. Show that $* : \mathfrak{g}_{\mathbb{C}} \to \mathfrak{g}_{\mathbb{C}}$, $Z = X + iY \mapsto Z^* = -X + iY$ defines an involution on $\mathfrak{g}_{\mathbb{C}}$, i.e. it satisfies
$$(Z^*)^* = Z \qquad \text{and} \qquad [Z_1, Z_2]^* = [Z_2^*, Z_1^*]$$
for all $Z, Z_1, Z_2 \in \mathfrak{g}_{\mathbb{C}}$
2. Show that $\mathfrak{g} \mapsto (\mathfrak{g}_{\mathbb{C}}, *)$ is an isomorphism between the category of real Lie algebras and the category of involutive complex Lie algebras. What are the morphisms in those two categories? How does the functor $\mathfrak{g} \mapsto (\mathfrak{g}_{\mathbb{C}}, *)$ act on morphisms?

The *universal enveloping algebra* $\mathcal{U}(\mathfrak{g})$ of a Lie algebra $\mathfrak{g}$ can be constructed as the quotient $\mathcal{T}(\mathfrak{g})/\mathcal{J}$ of the tensor algebra $\mathcal{T}(\mathfrak{g})$ over $\mathfrak{g}$ by the ideal $\mathcal{J}$ generated by

$$\{X\otimes Y - Y\otimes X - [X, Y] | X, Y \in \mathfrak{g}\}.$$

The universal enveloping algebra is characterized by a universal property. Composing the embedding $\imath : \mathfrak{g} \to \mathcal{T}(\mathfrak{g})$ with the canonical projection $p : \mathcal{T}(\mathfrak{g}) \to \mathcal{T}(\mathfrak{g})/\mathcal{J}$ we get an embedding $\imath' = p\circ\imath : \mathfrak{g} \to \mathcal{U}(\mathfrak{g})$ of $\mathfrak{g}$ into its enveloping algebra. For every algebra $\mathcal{A}$ and every Lie algebra homomorphism $R : \mathfrak{g} \to \mathcal{A}$ there exists a unique algebra homomorphism $\mathcal{U}(R) : \mathcal{U}(\mathfrak{g}) \to \mathcal{A}$ such that the following diagram commutes,

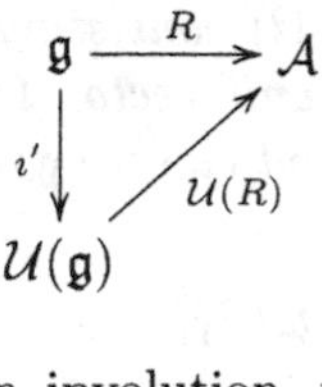

i.e. $\mathcal{U}(R)\circ\imath' = R$. If $\mathfrak{g}$ has an involution, then it can be extended to an involution of $\mathcal{U}(g)$.

The enveloping algebra $\mathcal{U}(\mathfrak{g})$ becomes a bialgebra, if we extend the Lie algebra homomorphism

$$\begin{aligned} &\varepsilon : \mathfrak{g} \to \mathbb{C}, && \varepsilon(X) = 0, \\ &\Delta : \mathfrak{g} \to \mathcal{U}(\mathfrak{g})\otimes\mathcal{U}(\mathfrak{g}), && \Delta(X) = X\otimes\mathbf{1} + \mathbf{1}\otimes X, \end{aligned}$$

to $\mathcal{U}(\mathfrak{g})$. We will denote the coproduct $\mathcal{U}(\Delta)$ and the counit $\mathcal{U}(\varepsilon)$ again by Δ and ε. It is even a Hopf algebra with the antipode $S : \mathcal{U}(\mathfrak{g}) \to \mathcal{U}(\mathfrak{g})$ given by $S(X) = -X$ on $\mathfrak{g}$ and extended as an anti-homomorphism.

Exercise 1.33. Let $\mathfrak{g}$ be a real Lie algebra and $\mathcal{U} = \mathcal{U}(\mathfrak{g}_\mathbb{C})$ the enveloping algebra of its complexification.

1. Let $(j_{st})_{0\leq s\leq t}$ be a Lévy process on $\mathcal{U}$. Show that its restriction to g is a Lévy process on $\mathfrak{g}$.
2. Let $(k_{st})_{0\leq s\leq t}$ now be a Lévy process on $\mathfrak{g}$. Verify that its extension to $\mathcal{U}$ is a Lévy process on $\mathcal{U}$.
3. Show that this establishes a one-to-one correspondence between Lévy processes on a real Lie algebra and Lévy processes on its universal enveloping algebra.

We will now show that Lévy processes on real Lie algebras are the same as factorizable representation of current algebras.

Let $\mathfrak{g}$ be a real Lie algebra and $(\mathbb{T}, \mathcal{T}, \mu)$ a measure space (e.g. the real line $\mathbb{R}$ with the Lebesgue measure λ). Then the set of $\mathfrak{g}$-valued step functions

$$\mathfrak{g}^I = \left\{ X = \sum_{i=1}^n X_i \mathbf{1}_{M_i}; X_i \in \mathfrak{g}, M_i \in \mathcal{T}, \mu(M_i) < \infty, M_i \subseteq I, n \in \mathbb{N} \right\}.$$

on $I \subseteq \mathbb{T}$ is again a real Lie algebra with the pointwise Lie bracket. For $I_1 \subseteq I_2$ we have an inclusion $i_{I_1,I_2} : \mathfrak{g}^{I_1} \to \mathfrak{g}^{I_2}$, simply extending the functions as zero outside I_1. Furthermore, for disjoint subsets $I_1, I_2 \in \mathcal{T}$, $\mathfrak{g}^{I_1\cup I_2}$ is equal to the direct sum $\mathfrak{g}^{I_1} \oplus \mathfrak{g}^{I_2}$. If π be a representation of $\mathfrak{g}^\mathbb{T}$ and $I \in \mathcal{T}$, then have also a representation $\pi^I = \pi \circ i_{I,\mathbb{T}}$ of $\mathfrak{g}^I$

Recall that for two representation ρ_1, ρ_2 of two Lie algebras $\mathfrak{g}_1$ and $\mathfrak{g}_2$, acting on (pre-) Hilbert spaces H_1 and H_2, we can define a representation $\rho = (\rho_1 \otimes \rho_2)$ of $\mathfrak{g}_1 \oplus \mathfrak{g}_1$ acting on $H_1 \otimes H_2$ by

$$(\rho_1 \otimes \rho_2)(X_1 + X_2) = \rho_1(X_1) \otimes \mathbf{1} + \mathbf{1} \otimes \rho_2(X_2),$$

for $X_1 \in \mathfrak{g}_1$, $X_2 \in \mathfrak{g}_2$.

Definition 1.34. *A triple (π, D, Ω) consisting of a representation π of $\mathfrak{g}^\mathbb{T}$ by anti-hermitian operators and a unit vector $\Omega \in D$ is called a* factorizable representation *of the simple current algebra $\mathfrak{g}^\mathbb{T}$, if the following conditions are satisfied.*

1. *(Factorization property) For all $I_1, I_2 \in \mathcal{T}$, $I_1 \cap I_2 = \emptyset$, we have*

$$(\pi^{I_1\cup I_2}, D, \Omega) \cong (\pi^{I_1} \otimes \pi^{I_2}, D \otimes D, \Omega \otimes \Omega).$$

2. *(Invariance) The linear functional $\varphi_I : \mathcal{U}(\mathfrak{g}) \to$ determined by*

$$\varphi_I(X^n) = \langle \Omega, \pi(X\mathbf{1}_I)^n \Omega \rangle$$

for $X \in \mathfrak{g}$, $I \in \mathcal{T}$ depends only on $\mu(I)$.

3. *(Weak continuity) For any sequence* $(I_k)_{k\in\mathbb{N}}$ *with* $\lim_{k\to\infty}\mu(I_k)=0$ *we have* $\lim_{k\to\infty}\varphi_{I_k}(u)=\varepsilon(u)$ *for all* $u\in\mathcal{U}(\mathfrak{g})$.

Proposition 1.35. *Let* $\mathfrak{g}$ *be a real Lie algebra and* $(\mathbb{T},\mathcal{T},\mu)=(\mathbb{R}_+,\mathcal{B}(\mathbb{R}_+),\lambda)$. *Then we have a one-to-one correspondence between factorizable representations of* $\mathfrak{g}^{\mathbb{R}_+}$ *and Lévy processes on* $\mathfrak{g}$.

The relation which is used to switch from one to the other is

$$\pi(X\mathbf{1}_{[s,t[})=j_{st}(X)$$

for $0\leq s\leq t$ and $X\in\mathfrak{g}$.

Proposition 1.36. *Let* $\mathfrak{g}$ *be a real Lie algebra and* $(\mathbb{T},\mathcal{T},\mu)$ *a measure space without atoms. Then all factorizable representations of* $\mathfrak{g}^{\mathbb{T}}$ *are characterized by generators or equivalently by Schürmann triples on* $\mathcal{U}(\mathfrak{g}_{\mathbb{C}})$. *They have a realization on the symmetric Fock space* $\Gamma\big(L^2(\mathbb{T},\mathcal{T},\mu)\big)$ *determined by*

$$\pi(X\mathbf{1}_I)=A^*\big(\mathbf{1}_I\times\rho(X)\big)+\Lambda\big(\mathbf{1}_I\otimes\rho(X)\big)+A\big(\mathbf{1}_I\otimes\eta(X^*)\big)+\mu(I)L(X)$$

for $I\in\mathcal{T}$ *with* $\mu(I)<\infty$ *and* $X\in\mathfrak{g}$.

The Quantum Azéma Martingale

Let $q\in\mathbb{C}$ and $\mathcal{B}_q$ the involutive bialgebra with generators x,x^*,y,y^* and relations

$$\begin{aligned} yx=qxy,&\qquad x^*y=qyx^*,\\ \Delta(x)=x\otimes y+\mathbf{1}\otimes x,&\qquad \Delta(y)=y\otimes y,\\ \varepsilon(x)=0,&\qquad \varepsilon(y)=1. \end{aligned}$$

Proposition 1.37. *There exists a unique Schürmann triple on* $\mathcal{B}_q$ *acting on* $D=\mathbb{C}$ *with*

$$\begin{aligned} \rho(y)&=q,\ \ \rho(x)=0,\\ \eta(y)&=0,\ \ \eta(x)=1,\\ L(y)&=0,\ \ L(x)=0. \end{aligned}$$

Let $(j_{st})_{0\leq s\leq t}$ be the associated Lévy process on $\mathcal{B}_q$ and set $Y_t=j_{0t}(y)$, $X_t=j_{0t}(x)$, and $X_t^*=j_{0t}(x^*)$. These operator processes are determined by the quantum stochastic differential equations

$$\mathrm{d}Y_t=(q-1)Y_t\mathrm{d}\Lambda_t, \tag{1.6}$$
$$\mathrm{d}X_t=\mathrm{d}A_t^*+(q-1)X_t\mathrm{d}\Lambda_t, \tag{1.7}$$
$$\mathrm{d}X_t^*=\mathrm{d}A_t+(\overline{q}-1)X_t\mathrm{d}\Lambda_t, \tag{1.8}$$

with initial conditions $Y_0=\mathbf{1}$, $X_0=X_0^*=0$. This process is the *quantum Azéma martingale* introduced by Parthasarathy [Par90], see also [Sch91a]. The first Equation (1.6) can be solved explicitely, the operator process $(Y_t)_{t\geq 0}$ is the second quantization of multiplication by q, i.e.,

$$Y_t = \Gamma_t(q), \qquad \text{for } t \geq 0$$

Its action on exponential vectors is given by

$$Y_t\mathcal{E}(f) = \mathcal{E}\big(qf\mathbf{1}_{[0,t[} + f\mathbf{1}_{[t,+\infty[}\big).$$

The hermitian operator process $(Z_t)_{t\geq 0}$ defined by $Z_t = X_t + X_t^*$ has as classical version the classical Azéma martingale $(M_t)_{t\geq 0}$ introduced by Azéma and Emery, cf. [Eme89], i.e. is has the same joint moments,

$$\langle \Omega, Z_{t_1}^{n_1} \cdots Z_{t_k}^{n_k} \Omega \rangle = E\left(M_{t_1}^{n_1} \cdots M_{t_k}^{n_k}\right)$$

for all $n_1, \ldots, n_k \in \mathbb{N}$, $t_1, \ldots, t_k \in \mathbb{R}_+$. This was the first example of a classical normal martingale having the so-called *chaotic representation property*, which is not a classical Lévy process.

2 Lévy Processes and Dilations of Completely Positive Semigroups

In this section we will show how Lévy process can be used to construct dilations of quantum dynamical semigroups on the matrix algebra $\mathcal{M}_d$. That unitary cocycles on the symmetric Fock space tensor a finite-dimensional initial space can be interpreted as a Lévy process on a certain involutive bialgebra, was first observed in [Sch90]. For more details on quantum dynamical semigroups and their dilations, see [Bha01, Bha05] and the references therein.

2.1 The Non-Commutative Analogue of the Algebra of Coefficients of the Unitary Group

For $d \in \mathbb{N}$ we denote by $\mathcal{U}_d$ the free non-commutative (!) $*$-algebra generated by indeterminates u_{ij}, u_{ij}^*, $i,j = 1, \ldots, d$ with the relations

$$\sum_{j=1}^{d} u_{kj}u_{\ell j}^* = \delta_{k\ell},$$

$$\sum_{j=1}^{d} u_{jk}^* u_{j\ell} = \delta_{k\ell},$$

The $*$-algebra $\mathcal{U}_d$ is turned into a $*$-bialgebra, if we put

$$\Delta(u_{k\ell}) = \sum_{j=1}^{d} u_{kj} \otimes u_{j\ell},$$

$$\varepsilon(u_{k\ell}) = \delta_{k\ell}.$$

This $*$-bialgebra has been investigated by Glockner and von Waldenfels, see [GvW89]. If we assume that the generators u_{ij}, u_{ij}^* commute, we obtain the coefficient algebra of the unitary group $U(d)$. This is why $\mathcal{U}_d$ is often called the *non-commutative analogue of the algebra of coefficients of the unitary group.* It is isomorphic to the $*$-algebra generated by the mappings

$$\xi_{k\ell} : \mathcal{U}(\mathbb{C}^d \otimes H) \to \mathcal{B}(H)$$

with

$$\xi_{k\ell}(U) = P_k U P_\ell^* = U_{k\ell}$$

for $U \in \mathcal{U}(\mathbb{C}^d \otimes H) \subseteq \mathcal{M}_d\big(\mathcal{B}(H)\big)$, where H is an infinite-dimensional, separable Hilbert space and $\mathcal{U}(\mathbb{C}^d \otimes H)$ denotes the unitary group of operators on $\mathbb{C}^d \otimes H$. Moreover $\mathcal{B}(H)$ denotes the $*$-algebra of bounded operators on H, $\mathcal{M}_d\big(\mathcal{B}(H)\big)$ the $*$-algebra of $d \times d$-matrices with elements from $\mathcal{B}(H)$ and $P_k : \mathbb{C}^d \otimes H \to H$ the projection on the k-th component.

Proposition 2.1. *1. On $\mathcal{U}_1$ a faithful Haar measure is given by $\lambda(u^n) = \delta_{0,n}$, $n \in \mathbb{Z}$.*
2. On $\mathcal{U}_1$ an antipode is given by setting $S(x) = x^*$ and extending S as a $*$-algebra homomorphism.
3. For $d > 1$ the bialgebra $\mathcal{U}_d$ does not possess an antipode.

Exercise 2.2. Recall that a (two-sided) *Haar measure* on a bialgebra $\mathcal{B}$ is a normalized linear functional λ satisfying

$$\lambda \star \varphi = \varphi(1)\lambda = \varphi \star \lambda$$

for all linear functionals φ on $\mathcal{B}$.
Verify (1) and (2).

Proof. Let us prove (3). We suppose that an antipode exists. Then

$$\begin{aligned} u_{\ell k}^* &= \sum_{n=1}^d \sum_{j=1}^d S(u_{kj}) u_{jn} u_{\ell n}^* \\ &= \sum_{j=1}^d S(u_{kj}) \sum_{n=1}^d u_{jn} u_{\ell n}^* \\ &= \sum_{j=1}^d S(u_{kj}) \delta_{j\ell} = S(u_{k\ell}). \end{aligned}$$

Similarly, one proves that $S(u_{k\ell}^*) = u_{lk}$. Since S is an antipode, it has to be an algebra anti-homomorphisms. Therefore,

$$S\left(\sum_{j=1}^d u_{kj} u_{\ell j}^*\right) = \sum_{j=1}^d S(u_{\ell j}^*) S(u_{kj}) = \sum_{j=1}^d u_{j\ell} u_{jk}^*,$$

which is not equal to $\delta_{k\ell}$, if $d > 1$. □

Remark 2.3. Since $\mathcal{U}_d$ does not have an antipode for $d > 1$, it is not a compact quantum group (for $d = 1$, of course, its C^*-completion is the compact quantum group of continuous functions on the circle S^1). We do not know, if $\mathcal{U}_d$ has a Haar measure for $d > 1$.

We have $\mathcal{U}_n = \mathbb{C}1 \oplus \mathcal{U}_n^0$, where $\mathcal{U}_n^0 = K_1 = \ker \varepsilon$ is the ideal generated by $\hat{u}_{ij} = u_{ij} - \delta_{ij}1$, $i, j = 1, \ldots n$, and their adjoints. The defining relations become

$$-\sum_{j=1}^{d} \hat{u}_{ij}\hat{u}_{kj}^* = \hat{u}_{ik} + \hat{u}_{ki}^* = -\sum_{j=1}^{d} \hat{u}_{ji}^*\hat{u}_{jk}, \tag{2.1}$$

for $i, k = 1, \ldots, n$, in terms of these generators. We shall also need the ideals

$$K_2 = \operatorname{span}\{ab|a, b \in K_1\} \quad \text{and} \quad K_3 = \operatorname{span}\{abc|a, b, c \in K_1\}.$$

2.2 An Example of a Lévy Process on $\mathcal{U}_d$

A one-dimensional representation $\sigma : \mathcal{U}_d \to \mathbb{C}$ is determined by the matrix $w = (w_{ij})_{1\leq i,j\leq d} \in \mathcal{M}_d$, $w_{ij} = \sigma(u_{ij})$. The relations in $\mathcal{U}_d$ imply that w is unitary. For $\ell = (\ell_{ij}) \in \mathcal{M}_d$ we can define a σ-cocycle (or σ-derivation) as follows. We set

$$\begin{aligned} \eta_\ell(u_{ij}) &= \ell_{ij}, \\ \eta_\ell(u_{ij}^*) &= -(w^*\ell)_{ji} = -\sum_{k=1}^{d} \overline{w}_{kj}\ell_{ki}, \end{aligned}$$

on the generators and require η_ℓ to satisfy

$$\eta_\ell(ab) = \sigma(a)\eta_\ell(b) + \eta_\ell(a)\varepsilon(b)$$

for $a, b \in \mathcal{U}_d$. The hermitian linear functional $L_{w,\ell} : \mathcal{U}_d \to \mathbb{C}$ with

$$\begin{aligned} L_{w,\ell}(\mathbf{1}) &= 0, \\ L_{w,\ell}(u_{ij}) &= \overline{L_{w,\ell}(u_{ij}^*)} = -\frac{1}{2}(\ell^*\ell)_{ij} = -\frac{1}{2}\sum_{k=1}^{d} \overline{\ell}_{ki}\ell_{kj}, \\ L_{w,\ell}(ab) &= \varepsilon(a)L_{w,\ell}(b) + \overline{\eta_\ell(a^*)}\eta_\ell(b) + L_{w,\ell}(a)\varepsilon(b) \end{aligned}$$

for $a, b \in \mathcal{U}_d$, can be shown to be a generator with Schürmann triple $(\sigma, \eta_\ell, L_{w,\ell})$. The generator $L_{w,\ell}$ is Gaussian if and only if w is the identity matrix.

The associated Lévy process on $\mathcal{U}_d$ is determined by the quantum stochastic differential equations

$$\mathrm{d}j_{st}(u_{ij}) = \sum_{k=1}^{d} j_{st}(u_{ik}) \left(\ell_{kj}\mathrm{d}A_t^* + (w_{kj} - \delta_{kj})\mathrm{d}\Lambda_t - \sum_{n=1}^{d} w_{nj}\overline{\ell}_{nk}\mathrm{d}A_t - \frac{1}{2}\sum_{n=1}^{d} \overline{\ell}_{nk}\ell_{nj}\mathrm{d}t \right),$$

on $\Gamma\big(L^2(\mathbb{R}_+,\mathbb{C})\big)$ with initial conditions $j_{ss}(u_{ij}) = \delta_{ij}$.

We define an operator process $(U_{st})_{0\le s\le t}$ in $\mathcal{M}_d \otimes \mathcal{B}\Big(\Gamma\big(L^2(\mathbb{R}_+,\mathbb{C})\big)\Big) \cong \mathcal{B}\Big(\mathbb{C}^d \otimes \Gamma\big(L^2(\mathbb{R}_+,\mathbb{C})\big)\Big)$ by

$$U_{st} = \big(j_{st}(u_{ij})\big)_{1\le i,j\le d},$$

for $0 \le s \le t$. Then $(U_{st})_{0\le s\le t}$ is a unitary operator process and satisfies the quantum stochastic differential equation

$$\mathrm{d}U_{st} = U_{st}\left(\ell \mathrm{d}A_t^* + (w-\mathbf{1})\mathrm{d}\Lambda_t - \ell^* w \mathrm{d}A_t - \frac{1}{2}\ell^*\ell \mathrm{d}t\right)$$

with initial condition $U_{ss} = \mathbf{1}$. The increment property of $(j_{st})_{0\le s\le t}$ implies that $(U_{st})_{0\le s\le t}$ satisfies

$$U_{0s}U_{s,s+t} = U_{0,s+t} \tag{2.2}$$

for all $0 \le s \le t$.

Let $S_t : L^2(\mathbb{R}_+,\mathcal{K}) \to L^2(\mathbb{R}_+,\mathcal{K})$ be the shift operator,

$$S_t f(s) = \begin{cases} f(s-t) & \text{if } s \ge t, \\ 0 & \text{else,} \end{cases}$$

for $f \in L^2(\mathbb{R}_+,\mathcal{K})$, and define $W_t : \Gamma\big(L^2(\mathbb{R}_+,\mathcal{K})\big) \otimes \Gamma\big(L^2([0,t[,\mathcal{K})\big) \to \Gamma\big(L^2(\mathbb{R}_+,\mathcal{K})\big)$ by

$$W_t\big(\mathcal{E}(f)\otimes\mathcal{E}(g)\big) = \mathcal{E}(g + S_t f),$$

on exponential vectors $\mathcal{E}(f), \mathcal{E}(g)$ of functions $f \in L^2(\mathbb{R}_+,\mathcal{K})$, $g \in L^2([0,t[,\mathcal{K})$. Then the *CCR flow* $\gamma_t : \mathcal{B}\Big(\Gamma\big(L^2(\mathbb{R}_+,\mathcal{K})\big)\Big)$ is defined by

$$\gamma_t(Z) = W_t(Z\otimes\mathbf{1})W_t^*,$$

for $Z \in \mathcal{B}\Big(\Gamma\big(L^2(\mathbb{R}_+,\mathcal{K})\big)\Big)$. On $\mathcal{B}\Big(\mathbb{C}^d \otimes \Gamma\big(L^2(\mathbb{R}_+,\mathcal{K})\big)\Big)$ we have the E_0-semigroup $(\tilde{\gamma}_t)_{t\ge 0}$ with $\tilde{\gamma}_t = \mathrm{id}\otimes\gamma_t$.

We have $U_{s,s+t} = \tilde{\gamma}_s(U_{0t})$ for all $s,t\ge 0$ and therefore increment property (2.2) implies that $(U_t)_{t\ge0}$ with $U_t = U_{0t}$, $t \ge 0$, is a left cocycle of $(\tilde{\gamma}_t)_{t\ge0}$, i.e.

$$U_{s+t} = U_s\tilde{\gamma}_s(U_t),$$

for all $s,t\ge 0$. One can check that $(U_t)_{t\ge0}$ is also local and continuous, i.e. an *HP-cocycle*, see [Lin05, Bha05].

Therefore we can define a new E_0-semigroup $(\eta_t)_{t\ge0}$ on the algebra $\mathcal{B}\Big(\mathbb{C}^d\otimes \Gamma\big(L^2(\mathbb{R}_+,\mathcal{K})\big)\Big)$ by

$$\eta_t(Z) = U_t\tilde{\gamma}_t(Z)U_t^*, \tag{2.3}$$

for $Z \in \mathcal{B}\Big(\mathbb{C}^d \otimes \Gamma\big(L^2(\mathbb{R}_+,\mathcal{K})\big)\Big)$ and $t \ge 0$.

Let $\{e_1, \ldots, e_d\}$ be the standard basis of $\mathbb{C}^d$ and denote by $\mathbb{E}_0$ the conditional expectation from $\mathcal{B}\Big(\mathbb{C}^d \otimes \Gamma\big(L^2(\mathbb{R}_+, \mathcal{K})\big)\Big)$ to $\mathcal{B}(\mathbb{C}^d) \cong \mathcal{M}_d$ determined by

$$\big(\mathbb{E}_0(Z)\big)_{ij} = \langle e_i \otimes \Omega, Z e_j \otimes \Omega\rangle$$

for $Z \in \mathcal{B}\Big(\mathbb{C}^d \otimes \Gamma\big(L^2(\mathbb{R}_+, \mathcal{K})\big)\Big)$. Then

$$\tau_t = \mathbb{E}_0\big(\eta_t(X \otimes \mathbf{1})\big) \tag{2.4}$$

defines a quantum dynamical semigroup on $\mathcal{M}_d$. It acts on the matrix units E_{ij} by

$$\begin{aligned}
&\tau_t(E_{ij}) \\
&= \begin{pmatrix} \langle e_1 \otimes \Omega, U_t(E_{ij} \otimes \mathbf{1})U_t^* e_1 \otimes \Omega\rangle & \cdots & \langle e_1 \otimes \Omega, U_t(E_{ij} \otimes \mathbf{1})U_t^* e_d \otimes \Omega\rangle \\ \vdots & & \vdots \\ \langle e_d \otimes \Omega, U_t(E_{ij} \otimes \mathbf{1})U_t^* e_1 \otimes \Omega\rangle & \cdots & \langle e_d \otimes \Omega, U_t(E_{ij} \otimes \mathbf{1})U_t^* e_d \otimes \Omega\rangle \end{pmatrix} \\
&= \varphi_t \begin{pmatrix} u_{1i}u_{1j}^* & u_{1i}u_{2j}^* & \cdots & u_{1i}u_{dj}^* \\ u_{2i}u_{1j}^* & u_{2i}u_{2j}^* & \cdots & u_{2i}u_{dj}^* \\ \vdots & \vdots & & \vdots \\ u_{di}u_{1j}^* & u_{di}u_{2j}^* & \cdots & u_{di}u_{dj}^* \end{pmatrix},
\end{aligned}$$

and therefore the generator $\mathcal{L}$ of $(\tau_t)_{t\geq 0}$ is given by

$$\mathcal{L}(E_{ij}) = \big(L_{w,\ell}(u_{ki}u_{mj}^*)\big)_{1\leq k,m\leq d},$$

for $1 \leq i, j \leq d$.

Lemma 2.4. *The generator $\mathcal{L}$ of $(\tau_t)_{t\geq 0}$ is given by*

$$\mathcal{L}(X) = \ell^* w X w^* \ell - \frac{1}{2}\{X, \ell^*\ell\}$$

for $X \in \mathcal{M}_d$.

Proof. We have, of course, $\frac{\mathrm{d}}{\mathrm{d}t}\big|_{t=0} \varphi_t(u_{ki}u_{mj}^*) = L_{w,\ell}(u_{ki}u_{mj}^*)$. Using (1.3) and the definition of the Schürmann triple, we get

$$\begin{aligned}
L_{w,\ell}(u_{ki}u_{mj}^*) &= \varepsilon(u_{ki})L_{w,\ell}(u_{mj}^*) + \overline{\eta_\ell(u_{ki}^*)}\eta_\ell(u_{mj}^*) + L_{w,\ell}(u_{ki})\varepsilon(u_{mj}^*) \\
&= -\frac{1}{2}\delta_{ki}\overline{(\ell^*\ell)_{mj}} + (\ell^* w)_{ki}(w^*\ell)_{jm} - \frac{1}{2}(\ell^*\ell)_{ki}\delta_{mj}.
\end{aligned}$$

Writing this in matrix form, we get

$$\big(L_{w,\ell}(u_{ki}u_{mj}^*)\big)_{1\leq k,m\leq d} = -\frac{1}{2}E_{ij}\ell^*\ell + \ell^* w E_{ij} w^*\ell - \frac{1}{2}\ell^*\ell E_{ij},$$

and therefore the formula given in the Lemma. □

2.3 Classification of Generators on $\mathcal{U}_d$

In this section we shall classify all Lévy processes on $\mathcal{U}_d$, see also [Sch97] and [Fra00, Section 4].

The functionals $D_{ij} : \mathcal{U}_d \to \mathbb{C}$, $i, j = 1, \ldots, d$ defined by

$$\begin{aligned} D_{ij}(\hat{u}_{kl}) &= D_{ij}(u_{kl}) = i\delta_{ik}\delta_{jl}, \\ D_{ij}(\hat{u}^*_{kl}) &= D_{ij}(u^*_{kl}) = -D_{ij}(\hat{u}_{lk}) = -i\delta_{il}\delta_{jk}, \\ D_{ij}(u) &= 0 \quad \text{if } u \notin \operatorname{span}\{\hat{u}_{ij}, \hat{u}^*_{ij}; i, j = 1, \ldots, d\}, \end{aligned}$$

for $i, j, k, l = 1, \ldots, d$, are drift generators, since they are hermitian and form Schürmann triples together with the zero cocycle $\eta = 0$ and the trivial representation $\rho = \varepsilon$.

Let $A = (a_{jk}) \in \mathcal{M}_d(\mathbb{C})$ be a complex $d \times d$-matrix. It is not difficult to see that the triples $(\varepsilon, \eta_A : \mathcal{U}_d \to \mathbb{C}, G_A)$, $i, j = 1, \ldots, d$ defined by

$$\begin{aligned} \eta_A(\hat{u}_{jk}) &= \eta_A(u_{jk}) = a_{jk}, \\ \eta_A(\hat{u}^*_{jk}) &= \eta_A(u^*_{jk}) = -\eta_A(u_{kj}) = -a_{kj}, \\ \eta_A(1) &= \eta_A(uv) = 0 \quad \text{for } u, v \in \mathcal{U}^0_d, \end{aligned}$$

and

$$\begin{aligned} G_A(1) &= G_A(\hat{u}_{jk} - \hat{u}^*_{kj}) = 0, \qquad \text{for } j, k = 1, \ldots, d, \\ G_A(\hat{u}_{jk} + \hat{u}^*_{kj}) &= -G_A\left(\sum_{l=1}^d \hat{u}^*_{lj}\hat{u}_{lk}\right) = -\sum_{l=1}^d \overline{a_{lj}}a_{lk} = -(A^*A)_{jk}, \\ &\qquad\qquad \text{for } j, k = 1, \ldots, d, \\ G_A(uv) &= \langle \eta_A(u^*), \eta_A(v)\rangle = \overline{\eta_A(u^*)}\eta_A(v), \end{aligned}$$

for $u, v \in \mathcal{U}^0_d$, are Schürmann triples. Furthermore, we have $\eta_A|_{K_2} = 0$ and $G_A|_{K_3} = 0$, i.e. the generators G_A are Gaussian. On the elements $\hat{u}_{jk}, \hat{u}^*_{jk}$, $j, k = 1, \ldots, d$, this gives

$$\begin{aligned} G_A(\hat{u}_{jk}) &= -\frac{1}{2}(A^*A)_{jk} \\ G_A(\hat{u}^*_{jk}) &= -\frac{1}{2}(A^*A)_{kj} \\ G_A(\hat{u}_{jk}\hat{u}_{lm}) &= \overline{\eta_A(\hat{u}^*_{jk})}\eta_A(\hat{u}_{lm}) = -\overline{a_{kj}}a_{lm}, \\ G_A(\hat{u}_{jk}\hat{u}^*_{lm}) &= \overline{a_{kj}}a_{ml} \\ G_A(\hat{u}^*_{jk}\hat{u}_{lm}) &= \overline{a_{jk}}a_{lm} \\ G_A(\hat{u}^*_{jk}\hat{u}^*_{lm}) &= -\overline{a_{jk}}a_{ml} \end{aligned}$$

for $j, k, l, m = 1, \ldots, d$.

Let us denote the standard basis of $\mathcal{M}_d(\mathbb{C})$ by E_{jk}, $j, k = 1, \ldots, d$. We define the functionals $G_{jk,lm} : \mathcal{U}_d \to \mathbb{C}$ by

$$
\begin{aligned}
G_{jk,lm}(1) &= 0, \\
G_{jk,lm}(\hat{u}_{rs}) &= -\frac{1}{2}\delta_{kr}\delta_{jl}\delta_{ms} = -\frac{1}{2}(E_{jk}^*E_{lm})_{rs}, \qquad \text{for } r,s=1,\dots,d, \\
G_{jk,lm}(\hat{u}_{rs}^*) &= -\frac{1}{2}\delta_{ks}\delta_{jl}\delta_{mr} = -\frac{1}{2}(E_{jk}^*E_{lm})_{sr}, \qquad \text{for } r,s=1,\dots,d, \\
G_{jk,lm}(uv) &= \langle \eta_{E_{jk}}(u^*), \eta_{E_{lm}}(v)\rangle = \overline{\eta_{E_{jk}}(u^*)}\eta_{E_{lm}}(v),
\end{aligned}
$$

for $u, v \in \mathcal{U}_n^0$, $j,k,l,m = 1,\dots,d$.

Theorem 2.5. *A generator $L : \mathcal{U}_d \to \mathbb{C}$ is Gaussian, if and only if it is of the form*

$$
L = \sum_{j,k,l,m=1}^{d} \sigma_{jk,lm} G_{jk,lm} + \sum_{j,k=1}^{d} b_{jk} D_{jk},
$$

with a hermitian $d \times d$-matrix (b_{jk}) and a positive semi-definite $d^2 \times d^2$-matrix $(\sigma_{jk,lm})$. It is a drift, if and only if $\sigma_{jk,lm} = 0$ for $j,k,l,m = 1,\dots,d$.

Proof. Applying L to Equation (2.1), we see that $L(\hat{u}_{jk}) = -L(\hat{u}_{kj}^*)$ has to hold for a drift generator. By the hermitianity we get $L(\hat{u}_{jk}) = \overline{L(\hat{u}_{jk}^*)}$, and thus a drift generator L has to be of the form $\sum_{j,k=1}^{n} b_{ij} D_{ij}$ with a hermitian $d \times d$-matrix (b_{ij}).

Let (ρ, η, L) be a Schürmann triple with a Gaussian generator L. Then we have $\rho = \varepsilon\,\mathrm{id}$, and $\eta(1) = 0$, $\eta|_{K_2} = 0$. By applying η to Equation (2.1), we get $\eta(\hat{u}_{ij}^*) = -\eta(\hat{u}_{ji})$. Therefore $\eta(\mathcal{U}_d)$ has at most dimension d^2 and the Schürmann triple (ρ, η, L) can be realized on the Hilbert space $\mathcal{M}_d(\mathbb{C})$ (where the inner product is defined by $\langle A, B\rangle = \sum_{j,k=1}^{d} \overline{a_{jk}} b_{jk}$ for $A = (a_{jk}), B = (b_{jk}) \in \mathcal{M}_d(\mathbb{C})$). We can write η as

$$
\eta = \sum_{j,k=1}^{d} \eta_{A_{jk}} E_{jk} \tag{2.5}
$$

where the matrices A_{jk} are defined by $(A_{jk})_{lm} = \langle E_{lm}, \eta(\hat{u}_{jk})\rangle$, for $j,k,l,m = 1,\dots,d$.

Then we get

$$
\begin{aligned}
L(\hat{u}^{\bullet_1}_{rs}\hat{u}^{\bullet_2}_{tu}) &= \langle \eta\big((\hat{u}^{\bullet_1}_{rs})^*\big), \eta(\hat{u}^{\bullet_2}_{tu})\rangle \\
&= \sum_{j,k,l,m=1}^{d} \langle \eta_{A_{jk}}\big((\hat{u}^{\bullet_1}_{rs})^*\big)E_{jk}), \eta_{A_{lm}}(\hat{u}^{\bullet_2}_{tu})E_{lm})\rangle \\
&= \sum_{j,k=1}^{d} \overline{\eta_{A_{jk}}\big((\hat{u}^{\bullet_1}_{rs})^*\big)}\eta_{A_{jk}}(\hat{u}^{\bullet_2}_{tu}) \\
&= \begin{cases} \sum_{j,k=1}^{d} -\overline{(A_{jk})_{sr}}(A_{jk})_{tu} & \text{if } \hat{u}^{\bullet_1} = \hat{u}^{\bullet_2} = \hat{u} \\ \sum_{j,k=1}^{d} \overline{(A_{jk})_{sr}}(A_{jk})_{ut} & \text{if } \hat{u}^{\bullet_1} = \hat{u}, \hat{u}^{\bullet_2} = \hat{u}^* \\ \sum_{j,k=1}^{n} \overline{(A_{jk})_{rs}}(A_{jk})_{tu} & \text{if } \hat{u}^{\bullet_1} = \hat{u}^*, \hat{u}^{\bullet_2} = \hat{u} \\ \sum_{j,k=1}^{d} -\overline{(A_{jk})_{rs}}(A_{jk})_{ut} & \text{if } \hat{u}^{\bullet_1} = \hat{u}^{\bullet_2} = \hat{u}^* \end{cases} \\
&= \sum_{l,m,p,q=1}^{d} \sigma_{lm,pq}G_{jk,lm}(\hat{u}^{\bullet_1}_{rs}\hat{u}^{\bullet_2}_{tu})
\end{aligned}
$$

$$
= \left(\sum_{j,k,l,m=1}^{d} \sigma_{jk,lm}G_{jk,lm} + \sum_{jmj=1}^{n} b_{jk}D_{jk}\right)(\hat{u}^{\bullet_1}_{rs}\hat{u}^{\bullet_2}_{tu})
$$

for $r,s,t,u = 1,\ldots,d$, where $\sigma = (\sigma_{lm,pq}) \in \mathcal{M}_{d^2}(\mathbb{C})$ is the positive semi-definite matrix defined by

$$
\sigma_{lm,pq} = \sum_{j,k=1}^{d} \overline{(A_{jk})_{lm}}(A_{jk})_{pq}
$$

for $l,m,p,q = 1,\ldots,d$.

Setting $b_{jk} = -\frac{i}{2}L(\hat{u}_{jk} - \hat{u}^*_{kj})$, for $j,k = 1,\ldots,d$, we get

$$
\begin{aligned}
L(\hat{u}_{rs}) &= L\left(\frac{\hat{u}_{rs}+\hat{u}^*_{sr}}{2} + \frac{\hat{u}_{rs}-\hat{u}^*_{sr}}{2}\right) = -\frac{1}{2}\sum_{p=1}^{d}\langle \eta(\hat{u}_{pr}), \eta(\hat{u}_{ps})\rangle + ib_{rs} \\
&= -\frac{1}{2}\sum_{j,k=1}^{n}(A^*_{jk}A_{jk})_{rs} + ib_{rs} \\
&= \left(\sum_{j,k,l,m=1}^{d} \sigma_{jk,lm}G_{jk,lm} + \sum_{jmj=1}^{d} b_{jk}D_{jk}\right)(\hat{u}_{rs})
\end{aligned}
$$

$$L(\hat{u}_{sr}^*) = L\left(\frac{\hat{u}_{rs}+\hat{u}_{sr}^*}{2} - \frac{\hat{u}_{rs}-\hat{u}_{sr}^*}{2}\right) = -\frac{1}{2}\sum_{p=1}^d \langle \eta(\hat{u}_{pr}), \eta(\hat{u}_{ps})\rangle - ib_{rs}$$

$$= -\frac{1}{2}\sum_{j,k=1}^d (A_{jk}^* A_{jk})_{rs} - ib_{rs}$$

$$= \left(\sum_{j,k,l,m=1}^d \sigma_{jk,lm} G_{jk,lm} + \sum_{jmj=1}^d b_{jk} D_{jk}\right)(\hat{u}_{sr}^*)$$

where we used Equation (2.1) for evaluating $L(\hat{u}_{rs} + \hat{u}_{sr}^*)$. Therefore we have $L = \sum_{j,k,l,m=1}^d \sigma_{jk,lm} G_{jk,lm} + \sum_{jmj=1}^n b_{jk} D_{jk}$, since both sides vanish on K_3 and on **1**. The matrix (b_{jk}) is hermitian, since L is hermitian,

$$\overline{b_{jk}} = \frac{i}{2}\overline{L(\hat{u}_{jk} - \hat{u}_{kj}^*)} = \frac{i}{2}L(\hat{u}_{jk}^* - \hat{u}_{kj}) = b_{kj},$$

for $j, k = 1, \ldots, d$.

Conversely, let $L = \sum_{i,j=1}^d \sigma_{jk,lm} G_{jk,lm} + \sum_{j,k=1}^d b_{jk} D_{jk}$ with a positive semi-definite $d^2 \times d^2$-matrix $(\sigma_{jk,lm})$ and a hermitian $d \times d$-matrix (b_{jk}). Then we can choose a matrix $M = (m_{kl,ml}) \in \mathcal{M}_{d^2}(\mathbb{C})$ such that $\sum_{p,q=1}^d \overline{m_{pq,jk}} m_{pq,lm} = \sigma_{jk,lm}$ for all $i, j, r, s = 1, \ldots, d$. We define $\eta : \mathcal{U}_d \to \mathbb{C}^{d^2}$ by the matrices A_{jk} with components $(A_{jk})_{lm} = m_{jk,lm}$ as in Equation (2.5). It is not difficult to see that $(\varepsilon\, \mathrm{id}_{\mathbb{C}^{d^2}}, \eta, L)$ is a Schürmann triple and L therefore a Gaussian generator. □

We can give the generators of a Gaussian Lévy process on $\mathcal{U}_n$ also in the following form, cf. [Sch93, Theorem 5.1.12]

Proposition 2.6. *Let $L^1, \ldots, L^n, M \in \mathcal{M}_d(\mathbb{C})$, with $M^* = M$, and let H be an n-dimensional Hilbert space with orthonormal basis $\{e_1, \ldots, e_n\}$. Then there exists a unique Gaussian Schürmann triple (ρ, η, L) with*

$$\rho = \varepsilon \mathrm{id}_H,$$
$$\eta(u_{jk}) = \sum_{\nu=1}^n L_{jk}^\nu e_\nu,$$
$$\eta(u_{jk}^*) = -\eta(u_{kj}),$$
$$L(u_{jk}) = \frac{1}{2}\sum_{r=1}^d \langle \eta(u_{jr}^*), \eta(u_{kr})\rangle + iM_{jk}$$

for $1 \le j, k \le d$.

The following theorem gives a classification of all Lévy processes on $\mathcal{U}_d$.

Theorem 2.7. *Let H be a Hilbert space, U a unitary operator on $H \otimes \mathbb{C}^d$, $A = (a_{jk})$ an element of the Hilbert space $H \otimes \mathcal{M}_d(\mathbb{C})$ and $\lambda = (\lambda_{jk}) \in \mathcal{M}_d(\mathbb{C})$ a hermitian matrix. Then there exists a unique Schürmann triple (ρ, η, L) on H such that*

$$\rho(u_{jk}) = P_j U P_k^*, \tag{2.6a}$$

$$\eta(u_{jk}) = a_{jk}, \tag{2.6b}$$

$$L(u_{jk} - u_{kj}^*) = 2i\lambda_{jk}, \tag{2.6c}$$

for $j, k = 1 \ldots, d$, where $P_j : H \otimes \mathbb{C}^d \to H \otimes \mathbb{C}e_j \cong H$ projects a vector with entries in H to its jth component.

Furthermore, all Schürmann triples on $\mathcal{U}_n$ are of this form.

Proof. Let us first show that all Schürmann triples are of the form given in the theorem. If (ρ, η, L) is a Schürmann triple, then we can use the Equations (2.6) to define U, A, and λ. The defining relations of $\mathcal{U}_d$ imply that U is unitary, since

$$U^*UP_l^* = \sum_{j=1}^d \sum_{k=1}^d P_j U^* P_k^* P_k U P_l^* = \sum_{j=1}^d \sum_{k=1}^d \rho(u_{kj}^* u_{kl}) = \sum_{j=1}^d \delta_{jl}\rho(\mathbf{1}) = \mathrm{id}_{H \otimes e_l},$$

$$UU^*P_l^* = \sum_{j=1}^d \sum_{k=1}^d P_j U P_k^* P_k U^* P_l^* = \sum_{j=1}^d \sum_{k=1}^d \rho(u_{jk}^* u_{lk}) = \sum_{j=1}^d \delta_{jl}\rho(\mathbf{1}) = \mathrm{id}_{H \otimes e_l},$$

for $l = 1, \ldots, d$, where $e_1, \ldots, e_d$ denotes the standard basis of $\mathbb{C}^d$. The hermitianity of λ is an immediate consequence of the hermitianity of L.

Conversely, let U, A, and λ be given. Then there exists a unique representation ρ on H such that $\rho(u_{jk}) = P_j U P_k^*$, for $j, k, = 1, \ldots, d$, since the unitarity of U implies that the defining relations of $\mathcal{U}_n$ are satisfied. We can set $\eta(\hat{u}_{jk}) = a_{jk}$, and extend via $\eta(u_{ki}^*) = -\eta\left(\hat{u}_{ik} + \sum_{j=1}^d \hat{u}_{ji}^* \hat{u}_{jk}\right) = -a_{ik} - \sum_{j=1}^d \rho(\hat{u}_{ji})^* a_{jk}$, for $i, k = 1, \ldots, d$ and $\eta(uv) = \rho(u)\eta(v) + \eta(u)\varepsilon(v)$ (i.e. Equation (1.2), for $u, v \in \mathcal{U}_d$, in this way we obtain the unique (ρ, ε)-cocycle with $\eta(\hat{u}_{jk}) = a_{jk}$. Then we set $L(u_{jk}) = i\lambda_{jk} - \frac{1}{2}\sum_{l=1}^d \langle a_{lj}, a_{lk}\rangle$ and $L(u_{kj}^*) = -i\lambda_{jk} - \frac{1}{2}\sum_{l=1}^d \langle a_{lj}, a_{lk}\rangle$, for $j, k = 1, \ldots, d$, and use Equation (1.3) to extend it to all of $\mathcal{U}_d$. This extension is again unique, because the Relation (2.1) implies $L(u_{jk} + u_{kj}^*) = -\sum_{l=1}^d \langle a_{lj}, a_{lk}\rangle$, and this together with $L(u_{jk} - u_{kj}^*) = 2i\lambda_{jk}$ determines L on the generators u_{jk}, u_{jk}^* of $\mathcal{U}_d$. But once L is defined on the generators, it is determined on all of $\mathcal{U}_d$ thanks to Equation (1.3). □

2.4 Dilations of Completely Positive Semigroups on $\mathcal{M}_d$

Let $(\tau_t)_{t\geq 0}$ be a *quantum dynamical semigroup* on $\mathcal{M}_d$, i.e. a weakly continuous semigroup of completely positive maps $\tau_t : \mathcal{M}_d \to \mathcal{M}_d$.

Definition 2.8. *A semigroup $(\theta_t)_{t\geq 0}$ of not necessarily unital endomorphisms of $\mathcal{B}(\mathcal{H})$ with $\mathbb{C}^d \subseteq \mathcal{H}$ is called a* dilation *of $(\tau_t)_{t\geq 0})$, if*

$$\tau_t(X) = P\theta_t(X)P$$

holds for all $t \geq 0$ and all $X \in \mathcal{M}_d = \mathcal{B}(\mathbb{C}^d) = P\mathcal{B}(\mathcal{H})P$. Here P is the orthogonal projection from $\mathcal{H}$ to $\mathbb{C}^d$.

Example 2.9. We can use the construction in Section 2.2 to get an example. Let $(\tau_t)_{t\geq 0}$ be the semigroup defined in (2.4). We identify $\mathbb{C}^d$ with the subspace $\mathbb{C}^d \otimes \Omega \subseteq \mathbb{C}^d \otimes \Gamma\big(L^2(\mathbb{R}_+, \mathcal{K})\big)$. The orthogonal projection $P : \mathcal{H} \to \mathbb{C}^d$ is given by $P = \mathrm{id}_{\mathbb{C}^d} \otimes P_\Omega$, where P_Ω denotes the projection onto the vacuum vector. Furthermore, we consider $\mathcal{M}_d$ as a subalgebra of $\mathcal{B}\Big(\mathbb{C}^d \otimes \Gamma\big(L^2(\mathbb{R}_+, \mathcal{K})\big)\Big)$ by letting a matrix $X \in \mathcal{M}_d$ act on $v \otimes w \in \mathbb{C}^d \otimes \Gamma\big(L^2(\mathbb{R}_+, \mathcal{K})\big)$ as $X \otimes P_\Omega$.

Note that we have

$$\mathbb{E}_0(X) \otimes P_\Omega = PXP$$

for all $X \in \mathcal{B}\Big(\mathbb{C}^d \otimes \Gamma\big(L^2(\mathbb{R}_+, \mathcal{K})\big)\Big)$.

Then the semigroup $(\eta_t)_{t\geq 0}$ defined in (2.3) is a dilation of $(\tau_t)_{t\geq 0}$, since

$$\begin{aligned} P\eta_t(X \otimes P_\Omega)P &= PU_t\tilde{\gamma}_t(X \otimes P_\Omega)U_t^*P = PU_t(X \otimes \mathrm{id}_{\Gamma(L^2([0,t],\mathcal{K}))} \otimes P_\Omega)U_t^*P \\ &= PU_t(X \otimes \mathbf{1})U_t^*P = \tau_t(X) \otimes P_\Omega \end{aligned}$$

for all $X \in \mathcal{M}_d$. Here we used that fact that the HP-cocycle $(U_t)_{t\geq 0}$ is adapted.

Definition 2.10. *A dilation $(\theta_t)_{t\geq 0}$ on $\mathcal{H}$ of a quantum dynamical semigroup $(\tau_t)_{t\geq 0}$ on $\mathbb{C}^d$ is called* minimal, *if the subspace generated by the $\theta_t(X)$ from $\mathbb{C}^d$ is dense in $\mathcal{H}$, i.e. if*

$$\overline{\mathrm{span}\,\{\theta_{t_1}(X_1)\cdots\theta_{t_n}(X_n)v | t_1, \ldots, t_n \geq 0, X_1, \ldots, X_n \in \mathcal{M}_d, v \in \mathbb{C}^d, n \in \mathbb{N}\}}$$

is equal to $\mathcal{H}$.

Lemma 2.11. *It is sufficient to consider ordered times $t_1 \geq t_2 \geq \cdots \geq t_n \geq 0$, since*

$$\begin{aligned} &\mathrm{span}\,\big\{\theta_{t_1}(X_1)\cdots\theta_{t_n}(X_n)v | t_1 \geq \ldots \geq t_n \geq 0, X_1, \ldots, X_n \in \mathcal{M}_d, v \in \mathbb{C}^d\big\} \\ &= \mathrm{span}\,\big\{\theta_{t_1}(X_1)\cdots\theta_{t_n}(X_n)v | t_1, \ldots, t_n \geq 0, X_1, \ldots, X_n \in \mathcal{M}_d, v \in \mathbb{C}^d\big\} \end{aligned}$$

Proof. See [Bha01, Section 3] □

Example 2.12. We will now show that the dilation from Example 2.9 is not minimal, if w and ℓ not linearly independent.

Due to the adaptedness of the HP-cocycle $(U_t)_{t\geq 0}$, we can write

$$\eta_t(X \otimes P_\Omega) = U_t \tilde{\gamma}_t(X \otimes P_\Omega) U_t^* = \big(U_t(X \otimes \mathbf{1})U_t^*\big) \otimes P_\Omega$$

on $\Gamma\big(L^2([0,t],\mathcal{K})\big) \otimes \Gamma\big(L^2([t,\infty[,\mathcal{K})\big)$. Let

$$\hat{\eta}_t(X) = \eta_t(X \otimes \mathbf{1}) = U_t(X \otimes \mathbf{1})U_t^*$$

for $X \in \mathcal{M}_d$ and $t \geq 0$, then we have

$$\hat{\eta}_{t_1}(X_1) \cdots \hat{\eta}_{t_n}(X_n) v = \eta_{t_1}(X_1 \otimes P_\Omega) \cdots \eta_{t_n}(X_n \otimes P_\Omega) v$$

for $v \in \mathbb{C}^d \otimes \Omega$, $n \in \mathbb{N}$, $t_1 \geq \cdots \geq t_n \geq 0$, $X_1, \ldots, X_n \mathcal{M}_d$, i.e. time-ordered products of the $\hat{\eta}_t(X)$ generate the same subspace from $\mathbb{C}^d \otimes \Omega$ as the $\eta_t(X \otimes P_\Omega)$. Using the quantum Itô formula, one can show that the operators $\hat{\eta}_t(X)$, $X \in \mathcal{M}_d$ satisfy the quantum stochastic differential equation.

$$\hat{\eta}_t(X) = U_t(X \otimes \mathbf{1})U_t^* = X \otimes \mathbf{1} + \int_0^t U_s(wXw^* - X)U_s^* \mathrm{d}\Lambda_s, \qquad t \geq 0,$$

if $\ell = \lambda w$.

Since the quantum stochastic differential equation for $\hat{\eta}_t(X)$ has no creation part, these operators leave $\mathbb{C}^d \otimes \Omega$ invariant. More precisely, the subspace

$$\big\{\hat{\eta}_{t_1}(X_1) \cdots \hat{\eta}_{t_n}(X_n) v \otimes \Omega | t_1 \geq \ldots \geq t_n \geq 0, X_1, \ldots, X_n \in \mathcal{M}_d, v \in \mathbb{C}^d\big\}$$

is equal to $\mathbb{C}^d \otimes \Omega$, and therefore the dilation $(\eta_t)_{t\geq 0}$ is not minimal, if w and ℓ are not linearly independent. Note that in this case the quantum dynamical semigroup is also trivial, i.e. $\tau_t = \mathrm{id}$ for all $t \geq 0$, since its generator vanishes.

One can show that the converse is also true, if w and ℓ are linearly independent, then the dilation $(\eta_t)_{t\geq 0}$ is minimal.

The general form of the generator of a quantum dynamical semigroup on $\mathcal{M}_d$ was determined by [GKS76, Lin76].

Theorem 2.13. *Let $(\tau_t)_{t\geq 0}$ be a quantum dynamical semigroup on $\mathcal{M}_d$. Then there exist matrices $M, L^1, \ldots, L^n \in \mathcal{M}_d$, with $M^* = M$, such that the generator $\mathcal{L} = \frac{\mathrm{d}}{\mathrm{d}t}\tau_t$ is given by*

$$\mathcal{L}(X) = i[M, X] + \sum_{k=1}^n \left((L^k)^* X L^k - \frac{1}{2}\left\{X, (L^k)^* L^k\right\}\right)$$

for $X \in \mathcal{M}_d$.

Note that $M, L^1, \ldots, L^n \in \mathcal{M}_d$ are not uniquely determined by $(\tau_t)_{t\geq 0}$.

Proposition 2.6 allows us to associate a Lévy process on $\mathcal{U}_d$ to $L^1, \ldots, L^n$, M. It turns out that the cocycle constructed from this Lévy process as in Section 2.2 dilates the quantum dynamical semigroup whose generator $\mathcal{L}$ is given by $L^1, \ldots, L^n, M$.

Proposition 2.14. *Let $n \in \mathbb{N}$, $M, L^1, \cdots, L^n \in \mathcal{M}_d$, $M^* = M$, and let $(j_{st})_{0\leq s\leq t}$ be the Lévy process on $\mathcal{U}_d$ over $\Gamma\big(L^2(\mathbb{R}_+, \mathbb{C}^n)\big)$, whose Schürmann triple is constructed from $M, L^1, \cdots, L^n$ as in Proposition 2.6. Then the semigroup $(\eta_t)_{t\geq 0}$ defined from the unitary cocycle*

$$U_t = \begin{pmatrix} j_{0t}(u_{11}) & \cdots & j_{0t}(u_{1d}) \\ \vdots & & \vdots \\ j_{0t}(u_{d1}) & \cdots & j_{0t}(u_{dd}) \end{pmatrix}$$

as in (2.3) is a dilation of the quantum dynamical semigroup $(\tau_t)_{t\geq 0}$ with generator

$$\mathcal{L}(X) = i[M, X] + \sum_{k=1}^{n} \left((L^k)^* X L^k - \frac{1}{2}\left\{X, (L^k)^* L^k\right\} \right)$$

for $X \in \mathcal{M}_d$.

Proof. The calculation is similar to the one in Section 2.2. □

We denote this dilation by $(\eta_t)_{t\geq 0}$ and define again $\hat{\eta}_t : \mathcal{M}_d \to \mathcal{B}\Big(C^d \otimes \Gamma\big(L^2([0,t], \mathcal{K})\big)\Big)$, $t \geq 0$ by

$$\hat{\eta}_t(X) = \eta_t(X \otimes \mathbf{1}) = U_t(X \otimes \mathbf{1})U_t^*$$

for $X \in \mathcal{M}_d$.

Denote by $\mathcal{Q}_d$ the subalgebra of $\mathcal{U}_d$ generated by $u_{ij}u_{k\ell}^*$, $1 \leq i, j, k, \ell \leq d$. This is even a subbialgebra, since

$$\Delta(u_{ij}u_{k\ell}^*) = \sum_{r,s=1}^{d} u_{ir}u_{ks}^* \otimes u_{rj}u_{s\ell}^*$$

for all $1 \leq i, j, k, \ell \leq d$.

Lemma 2.15. *Let $\eta : \mathcal{U}_d \to H$ be the cocycle associated to $L^1, \ldots, L^n, M \in \mathcal{M}_d(\mathbb{C})$, with $M^* = M$, in Proposition 2.6.*

(a) η is surjective, if and only if $L^1, \ldots, L^n$ are linearly independent.
(b) $\eta|_{\mathcal{Q}_d}$ is surjective, if and only if $I, L^1, \ldots, L^n$ are linearly independent, where I denotes the identity matrix.

Proof. (a) $\mathcal{U}_d$ is generated by $\{u_{ij} | 1 \leq i, j \leq d\}$, so by Lemma I.1.26 we have $\eta(\mathcal{U}_d) = \text{span}\{\eta(u_{ij}) | 1 \leq i, j \leq d\}$.
Denote by $\Lambda_1 : H \to \text{span}\left\{(L^1)^*, \ldots, (L^n)^*\right\} \subseteq \mathcal{M}_d(\mathbb{C})$ the linear map defined by $\Lambda_1(e_\nu) = (L^\nu)^*$, $\nu = 1, \ldots, n$. Then we have $\ker \Lambda_1 = \eta(\mathcal{U}_d)^\perp$, since

$$\langle v, \eta(u_{ij})\rangle = \sum_{\nu=1}^{d} \overline{v_\nu} L^\nu_{ij} = \overline{\Lambda_1(v)_{ji}}, \qquad 1 \le i,j \le d,$$

for $v = \sum_{\nu=1}^n v_\nu e_\nu \in H$. The map Λ_1 is injective, if and only if $L^1, \ldots, L^n$ are linearly independent. Since $\ker \Lambda_1 = \eta(\mathcal{U}_d)^\perp$, this is also equivalent to the surjectivity of $\eta|_{\mathcal{U}_d}$.

(b) We have $\eta(\mathcal{Q}_d) = \mathrm{span}\{\eta(u_{ij}u^*_{k\ell}) | 1 \le i,j,k,\ell \le d\}$.
Denote by $\Lambda_2 : H \to \mathrm{span}\{(L^1)^* \otimes I - I \otimes (L^1)^*, \ldots, (L^n)^* \otimes I - I \otimes (L^n)^*\} \subseteq \mathcal{M}_d(\mathbb{C}) \otimes \mathcal{M}_d(\mathbb{C})$ the linear map defined by $\Lambda_1(e_\nu) = (L^\nu)^* \otimes I - I \otimes (L^\nu)^*$, $\nu = 1, \ldots, n$. Then we have $\ker \Lambda_2 = \eta(\mathcal{Q}_d)^\perp$, since

$$\begin{aligned}
\langle v, \eta(u_{ij}u^*_{k\ell})\rangle &= \langle v, \varepsilon(u_{ij})\eta(u^*_{k\ell}) + \eta(u_{ij}\varepsilon(u^*_{k\ell})\rangle \\
&= \langle v, -\delta_{ij}\eta(u_{\ell k}) + \eta(u_{ij}\delta_{k\ell})\rangle \\
&= \sum_{\nu=1}^{d} \overline{v_\nu}(L^\nu_{ij}\delta_{k\ell} - \delta_{ij}L^\nu_{\ell k}) \\
&= \overline{\Lambda_2(v)_{ji,k\ell}}, \qquad 1 \le i,j,k,\ell \le d,
\end{aligned}$$

for $v = \sum_{\nu=1}^n v_\nu e_\nu \in H$.
The map Λ_2 is injective, if and only if $L^1, \ldots, L^n$ are linearly independent and $I \notin \mathrm{span}\{L^1, \ldots, L^n\}$, i.e. iff $I, L^1, \ldots, L^n$ are linearly independent. Since $\ker \Lambda_2 = \eta(\mathcal{Q}_d)^\perp$, it follows that this is equivalent to the surjectivity of $\eta|_{\mathcal{Q}_d}$.

□

Bhat [Bha01, Bha05] has given a necessary and sufficient condition for the minimality of dilations of the form we are considering.

Theorem 2.16. *[Bha01, Theorem 9.1] The dilation $(\eta_t)_{t\ge0}$ is minimal if and only if $I, L^1, \ldots, L^n$ are linearly independent.*

Remark 2.17. The preceding arguments show that the condition in Bhat's theorem is necessary. Denote by $\mathcal{H}_0$ the subspace of $\Gamma\big(L^2(\mathbb{R}_+, \mathbb{C}^n)\big)$, which is generated by operators of form $j_{st}(u_{ij}u^*_{k\ell})$. By Theorem 1.20, this subspace is dense in $\Gamma\big(L^2(\mathbb{R}_+, \eta(\mathcal{Q}_d)\big)$. Therefore the subspace generated by elements of the $\hat{\eta}_t(X) = U_t(X \otimes \mathbf{1})U^*_t$ from $\mathbb{C}^d \otimes \Omega$ is contained in $\mathbb{C}^d \otimes \mathcal{H}_0$. If η is minimal, then this subspace in dense in $\mathbb{C}^d \otimes \Gamma\big(L^2(\mathbb{R}_+, \mathbb{C}^n)\big)$. But this can only happen if $\mathcal{H}_0$ is dense in $\Gamma\big(L^2(\mathbb{R}_+, \mathbb{C}^n)\big)$. This implies $\eta(\mathcal{Q}_d) = \mathbb{C}^d$ and therefore that $I, L^1, \ldots, L^n$ are linearly independent.

Bhat's theorem is actually more general, it also applies to dilations of quantum dynamical semigroups on the algebra of bounded operators on an infinite-dimensional separable Hilbert space, whose generator involves infinitely many L's, see [Bha01, Bha05].

3 The Five Universal Independences

In classical probability theory there exists only one canonical notion of independence. But in quantum probability many different notions of independence have been used, e.g., to obtain central limit theorems or to develop a quantum stochastic calculus. If one requires that the joint law of two independent random variables should be determined by their marginals, then an independence gives rise to a product. Imposing certain natural condition, e.g., that functions of independent random variables should again be independent or an associativity property, it becomes possible to classify all possible notions of independence. This program has been carried out in recent years by Schürmann [Sch95a], Speicher [Spe97], Ben Ghorbal and Schürmann [BGS99][BGS02], and Muraki [Mur03, Mur02]. In this section we will present the results of these classifications. Furthermore we will formulate a category theoretical approach to the notion of independence and show that boolean, monotone, and anti-monotone independence can be reduced to tensor independence in a similar way as the bosonization of Fermi independence [HP86] or the symmetrization of [Sch93, Section 3].

3.1 Preliminaries on Category Theory

We recall the basic definitions and properties from category theory that we shall use. For a thorough introduction, see, e.g., [Mac98].

Definition 3.1. *A* category $\mathcal{C}$ *consists of*

(a) a class $\mathrm{Ob}\,\mathcal{C}$ *of* objects *denoted by* $A, B, C, \ldots$,
(b) a class $\mathrm{Mor}\,\mathcal{C}$ *of* morphism *(or* arrows*) denoted by* $f, g, h, \ldots$,
(c) mappings $\mathrm{tar}, \mathrm{src} : \mathrm{Mor}\,\mathcal{C} \to \mathrm{Ob}\,\mathcal{C}$ *assigning to each morphism* f *its* source *(or* domain*)* $\mathrm{src}(f)$ *and its* target *(or* codomain*)* $\mathrm{tar}(f)$*. We will say that* f *is a morphism in* $\mathcal{C}$ *from* A *to* B *or write "*$f : A \to B$ *is a morphism in* $\mathcal{C}$*" if* f *is a morphism in* $\mathcal{C}$ *with source* $\mathrm{src}(f) = A$ *and target* $\mathrm{tar}(f) = B$,
(d) a composition $(f, g) \mapsto g \circ f$ *for pairs of morphisms* f, g *that satisfy* $\mathrm{src}(g) = \mathrm{tar}(f)$,
(e) and a map $\mathrm{id} : \mathrm{Ob}\,\mathcal{C} \to \mathrm{Mor}\,\mathcal{C}$ *assigning to an object* A *of* $\mathcal{C}$ *the* identity morphism $\mathrm{id}_A : A \to A$,

such that the

(1) associativity property*: for all morphisms* $f : A \to B$, $g : B \to C$, *and* $h : C \to D$ *of* $\mathcal{C}$*, we have*

$$(h \circ g) \circ f = h \circ (g \circ f),$$

and the
(2) identity property*:* $\mathrm{id}_{\mathrm{tar}(f)} \circ f = f$ *and* $f \circ \mathrm{id}_{\mathrm{src}(f)} = f$ *holds for all morphisms* f *of* $\mathcal{C}$,

are satisfied.

Let us emphasize that it is not so much the objects, but the morphisms that contain the essence of a category (even though categories are usually named after their objects). Indeed, it is possible to define categories without referring to the objects at all, see the definition of "arrows-only metacategories" in [Mac98, Page 9]. The objects are in one-to-one correspondence with the identity morphisms, in this way $\mathrm{Ob}\,\mathcal{C}$ can always be recovered from $\mathrm{Mor}\,\mathcal{C}$.

We give an example.

Example 3.2. Let $\mathrm{Ob}\,\mathfrak{Set}$ be the class of all sets (of a fixed universe) and $\mathrm{Mor}\,\mathfrak{Set}$ the class of total functions between them. Recall that a *total function* (or simply *function*) is a triple (A, f, B), where A and B are sets, and $f \subseteq A \times B$ is a subset of the cartesian product of A and B such that for a given $x \in A$ there exists a unique $y \in B$ with $(x, y) \in f$. Usually one denotes this unique element by $f(x)$, and writes $x \mapsto f(x)$ to indicate $\big(x, f(x)\big) \in f$. The triple (A, f, B) can also be given in the form $f : A \to B$. We define

$$\mathrm{src}\big((A, f, B)\big) = A, \qquad \text{and} \qquad \mathrm{tar}\big((A, f, B)\big) = B.$$

The composition of two morphisms (A, f, B) and (B, g, C) is defined as

$$(B, g, C) \circ (A, f, B) = (A, g \circ f, C),$$

where $g \circ f$ is the usual composition of the functions f and g, i.e.

$$g \circ f = \{(x, z) \in A \times C;\ \text{there exists a } y \in B \text{ s.t. } (x, y) \in f \text{ and } (y, z) \in g\}.$$

The identity morphism assigned to an object A is given by (A, id_A, A), where $\mathrm{id}_A \subseteq A \times A$ is the identity function, $\mathrm{id}_A = \{(x, x); x \in A\}$. It is now easy to check that these definitions satisfy the associativity property and the identity property, and therefore define a category. We shall denote this category by $\mathfrak{Set}$.

Definition 3.3. *Let $\mathcal{C}$ be a category. A morphism $f : A \to B$ in $\mathcal{C}$ is called an* isomorphism *(or* invertible*), if there exists a morphism $g : B \to A$ in $\mathcal{C}$ such that $g \circ f = \mathrm{id}_A$ and $f \circ g = \mathrm{id}_B$. Such a morphism g is uniquely determined, if it exists, it is called the* inverse *of f and denoted by $g = f^{-1}$. Objects A and B are called* isomorphic, *if there exists an isomorphism $f : A \to B$.*

Morphisms f with $\mathrm{tar}(f) = \mathrm{src}(f) = A$ *are called* endomorphisms *of A. Isomorphic endomorphism are called* automorphisms.

For an arbitrary pair of objects $A, B \in \mathrm{Ob}\,\mathcal{C}$ we define $\mathrm{Mor}_{\mathcal{C}}(A, B)$ to be the collection of morphisms from A to B, i.e.

$$\mathrm{Mor}_{\mathcal{C}}(A, B) = \{f \in \mathrm{Mor}\,\mathcal{C}; \mathrm{src}(f) = A \text{ and } \mathrm{tar}(f) = B\}.$$

Often the collections $\mathrm{Mor}_{\mathcal{C}}(A,B)$ are also denoted by $\hom_{\mathcal{C}}(A,B)$ and called the *hom-sets* of $\mathcal{C}$. In particular, $\mathrm{Mor}_{\mathcal{C}}(A,A)$ contains exactly the endomorphisms of A, they form a semigroup with identity element with respect to the composition of $\mathcal{C}$ (if $\mathrm{Mor}_{\mathcal{C}}(A,A)$ is a set).

Compositions and inverses of isomorphisms are again isomorphisms. The automorphisms of an object form a group (if they form a set).

Example 3.4. Let $(G,\circ,e)$ be a semigroup with identity element e. Then $(G,\circ,e)$ can be viewed as a category. The only object of this category is G itself, and the morphisms are the elements of G. The identity morphism is e and the composition is given by the composition of G.

Definition 3.5. *For every category $\mathcal{C}$ we can define its* dual or opposite category $\mathcal{C}^{\mathrm{op}}$. *It has the same objects and morphisms, but target and source are interchanged, i.e.*

$$\mathrm{tar}_{\mathcal{C}^{\mathrm{op}}}(f) = \mathrm{src}_{\mathcal{C}}(f) \text{ and } \mathrm{src}_{\mathcal{C}^{\mathrm{op}}}(f) = \mathrm{tar}_{\mathcal{C}}(f)$$

and the composition is defined by $f \circ_{\mathrm{op}} g = g \circ f$. We obviously have $\mathcal{C}^{\mathrm{op}\,\mathrm{op}} = \mathcal{C}$.

Dualizing, i.e. passing to the opposite category, is a very useful concept in category theory. Whenever we define something in a category, like an epimorphism, a terminal object, a product, etc., we get a definition of a "co-something", if we take the corresponding definition in the opposite category. For example, an *epimorphism* or *epi* in $\mathcal{C}$ is a morphism in $\mathcal{C}$ which is right cancellable, i.e. $h \in \mathrm{Mor}\,\mathcal{C}$ is called an epimorphism, if for any morphisms $g_1, g_2 \in \mathrm{Mor}\,\mathcal{C}$ the equality $g_1 \circ h = g_2 \circ h$ implies $g_1 = g_2$. The dual notion of a epimorphism is a morphism, which is an epimorphism in the category $\mathcal{C}^{\mathrm{op}}$, i.e. a morphism that is left cancellable. It could therefore be called a "co-epimorphism", but the generally accepted name is *monomorphism* or *monic*. The same technique of dualizing applies not only to definitions, but also to theorems. A morphism $r : B \to A$ in $\mathcal{C}$ is called a *right inverse* of $h : A \to B$ in $\mathcal{C}$, if $h \circ r = \mathrm{id}_B$. If a morphism has a right inverse, then it is necessarily an epimorphism, since $g_1 \circ g = g_2 \circ h$ implies $g_1 = g_1 \circ g \circ r = g_2 \circ h \circ r = g_2$, if we compose both sides of the equality with a right inverse r of h. Dualizing this result we see immediately that a morphism $f : A \to B$ that has a *left inverse* (i.e. a morphism $l : B \to A$ such that $l \circ f = \mathrm{id}_A$) is necessarily a monomorphism. Left inverses are also called *retractions* and right inverses are also called *sections*. Note that one-sided inverses are usually not unique.

Definition 3.6. *A category $\mathcal{D}$ is called a* subcategory *of the category $\mathcal{C}$, if*

(1) the objects of $\mathcal{D}$ form a subclass of $\mathrm{Ob}\,\mathcal{C}$, and the morphisms of $\mathcal{D}$ form a subclass of $\mathrm{Mor}\,\mathcal{C}$,
(2) for any morphism f of $\mathcal{D}$, the source and target of f in $\mathcal{C}$ are objects of $\mathcal{D}$ and agree with the source and target taken in $\mathcal{D}$,
(3) for every object D of $\mathcal{D}$, the identity morphism id_D of $\mathcal{C}$ is a morphism of $\mathcal{D}$, and

(4) for any pair $f: A \to B$ and $g: B \to C$ in $\mathcal{D}$, the composition $g \circ f$ in $\mathcal{C}$ is a morphism of $\mathcal{D}$ and agrees with the composition of f and g in $\mathcal{D}$.

A subcategory $\mathcal{D}$ of $\mathcal{C}$ is called full, *if for any two objects $A, B \in \mathrm{Ob}\,\mathcal{D}$ all $\mathcal{C}$-morphisms from A to B belong also to $\mathcal{D}$, i.e. if*

$$\mathrm{Mor}_{\mathcal{D}}(A,B) = \mathrm{Mor}_{\mathcal{C}}(A,B).$$

Remark 3.7. If D is an object of $\mathcal{D}$, then the identity morphism of D in $\mathcal{D}$ is the same as that in $\mathcal{C}$, since the identity element of a semigroup is unique, if it exists.

Exercise 3.8. Let $(G, \circ, e)$ be a unital semigroup. Show that a subsemigroup G_0 of G defines a subcategory of $(G, \circ, e)$ (viewed as a category), if and only if $e \in G_0$.

Definition 3.9. *Let $\mathcal{C}$ and $\mathcal{D}$ be two categories. A* covariant functor *(or simply* functor*) $T : \mathcal{C} \to \mathcal{D}$ is a map for objects and morphisms, every object $A \in \mathrm{Ob}\,\mathcal{C}$ is mapped to an object $T(A) \in \mathrm{Ob}\,\mathcal{D}$, and every morphism $f : A \to B$ in $\mathcal{C}$ is mapped to a morphism $T(f) : T(A) \to T(B)$ in $\mathcal{D}$, such that the identities and the composition are respected, i.e. such that*

$$\begin{aligned} T(\mathrm{id}_A) &= \mathrm{id}_{T(A)}, && \textit{for all } A \in \mathrm{Ob}\,\mathcal{C} \\ T(g \circ f) &= T(g) \circ T(f), && \textit{whenever } g \circ f \textit{ is defined in } \mathcal{C}. \end{aligned}$$

We will denote the collection of all functors between two categories $\mathcal{C}$ and $\mathcal{D}$ by $\mathrm{Funct}(\mathcal{C}, \mathcal{D})$.

A contravariant functor *$T : \mathcal{C} \to \mathcal{D}$ maps an object $A \in \mathrm{Ob}\,\mathcal{C}$ to an object $T(A) \in \mathrm{Ob}\,\mathcal{D}$, and a morphism $f : A \to B$ in $\mathcal{C}$ to a morphism $T(f) : T(B) \to T(A)$ in $\mathcal{D}$, such such that*

$$\begin{aligned} T(\mathrm{id}_A) &= \mathrm{id}_{T(A)}, && \textit{for all } A \in \mathrm{Ob}\,\mathcal{C} \\ T(g \circ f) &= T(f) \circ T(g), && \textit{whenever } g \circ f \textit{ is defined in } \mathcal{C}. \end{aligned}$$

Example 3.10. Let $\mathcal{C}$ be a category. The *identity functor* $\mathrm{id}_{\mathcal{C}} : \mathcal{C} \to \mathcal{C}$ is defined by $\mathrm{id}_{\mathcal{C}}(A) = A$ and $\mathrm{id}_{\mathcal{C}}(f) = f$.

Example 3.11. The *inclusion* of a subcategory $\mathcal{D}$ of $\mathcal{C}$ into $\mathcal{C}$ also defines a functor, we can denote it by $\subseteq : \mathcal{D} \to \mathcal{C}$ or by $\mathcal{D} \subseteq \mathcal{C}$.

Example 3.12. The functor $\mathrm{op} : \mathcal{C} \to \mathcal{C}^{\mathrm{op}}$ that is defined as the identity map on the objects and morphisms is a contravariant functor. This functor allows to obtain covariant functors from contravariant ones. Let $T : \mathcal{C} \to \mathcal{D}$ be a contravariant functor, then $T \circ \mathrm{op} : \mathcal{C}^{\mathrm{op}} \to \mathcal{D}$ and $\mathrm{op} \circ T : \mathcal{C} \to \mathcal{D}^{\mathrm{op}}$ are covariant.

Example 3.13. Let G and H be unital semigroups, then the functors $T : G \to H$ are precisely the identity preserving semigroup homomorphisms from G to H.

Functors can be composed, if we are given two functors $S : \mathcal{A} \to \mathcal{B}$ and $T : \mathcal{B} \to \mathcal{C}$, then the composition $T \circ S : \mathcal{A} \to \mathcal{C}$,

$$\begin{aligned} (T \circ S)(A) &= T(S(A)), \qquad \text{for } A \in \mathrm{Ob}\,\mathcal{A}, \\ (T \circ S)(f) &= T(S(f)), \qquad \text{for } f \in \mathrm{Mor}\,\mathcal{A}, \end{aligned}$$

is again a functor. The composite of two covariant or two contravariant functors is covariant, whereas the composite of a covariant and a contravariant functor is contravariant. The identity functor obviously is an identity w.r.t. to this composition. Therefore we can define categories of categories, i.e. categories whose objects are categories and whose morphisms are the functors between them.

Definition 3.14. *Let $\mathcal{C}$ and $\mathcal{D}$ be two categories and let $S, T : \mathcal{C} \to \mathcal{D}$ be two functors between them. A* natural transformation *(or* morphism of functors*) $\eta : S \to T$ assigns to every object $A \in \mathrm{Ob}\,\mathcal{C}$ of $\mathcal{C}$ a morphism $\eta_A : S(A) \to T(A)$ such that the diagram*

$$\begin{array}{ccc} S(A) & \xrightarrow{\eta_A} & T(A) \\ {\scriptstyle S(f)}\downarrow & & \downarrow{\scriptstyle T(f)} \\ S(B) & \xrightarrow{\eta_B} & T(B) \end{array}$$

is commutative for every morphisms $f : A \to B$ in $\mathcal{C}$. The morphisms η_A, $A \in \mathrm{Ob}\,\mathcal{C}$ are called the components *of η. If every component η_A of $\eta : S \to T$ is an isomorphism, then $\eta : S \to T$ is called a* natural isomomorphism *(or a* natural equivalence*), in symbols this is expressed as $\eta : S \cong T$.*

We will denote the collection of all natural transformations between two functors $S, T : \mathcal{C} \to \mathcal{D}$ by $\mathrm{Nat}(S, T)$.

Exercise 3.15. Let G_1 and G_2 be two groups (regarded as categories as in Example 3.4). $S, T : G_1 \to G_2$ are functors, if they are group homomorphisms, see Example 3.13. Show that there exists a natural transformation $\eta : S \to T$ if and only if S and T are conjugate, i.e. if there exists an element $h \in G$ such that $T(g) = hS(g)h^{-1}$ for all $g \in G_1$.

Definition 3.16. *Natural transformations can also be composed. Let $S, T, U : \mathcal{B} \to \mathcal{C}$ and let $\eta : S \to T$ and $\vartheta : T \to U$ be two natural transformations. Then we can define a natural transformation $\vartheta \cdot \eta : S \to U$, its components are simply $(\vartheta \cdot \eta)_A = \vartheta_A \circ \eta_A$. To show that this defines indeed a natural transformation, take a morphism $f : A \to B$ of $\mathcal{B}$. Then the following diagram is commutative, because the two trapezia are.*

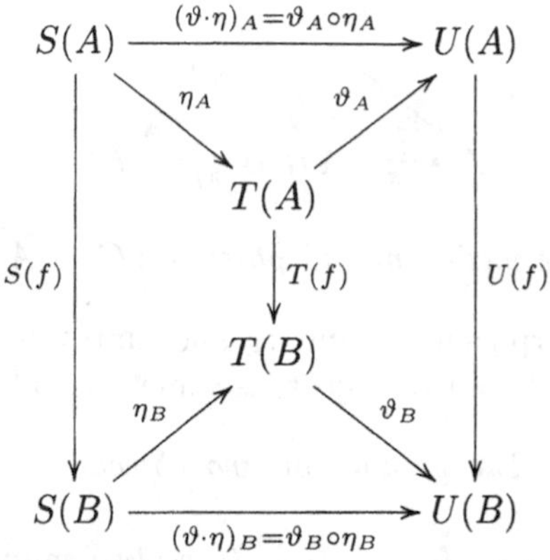

For a given functor $S : \mathcal{B} \to \mathcal{C}$ there exists also the identical natural transformation $\mathrm{id}_S : S \to S$ *that maps $A \in \mathrm{Ob}\,\mathcal{B}$ to $\mathrm{id}_{S(A)} \in \mathrm{Mor}\,\mathcal{C}$, it is easy to check that it behaves as a unit for the composition defined above.*

Therefore we can define the functor category $\mathcal{C}^{\mathcal{B}}$ *that has the functors from $\mathcal{B}$ to $\mathcal{C}$ as objects and the natural transformations between them as morphisms.*

Remark 3.17. Note that a natural transformation $\eta : S \to T$ has to be defined as the triple $(S, (\eta_A)_A, T)$ consisting of its the source S, its components $(\eta_A)_A$ and its target T. The components $(\eta_A)_A$ do not uniquely determine the functors S and T, they can also belong to a natural transformation between another pair of functors (S', T').

Definition 3.18. *Two categories $\mathcal{B}$ and $\mathcal{C}$ can be called* isomorphic, *if there exists an invertible functor $T : \mathcal{B} \to \mathcal{C}$. A useful weaker notion is that of* equivalence *or* categorical equivalence. *Two categories $\mathcal{B}$ and $\mathcal{C}$ are equivalent, if there exist functors $F : \mathcal{B} \to \mathcal{C}$ and $G : \mathcal{C} \to \mathcal{B}$ and natural isomorphisms $G \circ F \cong \mathrm{id}_{\mathcal{B}}$ and $F \circ G \cong \mathrm{id}_{\mathcal{C}}$.*

We will look at products and coproducts of objects in a category. The idea of the product of two objects is an abstraction of the Cartesian product of two sets. For any two sets M_1 and M_2 their Cartesian product $M_1 \times M_2$ has the property that for any pair of maps (f_1, f_2), $f_1 : N \to M_1$, $f_2 : N \to M_2$, there exists a unique map $h : N \to M_1 \times M_2$ such that $f_i = p_i \circ h$ for $i = 1, 2$, where $p_i : M_1 \times M_2 \to M_i$ are the canonical projections $p_i(m_1, m_2) = m_i$. Actually, the Cartesian product $M_1 \times M_2$ is characterized by this property up to isomorphism (of the category $\mathfrak{Set}$, i.e. set-theoretical bijection).

Definition 3.19. *A triple $(A \,\Pi\, B, \pi_A, \pi_B)$ is called a* product *(or* binary product*) of the objects A and B in the category $\mathcal{C}$, if for any object $C \in \mathrm{Ob}\,\mathcal{C}$ and any morphisms $f : C \to A$ and $g : C \to B$ there exists a unique morphism h such that the following diagram commutes,*

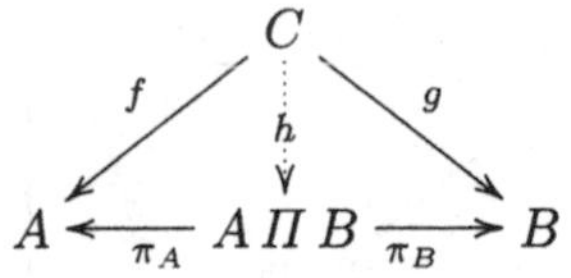

We will also denote the mediating morphism $h : C \to A \mathit{\Pi} B$ *by* $[f, g]$.

Often one omits the morphisms π_A and π_B and simply calls $A \mathit{\Pi} B$ the product of A and B. The product of two objects is sometimes also denoted by $A \times B$.

Proposition 3.20. *(a) The product of two objects is unique up to isomorphism, if it exists.*

(b) Let $f_1 : A_1 \to B_1$ *and* $f_2 : A_2 \to B_2$ *be two morphisms in a category* $\mathcal{C}$ *and assume that the products* $A_1 \mathit{\Pi} A_2$ *and* $B_1 \mathit{\Pi} B_2$ *exist in* $\mathcal{C}$*. Then there exists a unique morphism* $f_1 \mathit{\Pi} f_2 : A_1 \mathit{\Pi} A_2 \to B_1 \mathit{\Pi} B_2$ *such that the following diagram commutes,*

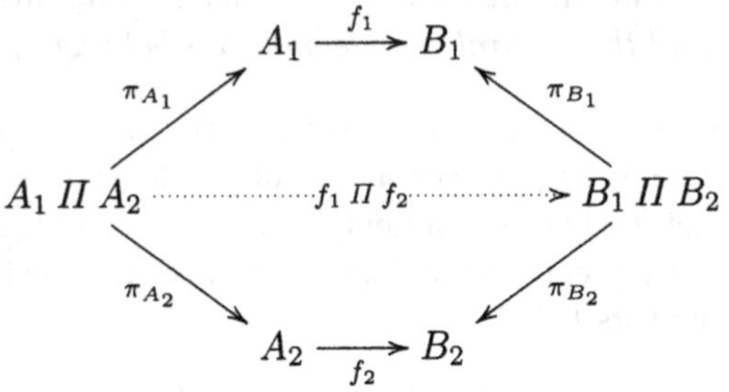

(c) Let $A_1, A_2, B_1, B_2, C_1, C_2$ *be objects of a category* $\mathcal{C}$ *and suppose that the products* $A_1 \mathit{\Pi} A_2$*,* $B_1 \mathit{\Pi} B_2$ *and* $C_1 \mathit{\Pi} C_2$ *exist in* $\mathcal{C}$*. Then we have*

$$\mathrm{id}_{A_1} \mathit{\Pi}\, \mathrm{id}_{A_2} = \mathrm{id}_{A_1 \mathit{\Pi} A_2} \text{ and } (g_1 \mathit{\Pi} g_2) \circ (f_1 \mathit{\Pi} f_2) = (g_1 \circ f_1) \mathit{\Pi} (g_2 \circ f_2)$$

for all morphisms $f_i : A_i \to B_i$*,* $g_i : B_i \to C_i$*,* $i = 1, 2$*.*

Proof. (a) Suppose we have two candidates (P, π_A, π_B) and (P', π'_A, π'_B) for the product of A and B, we have to show that P and P' are isomorphic. Applying the defining property of the product to (P, π_A, π_B) with $C = P'$ and to (P', π'_A, π'_B) with $C = P$, we get the following two commuting diagrams,

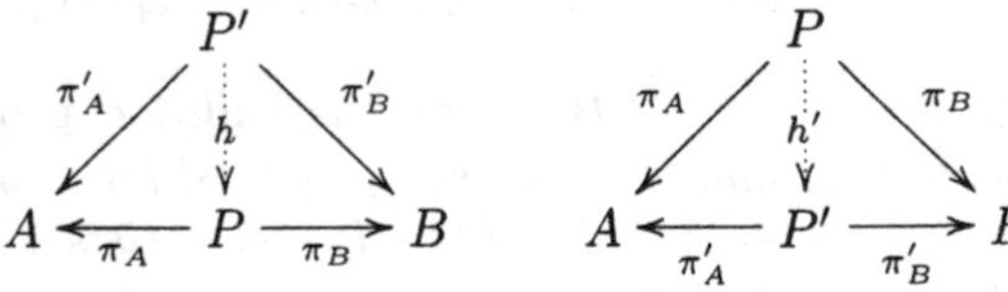

We get $\pi_A \circ h \circ h' = \pi'_A \circ h' = \pi_A$ and $\pi_B \circ h \circ h' = \pi'_B \circ h' = \pi_B$, i.e. the diagram

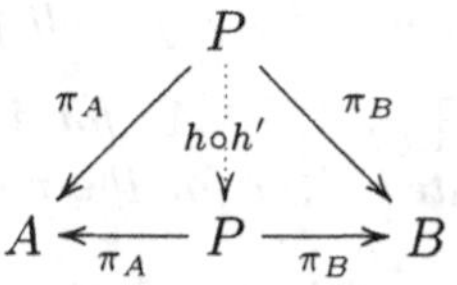

is commutative. It is clear that this diagram also commutes, if we replace $h \circ h'$ by id_P, so the uniqueness implies $h \circ h' = \mathrm{id}_P$. Similarly one proves $h' \circ h = \mathrm{id}_{P'}$, so that $h : P' \to P$ is the desired isomorphism.

(b) The unique morphism $f_1 \mathit{\Pi} f_2$ exists by the defining property of the product of B_1 and B_2, as we can see from the diagram

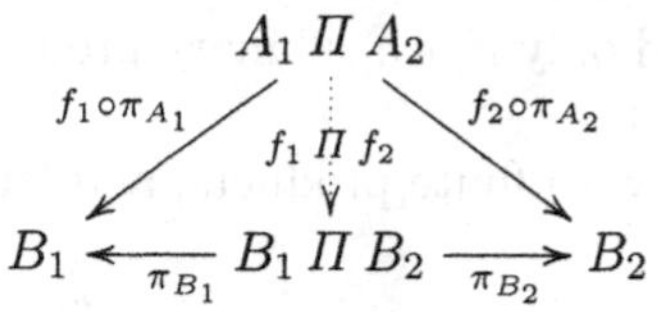

(c) Both properties follow from the uniqueness of the mediating morphism in the defining property of the product. To prove $\mathrm{id}_{A_1} \mathit{\Pi}\, \mathrm{id}_{A_2} = \mathrm{id}_{A_1 \mathit{\Pi} A_2}$ one has to show that both expressions make the diagram

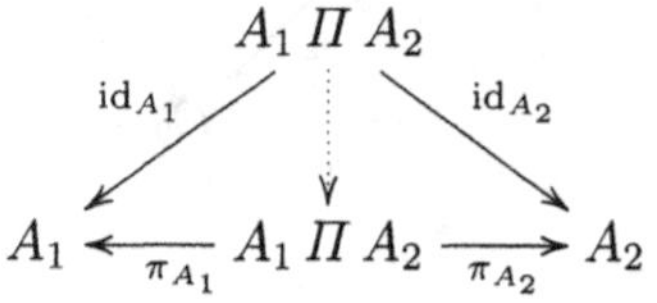

commutative, for the the second equality one checks that $(g_1 \mathit{\Pi} g_2) \circ (f_1 \mathit{\Pi} f_2)$ and $(g_1 \circ f_1) \mathit{\Pi} (g_2 \circ f_2)$ both make the diagram

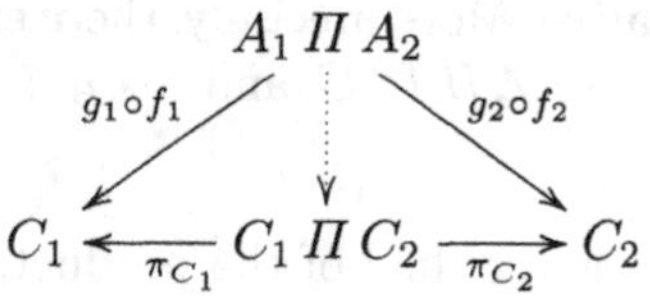

commutative.

□

The notion of product extends also to more then two objects.

Definition 3.21. *Let $(A_i)_{i \in I}$ be a family of objects of a category $\mathcal{C}$, indexed by some set I. The pair $\left(\prod_{i\in I} A_i, \left(\pi_j : \prod_{i\in I} A_i \to A_j\right)_{j\in I}\right)$ consisting of an object $\prod_{i\in I} A_i$ of $\mathcal{C}$ and a family of morphisms $\left(\pi_j : \prod_{i\in I} A_i \to A_j\right)_{j\in I}$ of $\mathcal{C}$ is a* product *of the family $(A_i)_{i\in I}$ if for any object C and any family of morphisms $(f_i : C \to A_i)_{i\in I}$ there exists a unique morphism $h : C \to \prod_{i\in I} A_i$ such that*

$$\pi_j \circ h = f_j, \quad \textit{for all } j \in I$$

holds. The morphism $\pi_j : \prod_{i\in I} A_i \to A_j$ *for* $j \in I$ *is called the* j*th* product projection. *We will also write* $[f_i]_{i\in I}$ *for the morphism* $h : C \to \prod_{i\in I} A_i$.

An object T of a category $\mathcal{C}$ is called *terminal*, if for any object C of $\mathcal{C}$ there exists a unique morphism from C to T. A terminal object is unique up to isomorphism, if it exists. A product of the empty family is a terminal object.

Exercise 3.22. (a) We say that a category $\mathcal{C}$ has finite products if for any family of objects indexed by a finite set there exists a product. Show that this is the case if and only if it has binary products for all pairs of objects and a terminal object.
(b) Let $\mathcal{C}$ be a category with finite products, and let

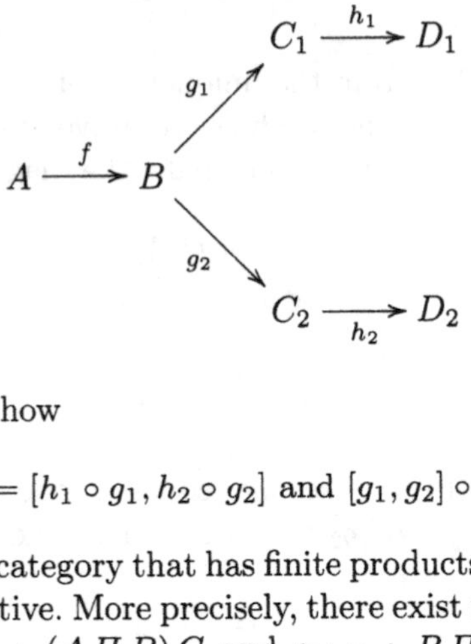

be morphisms in $\mathcal{C}$. Show

$$(h_1\, \Pi\, h_2) \circ [g_1, g_2] = [h_1 \circ g_1, h_2 \circ g_2] \text{ and } [g_1, g_2] \circ f = [g_1 \circ f, g_2 \circ f].$$

Remark 3.23. Let $\mathcal{C}$ be a category that has finite products. Then the product is associative and commutative. More precisely, there exist natural isomorphisms $\alpha_{A,B,C} : A\,\Pi\,(B\,\Pi\,C) \to (A\,\Pi\,B)C$ and $\gamma_{A,B} : B\,\Pi\,A \to A\,\Pi\,B$ for all objects $A, B, C \in \mathrm{Ob}\,\mathcal{C}$.

The notion *coproduct* is the dual of the product, i.e.

$$\left(\coprod_{i\in I} A_i, \left(\imath_j : A_j \to \coprod_{i\in I} A_i\right)_{j\in I}\right)$$

is called a coproduct of the family $(A_i)_{i\in I}$ of objects in $\mathcal{C}$, if it is a product of the same family in the category $\mathcal{C}^{\mathrm{op}}$. Formulated in terms of objects and morphisms of $\mathcal{C}$ only, this amounts to the following.

Definition 3.24. *Let* $(A_i)_{i\in I}$ *be a family of objects of a category* $\mathcal{C}$*, indexed by some set* I*. The pair* $\left(\coprod_{i\in I} A_i, \left(\imath_j : A_k \to \prod_{i\in I} A_i\right)_{j\in I}\right)$ *consisting of an object* $\coprod_{i\in I} A_i$ *of* $\mathcal{C}$ *and a family of morphisms* $\left(\imath_j : A_j \to \coprod_{i\in I} A_i\right)_{j\in I}$ *of* $\mathcal{C}$

is a coproduct of the family $(A_i)_{i\in I}$ *if for any object* C *and any family of morphisms* $(f_i : A_i \to C)_{i\in I}$ *there exists a unique morphism* $h : \coprod_{i\in I} A_i \to C$ *such that*

$$h \circ \imath_j = f_j, \qquad \textit{for all } j \in I$$

holds. The morphism $\imath_j : A_j \to \coprod_{i\in I} A_i$ *for* $j \in I$ *is called the* j*th* coproduct injection. *We will write* $[f_i]_{i\in I}$ *for the morphism* $h : \coprod_{i\in I} A_i \to C$.

A coproduct of the empty family in $\mathcal{C}$ is an *initial object*, i.e. an object I such that for any object A of $\mathcal{C}$ there exists exactly one morphism from I to A.

It is straightforward to translate Proposition 3.20 to its counterpart for the coproduct.

Example 3.25. In the trivial unital semigroup $(G = \{e\}, \cdot, e)$, viewed as a category (note that is is isomorphic to the discrete category over a set with one element) its only object G is a terminal and initial object, and also a product and coproduct for any family of objects. The product projections and coproduct injections are given by the unique morphism e of this category.

In any other unital semigroup there exist no initial or terminal objects and no binary or higher products or coproducts.

Example 3.26. In the category $\mathfrak{Set}$ a binary product of two sets A and B is given by their Cartesian product $A\times B$ (together with the obvious projections) and any set with one element is terminal. A coproduct of A and B is defined by their disjoint union $A\dot{\cup}B$ (together with the obvious injections) and the empty set is an initial object. Recall that we can define the disjoint union as $A\dot{\cup}B = (A \times \{A\}) \cup (B \times \{B\})$.

Exercise 3.27. Let $\mathfrak{Vek}$ be the category that has as objects all vector spaces (over some field $\mathbb{K}$) and as morphisms the $\mathbb{K}$-linear maps between them. The trivial vector space $\{0\}$ is an initial and terminal object in this category. Show that the direct sum of (finitely many) vector spaces is a product and a coproduct in this category.

The following example shall be used throughout this section and the following.

Example 3.28. The coproduct in the category of unital algebras $\mathfrak{Alg}$ is the *free product* of $*$-algebras *with* identification of the units. Let us recall its defining universal property. Let $\{\mathcal{A}_k\}_{k\in I}$ be a family of unital $*$-algebras and $\coprod_{k\in I} \mathcal{A}_k$ their free product, with canonical inclusions $\{i_k : \mathcal{A}_k \to \coprod_{k\in I} \mathcal{A}_k\}_{k\in I}$. If $\mathcal{B}$ is any unital $*$-algebra, equipped with unital $*$-algebra homomorphisms $\{i'_k : \mathcal{A}_k \to \mathcal{B}\}_{k\in I}$, then there exists a unique unital $*$-algebra homomorphism $h : \coprod_{k\in I} \mathcal{A}_k \to \mathcal{B}$ such that

$$h \circ i_k = i'_k, \qquad \text{for all} \quad k \in I.$$

It follows from the universal property that for any pair of unital $*$-algebra homomorphisms $j_1 : \mathcal{A}_1 \to \mathcal{B}_1$, $j_2 : \mathcal{A}_2 \to \mathcal{B}_2$ there exists a unique unital $*$-algebra homomorphism $j_1 \coprod j_2 : \mathcal{A}_1 \coprod \mathcal{A}_2 \to \mathcal{B}_1 \coprod \mathcal{B}_2$ such that the diagram

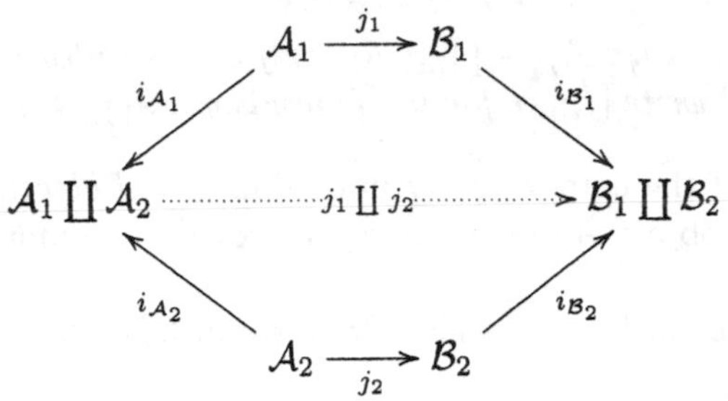

commutes.

The free product $\coprod_{k \in I} \mathcal{A}_k$ can be constructed as a sum of tensor products of the $\mathcal{A}_k$, where neighboring elements in the product belong to different algebras. For simplicity, we illustrate this only for the case of the free product of two algebras. Let

$$\mathbb{A} = \bigcup_{n \in \mathbb{N}} \{\epsilon \in \{1,2\}^n | \epsilon_1 \neq \epsilon_2 \neq \cdots \neq \epsilon_n\}$$

and decompose $\mathcal{A}_i = \mathbb{C}\mathbf{1} \oplus \mathcal{A}_i^0$, $i = 1, 2$, into a direct sum of vector spaces. As a coproduct $\mathcal{A}_1 \coprod \mathcal{A}_2$ is unique up to isomorphism, so the construction does not depend on the choice of the decompositions.

Then $\mathcal{A}_1 \coprod \mathcal{A}_2$ can be constructed as

$$\mathcal{A}_1 \coprod \mathcal{A}_2 = \bigoplus_{\epsilon \in \mathbb{A}} \mathcal{A}^\epsilon,$$

where $\mathcal{A}^\emptyset = \mathbb{C}$, $\mathcal{A}^\epsilon = \mathcal{A}^0_{\epsilon_1} \otimes \cdots \otimes \mathcal{A}^0_{\epsilon_n}$ for $\epsilon = (\epsilon_1, \ldots, \epsilon_n)$. The multiplication in $\mathcal{A}_1 \coprod \mathcal{A}_2$ is inductively defined by

$$(a_1 \otimes \cdots \otimes a_n) \cdot (b_1 \otimes \cdots \otimes b_m) = \begin{cases} a_1 \otimes \cdots \otimes (a_n \cdot b_1) \otimes \cdots \otimes b_m & \text{if } \epsilon_n = \delta_1, \\ a_1 \otimes \cdots \otimes a_n \otimes b_1 \otimes \cdots \otimes b_m & \text{if } \epsilon_n \neq \delta_1, \end{cases}$$

for $a_1 \otimes \cdots \otimes a_n \in \mathcal{A}^\epsilon$, $b_1 \otimes \cdots \otimes b_m \in \mathcal{A}^\delta$. Note that in the case $\epsilon_n = \delta_1$ the product $a_n \cdot b_1$ is not necessarily in $\mathcal{A}^0_{\epsilon_n}$, but is in general a sum of a multiple of the unit of $\mathcal{A}_{\epsilon_n}$ and an element of $\mathcal{A}^0_{\epsilon_n}$. We have to identify $a_1 \otimes \cdots a_{n-1} \otimes 1 \otimes b_2 \otimes \cdots b_m$ with $a_1 \otimes \cdots \otimes a_{n-1} \cdot b_2 \otimes \cdots b_m$.

Since $\coprod$ is the coproduct of a category, it is commutative and associative in the sense that there exist natural isomorphisms

$$\gamma_{\mathcal{A}_1, \mathcal{A}_2} : \mathcal{A}_1 \coprod \mathcal{A}_2 \stackrel{\cong}{\to} \mathcal{A}_2 \coprod \mathcal{A}_1, \tag{3.1}$$

$$\alpha_{\mathcal{A}_1, \mathcal{A}_2, \mathcal{A}_3} : \mathcal{A}_1 \coprod \left(\mathcal{A}_2 \coprod \mathcal{A}_3\right) \stackrel{\cong}{\to} \left(\mathcal{A}_1 \coprod \mathcal{A}_2\right) \coprod \mathcal{A}_3$$

for all unital $*$-algebras $\mathcal{A}_1, \mathcal{A}_2, \mathcal{A}_3$. Let $i_\ell : \mathcal{A}_\ell \to \mathcal{A}_1 \coprod \mathcal{A}_2$ and $i'_\ell : \mathcal{A}_\ell \to \mathcal{A}_2 \coprod \mathcal{A}_1$, $\ell = 1, 2$ be the canonical inclusions. The commutativity constraint $\gamma_{\mathcal{A}_1,\mathcal{A}_2} : \mathcal{A}_1 \coprod \mathcal{A}_2 \to \mathcal{A}_2 \coprod \mathcal{A}_1$ maps an element of $\mathcal{A}_1 \coprod \mathcal{A}_2$ of the form $i_1(a_1)i_2(b_1)\cdots i_2(b_n)$ with $a_1, \ldots, a_n \in \mathcal{A}_1$, $b_1, \ldots, b_n \in \mathcal{A}_2$ to

$$\gamma_{\mathcal{A}_1,\mathcal{A}_2}\big(i_1(a_1)i_2(b_1)\cdots i_2(b_n)\big) = i'_1(a_1)i'_2(b_1)\cdots i'_2(b_n) \in \mathcal{A}_2 \coprod \mathcal{A}_1.$$

Exercise 3.29. We also consider non-unital algebras. Show that the *free product* of $*$-algebras *without* identification of units is a coproduct in the category $\mathfrak{nuAlg}$ of non-unital (or rather not necessarily unital) algebras. Give an explicit construction for the free product of two non-unital algebras.

Exercise 3.30. Show that the following defines a a functor from the category of non-unital algebras $\mathfrak{nuAlg}$ to the category of unital algebras $\mathfrak{Alg}$. For an algebra $\mathcal{A} \in \mathrm{Ob}\,\mathfrak{nuAlg}$, $\tilde{\mathcal{A}}$ is equal to $\tilde{\mathcal{A}} = \mathbb{C}\mathbf{1} \oplus \mathcal{A}$ as a vector space and the multiplication is defined by

$$(\lambda\mathbf{1} + a)(\lambda'\mathbf{1} + a') = \lambda\lambda'\mathbf{1} + \lambda' a + \lambda a' + aa'$$

for $\lambda, \lambda' \in \mathbb{C}$, $a, a' \in \mathcal{A}$. We will call $\tilde{\mathcal{A}}$ the *unitization* of $\mathcal{A}$. Note that $\mathcal{A} \cong 0\mathbf{1} + \mathcal{A} \subseteq \tilde{\mathcal{A}}$ is not only a subalgebra, but even an ideal in $\tilde{\mathcal{A}}$.

How is the functor defined on the morphisms?

Show that the following relation holds between the free product *with* identification of units $\coprod_{\mathfrak{Alg}}$ and the free product *without* identification of units $\coprod_{\mathfrak{nuAlg}}$,

$$\widetilde{\mathcal{A}_1 \coprod_{\mathfrak{nuAlg}} \mathcal{A}_2} \cong \tilde{\mathcal{A}}_1 \coprod_{\mathfrak{Alg}} \tilde{\mathcal{A}}_2$$

for all $\mathcal{A}_1, \mathcal{A}_2 \in \mathrm{Ob}\,\mathfrak{nuAlg}$.

Note furthermore that the range of this functor consists of all algebras that admit a decomposition of the form $\mathcal{A} = \mathbb{C}\mathbf{1} \oplus \mathcal{A}_0$, where $\mathcal{A}_0$ is a subalgebra. This is equivalent to having a one-dimensional representation. The functor is not surjective, e.g., the algebra $\mathcal{M}_2$ of 2×2-matrices can not be obtained as a unitization of some other algebra.

Let us now come to the definition of a tensor category.

Definition 3.31. *A category* $(\mathcal{C}, \Box)$ *equipped with a bifunctor* $\Box : \mathcal{C} \times \mathcal{C} \to \mathcal{C}$, *called* tensor product, *that is associative up to a natural isomorphism*

$$\alpha_{A,B,C} : A\Box(B\Box C) \stackrel{\cong}{\to} (A\Box B)\Box C, \qquad \text{for all } A, B, C \in \mathrm{Ob}\,\mathcal{C},$$

and an element E that is, up to natural isomorphisms

$$\lambda_A : E\Box A \stackrel{\cong}{\to} A, \quad \text{and} \quad \rho_A : A\Box E \stackrel{\cong}{\to} A, \quad \text{for all } A \in \mathrm{Ob}\,\mathcal{C},$$

a unit for $\Box$, *is called a* tensor category *or* monoidal category, *if the* pentagon axiom

$$\begin{array}{ccccc} & & (A\Box B)\Box(C\Box D) & & \\ & \overset{\alpha_{A,B,C\Box D}}{\nearrow} & & \overset{\alpha_{A\Box B,C,D}}{\searrow} & \\ A\Box(B\Box(C\Box D)) & & & & ((A\Box B)\Box C)\Box D \\ \Big\downarrow {\scriptstyle \mathrm{id}_A\Box\alpha_{B,C,D}} & & & & \Big\uparrow {\scriptstyle \alpha_{A,B,C}\Box\mathrm{id}_D} \\ A\Box((B\Box C)\Box D) & & \xrightarrow[\alpha_{A,B\Box C,D}]{} & & (A\Box(B\Box C))\Box D \end{array}$$

and the triangle axiom

$$\begin{array}{ccc} A\Box(E\Box C) & \xrightarrow{\alpha_{A,E,C}} & (A\Box E)\Box C \\ & {\scriptstyle \mathrm{id}_A\Box\lambda_C}\searrow \quad \swarrow{\scriptstyle \rho_A\Box\mathrm{id}_C} & \\ & A\Box C & \end{array}$$

are satisfied for all objects A, B, C, D *of* $\mathcal{C}$.

If a category has products or coproducts for all finite sets of objects, then the universal property guarantees the existence of the isomorphisms α, λ, and ρ that turn it into a tensor category.

A functor between tensor categories, that behaves "nicely" with respect to the tensor products, is called a *tensor functor* or *monoidal functor*, see, e.g., Section XI.2 in MacLane[Mac98].

Definition 3.32. *Let* $(\mathcal{C},\Box)$ *and* $(\mathcal{C}',\Box')$ *be two tensor categories. A* cotensor functor *or* comonoidal functor $F : (\mathcal{C},\Box) \to (\mathcal{C}',\Box')$ *is an ordinary functor* $F : \mathcal{C} \to \mathcal{C}'$ *equipped with a morphism* $F_0 : F(E_\mathcal{C}) \to E_{\mathcal{C}'}$ *and a natural transformation* $F_2 : F(\,\cdot\,\Box\,\cdot\,) \to F(\,\cdot\,)\Box' F(\,\cdot\,)$, *i.e. morphisms* $F_2(A,B) : F(A\Box B) \to F(A)\Box' F(B)$ *for all* $A, B \in \mathrm{Ob}\,\mathcal{C}$ *that are natural in* A *and* B, *such that the diagrams*

$$\begin{array}{ccc} F\big(A\Box(B\Box C)\big) & \xrightarrow{F(\alpha_{A,B,C})} & F\big((A\Box B)\Box C\big) \\ \Big\downarrow {\scriptstyle F_2(A,B\Box C)} & & \Big\downarrow {\scriptstyle F_2(A\Box B,C)} \\ F(A)\Box' F(B\Box C) & & F(A\Box B)\Box' F(C) \\ \Big\downarrow {\scriptstyle \mathrm{id}_{F(A)}\Box' F_2(B,C)} & & \Big\downarrow {\scriptstyle F_2(A,B)\Box'\mathrm{id}_{F(C)}} \\ F(A)\Box'\big(F(B)\Box' F(C)\big) & \xrightarrow[\alpha'_{F(A),F(B),F(C)}]{} & \big(F(A)\Box' F(B)\big)\Box' F(C) \end{array} \tag{3.2}$$

$$\begin{array}{ccc} F(B\Box E_\mathcal{C}) & \xrightarrow{F_2(B,E_\mathcal{C})} & F(B)\Box' F(E_\mathcal{C}) \\ \Big\downarrow {\scriptstyle F(\rho_B)} & & \Big\downarrow {\scriptstyle \mathrm{id}_B\Box' F_0} \\ F(B) & \xleftarrow[\rho'_{F(B)}]{} & F(B)\Box' E_{\mathcal{C}'} \end{array} \tag{3.3}$$

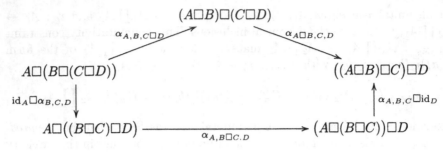

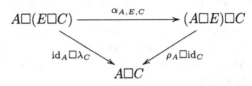

$$\begin{array}{ccc} F(E_{\mathcal{C}}\Box B) & \xrightarrow{F_2(E_{\mathcal{C}},B)} & F(E_{\mathcal{C}})\Box' F(B) \\ {\scriptstyle F(\lambda_B)}\downarrow & & \downarrow{\scriptstyle F_0\Box' \mathrm{id}_B} \\ F(B) & \xleftarrow[\lambda'_{F(B)}]{} & E_{\mathcal{C}'}\Box' F(B) \end{array} \tag{3.4}$$

commute for all $A, B, C \in \mathrm{Ob}\,\mathcal{C}$.

We have reversed the direction of F_0 and F_2 in our definition. In the case of a strong tensor functor, i.e. when all the morphisms are isomorphisms, our definition of a cotensor functor is equivalent to the usual definition of a tensor functor as, e.g., in MacLane[Mac98].

The conditions are exactly what we need to get morphisms

$$F_n(A_1, \ldots, A_n) : F(A_1 \Box \cdots \Box A_n) \to F(A_1) \Box' \cdots \Box' F(A_n)$$

for all finite sets $\{A_1, \ldots, A_n\}$ of objects of $\mathcal{C}$ such that, up to these morphisms, the functor $F : (\mathcal{C}, \Box) \to (\mathcal{C}', \Box')$ is a homomorphism.

3.2 Classical Stochastic Independence and the Product of Probability Spaces

Two random variables $X_1 : (\Omega, \mathcal{F}, P) \to (E_1, \mathcal{E}_1)$ and $X_2 : (\Omega, \mathcal{F}, P) \to (E_2, \mathcal{E}_2)$, defined on the same probability space $(\Omega, \mathcal{F}, P)$ and with values in two possibly distinct measurable spaces $(E_1, \mathcal{E}_1)$ and $(E_2, \mathcal{E}_2)$, are called *stochastically independent* (or simply *independent*) w.r.t. P, if the σ-algebras $X_1^{-1}(\mathcal{E}_1)$ and $X_2^{-1}(\mathcal{E}_2)$ are independent w.r.t. P, i.e. if

$$P\big((X_1^{-1}(M_1) \cap X_2^{-1}(M_2)\big) = P\big((X_1^{-1}(M_1)\big)P\big(X_2^{-1}(M_2)\big)$$

holds for all $M_1 \in \mathcal{E}_1$, $M_2 \in \mathcal{E}_2$. If there is no danger of confusion, then the reference to the measure P is often omitted.

This definition can easily be extended to arbitrary families of random variables. A family $\big(X_j : (\Omega, \mathcal{F}, P) \to (E_j, \mathcal{E}_j)\big)_{j\in J}$, indexed by some set J, is called independent, if

$$P\left(\bigcap_{k=1}^{n} X_{j_k}^{-1}(M_{j_k})\right) = \prod_{k=1}^{n} P\big(X_{j_k}^{-1}(M_{j_k})\big)$$

holds for all $n \in \mathbb{N}$ and all choices of indices $k_1, \ldots, k_n \in J$ with $j_k \neq j_\ell$ for $j \neq \ell$, and all choices of measurable sets $M_{j_k} \in \mathcal{E}_{j_k}$.

There are many equivalent formulations for independence, consider, e.g., the following proposition.

Proposition 3.33. *Let X_1 and X_2 be two real-valued random variables. The following are equivalent.*

(i) X_1 *and* X_2 *are independent.*
(ii) For all bounded measurable functions f_1, f_2 *on* $\mathbb{R}$ *we have*

$$\mathbb{E}\big(f_1(X_1)f_2(X_2)\big) = \mathbb{E}\big(f_1(X_1)\big)\mathbb{E}\big(f_2(X_2)\big).$$

(iii) The probability space $(\mathbb{R}^2, \mathcal{B}(\mathbb{R}^2), P_{(X_1,X_2)})$ *is the product of the probability spaces* $(\mathbb{R}, \mathcal{B}(\mathbb{R}), P_{X_1})$ *and* $(\mathbb{R}, \mathcal{B}(\mathbb{R}), P_{X_2})$, *i.e.*

$$P_{(X_1,X_2)} = P_{X_1} \otimes P_{X_2}.$$

We see that stochastic independence can be reinterpreted as a rule to compute the joint distribution of two random variables from their marginal distribution. More precisely, their joint distribution can be computed as a product of their marginal distributions. This product is associative and can also be iterated to compute the joint distribution of more than two independent random variables.

The classifications of independence for non-commutative probability spaces [Spe97, BGS99, BG01, Mur03, Mur02] that we are interested in are based on redefining independence as a product satisfying certain natural axioms.

3.3 Definition of Independence in the Language of Category Theory

We will now define the notion of independence in the language of category theory. The usual notion of independence for classical probability theory and the independences classified in [Spe97, BGS99, BG01, Mur03, Mur02] will then be instances of this general notion obtained by considering the category of classical probability spaces or categories of algebraic probability spaces.

In order to define a notion of independence we need less than a (co-) product, but a more than a tensor product. What we need are inclusions or projections that allow us to view the objects A, B as subsystems of their product $A\Box B$.

Definition 3.34. *A* tensor category with projections $(\mathcal{C}, \Box, \pi)$ *is a tensor category* $(\mathcal{C}, \Box)$ *equipped with two natural transformations* $\pi_1 : \Box \to P_1$ *and* $\pi_2 : \Box \to P_2$, *where the bifunctors* $P_1, P_2 : \mathcal{C} \times \mathcal{C} \to \mathcal{C}$ *are defined by* $P_1(B_1, B_2) = B_1$, $P_2(B_1, B_2) = B_2$, *on pairs of objects* B_1, B_2 *of* $\mathcal{C}$, *and similarly on pairs of morphisms. In other words, for any pair of objects* B_1, B_2 *there exist two morphisms* $\pi_{B_1} : B_1\Box B_2 \to B_1$, $\pi_{B_2} : B_1\Box B_2 \to B_2$, *such that for any pair of morphisms* $f_1 : A_1 \to B_1$, $f_2 : A_2 \to B_2$, *the following diagram commutes,*

$$\begin{array}{ccccc}
A_1 & \xleftarrow{\pi_{A_1}} & A_1\Box A_2 & \xrightarrow{\pi_{A_2}} & A_2 \\
\downarrow{\scriptstyle f_1} & & \downarrow{\scriptstyle f_1\Box f_2} & & \downarrow{\scriptstyle f_2} \\
B_1 & \xleftarrow[\pi_{B_1}]{} & B_1\Box B_2 & \xrightarrow[\pi_{B_2}]{} & B_2.
\end{array}$$

Similarly, a tensor product with inclusions $(\mathcal{C}, \Box, i)$ *is a tensor category* $(\mathcal{C}, \Box)$ *equipped with two natural transformations* $i_1 : P_1 \to \Box$ *and* $i_2 : P_2 \to \Box$, *i.e. for any pair of objects* B_1, B_2 *there exist two morphisms* $i_{B_1} : B_1 \to B_1 \Box B_2$, $i_{B_2} : B_2 \to B_1 \Box B_2$, *such that for any pair of morphisms* $f_1 : A_1 \to B_1$, $f_2 : A_2 \to B_2$, *the following diagram commutes,*

$$\begin{array}{ccccc} A_1 & \xrightarrow{i_{A_1}} & A_1 \Box A_2 & \xleftarrow{i_{A_2}} & A_2 \\ \downarrow{\scriptstyle f_1} & & \downarrow{\scriptstyle f_1 \Box f_2} & & \downarrow{\scriptstyle f_2} \\ B_1 & \xrightarrow{i_{B_1}} & B_1 \Box B_2 & \xleftarrow{i_{B_2}} & B_2. \end{array}$$

In a tensor category with projections or with inclusions we can define a notion of independence for morphisms.

Definition 3.35. *Let* $(\mathcal{C}, \Box, \pi)$ *be a tensor category with projections. Two morphism* $f_1 : A \to B_1$ *and* $f_2 : A \to B_2$ *with the same source* A *are called* independent *(with respect to* $\Box$*), if there exists a morphism* $h : A \to B_1 \Box B_2$ *such that the diagram*

$$\begin{array}{ccccc} & \swarrow^{f_1} & A & \searrow^{f_2} & \\ & & \downarrow{\scriptstyle h} & & \\ B_1 & \xleftarrow[\pi_{B_1}]{} & B_1 \Box B_2 & \xrightarrow[\pi_{B_2}]{} & B_2 \end{array} \qquad (3.5)$$

commutes.

In a tensor category with inclusions $(\mathcal{C}, \Box, i)$, *two morphisms* $f_1 : B_1 \to A$ *and* $f_2 : B_2 \to A$ *with the same target* B *are called independent, if there exists a morphism* $h : B_1 \Box B_2 \to A$ *such that the diagram*

$$\begin{array}{ccccc} & \nearrow^{f_1} & A & \nwarrow^{f_2} & \\ & & \uparrow{\scriptstyle h} & & \\ B_1 & \xrightarrow[i_{B_1}]{} & B_1 \Box B_2 & \xleftarrow[i_{B_2}]{} & B_2 \end{array} \qquad (3.6)$$

commutes.

This definition can be extended in the obvious way to arbitrary sets of morphisms.

If $\Box$ is actually a product (or coproduct, resp.), then the universal property in Definition 3.19 implies that for all pairs of morphisms with the same source (or target, resp.) there exists even a unique morphism that makes diagram (3.5) (or (3.6), resp.) commuting. Therefore in that case all pairs of morphism with the same source (or target, resp.) are independent.

We will now consider several examples. We will show that for the category of classical probability spaces we recover usual stochastic independence, if we take the product of probability spaces, cf. Proposition 3.36.

Example: Independence in the Category of Classical Probability Spaces

The category $\mathfrak{Meas}$ of measurable spaces consists of pairs $(\Omega, \mathcal{F})$, where Ω is a set and $\mathcal{F} \subseteq \mathcal{P}(\Omega)$ a σ-algebra. The morphisms are the measurable maps. This category has a product,

$$(\Omega_1, \mathcal{F}_1) \,\Pi\, (\Omega_2, \mathcal{F}_2) = (\Omega_1 \times \Omega_2, \mathcal{F}_1 \otimes \mathcal{F}_2)$$

where $\Omega_1 \times \Omega_2$ is the Cartesian product of Ω_1 and Ω_2, and $\mathcal{F}_1 \otimes \mathcal{F}_2$ is the smallest σ-algebra on $\Omega_1 \times \Omega_2$ such that the canonical projections $p_1 : \Omega_1 \times \Omega_2 \to \Omega_1$ and $p_2 : \Omega_1 \times \Omega_2 \to \Omega_2$ are measurable.

The category of probability spaces $\mathfrak{Prob}$ has as objects triples $(\Omega, \mathcal{F}, P)$ where $(\Omega, \mathcal{F})$ is a measurable space and P a probability measure on $(\Omega, \mathcal{F})$. A morphism $X : (\Omega_1, \mathcal{F}_1, P_1) \to (\Omega_1, \mathcal{F}_2, P_2)$ is a measurable map $X : (\Omega_1, \mathcal{F}_1) \to (\Omega_1, \mathcal{F}_2)$ such that

$$P_1 \circ X^{-1} = P_2.$$

This means that a random variable $X : (\Omega, \mathcal{F}, P) \to (E, \mathcal{E})$ automatically becomes a morphism, if we equip $(E, \mathcal{E})$ with the measure

$$P_X = P \circ X^{-1}$$

induced by X.

This category does not have universal products. But one can check that the product of measures turns $\mathfrak{Prob}$ into a tensor category,

$$(\Omega_1, \mathcal{F}_1, P_1) \otimes (\Omega_2, \mathcal{F}_2, P_2) = (\Omega_1 \times \Omega_2, \mathcal{F}_1 \otimes \mathcal{F}_2, P_1 \otimes P_2),$$

where $P_1 \otimes P_2$ is determined by

$$(P_1 \otimes P_2)(M_1 \times M_2) = P_1(M_1)P_2(M_2),$$

for all $M_1 \in \mathcal{F}_1$, $M_2 \in \mathcal{F}_2$. It is even a tensor category with projections in the sense of Definition 3.34 with the canonical projections $p_1 : (\Omega_1 \times \Omega_2, \mathcal{F}_1 \otimes \mathcal{F}_2, P_1 \otimes P_2) \to (\Omega_1, \mathcal{F}_1, P_1)$, $p_2 : (\Omega_1 \times \Omega_2, \mathcal{F}_1 \otimes \mathcal{F}_2, P_1 \otimes P_2) \to (\Omega_2, \mathcal{F}_2, P_2)$ given by $p_1\big((\omega_1, \omega_2)\big) = \omega_1$, $p_2\big((\omega_1, \omega_2)\big) = \omega_2$ for $\omega_1 \in \Omega_1$, $\omega_2 \in \Omega_2$.

The notion of independence associated to this tensor product with projections is exactly the one used in probability.

Proposition 3.36. *Two random variables $X_1 : (\Omega, \mathcal{F}, P) \to (E_1, \mathcal{E}_1)$ and $X_2 : (\Omega, \mathcal{F}, P) \to (E_2, \mathcal{E}_2)$, defined on the same probability space $(\Omega, \mathcal{F}, P)$ and with values in measurable spaces $(E_1, \mathcal{E}_1)$ and $(E_2, \mathcal{E}_2)$, are stochastically independent, if and only if they are independent in the sense of Definition 3.35 as morphisms $X_1 : (\Omega, \mathcal{F}, P) \to (E_1, \mathcal{E}_1, P_{X_1})$ and $X_2 : (\Omega, \mathcal{F}, P) \to (E_2, \mathcal{E}_2, P_{X_2})$ of the tensor category with projections $(\mathfrak{Prob}, \otimes, p)$.*

Proof. Assume that X_1 and X_2 are stochastically independent. We have to find a morphism $h : (\Omega, \mathcal{F}, P) \to (E_1 \times E_2, \mathcal{E}_1 \otimes \mathcal{E}_2, P_{X_1} \otimes P_{X_2})$ such that the diagram

$$\begin{array}{ccccc} & & (\Omega, \mathcal{F}, P) & & \\ & \swarrow^{X_1} & \downarrow h & {}^{X_2}\searrow & \\ (E_1, \mathcal{E}_1, P_{X_1}) & \xleftarrow[p_{E_1}]{} & (E_1 \times E_2, \mathcal{E}_1 \otimes \mathcal{E}_2, P_{X_1} \otimes P_{X_2}) & \xrightarrow[p_{E_2}]{} & (E_2, \mathcal{E}_2, P_{X_2}) \end{array}$$

commutes. The only possible candidate is $h(\omega) = \big(X_1(\omega), X_2(\omega)\big)$ for all $\omega \in \Omega$, the unique map that completes this diagram in the category of measurable spaces and that exists due to the universal property of the product of measurable spaces. This is a morphism in $\mathfrak{Prob}$, because we have

$$\begin{aligned} P\big(h^{-1}(M_1 \times M_2)\big) &= P\big(X_1^{-1}(M_1) \cap X_2^{-1}(M_2)\big) = P\big(X_1^{-1}(M_1)\big)P\big(X_2^{-1}(M_2)\big) \\ &= P_{X_1}(M_1)P_{X_2}(M_2) = (P_{X_1} \otimes P_{X_2})(M_1 \times M_2) \end{aligned}$$

for all $M_1 \in \mathcal{E}_1$, $M_2 \in \mathcal{E}_2$, and therefore

$$P \circ h^{-1} = P_{X_1} \otimes P_{X_2}.$$

Conversely, if X_1 and X_2 are independent in the sense of Definition 3.35, then the morphism that makes the diagram commuting has to be again $h : \omega \mapsto \big(X_1(\omega), X_2(\omega)\big)$. This implies

$$P_{(X_1, X_2)} = P \circ h^{-1} = P_{X_1} \otimes P_{X_2}$$

and therefore

$$P\big(X_1^{-1}(M_1) \cap X_2^{-1}(M_2)\big) = P\big(X_1^{-1}(M_1)\big)P\big(X_2^{-1}(M_2)\big)$$

for all $M_1 \in \mathcal{E}_1$, $M_2 \in \mathcal{E}_2$. □

Example: Tensor Independence in the Category of Algebraic Probability Spaces

By the *category of algebraic probability spaces* $\mathfrak{AlgProb}$ we denote the category of associative unital algebras over $\mathbb{C}$ equipped with a unital linear functional. A morphism $j : (\mathcal{A}_1, \varphi_1) \to (\mathcal{A}_2, \varphi_2)$ is a quantum random variable, i.e. an algebra homomorphism $j : \mathcal{A}_1 \to \mathcal{A}_2$ that preserves the unit and the functional, i.e. $j(\mathbf{1}_{\mathcal{A}_1}) = \mathbf{1}_{\mathcal{A}_2}$ and $\varphi_2 \circ j = \varphi_1$.

The tensor product we will consider on this category is just the usual tensor product $(\mathcal{A}_1 \otimes \mathcal{A}_2, \varphi_1 \otimes \varphi_2)$, i.e. the algebra structure of $\mathcal{A}_1 \otimes \mathcal{A}_2$ is defined by

$$\begin{aligned} \mathbf{1}_{\mathcal{A}_1 \otimes \mathcal{A}_2} &= \mathbf{1}_{\mathcal{A}_1} \otimes \mathbf{1}_{\mathcal{A}_2}, \\ (a_1 \otimes a_2)(b_1 \otimes b_2) &= a_1 b_1 \otimes a_2 b_2, \end{aligned}$$

and the new functional is defined by

$$(\varphi_1 \otimes \varphi_2)(a_1 \otimes a_2) = \varphi_1(a_1)\varphi_2(a_2),$$

for all $a_1, b_1 \in \mathcal{A}_1$, $a_2, b_2 \in \mathcal{A}_2$.

This becomes a tensor category with inclusions with the inclusions defined by

$$i_{\mathcal{A}_1}(a_1) = a_1 \otimes \mathbf{1}_{\mathcal{A}_2},$$
$$i_{\mathcal{A}_2}(a_2) = \mathbf{1}_{\mathcal{A}_1} \otimes a_2,$$

for $a_1 \in \mathcal{A}_1$, $a_2 \in \mathcal{A}_2$.

One gets the category of $*$-algebraic probability spaces, if one assumes that the underlying algebras have an involution and the functional are states, i.e. also positive. Then an involution is defined on $\mathcal{A}_1 \otimes \mathcal{A}_2$ by $(a_1 \otimes a_2)^* = a_1^* \otimes a_2^*$ and $\varphi_1 \otimes \varphi_2$ is again a state.

The notion of independence associated to this tensor product with inclusions by Definition 3.35 is the usual notion of *Bose* or *tensor independence* used in quantum probability, e.g., by Hudson and Parthasarathy.

Proposition 3.37. *Two quantum random variables $j_1 : (\mathcal{B}_1, \psi_1) \to (\mathcal{A}, \varphi)$ and $j_2 : (\mathcal{B}_2, \psi_2) \to (\mathcal{A}, \varphi)$, defined on algebraic probability spaces $(\mathcal{B}_1, \psi_1)$, $(\mathcal{B}_2, \psi_2)$ and with values in the same algebraic probability space $(\mathcal{A}, \varphi)$ are independent if and only if the following two conditions are satisfied.*

(i) The images of j_1 and j_2 commute, i.e.

$$\big[j_1(a_1), j_2(a_2)\big] = 0,$$

for all $a_1 \in \mathcal{A}_1$, $a_2 \in \mathcal{A}_2$.

(ii) φ satisfies the factorization property

$$\varphi\big(j_1(a_1)j_2(a_2)\big) = \varphi\big(j_1(a_1)\big)\varphi\big(j_2(a_2)\big),$$

for all $a_1 \in \mathcal{A}_1$, $a_2 \in \mathcal{A}_2$.

We will not prove this Proposition since it can be obtained as a special case of Proposition 3.38, if we equip the algebras with the trivial $\mathbb{Z}_2$-grading $\mathcal{A}^{(0)} = \mathcal{A}$, $\mathcal{A}^{(1)} = \{0\}$.

Example: Fermi Independence

Let us now consider the category of $\mathbb{Z}_2$-graded algebraic probability spaces $\mathbb{Z}_2$-$\mathfrak{AlgProb}$. The objects are pairs $(\mathcal{A}, \varphi)$ consisting of a $\mathbb{Z}_2$-graded unital algebra $\mathcal{A} = \mathcal{A}^{(0)} \oplus \mathcal{A}^{(1)}$ and an even unital functional φ, i.e. $\varphi|_{\mathcal{A}^{(1)}} = 0$. The morphisms are random variables that don't change the degree, i.e., for $j : (\mathcal{A}_1, \varphi_1) \to (\mathcal{A}_2, \varphi_2)$, we have

$$j(\mathcal{A}_1^{(0)}) \subseteq \mathcal{A}_2^{(0)} \quad \text{and} \quad j(\mathcal{A}_1^{(1)}) \subseteq \mathcal{A}_2^{(1)}.$$

The tensor product $(\mathcal{A}_1 \otimes_{\mathbb{Z}_2} \mathcal{A}_2, \varphi_1 \otimes \varphi_2) = (\mathcal{A}_1, \varphi_1) \otimes_{\mathbb{Z}_2} (\mathcal{A}_2, \varphi_2)$ is defined as follows. The algebra $\mathcal{A}_1 \otimes_{\mathbb{Z}_2} \mathcal{A}_2$ is the graded tensor product of $\mathcal{A}_1$ and $\mathcal{A}_2$, i.e. $(\mathcal{A}_1 \otimes_{\mathbb{Z}_2} \mathcal{A}_2)^{(0)} = \mathcal{A}_1^{(0)} \otimes \mathcal{A}_2^{(0)} \oplus \mathcal{A}_1^{(1)} \otimes \mathcal{A}_2^{(1)}$, $(\mathcal{A}_1 \otimes_{\mathbb{Z}_2} \mathcal{A}_2)^{(1)} = \mathcal{A}_1^{(1)} \otimes \mathcal{A}_2^{(0)} \oplus \mathcal{A}_1^{(0)} \otimes \mathcal{A}_2^{(1)}$, with the algebra structure given by

$$\mathbf{1}_{\mathcal{A}_1 \otimes_{\mathbb{Z}_2} \mathcal{A}_2} = \mathbf{1}_{\mathcal{A}_1} \otimes \mathbf{1}_{\mathcal{A}_2},$$
$$(a_1 \otimes a_2) \cdot (b_1 \otimes b_2) = (-1)^{\deg a_2 \deg b_1} a_1 b_1 \otimes a_2 b_2,$$

for all homogeneous elements $a_1, b_1 \in \mathcal{A}_1$, $a_2, b_2 \in \mathcal{A}_2$. The functional $\varphi_1 \otimes \varphi_2$ is simply the tensor product, i.e. $(\varphi_1 \otimes \varphi_2)(a_1 \otimes a_2) = \varphi_1(a_1) \otimes \varphi_2(a_2)$ for all $a_1 \in \mathcal{A}_1$, $a_2 \in \mathcal{A}_2$. It is easy to see that $\varphi_1 \otimes \varphi_2$ is again even, if φ_1 and φ_2 are even. The inclusions $i_1 : (\mathcal{A}_1, \varphi_1) \to (\mathcal{A}_1 \otimes_{\mathbb{Z}_2} \mathcal{A}_2, \varphi_1 \otimes \varphi_2)$ and $i_2 : (\mathcal{A}_2, \varphi_2) \to (\mathcal{A}_1 \otimes_{\mathbb{Z}_2} \mathcal{A}_2, \varphi_1 \otimes \varphi_2)$ are defined by

$$i_1(a_1) = a_1 \otimes \mathbf{1}_{\mathcal{A}_2} \quad \text{and } i_2(a_2) = \mathbf{1}_{\mathcal{A}_1} \otimes a_2,$$

for $a_1 \in \mathcal{A}_1$, $a_2 \in \mathcal{A}_2$.

If the underlying algebras are assumed to have an involution and the functionals to be states, then the involution on the $\mathbb{Z}_2$-graded tensor product is defined by $(a_1 \otimes a_2)^* = (-1)^{\deg a_1 \deg a_2} a_1^* \otimes a_2^*$, this gives the category of $\mathbb{Z}_2$-graded $*$-algebraic probability spaces.

The notion of independence associated to this tensor category with inclusions is called *Fermi independence* or *anti-symmetric independence.*

Proposition 3.38. *Two random variables $j_1 : (\mathcal{B}_1, \psi_1) \to (\mathcal{A}, \varphi)$ and $j_2 : (\mathcal{B}_2, \psi_2) \to (\mathcal{A}, \varphi)$, defined on two $\mathbb{Z}_2$-graded algebraic probability spaces $(\mathcal{B}_1, \psi_1)$, $(\mathcal{B}_2, \psi_2)$ and with values in the same $\mathbb{Z}_2$-algebraic probability space $(\mathcal{A}, \varphi)$ are independent if and only if the following two conditions are satisfied.*

(i) The images of j_1 and j_2 satisfy the commutation relations

$$j_2(a_2) j_1(a_1) = (-1)^{\deg a_1 \deg a_2} j_1(a_1) j_2(a_2)$$

for all homogeneous elements $a_1 \in \mathcal{B}_1$, $a_2 \in \mathcal{B}_2$.

(ii) φ satisfies the factorization property

$$\varphi\big(j_1(a_1) j_2(a_2)\big) = \varphi\big(j_1(a_1)\big)\varphi\big(j_2(a_2)\big),$$

for all $a_1 \in \mathcal{B}_1$, $a_2 \in \mathcal{B}_2$.

Proof. The proof is similar to that of Proposition 3.36, we will only outline it. It is clear that the morphism $h : (\mathcal{B}_1, \psi_1) \otimes_{\mathbb{Z}_2} (\mathcal{B}_2, \psi_2) \to (\mathcal{A}, \varphi)$ that makes the diagram in Definition 3.35 commuting, has to act on elements of $\mathcal{B}_1 \otimes \mathbf{1}_{\mathcal{B}_2}$ and $\mathbf{1}_{\mathcal{B}_1} \otimes \mathcal{B}_2$ as

$$h(b_1 \otimes \mathbf{1}_{\mathcal{B}_2}) = j_1(b_1) \quad \text{and} \quad h(\mathbf{1}_{\mathcal{B}_1} \otimes b_2) = j_2(b_2).$$

This extends to a homomorphism from $(\mathcal{B}_1,\psi_1)\otimes_{\mathbb{Z}_2}(\mathcal{B}_2,\psi_2)$ to $(\mathcal{A},\varphi)$, if and only if the commutation relations are satisfied. And the resulting homomorphism is a quantum random variable, i.e. satisfies $\varphi\circ h=\psi_1\otimes\psi_2$, if and only if the factorization property is satisfied. □

Example: Free Independence

We will now introduce another tensor product with inclusions for the category of algebraic probability spaces $\mathfrak{AlgProb}$. On the algebras we take simply the free product of algebras with identifications of units introduced in Example 3.28. This is the coproduct in the category of algebras, therefore we also have natural inclusions. It only remains to define a unital linear functional on the free product of the algebras.

Voiculescu's[VDN92] *free product* $\varphi_1*\varphi_2$ of two unital linear functionals $\varphi_1:\mathcal{A}_1\to\mathbb{C}$ and $\varphi_2:\mathcal{A}_2\to\mathbb{C}$ can be defined recursively by

$$(\varphi_1*\varphi_2)(a_1a_2\cdots a_m)=\sum_{I\subsetneqq\{1,\ldots,m\}}(-1)^{m-\sharp I+1}(\varphi_1*\varphi_2)\left(\overrightarrow{\prod_{k\in I}}a_k\right)\prod_{k\notin I}\varphi_{\epsilon_k}(a_k)$$

for a typical element $a_1a_2\cdots a_m\in\mathcal{A}_1\coprod\mathcal{A}_2$, with $a_k\in\mathcal{A}_{\epsilon_k}$, $\epsilon_1\neq\epsilon_2\neq\cdots\neq\epsilon_m$, i.e. neighboring a's don't belong to the same algebra. $\sharp I$ denotes the number of elements of I and $\overrightarrow{\prod}_{k\in I}a_k$ means that the a's are to be multiplied in the same order in which they appear on the left-hand-side. We use the convention $(\varphi_1*\varphi_2)\left(\overrightarrow{\prod}_{k\in\emptyset}a_k\right)=1$.

It turns out that this product has many interesting properties, e.g., if φ_1 and φ_2 are states, then their free product is a again a state. For more details, see [BNT05] and the references given there.

Examples: Boolean, Monotone, and Anti-monotone Independence

Ben Ghorbal and Schürmann[BG01, BGS99] and Muraki[Mur03] have also considered the category of non-unital algebraic probability $\mathfrak{nuAlgProb}$ consisting of pairs $(\mathcal{A},\varphi)$ of a not necessarily unital algebra $\mathcal{A}$ and a linear functional φ. The morphisms in this category are algebra homomorphisms that leave the functional invariant. On this category we can define three more tensor products with inclusions corresponding to the boolean product $\diamond$, the monotone product $\triangleright$ and the anti-monotone product $\triangleleft$ of states. They can be defined by

$$\varphi_1\diamond\varphi_2(a_1a_2\cdots a_m)=\prod_{k=1}^m\varphi_{\epsilon_k}(a_k),$$

$$\varphi_1\triangleright\varphi_2(a_1a_2\cdots a_m)=\varphi_1\left(\overrightarrow{\prod_{k:\epsilon_k=1}}a_k\right)\prod_{k:\epsilon_k=2}\varphi_2(a_k),$$

$$\varphi_1\triangleleft\varphi_2(a_1a_2\cdots a_m)=\prod_{k:\epsilon_k=1}\varphi_1(a_k)\,\varphi_2\left(\overrightarrow{\prod_{k:\epsilon_k=2}}a_k\right),$$

for $\varphi_1 : \mathcal{A}_1 \to \mathbb{C}$ and $\varphi_2 : \mathcal{A}_2 \to \mathbb{C}$ and a typical element $a_1 a_2 \cdots a_m \in \mathcal{A}_1 \coprod \mathcal{A}_2$, $a_k \in \mathcal{A}_{\epsilon_k}$, $\epsilon_1 \neq \epsilon_2 \neq \cdots \neq \epsilon_m$, i.e. neighboring a's don't belong to the same algebra. Note that for the algebras and the inclusions we use here the free product without units, the coproduct in the category of not necessarily unital algebras.

The monotone and anti-monotone product are not commutative, but related by

$$\varphi_1 \triangleright \varphi_2 = (\varphi_2 \triangleleft \varphi_1) \circ \gamma_{\mathcal{A}_1,\mathcal{A}_2},$$

for all linear functionals $\varphi_1 : \mathcal{A}_1 \to \mathbb{C}$, $\varphi_2 : \mathcal{A}_2 \to \mathbb{C}$, where $\gamma_{\mathcal{A}_1,\mathcal{A}_2} : \mathcal{A}_1 \coprod \mathcal{A}_2 \to \mathcal{A}_2 \coprod \mathcal{A}_1$ is the commutativity constraint (for the commutativity constraint for the free product of unital algebras see Equation (3.1)). The boolean product is commutative, i.e. it satisfies

$$\varphi_1 \diamond \varphi_2 = (\varphi_2 \diamond \varphi_1) \circ \gamma_{\mathcal{A}_1,\mathcal{A}_2},$$

for all linear functionals $\varphi_1 : \mathcal{A}_1 \to \mathbb{C}$, $\varphi_2 : \mathcal{A}_2 \to \mathbb{C}$.

Exercise 3.39. The boolean, the monotone and the anti-monotone product can also be defined for unital algebras, if they are in the range of the unitization functor introduced in Exercise 3.30.

Let $\varphi_1 : \mathcal{A}_1 \to \mathbb{C}$ and $\varphi_2 : \mathcal{A}_2 \to \mathbb{C}$ be two unital functionals on algebras $\mathcal{A}_1$, $\mathcal{A}_2$, which can be decomposed as $\mathcal{A}_1 = \mathbb{C}\mathbf{1} \oplus \mathcal{A}_1^0$, $\mathcal{A}_2 = \mathbb{C}\mathbf{1} \oplus \mathcal{A}_2^0$. Then we define the boolean, monotone, or anti-monotone product of φ_1 and φ_2 as the unital extension of the boolean, monotone, or anti-monotone product of their restrictions $\varphi_1|_{\mathcal{A}_1^0}$ and $\varphi_2|_{\mathcal{A}_2^0}$.

Show that this leads to the following formulas.

$$\varphi_1 \diamond \varphi_2(a_1 a_2 \cdots a_n) = \prod_{i=1}^{n} \varphi_{\epsilon_i}(a_i),$$

$$\varphi_1 \triangleright \varphi_2(a_1 a_2 \cdots a_n) = \varphi_1\left(\prod_{i:\epsilon_i=1} a_i\right) \prod_{i:\epsilon_i=2} \varphi_2(a_i),$$

$$\varphi_1 \triangleleft \varphi_2(a_1 a_2 \cdots a_n) = \prod_{i:\epsilon_i=1} \varphi_1(a_i) \varphi_2\left(\prod_{i:\epsilon_i=2} a_i\right),$$

for $a_1 a_2 \cdots a_n \in \mathcal{A}_1 \coprod \mathcal{A}_2$, $a_i \in \mathcal{A}_{\epsilon_i}^0$, $\epsilon_1 \neq \epsilon_2 \neq \cdots \neq \epsilon_n$. We use the convention that the empty product is equal to the unit element.

These products can be defined in the same way for $*$-algebraic probability spaces, where the algebras are unital $*$-algebras having such a decomposition $\mathcal{A} = \mathbb{C}\mathbf{1} \oplus \mathcal{A}_0$ and the functionals are states. To check that $\varphi_1 \diamond \varphi_2, \varphi_1 \triangleright \varphi_2, \varphi_1 \triangleleft \varphi_2$ are again states, if φ_1 and φ_2 are states, one can verify that the following constructions give their GNS representations. Let (π_1, H_1, ξ_1) and (π_2, H_2, ξ_2) denote the GNS representations of $(\mathcal{A}_1, \varphi_1)$ and $(\mathcal{A}_2, \varphi_2)$. The GNS representations of $(\mathcal{A}_1 \coprod \mathcal{A}_2, \varphi_1 \diamond \varphi_2)$, $(\mathcal{A}_1 \coprod \mathcal{A}_2, \varphi_1 \triangleright \varphi_2)$, and $(\mathcal{A}_1 \coprod \mathcal{A}_2, \varphi_1 \lhd \varphi_2)$ can all be defined on the Hilbert space $H = H_1 \otimes H_2$ with the state vector $\xi = \xi_1 \otimes \xi_2$. The representations are defined by $\pi(\mathbf{1}) = \mathrm{id}$ and

$$\begin{array}{lll} \pi|_{\mathcal{A}_1^0} = \pi_1 \otimes P_2, & \pi|_{\mathcal{A}_2^0} = P_1 \otimes \pi_2, & \text{for } \varphi_1 \diamond \varphi_2, \\ \pi|_{\mathcal{A}_1^0} = \pi_1 \otimes P_2, & \pi|_{\mathcal{A}_2^0} = \mathrm{id}_{H_2} \otimes \pi_2, & \text{for } \varphi_1 \rhd \varphi_2, \\ \pi|_{\mathcal{A}_1^0} = \pi_1 \otimes \mathrm{id}_{H_2}, & \pi|_{\mathcal{A}_2^0} = P_1 \otimes \pi_2, & \text{for } \varphi_1 \lhd \varphi_2, \end{array}$$

where P_1, P_2 denote the orthogonal projections $P_1 : H_1 \to \mathbb{C}\xi_1$, $P_2 : H_2 \to \mathbb{C}\xi_2$. For the boolean case, $\xi = \xi_1 \otimes \xi_2 \in H_1 \otimes H_2$ is not cyclic for π, only the subspace $\mathbb{C}\xi \oplus H_1^0 \oplus H_2^0$ can be generated from ξ.

3.4 Reduction of an Independence

For a reduction of independences we need a little bit more than a cotensor functor.

Definition 3.40. *Let $(\mathcal{C}, \Box, i)$ and $(\mathcal{C}', \Box', i')$ be two tensor categories with inclusions and assume that we are given functors $I : \mathcal{C} \to \mathcal{D}$ and $I' : \mathcal{C}' \to \mathcal{D}$ to some category $\mathcal{D}$. A* reduction (F, J) *of the tensor product $\Box$ to the tensor product $\Box'$ (w.r.t. $(\mathcal{D}, I, I')$)is a cotensor functor $F : (\mathcal{C}, \Box) \to (\mathcal{C}', \Box')$ and a natural transformation $J : I \to I' \circ F$, i.e. morphisms $J_A : A \to F(A)$ in $\mathcal{D}$ for all objects $A \in \mathrm{Ob}\,\mathcal{C}$ such that the diagram*

$$\begin{array}{ccc} I(A) & \xrightarrow{J_A} & I' \circ F(A) \\ {\scriptstyle I(f)}\downarrow & & \downarrow{\scriptstyle I' \circ F(f)} \\ I(B) & \xrightarrow[J_B]{} & I' \circ F(B) \end{array}$$

commutes for all morphisms $f : A \to B$ in $\mathcal{C}$.

In the simplest case, $\mathcal{C}$ will be a subcategory of $\mathcal{C}'$, I will be the inclusion functor from $\mathcal{C}$ into $\mathcal{C}'$, and I' the identity functor on $\mathcal{C}'$. Then such a reduction provides us with a system of inclusions $J_n(A_1, \ldots, A_n) = F_n(A_1, \ldots, A_n) \circ J_{A_1 \Box \cdots \Box A_n}$

$$J_n(A_1, \ldots, A_n) : A_1 \Box \cdots \Box A_n \to F(A_1) \Box' \cdots \Box' F(A_n)$$

with $J_1(A) = J_A$ that satisfies, e.g., $J_{n+m}(A_1, \ldots, A_{n+m}) = F_2\big(F(A_1)\Box' \cdots \Box' F(A_n), F(A_{n+1})\Box' \cdots \Box' F(A_{n+m})\big) \circ \big(J_n(A_1, \ldots, A_n) \Box J_m(A_{n+1}, \ldots, A_{n+m})\big)$ for all $n, m \in \mathbb{N}$ and $A_1, \ldots, A_{n+m} \in \mathrm{Ob}\,\mathcal{C}$.

A reduction between two tensor categories with projections would consist of a cotensor functor F and a natural transformation $P : F \to I'$.

In our applications we will also often encounter the case where $\mathcal{C}$ is not be a subcategory of $\mathcal{C}'$, but we have, e.g., a forgetful functor U from $\mathcal{C}$ to $\mathcal{C}'$ that "forgets" an additional structure that $\mathcal{C}$ has. An example for this situation is the reduction of Fermi independence to tensor independence in following subsection. Here we have to forget the $\mathbb{Z}_2$-grading of the objects of $\mathbb{Z}_2$-$\mathfrak{AlgProb}$ to get objects of $\mathfrak{AlgProb}$. In this situation a reduction of the tensor product with inclusions $\Box$ to the tensor product with inclusions $\Box'$ is a tensor function F from $(\mathcal{C}, \Box)$ to $(\mathcal{C}', \Box')$ and a natural transformation $J : U \to F$.

Example 3.41. The identity functor can be turned into a reduction from $(\mathfrak{Alg}, \coprod)$ to $(\mathfrak{Alg}, \otimes)$ (with the obvious inclusions).

The Symmetric Fock Space as a Tensor Functor

The category $\mathfrak{Vec}$ with the direct product $\oplus$ is of course a tensor category with inclusions and with projections, since the direct sum of vector spaces is both a product and a coproduct.

Not surprisingly, the usual tensor product of vector spaces is also a tensor product in the sense of category theory, but there are no canonical inclusions or projections. We can fix this by passing to the category $\mathfrak{Vek}_*$ of pointed vector spaces, whose objects are pairs (V, v) consisting of a vector space V and a non-zero vector $v \in V$. The morphisms $h : (V_1, v_1) \to (V_2, v_2)$ in this category are the linear maps $h : V_1 \to V_2$ with $h(v_1) = v_2$. In this category (equipped with the obvious tensor product $(V_1, v_1) \otimes (V_2, v_2) = (V_1 \otimes V_2, v_1 \otimes v_2)$) inclusions can be defined by $I_1 : V_1 \ni u \mapsto u \otimes v_2 \in V_1 \otimes V_2$ and $I_2 : V_1 \ni u \mapsto v_1 \otimes u \in V_1 \otimes V_2$.

Exercise 3.42. Show that in $(\mathfrak{Vek}_*, \otimes, I)$ all pairs of morphisms are independent, even though the tensor product is not a coproduct.

Proposition 3.43. *Take $\mathcal{D} = \mathfrak{Vek}$, $I = \mathrm{id}_{\mathfrak{Vek}}$, and $I' : \mathfrak{Vek}_* \to \mathfrak{Vek}$ the functor that forgets the fixed vector.*

The symmetric Fock space Γ is a reduction from $(\mathfrak{Vek}, \oplus, i)$ to $(\mathfrak{Vek}_*, \otimes, I)$ (w.r.t. $(\mathfrak{Vek}, \mathrm{id}_{\mathfrak{Vek}}, I')$).

We will not prove this proposition, we will only define all the natural transformations.

On the objects, Γ maps a vector space V to the pair $\big(\Gamma(V), \Omega\big)$ consisting of the algebraic symmetric Fock space

$$\Gamma(V) = \bigoplus_{n \in \mathbb{N}} V^{\otimes n}$$

and the vacuum vector Ω. The trivial vector space $\{0\}$ gets mapped to the field $\Gamma(\{0\}) = \mathbb{K}$ with the unit 1 as fixed vector. Linear maps $h : V_1 \to V_2$ get mapped to their second quantization $\Gamma(h) : \Gamma(V_1) \to \Gamma(V_2)$. $F_0 : \Gamma(\{0\}) = (\mathbb{K}, \mathbf{1}) \to (\mathbb{K}, \mathbf{1})$ is just the identity and F_2 is the natural isomorphism from $\Gamma(V_1 \oplus V_2)$ to $\Gamma(V_1) \otimes \Gamma(V_2)$ which acts on exponential vectors as

$$F_2 : \mathcal{E}(u_1 + u_2) \mapsto \mathcal{E}(u_1) \otimes \mathcal{E}(u_2)$$

for $u_1 \in V_1$, $u_2 \in V_2$.

The natural transformation $J : \mathrm{id}_{\mathfrak{Vec}} \to \Gamma$ finally is the embedding of V into $\Gamma(V)$ as one-particle space.

Example: Bosonization of Fermi Independence

We will now define the bosonization of Fermi independence as a reduction from $(\mathfrak{AlgProb}, \otimes, i)$ to $(\mathbb{Z}_2\text{-}\mathfrak{AlgProb}, \otimes_{\mathbb{Z}_2}, i)$. We will need the group algebra $\mathbb{C}\mathbb{Z}_2$ of $\mathbb{Z}_2$ and the linear functional $\varepsilon : \mathbb{C}\mathbb{Z}_2 \to \mathbb{C}$ that arises as the linear extension of the trivial representation of $\mathbb{Z}_2$, i.e.

$$\varepsilon(\mathbf{1}) = \varepsilon(g) = 1,$$

if we denote the even element of $\mathbb{Z}_2$ by $\mathbf{1}$ and the odd element by g.

The underlying functor $F : \mathbb{Z}_2\text{-}\mathfrak{AlgProb} \to \mathfrak{AlgProb}$ is given by

$$F : \begin{array}{ll} \mathrm{Ob}\, \mathbb{Z}_2\text{-}\mathfrak{AlgProb} \ni (\mathcal{A}, \varphi) & \mapsto (\mathcal{A} \otimes_{\mathbb{Z}_2} \mathbb{C}\mathbb{Z}_2, \varphi \otimes \varepsilon) \in \mathrm{Ob}\, \mathfrak{AlgProb}, \\ \mathrm{Mor}\, \mathbb{Z}_2\text{-}\mathfrak{AlgProb} \ni f & \mapsto f \otimes \mathrm{id}_{\mathbb{C}\mathbb{Z}_2} \in \mathrm{Mor}\, \mathfrak{AlgProb}. \end{array}$$

The unit element in both tensor categories is the one-dimensional unital algebra $\mathbb{C}\mathbf{1}$ with the unique unital functional on it. Therefore F_0 has to be a morphism from $F(\mathbb{C}\mathbf{1}) \cong \mathbb{C}\mathbb{Z}_2$ to $\mathbb{C}\mathbf{1}$. It is defined by $F_0(\mathbf{1}) = F_0(g) = \mathbf{1}$.

The morphism $F_2(\mathcal{A}, \mathcal{B})$ has to go from $F(\mathcal{A} \otimes_{\mathbb{Z}_2} \mathcal{B}) = (\mathcal{A} \otimes_{\mathbb{Z}_2} \mathcal{B}) \otimes \mathbb{C}\mathbb{Z}_2$ to $F(\mathcal{A}) \otimes F(\mathcal{B}) = (\mathcal{A} \otimes_{\mathbb{Z}_2} \mathbb{C}\mathbb{Z}_2) \otimes (\mathcal{B} \otimes_{\mathbb{Z}_2} \mathbb{C}\mathbb{Z}_2)$. It is defined by

$$a \otimes b \otimes \mathbf{1} \mapsto \begin{cases} (a \otimes \mathbf{1}) \otimes (b \otimes \mathbf{1}) & \text{if } b \text{ is even}, \\ (a \otimes g) \otimes (b \otimes \mathbf{1}) & \text{if } b \text{ is odd}, \end{cases}$$

and

$$a \otimes b \otimes g \mapsto \begin{cases} (a \otimes g) \otimes (b \otimes g) & \text{if } b \text{ is even}, \\ (a \otimes \mathbf{1}) \otimes (b \otimes g) & \text{if } b \text{ is odd}, \end{cases}$$

for $a \in \mathcal{A}$ and homogeneous $b \in \mathcal{B}$.

Finally, the inclusion $J_{\mathcal{A}} : \mathcal{A} \to \mathcal{A} \otimes_{\mathbb{Z}_2} \mathbb{C}\mathbb{Z}_2$ is defined by

$$J_{\mathcal{A}}(a) = a \otimes \mathbf{1}$$

for all $a \in \mathcal{A}$.

In this way we get inclusions $J_n = J_n(\mathcal{A}_1, \ldots, \mathcal{A}_n) = F_n(\mathcal{A}_1, \ldots, \mathcal{A}_n) \circ J_{\mathcal{A}_1 \otimes_{\mathbb{Z}_2} \cdots \otimes_{\mathbb{Z}_2} \mathcal{A}_n}$ of the graded tensor product $\mathcal{A}_1 \otimes_{\mathbb{Z}_2} \cdots \otimes_{\mathbb{Z}_2} \mathcal{A}_n$ into the usual tensor product $(\mathcal{A}_1 \otimes_{\mathbb{Z}_2} \mathbb{C}\mathbb{Z}_2) \otimes \cdots \otimes (\mathcal{A}_n \otimes_{\mathbb{Z}_2} \mathbb{C}\mathbb{Z}_2)$ which respect the states and allow to reduce all calculations involving the graded tensor product to calculations involving the usual tensor product on the bigger algebras $F(\mathcal{A}_1) = \mathcal{A}_1 \otimes_{\mathbb{Z}_2} \mathbb{C}\mathbb{Z}_2, \ldots, F(\mathcal{A}_n) = \mathcal{A}_n \otimes_{\mathbb{Z}_2} \mathbb{C}\mathbb{Z}_2$. These inclusions are determined by

$$J_n(\underbrace{\mathbf{1} \otimes \cdots \otimes \mathbf{1}}_{k-1 \text{ times}} \otimes a \otimes \underbrace{\mathbf{1} \otimes \cdots \otimes \mathbf{1}}_{n-k \text{ times}}) = \underbrace{\tilde{g} \otimes \cdots \otimes \tilde{g}}_{k-1 \text{ times}} \otimes \tilde{a} \otimes \underbrace{\tilde{\mathbf{1}} \otimes \cdots \otimes \tilde{\mathbf{1}}}_{n-k \text{ times}},$$

for $a \in \mathcal{A}_k$ odd, and

$$J_n(\underbrace{\mathbf{1} \otimes \cdots \otimes \mathbf{1}}_{k-1 \text{ times}} \otimes a \otimes \underbrace{\mathbf{1} \otimes \cdots \otimes \mathbf{1}}_{n-k \text{ times}}) = \underbrace{\tilde{\mathbf{1}} \otimes \cdots \otimes \tilde{\mathbf{1}}}_{k-1 \text{ times}} \otimes \tilde{a} \otimes \underbrace{\tilde{\mathbf{1}} \otimes \cdots \otimes \tilde{\mathbf{1}}}_{n-k \text{ times}},$$

for $a \in \mathcal{A}_k$ even, $1 \le k \le n$, where we used the abbreviations

$$\tilde{g} = \mathbf{1} \otimes g, \qquad \tilde{a} = a \otimes \mathbf{1}, \qquad \tilde{\mathbf{1}} = \mathbf{1} \otimes \mathbf{1}.$$

The Reduction of Boolean, Monotone, and Anti-Monotone Independence to Tensor Independence

We will now present the unification of tensor, monotone, anti-monotone, and boolean independence of Franz[Fra03b] in our category theoretical framework. It resembles closely the bosonization of Fermi independence in Subsection 3.4, but the group $\mathbb{Z}_2$ has to be replaced by the semigroup $M = \{\mathbf{1}, p\}$ with two elements, $\mathbf{1}\cdot\mathbf{1} = \mathbf{1}$, $\mathbf{1}\cdot p = p\cdot\mathbf{1} = p\cdot p = p$. We will need the linear functional $\varepsilon : \mathbb{C}M \to \mathbb{C}$ with $\varepsilon(\mathbf{1}) = \varepsilon(p) = 1$.

The underlying functor and the inclusions are the same for the reduction of the boolean, the monotone and the anti-monotone product. They map the algebra $\mathcal{A}$ of $(\mathcal{A}, \varphi)$ to the free product $F(\mathcal{A}) = \tilde{\mathcal{A}} \coprod \mathbb{C}M$ of the unitization $\tilde{\mathcal{A}}$ of $\mathcal{A}$ and the group algebra $\mathbb{C}M$ of M. For the unital functional $F(\varphi)$ we take the boolean product $\tilde{\varphi} \diamond \varepsilon$ of the unital extension $\tilde{\varphi}$ of φ with ε. The elements of $F(\mathcal{A})$ can be written as linear combinations of terms of the form

$$p^\alpha a_1 p \cdots p a_m p^\omega$$

with $m \in \mathbb{N}$, $\alpha, \omega \in \{0, 1\}$, $a_1, \ldots . a_m \in \mathcal{A}$, and $F(\varphi)$ acts on them as

$$F(\varphi)(p^\alpha a_1 p \cdots p a_m p^\omega) = \prod_{k=1}^{m} \varphi(a_k).$$

The inclusion is simply

$$J_\mathcal{A} : \mathcal{A} \ni a \mapsto a \in F(\mathcal{A}).$$

The morphism $F_0 : F(\mathbb{C}\mathbf{1}) = \mathbb{C}M \to \mathbb{C}\mathbf{1}$ is given by the trivial representation of M, $F_0(\mathbf{1}) = F_0(p) = \mathbf{1}$.

The only part of the reduction that is different for the three cases are the morphisms

$$F_2(\mathcal{A}_1, \mathcal{A}_2) : \mathcal{A}_1 \coprod \mathcal{A}_2 \to F(\mathcal{A}_1) \otimes F(\mathcal{A}_2) = (\tilde{\mathcal{A}}_1 \coprod \mathbb{C}M) \otimes (\tilde{\mathcal{A}}_2 \coprod \mathbb{C}M).$$

We set

$$F_2^{\mathrm{B}}(\mathcal{A}_1, \mathcal{A}_2)(a) = \begin{cases} a \otimes p & \text{if } a \in \mathcal{A}_1, \\ p \otimes a & \text{if } a \in \mathcal{A}_2, \end{cases}$$

for the boolean case,

$$F_2^{\mathrm{M}}(\mathcal{A}_1, \mathcal{A}_2)(a) = \begin{cases} a \otimes p & \text{if } a \in \mathcal{A}_1, \\ \mathbf{1} \otimes a & \text{if } a \in \mathcal{A}_2, \end{cases}$$

for the monotone case, and

$$F_2^{\mathrm{AM}}(\mathcal{A}_1, \mathcal{A}_2)(a) = \begin{cases} a \otimes \mathbf{1} & \text{if } a \in \mathcal{A}_1, \\ p \otimes a & \text{if } a \in \mathcal{A}_2, \end{cases}$$

for the anti-monotone case.

For the higher order inclusions $J_n^\bullet = F_n^\bullet(\mathcal{A}_1,\ldots,\mathcal{A}_n)\circ J_{\mathcal{A}_1\coprod\cdots\coprod\mathcal{A}_n}$, $\bullet\in\{\mathrm{B},\mathrm{M},\mathrm{AM}\}$, one gets

$$
\begin{aligned}
J_n^{\mathrm{B}}(a) &= p^{\otimes(k-1)}\otimes a\otimes p^{\otimes(n-k)},\\
J_n^{\mathrm{M}}(a) &= \mathbf{1}^{\otimes(k-1)}\otimes a\otimes p^{\otimes(n-k)},\\
J_n^{\mathrm{AM}}(a) &= p^{\otimes(k-1)}\otimes a\otimes \mathbf{1}^{\otimes(n-k)},
\end{aligned}
$$

if $a\in\mathcal{A}_k$.

One can verify that this indeed defines reductions (F^{B},J), (F^{M},J), and (F^{AM},J) from the categories $(\mathfrak{nuAlgProb},\diamond,i)$, $(\mathfrak{nuAlgProb},\triangleright,i)$, and $(\mathfrak{nuAlgProb},\triangleleft,i)$ to $(\mathfrak{AlgProb},\otimes,i)$. The functor $U:\mathfrak{nuAlgProb}\to\mathfrak{AlgProb}$ is the unitization of the algebra and the unital extension of the functional and the morphisms.

This reduces all calculations involving the boolean, monotone or anti-monotone product to the tensor product. These constructions can also be applied to reduce the quantum stochastic calculus on the boolean, monotone, and anti-monotone Fock space to the boson Fock space. Furthermore, they allow to reduce the theories of boolean, monotone, and anti-monotone Lévy processes to Schürmann's[Sch93] theory of Lévy processes on involutive bialgebras, see Franz[Fra03b] or Subsection 4.3.

Exercise 3.44. Construct a similar reduction for the category of unital algebras $\mathcal{A}$ having a decomposition $\mathcal{A}=\mathbb{C}\mathbf{1}\oplus\mathcal{A}_0$ and the boolean, monotone, or anti-monotone product defined for these algebras in Exercise 3.39

3.5 Classification of the Universal Independences

In the previous Subsection we have seen how a notion of independence can be defined in the language of category theory and we have also encountered several examples.

We are mainly interested in different categories of algebraic probability spaces. Their objects are pairs consisting of an algebra $\mathcal{A}$ and a linear functional φ on $\mathcal{A}$. Typically, the algebra has some additional structure, e.g., an involution, a unit, a grading, or a topology (it can be, e.g., a von Neumann algebra or a C^*-algebra), and the functional behaves nicely with respect to this additional structure, i.e., it is positive, unital, respects the grading, continuous, or normal. The morphisms are algebra homomorphisms, which leave the linear functional invariant, i.e., $j:(\mathcal{A},\varphi)\to(\mathcal{B},\psi)$ satisfies

$$\varphi=\psi\circ j$$

and behave also nicely w.r.t. additional structure, i.e., they can be required to be $*$-algebra homomorphisms, map the unit of $\mathcal{A}$ to the unit of $\mathcal{B}$, respect the grading, etc. We have already seen one example in Subsection 3.3.

The tensor product then has to specify a new algebra with a linear functional and inclusions for every pair of of algebraic probability spaces. If the category of algebras obtained from our algebraic probability space by forgetting the linear functional has a coproduct, then it is sufficient to consider the case where the new algebra is the coproduct of the two algebras.

Proposition 3.45. *Let $(\mathcal{C}, \Box, i)$ be a tensor category with inclusions and $F : \mathcal{C} \to \mathcal{D}$ a functor from $\mathcal{C}$ into another category $\mathcal{D}$ which has a coproduct $\coprod$ and an initial object $E_{\mathcal{D}}$. Then F is a tensor functor. The morphisms $F_2(A,B) : F(A) \coprod F(B) \to F(A \Box B)$ and $F_0 : E_{\mathcal{D}} \to F(E)$ are those guaranteed by the universal property of the coproduct and the initial object, i.e. $F_0 : E_{\mathcal{D}} \to F(E)$ is the unique morphism from $E_{\mathcal{D}}$ to $F(E)$ and $F_2(A,B)$ is the unique morphism that makes the diagram*

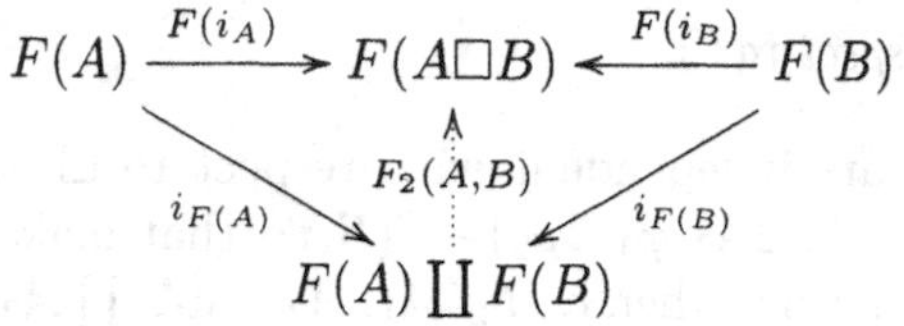

commuting.

Proof. Using the universal property of the coproduct and the definition of F_2, one shows that the triangles containing the $F(A)$ in the center of the diagram

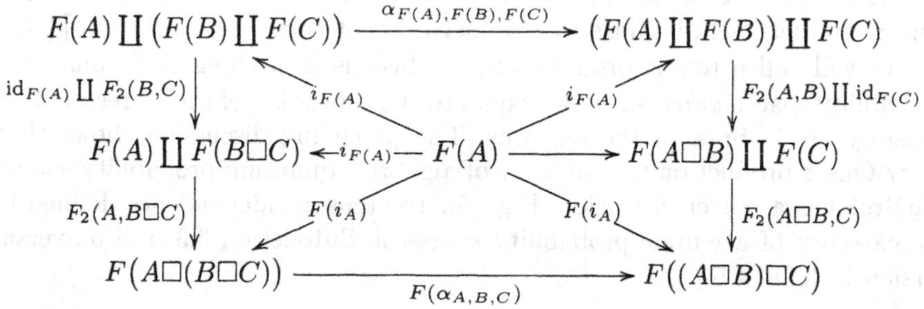

commute (where the morphism from $F(A)$ to $F(A \Box B) \coprod F(C)$ is given by $F(i_A) \coprod \mathrm{id}_{F(C)}$), and therefore that the morphisms corresponding to all the different paths form $F(A)$ to $F\big((A \Box B) \Box C\big)$ coincide. Since we can get similar diagrams with $F(B)$ and $F(C)$, it follows from the universal property of the triple coproduct $F(A) \coprod \big(F(B) \coprod F(C)\big)$ that there exists only a unique morphism from $F(A) \coprod (F(B) \coprod F(C))$ to $F\big((A \Box B) \Box C\big)$ and therefore that the whole diagram commutes.

The commutativity of the two diagrams involving the unit elements can be shown similarly. □

Let $\mathcal{C}$ now be a category of algebraic probability spaces and F the functor that maps a pair $(\mathcal{A}, \varphi)$ to the algebra $\mathcal{A}$, i.e., that "forgets" the linear functional φ. Suppose that $\mathcal{C}$ is equipped with a tensor product $\Box$ with inclusions and that $F(\mathcal{C})$ has a coproduct $\coprod$. Let $(\mathcal{A},\varphi)$, $(\mathcal{B},\psi)$ be two algebraic probability spaces in $\mathcal{C}$, we will denote the pair $(\mathcal{A},\varphi)\Box(\mathcal{B},\psi)$ also by $(\mathcal{A}\Box\mathcal{B},\varphi\Box\psi)$. By Proposition 3.45 we have morphisms $F_2(\mathcal{A},\mathcal{B}):\mathcal{A}\coprod\mathcal{B}\to\mathcal{A}\Box\mathcal{B}$ that define a natural transformation from the bifunctor $\coprod$ to the bifunctor $\Box$. With these morphisms we can define a new tensor product $\tilde{\Box}$ with inclusions by

$$(\mathcal{A},\varphi)\tilde{\Box}(\mathcal{B},\psi)=\left(\mathcal{A}\coprod\mathcal{B},(\varphi\Box\psi)\circ F_2(\mathcal{A},\mathcal{B})\right).$$

The inclusions are those defined by the coproduct.

Proposition 3.46. *If two random variables $f_1:(\mathcal{A}_1,\varphi_1)\to(\mathcal{B},\psi)$ and $f_1:(\mathcal{A}_1,\varphi_1)\to(\mathcal{B},\psi)$ are independent with respect to $\Box$, then they are also independent with respect to $\tilde{\Box}$.*

Proof. If f_1 and f_2 are independent with respect to $\Box$, then there exists a random variable $h:(\mathcal{A}_1\Box\mathcal{A}_2,\varphi_1\Box\varphi_2)\to(\mathcal{B},\psi)$ that makes diagram (3.6) in Definition 3.35 commuting. Then $h\circ F_2(\mathcal{A}_1,\mathcal{A}_2):(\mathcal{A}_1\coprod\mathcal{A}_2,\varphi_1\tilde{\Box}\varphi_2)\to(\mathcal{B},\psi)$ makes the corresponding diagram for $\tilde{\Box}$ commuting. □

The converse is not true. Consider the category of algebraic probability spaces with the tensor product, see Subsection 3.3, and take $B=\mathcal{A}_1\coprod\mathcal{A}_2$ and $\psi=(\varphi_1\otimes\varphi_2)\circ F_2(\mathcal{A}_1,\mathcal{A}_2)$. The canonical inclusions $i_{\mathcal{A}_1}:(\mathcal{A}_1,\varphi_1)\to(\mathcal{B},\psi)$ and $i_{\mathcal{A}_2}:(\mathcal{A}_2,\varphi_2)\to(\mathcal{B},\psi)$ are independent w.r.t. $\tilde{\otimes}$, but not with respect to the tensor product itself, because their images do not commute in $\mathcal{B}=\mathcal{A}_1\coprod\mathcal{A}_2$.

We will call a tensor product with inclusions in a category of quantum probability spaces *universal*, if it is equal to the coproduct of the corresponding category of algebras on the algebras. The preceding discussion shows that every tensor product on the category of algebraic quantum probability spaces $\mathfrak{AlgProb}$ has a universal version. E.g., for the tensor independence defined in the category of algebraic probability spaces in Subsection 3.3, the universal version is defined by

$$\varphi_1\tilde{\otimes}\varphi_2(a_1a_2\cdots a_m)=\varphi_1\left(\overrightarrow{\prod_{i:\epsilon_i=1}}a_i\right)\varphi_2\left(\overrightarrow{\prod_{i:\epsilon_i=2}}a_i\right)$$

for two unital functionals $\varphi_1:\mathcal{A}_1\to\mathbb{C}$ and $\varphi_2:\mathcal{A}_2\to\mathbb{C}$ and a typical element $a_1a_2\cdots a_m\in\mathcal{A}_1\coprod\mathcal{A}_2$, with $a_k\in\mathcal{A}_{\epsilon_k}$, $\epsilon_1\neq\epsilon_2\neq\cdots\neq\epsilon_m$, i.e. neighboring a's don't belong to the same algebra.

We will now reformulate the classification by Muraki[Mur03] and by Ben Ghorbal and Schürmann[BG01, BGS99] in terms of universal tensor products with inclusions for the category of algebraic probability spaces $\mathfrak{AlgProb}$.

In order to define a universal tensor product with inclusions on $\mathfrak{AlgProb}$ one needs a map that associates to a pair of unital functionals (φ_1,φ_2) on two algebras $\mathcal{A}_1$ and $\mathcal{A}_2$ a unital functional $\varphi_1\cdot\varphi_2$ on the free product $\mathcal{A}_1\coprod\mathcal{A}_2$ (with identification of the units) of $\mathcal{A}_1$ and $\mathcal{A}_2$ in such a way that the bifunctor

$$\Box : (\mathcal{A}_1, \varphi_1) \times (\mathcal{A}_2, \varphi_1) \mapsto (\mathcal{A}_1 \coprod \mathcal{A}_2, \varphi_1 \cdot \varphi_2)$$

satisfies all the necessary axioms. Since $\Box$ is equal to the coproduct $\coprod$ on the algebras, we don't have a choice for the isomorphisms α, λ, ρ implementing the associativity and the left and right unit property. We have to take the ones following from the universal property of the coproduct. The inclusions and the action of $\Box$ on the morphisms also have to be the ones given by the coproduct.

The associativity gives us the condition

$$\big((\varphi_1 \cdot \varphi_2) \cdot \varphi_3\big) \circ \alpha_{\mathcal{A}_1,\mathcal{A}_2,\mathcal{A}_3} = \varphi_1 \cdot (\varphi_2 \cdot \varphi_3), \tag{3.7}$$

for all $(\mathcal{A}_1, \varphi_1), (\mathcal{A}_2, \varphi_2), (\mathcal{A}_3, \varphi_3)$ in $\mathfrak{AlgProb}$. Denote the unique unital functional on $\mathbb{C}\mathbf{1}$ by δ, then the unit properties are equivalent to

$$(\varphi \cdot \delta) \circ \rho_{\mathcal{A}} = \varphi \quad \text{and} \quad (\delta \cdot \varphi) \circ \lambda_{\mathcal{A}} = \varphi,$$

for all $(\mathcal{A}, \varphi)$ in $\mathfrak{AlgProb}$. The inclusions are random variables, if and only if

$$(\varphi_1 \cdot \varphi_2) \circ i_{\mathcal{A}_1} = \varphi_1 \quad \text{and} \quad (\varphi_1 \cdot \varphi_2) \circ i_{\mathcal{A}_2} = \varphi_2 \tag{3.8}$$

for all $(\mathcal{A}_1, \varphi_1), (\mathcal{A}_2, \varphi_2)$ in $\mathfrak{AlgProb}$. Finally, from the functoriality of $\Box$ we get the condition

$$(\varphi_1 \cdot \varphi_2) \circ (j_1 \coprod j_2) = (\varphi_1 \circ j_1) \cdot (\varphi_2 \circ j_2) \tag{3.9}$$

for all pairs of morphisms $j_1 : (\mathcal{B}_1, \psi_1) \to (\mathcal{A}_1, \varphi_1)$, $j_2 : (\mathcal{B}_2, \psi_2) \to (\mathcal{A}_2, \varphi_2)$ in $\mathfrak{AlgProb}$.

Our Conditions (3.7), (3.8), and (3.9) are exactly the axioms (P2), (P3), and (P4) in Ben Ghorbal and Schürmann[BGS99], or the axioms (U2), the first part of (U4), and (U3) in Muraki[Mur03].

Theorem 3.47. *(Muraki [Mur03], Ben Ghorbal and Schürmann [BG01, BGS99]). There exist exactly two universal tensor products with inclusions on the category of algebraic probability spaces* $\mathfrak{AlgProb}$, *namely the universal version* $\tilde{\otimes}$ *of the tensor product defined in Section 3.3 and the one associated to the free product* $*$ *of states.*

For the classification in the non-unital case, Muraki imposes the additional condition

$$(\varphi_1 \cdot \varphi_2)(a_1 a_2) = \varphi_{\epsilon_1}(a_1)\varphi_{\epsilon_2}(a_2) \tag{3.10}$$

for all $(\epsilon_1, \epsilon_2) \in \{(1,2),(2,1)\}$, $a_1 \in \mathcal{A}_{\epsilon_1}$, $a_2 \in \mathcal{A}_{\epsilon_2}$.

Theorem 3.48. *(Muraki[Mur03]) There exist exactly five universal tensor products with inclusions satisfying (3.10) on the category of non-unital algebraic probability spaces* $\mathfrak{nuAlgProb}$, *namely the universal version* $\tilde{\otimes}$ *of the tensor product defined in Section 3.3 and the ones associated to the free product* $*$, *the boolean product* $\diamond$, *the monotone product* $\triangleright$ *and the anti-monotone product* $\triangleleft$.

The monotone and the anti-monotone are not symmetric, i.e. $(\mathcal{A}_1 \coprod \mathcal{A}_2, \varphi_1 \triangleright \varphi_2)$ and $(\mathcal{A}_2 \coprod \mathcal{A}_2, \varphi_2 \triangleright \varphi_1)$ are not isomorphic in general. Actually, the anti-monotone product is simply the mirror image of the monotone product,

$$(\mathcal{A}_1 \coprod \mathcal{A}_2, \varphi_1 \triangleright \varphi_2) \cong (\mathcal{A}_2 \coprod \mathcal{A}_1, \varphi_2 \triangleleft \varphi_1)$$

for all $(\mathcal{A}_1, \varphi_1), (\mathcal{A}_2, \varphi_2)$ in the category of non-unital algebraic probability spaces. The other three products are symmetric.

In the symmetric setting of Ben Ghorbal and Schürmann, Condition (3.10) is not essential. If one drops it and adds symmetry, one finds in addition the degenerate product

$$(\varphi_1 \bullet_0 \varphi_2)(a_1 a_2 \cdots a_m) = \begin{cases} \varphi_{\epsilon_1}(a_1) & \text{if } m = 1, \\ 0 & \text{if } m > 1. \end{cases}$$

and families

$$\varphi_1 \bullet_q \varphi_2 = q\big((q^{-1}\varphi_1) \cdot (q^{-1}\varphi_2)\big),$$

parametrized by a complex number $q \in \mathbb{C}\backslash\{0\}$, for each of the three symmetric products, $\bullet \in \{\tilde{\otimes}, *, \diamond\}$.

If one adds the condition that products of states are again states, then one can also show that the constant has to be equal to one.

Exercise 3.49. Consider the category of non-unital $*$-algebraic probability spaces, whose objects are pairs $(\mathcal{A}, \varphi)$ consisting of a not necessarily unital $*$-algebra $\mathcal{A}$ and a state $\varphi : \mathcal{A} \to \mathbb{C}$. Here a state is a linear functional $\varphi : \mathcal{A} \to \mathbb{C}$ whose unital extension $\tilde{\varphi} : \tilde{\mathcal{A}} \cong \mathbb{C}\mathbf{1} \oplus \mathcal{A} \to \mathbb{C}$, $\lambda\mathbf{1} + a \mapsto \tilde{\varphi}(\lambda\mathbf{1} + a) = \lambda + \varphi(a)$, to the unitization of $\mathcal{A}$ is a state.

Assume we have products $\cdot : S(\mathcal{A}_1) \times S(\mathcal{A}_2) \to S(\mathcal{A}_1 \coprod \mathcal{A}_2)$ of linear functionals on non-unital algebras $\mathcal{A}_1, \mathcal{A}_2$ that satisfy

$$(\varphi_1 \cdot \varphi_2)(a_1 a_2) = c_1 \varphi_1(a_1)\varphi_2(a_2),$$
$$(\varphi_1 \cdot \varphi_2)(a_2 a_1) = c_2 \varphi_1(a_1)\varphi_2(a_2),$$

for all linear functionals $\varphi_1 : \mathcal{A}_1 \to \mathbb{C}$, $\varphi_2 : \mathcal{A}_2 \to \mathbb{C}$, and elements $a_1 \in \mathcal{A}_1$, $a_2 \in \mathcal{A}_2$ with "universal" constants $c_1, c_2 \in \mathbb{C}$, i.e. constants that do not depend on the algebras, the functionals, or the algebra elements. That for every universal independence such constants have to exist is part of the proof of the classifications in [BG01, BGS99, Mur03].

Show that if the products of states are again states, then we have $c_1 = c_2 = 1$. Hint: Take for $\mathcal{A}_1$ and $\mathcal{A}_2$ the algebra of polynomials on $\mathbb{R}$ and for φ_1 and φ_2 evaluation in a point.

The proof of the classification of universal independences can be split into three steps.

Using the "universality" or functoriality of the product, one can show that there exist some "universal constants" - not depending on the algebras - and a formula for evaluating

$$(\varphi_1 \cdot \varphi_2)(a_1 a_2 \cdots a_m)$$

for $a_1 a_2 \cdots a_m \in \mathcal{A}_1 \coprod \mathcal{A}_2$, with $a_k \in \mathcal{A}_{\epsilon_k}$, $\epsilon_1 \neq \epsilon_2 \neq \cdots \neq \epsilon_m$, as a linear combination of products $\varphi_1(M_1)$, $\varphi_2(M_2)$, where M_1, M_2 are "sub-monomials" of $a_1 a_2 \cdots a_m$. Then in a second step it is shown by associativity that only products with *ordered* monomials M_1, M_2 contribute. This is the content of [BGS02, Theorem 5] in the commutative case and of [Mur03, Theorem 2.1] in the general case.

The third step, which was actually completed first in both cases, see [Spe97] and [Mur02], is to find the conditions that the universal constants have to satisfy, if the resulting product is associative. It turns out that the universal coefficients for $m > 5$ are already uniquely determined by the coefficients for $1 \leq m \leq 5$. Detailed analysis of the non-linear equations obtained for the coefficients of order up to five then leads to the classifications stated above.

4 Lévy Processes on Dual Groups

We now want to study quantum stochastic processes whose increments are free or independent in the sense of boolean, monotone, or anti-monotone independence. The approach based on bialgebras that we followed in the first Section works for the tensor product and fails in the other cases because the corresponding products are not defined on the tensor product, but on the free product of the algebra. The algebraic structure which has to replace bialgebras was first introduced by Voiculescu [Voi87, Voi90], who named them dual groups. In this section we will introduce these algebras and develop the theory of their Lévy processes. It turns out that Lévy processes on dual groups with boolean, monotonically, or anti-monotonically independent increments can be reduced to Lévy processes on involutive bialgebra. We do not know if this is also possible for Lévy processes on dual groups with free increments.

In the literature additive free Lévy processes have been studied most intensively, see, e.g., [GSS92, Bia98, Ans02, Ans03, BNT02b, BNT02a].

4.1 Preliminaries on Dual Groups

Denote by $\mathfrak{ComAlg}$ the category of commutative unital algebras and let $\mathcal{B} \in \mathrm{Ob}\,\mathfrak{ComAlg}$ be a commutative bialgebra. Then the mapping

$$\mathrm{Ob}\,\mathfrak{ComAlg} \ni \mathcal{A} \mapsto \mathrm{Mor}_{\mathfrak{ComAlg}}(\mathcal{B}, \mathcal{A})$$

can be understood as a functor from $\mathfrak{ComAlg}$ to the category of unital semigroups. The multiplication in $\mathrm{Mor}_{\mathfrak{Alg}}(\mathcal{B}, \mathcal{A})$ is given by the convolution, i.e.

$$f \star g = m_{\mathcal{A}} \circ (f \otimes g) \circ \Delta_{\mathcal{B}}$$

and the unit element is $\varepsilon_\mathcal{B}\mathbf{1}_\mathcal{A}$. A unit-preserving algebra homomorphism $h : \mathcal{A}_1 \to \mathcal{A}_2$ gets mapped to the unit-preserving semigroup homomorphism $\mathrm{Mor}_{\mathfrak{ComAlg}}(\mathcal{B}, \mathcal{A}_1) \ni f \to h \circ f \in \mathrm{Mor}_{\mathfrak{ComAlg}}(\mathcal{B}, \mathcal{A}_2)$, since

$$h \circ (f \star g) = (h \circ f) \star (h \circ g)$$

for all $\mathcal{A}_1, \mathcal{A}_2 \in \mathrm{Ob}\,\mathfrak{ComAlg}$, $h \in \mathrm{Mor}_{\mathfrak{ComAlg}}(\mathcal{A}_1, \mathcal{A}_2)$, $f, g \in \mathrm{Mor}_{\mathfrak{ComAlg}}(\mathcal{B}, \mathcal{A}_1)$.

If $\mathcal{B}$ is even a commutative Hopf algebra with antipode S, then $\mathrm{Mor}_{\mathfrak{ComAlg}}(\mathcal{B}, \mathcal{A})$ is a group with respect to the convolution product. The inverse of a homomorphism $f : \mathcal{B} \to \mathcal{A}$ with respect to the convolution product is given by $f \circ S$.

The calculation

$$\begin{aligned}(f \star g)(ab) &= m_\mathcal{A} \circ (f \otimes g) \circ \Delta_\mathcal{B}(ab) \\ &= f(a_{(1)}b_{(1)})g(a_{(2)}b_{(2)}) = f(a_{(1)})f(b_{(1)})g(a_{(2)})g(b_{(2)}) \\ &= f(a_{(1)})g(a_{(2)})f(b_{(1)})g(b_{(2)}) = (f \star g)(a)(f \star g)(b)\end{aligned}$$

shows that the convolution product $f \star g$ of two homomorphisms $f, g : \mathcal{B} \to \mathcal{A}$ is again a homomorphism. It also gives an indication why non-commutative bialgebras or Hopf algebras do not give rise to a similar functor on the category of non-commutative algebras, since we had to commute $f(b_{(1)})$ with $g(a_{(2)})$.

Zhang [Zha91], Berman and Hausknecht [BH96] showed that if one replaces the tensor product in the definition of bialgebras and Hopf algebras by the free product, then one arrives at a class of algebras that do give rise to a functor from the category of non-commutative algebras to the category of semigroups or groups.

A *dual group* [Voi87, Voi90] (called *H-algebra* or *cogroup* in the category of unital associative $*$-algebras in [Zha91] and [BH96], resp.) is a unital $*$-algebra $\mathcal{B}$ equipped with three unital $*$-algebra homomorphisms $\Delta : \mathcal{B} \to \mathcal{B} \coprod \mathcal{B}$, $S : \mathcal{B} \to \mathcal{B}$ and $\varepsilon : \mathcal{B} \to \mathbb{C}$ (also called *comultiplication*, *antipode*, and *counit*) such that

$$\left(\Delta \coprod \mathrm{id}\right) \circ \Delta = \left(\mathrm{id} \coprod \Delta\right) \circ \Delta, \tag{4.1}$$

$$\left(\varepsilon \coprod \mathrm{id}\right) \circ \Delta = \mathrm{id} = \left(\mathrm{id} \coprod \varepsilon\right) \circ \Delta, \tag{4.2}$$

$$m_\mathcal{B} \circ \left(S \coprod \mathrm{id}\right) \circ \Delta = \mathrm{id} = m_\mathcal{B} \circ \left(\mathrm{id} \coprod S\right) \circ \Delta, \tag{4.3}$$

where $m_\mathcal{B} : \mathcal{B} \coprod \mathcal{B} \to \mathcal{B}$, $m_\mathcal{B}(a_1 \otimes a_2 \otimes \cdots \otimes a_n) = a_1 \cdot a_2 \cdot \cdots \cdot a_n$, is the multiplication of $\mathcal{B}$. Besides the formal similarity, there are many relations between dual groups on the one side and Hopf algebras and bialgebras on the other side, cf. [Zha91]. For example, let $\mathcal{B}$ be a dual group with comultiplication Δ, and let $R : \mathcal{B} \coprod \mathcal{B} \to \mathcal{B} \otimes \mathcal{B}$ be the unique unital $*$-algebra homomorphism with

$$R_{\mathcal{B},\mathcal{B}} \circ i_1(b) = b \otimes \mathbf{1}, \qquad R_{\mathcal{B},\mathcal{B}} \circ i_2(b) = \mathbf{1} \otimes b,$$

for all $b \in \mathcal{B}$. Here $i_1, i_2 : \mathcal{B} \to \mathcal{B} \coprod \mathcal{B}$ denote the canonical inclusions of $\mathcal{B}$ into the first and the second factor of the free product $\mathcal{B} \coprod \mathcal{B}$. Then $\mathcal{B}$ is a bialgebra with the comultiplication $\overline{\Delta} = R_{\mathcal{B},\mathcal{B}} \circ \Delta$, see [Zha91, Theorem 4.2], but in general it is not a Hopf algebra.

We will not really work with dual groups, but the following weaker notion. A *dual semigroup* is a unital $*$-algebra $\mathcal{B}$ equipped with two unital $*$-algebra homomorphisms $\Delta : \mathcal{B} \to \mathcal{B} \coprod \mathcal{B}$ and $\varepsilon : \mathcal{B} \to \mathbb{C}$ such that Equations (4.1) and (4.2) are satisfied. The antipode is not used in the proof of [Zha91, Theorem 4.2], and therefore we also get an involutive bialgebra $(\mathcal{B}, \overline{\Delta}, \varepsilon)$ for every dual semigroup $(\mathcal{B}, \Delta, \varepsilon)$.

Note that we can always write a dual semigroup $\mathcal{B}$ as a direct sum $\mathcal{B} = \mathbb{C}\mathbf{1} \oplus \mathcal{B}^0$, where $\mathcal{B}^0 = \ker \varepsilon$ is even a $*$-ideal. Therefore it is in the range of the unitization functor and the boolean, monotone, and anti-monotone product can be defined for unital linear functionals on $\mathcal{B}$, cf. Exercise 3.39.

The comultiplication of a dual semigroup can also be used to define a convolution product. The *convolution* $j_1 \star j_2$ of two unital $*$-algebra homomorphisms $j_1, j_2 : \mathcal{B} \to \mathcal{A}$ is defined as

$$j_1 \star j_2 = m_{\mathcal{A}} \circ \left(j_1 \coprod j_2 \right) \circ \Delta.$$

As the composition of the three unital $*$-algebra homomorphisms $\Delta : \mathcal{B} \to \mathcal{B} \coprod \mathcal{B}$, $j_1 \coprod j_2 : \mathcal{B} \coprod \mathcal{B} \to \mathcal{A} \coprod \mathcal{A}$, and $m_{\mathcal{A}} : \mathcal{A} \coprod \mathcal{A} \to \mathcal{A}$, this is obviously again a unital $*$-algebra homomorphism. Note that this convolution can not be defined for arbitrary linear maps on $\mathcal{B}$ with values in some algebra, as for bialgebras, but only for unital $*$-algebra homomorphisms.

4.2 Definition of Lévy Processes on Dual Groups

Definition 4.1. *Let $j_1 : \mathcal{B}_1 \to (\mathcal{A}, \Phi), \ldots, j_n : \mathcal{B}_n \to (\mathcal{A}, \Phi)$ be quantum random variables over the same quantum probability space $(\mathcal{A}, \Phi)$ and denote their marginal distributions by $\varphi_i = \Phi \circ j_i$, $i = 1, \ldots, n$. The quantum random variables $(j_1, \ldots, j_n)$ are called tensor independent (respectively boolean independent, monotonically independent, anti-monotonically independent or free), if the state $\Phi \circ m_{\mathcal{A}} \circ (j_1 \coprod \cdots \coprod j_n)$ on the free product $\coprod_{i=1}^n \mathcal{B}_i$ is equal to the tensor product (boolean, monotone, anti-monotone, or free product, respectively) of $\varphi_1, \ldots, \varphi_n$.*

Note that tensor, boolean, and free independence do not depend on the order, but monotone and anti-monotone independence do. An n-tuple $(j_1, \ldots, j_n)$ of quantum random variables is monotonically independent, if and only if $(j_n, \ldots, j_1)$ is anti-monotonically independent.

We are now ready to define tensor, boolean, monotone, anti-monotone, and free Lévy processes on dual semigroups.

Definition 4.2. *[Sch95b] Let $(\mathcal{B}, \Delta, \varepsilon)$ be a dual semigroup. A quantum stochastic process $\{j_{st}\}_{0 \le s \le t \le T}$ on $\mathcal{B}$ over some quantum probability space $(\mathcal{A}, \Phi)$*

is called a tensor (resp. boolean, monotone, anti-monotone, or free) Lévy process on the dual semigroup $\mathcal{B}$, *if the following four conditions are satisfied.*

1. *(Increment property) We have*
$$j_{rs} \star j_{st} = j_{rt} \quad \textit{for all } 0 \le r \le s \le t \le T,$$
$$j_{tt} = \varepsilon \mathbf{1}_{\mathcal{A}} \quad \textit{for all } 0 \le t \le T.$$
2. *(Independence of increments) The family $\{j_{st}\}_{0 \le s \le t \le T}$ is tensor independent (resp. boolean, monotonically, anti-monotonically independent, or free) w.r.t. Φ, i.e. the n-tuple $(j_{s_1 t_2}, \ldots, j_{s_n t_n})$ is tensor independent (resp. boolean, monotonically, anti-monotonically independent, or free) for all $n \in \mathbb{N}$ and all $0 \le s_1 \le t_1 \le s_2 \le \cdots \le t_n \le T$.*
3. *(Stationarity of increments) The distribution $\varphi_{st} = \Phi \circ j_{st}$ of j_{st} depends only on the difference $t - s$.*
4. *(Weak continuity) The quantum random variables j_{st} converge to j_{ss} in distribution for $t \searrow s$.*

Remark 4.3. The independence property depends on the products and therefore for boolean, monotone and anti-monotone Lévy processes on the choice of a decomposition $\mathcal{B} = \mathbb{C}\mathbf{1} \oplus \mathcal{B}^0$. In order to show that the convolutions defined by $(\varphi_1 \diamond \varphi_2) \circ \Delta$, $(\varphi_1 \triangleright \varphi_2) \circ \Delta$, and $(\varphi_1 \triangleleft \varphi_2) \circ \Delta$ are associative and that the counit ε acts as unit element w.r.t. these convolutions, one has to use the universal property [BGS99, Condition (P4)], which in our setting is only satisfied for morphisms that respect the decomposition. Therefore we are forced to choose the decomposition given by $\mathcal{B}^0 = \ker \varepsilon$.

The marginal distributions $\varphi_{t-s} := \varphi_{st} = \Phi \circ j_{st}$ form again a convolution semigroup $\{\varphi_t\}_{t \in \mathbb{R}_+}$, with respect to the tensor (boolean, monotone, anti-monotone, or free respectively) convolution defined by $(\varphi_1 \tilde{\otimes} \varphi_2) \circ \Delta$ ($(\varphi_1 \diamond \varphi_2) \circ \Delta$, $(\varphi_1 \triangleright \varphi_2) \circ \Delta$, $(\varphi_1 \triangleleft \varphi_2) \circ \Delta$, or $(\varphi_1 * \varphi_2) \circ \Delta$, respectively). It has been shown that the generator $\psi : \mathcal{B} \to \mathbb{C}$,
$$\psi(b) = \lim_{t \searrow 0} \frac{1}{t} \big(\varphi_t(b) - \varepsilon(b) \big)$$
is well-defined for all $b \in \mathcal{B}$ and uniquely characterizes the semigroup $\{\varphi_t\}_{t \in \mathbb{R}_+}$, cf. [Sch95b, BGS99, Fra01].

Denote by S be the *flip* map $S : \mathcal{B} \coprod \mathcal{B} \to \mathcal{B} \coprod \mathcal{B}$, $S = m_{\mathcal{B} \coprod \mathcal{B}} \circ (i_2 \coprod i_1)$, where $i_1, i_2 : \mathcal{B} \to \mathcal{B} \coprod \mathcal{B}$ are the inclusions of $\mathcal{B}$ into the first and the second factor of the free product $\mathcal{B} \coprod \mathcal{B}$. The flip map S acts on $i_1(a_1) i_2(b_1) \cdots i_2(b_n) \in \mathcal{B} \coprod \mathcal{B}$ with $a_1, \ldots, a_n, b_1, \ldots, b_n \in \mathcal{B}$ as
$$S\Big(i_1(a_1) i_2(b_1) \cdots i_2(b_n)\Big) = i_2(a_1) i_1(b_1) \cdots i_1(b_n).$$
If $j_1 : \mathcal{B} \to \mathcal{A}_1$ and $j_2 : \mathcal{B} \to \mathcal{A}_2$ are two unital $*$-algebra homomorphisms, then we have $(j_2 \coprod j_1) \circ S = \gamma_{\mathcal{A}_1, \mathcal{A}_2} \circ (j_1 \coprod j_2)$. Like for bialgebras, the opposite comultiplication $\Delta^{\mathrm{op}} = S \circ \Delta$ of a dual semigroup $(\mathcal{B}, \Delta, \varepsilon)$ defines a new dual semigroup $(\mathcal{B}, \Delta^{\mathrm{op}}, \varepsilon)$.

Lemma 4.4. *Let $\{j_{st} : \mathcal{B} \to (\mathcal{A}, \Phi)\}_{0\le s\le t\le T}$ be a quantum stochastic process on a dual semigroup $(\mathcal{B}, \Delta, \varepsilon)$ and define its time-reversed process $\{j^{\rm op}_{st}\}_{0\le s\le t\le T}$ by*

$$j^{\rm op}_{st} = j_{T-t,T-s}$$

for $0 \le s \le t \le T$.

(i) The process $\{j_{st}\}_{0\le s\le t\le T}$ is a tensor (boolean, free, respectively) Lévy process on the dual semigroup $(\mathcal{B}, \Delta, \varepsilon)$ if and only if the time-reversed process $\{j^{\rm op}_{st}\}_{0\le s\le t\le T}$ is a tensor (boolean, free, respectively) Lévy process on the dual semigroup $(\mathcal{B}, \Delta^{\rm op}, \varepsilon)$.

(ii) The process $\{j_{st}\}_{0\le s\le t\le T}$ is a monotone Lévy process on the dual semigroup $(\mathcal{B}, \Delta, \varepsilon)$ if and only if the time-reversed process $\{j^{\rm op}_{st}\}_{0\le s\le t\le T}$ is an anti-monotone Lévy process on the dual semigroup $(\mathcal{B}, \Delta^{\rm op}, \varepsilon)$.

Proof. The equivalence of the stationarity and continuity property for the quantum stochastic processes $\{j_{st}\}_{0\le s\le t\le T}$ and $\{j^{\rm op}_{st}\}_{0\le s\le t\le T}$ is clear.

The increment property for $\{j_{st}\}_{0\le s\le t\le T}$ with respect to Δ is equivalent to the increment property of $\{j^{\rm op}_{st}\}_{0\le s\le t\le T}$ with respect to $\Delta^{\rm op}$, since

$$\begin{aligned} m_{\mathcal{A}} \circ \left(j^{\rm op}_{st} \coprod j^{\rm op}_{tu}\right) \circ \Delta^{\rm op} &= m_{\mathcal{A}} \circ \left(j_{T-t,T-s} \coprod j_{T-u,T-t}\right) \circ S \circ \Delta \\ &= m_{\mathcal{A}} \circ \gamma_{\mathcal{A},\mathcal{A}} \circ \left(j_{T-u,T-t} \coprod j_{T-t,T-s}\right) \circ \Delta \\ &= m_{\mathcal{A}} \circ \left(j_{T-u,T-t} \coprod j_{T-t,T-s}\right) \circ \Delta \end{aligned}$$

for all $0 \le s \le t \le u \le T$.

If $\{j_{st}\}_{0\le s\le t\le T}$ has monotonically independent increments, i.e. if the n-tuples $(j_{s_1t_2}, \ldots, j_{s_nt_n})$ are monotonically independent for all $n \in \mathbb{N}$ and all $0 \le s_1 \le t_1 \le s_2 \le \cdots \le t_n$, then the n-tuples $(j_{s_nt_n}, \ldots, j_{s_1t_1}) = (j^{\rm op}_{T-t_n,T-s_n}, \ldots, j^{\rm op}_{T-t_1,T-s_1})$ are anti-monotonically independent and therefore $\{j^{\rm op}_{st}\}_{0\le s\le t\le T}$ has anti-monotonically independent increments, and vice versa.

Since tensor and boolean independence and freeness do not depend on the order, $\{j_{st}\}_{0\le s\le t\le T}$ has tensor (boolean, free, respectively) independent increments, if and only $\{j^{\rm op}_{st}\}_{0\le s\le t\le T}$ has tensor (boolean, free, respectively) independent increments. □

Before we study boolean, monotone, and anti-monotone Lévy processes in more detail, we will show how the theory of tensor Lévy processes on dual semigroups reduces to the theory of Lévy processes on involutive bialgebras, see also [Sch95b]. If quantum random variables $j_1, \ldots, j_n$ are independent in the sense of Condition 2 in Definition 1.2, then they are also tensor independent in the sense of Definition 4.1. Therefore every Lévy process on the bialgebra $(\mathcal{B}, \overline{\Delta}, \varepsilon)$ associated to a dual semigroup $(\mathcal{B}, \Delta, \varepsilon)$ is automatically also a tensor Lévy process on the dual semigroup $(\mathcal{B}, \Delta, \varepsilon)$. To verify this, it is sufficient to note that the increment property in Definition 1.2 with respect to $\overline{\Delta}$ and the commutativity of the increments imply the increment property in Definition 4.2 with respect to Δ.

But tensor independence in general does not imply independence in the sense of Condition 2 in Definition 1.2, because the commutation relations are not necessarily satisfied. Therefore, in general, a tensor Lévy process on a dual semigroup $(\mathcal{B}, \Delta, \varepsilon)$ will *not* be a Lévy process on the involutive bialgebra $(\mathcal{B}, \overline{\Delta}, \varepsilon)$. But we can still associate an equivalent Lévy process on the involutive bialgebra $(\mathcal{B}, \overline{\Delta}, \varepsilon)$ to it. To do this, note that the convolutions of two unital functionals $\varphi_1, \varphi_2 : \mathcal{B} \to \mathbb{C}$ with respect to the dual semigroup structure and the tensor product and with respect to the bialgebra structure coincide, i.e.

$$(\varphi_1 \tilde{\otimes} \varphi_2) \circ \Delta = (\varphi_1 \otimes \varphi_2) \circ \overline{\Delta}.$$

for all unital functionals $\varphi_1, \varphi_2 : \mathcal{B} \to \mathbb{C}$. Therefore the semigroup of marginal distributions of a tensor Lévy process on the dual semigroup $(\mathcal{B}, \Delta, \varepsilon)$ is also a convolution semigroup of states on the involutive bialgebra $(\mathcal{B}, \overline{\Delta}, \varepsilon)$. It follows that there exists a unique (up to equivalence) Lévy process on the involutive bialgebra $(\mathcal{B}, \overline{\Delta}, \varepsilon)$ that has this semigroup as marginal distributions. It is easy to check that this process is equivalent to the given tensor Lévy process on the dual semigroup $(\mathcal{B}, \Delta, \varepsilon)$. We summarize our result in the following theorem.

Theorem 4.5. *Let $(\mathcal{B}, \Delta, \varepsilon)$ be a dual semigroup, and $(\mathcal{B}, \overline{\Delta}, \varepsilon)$ with $\overline{\Delta} = R_{\mathcal{B},\mathcal{B}} \circ \Delta$ the associated involutive bialgebra. The tensor Lévy processes on the dual semigroup $(\mathcal{B}, \Delta, \varepsilon)$ are in one-to-one correspondence (up to equivalence) with the Lévy processes on the involutive bialgebra $(\mathcal{B}, \overline{\Delta}, \varepsilon)$.*

Furthermore, every Lévy process on the involutive bialgebra $(\mathcal{B}, \overline{\Delta}, \varepsilon)$ is also a tensor Lévy process on the dual semigroup $(\mathcal{B}, \Delta, \varepsilon)$.

4.3 Reduction of Boolean, Monotone, and Anti-Monotone Lévy Processes to Lévy Processes on Involutive Bialgebras

In this subsection we will construct three involutive bialgebras for every dual semigroup $(\mathcal{B}, \Delta, \varepsilon)$ and establish a one-to-one correspondence between boolean, monotone, and anti-monotone Lévy processes on the dual semigroup $(\mathcal{B}, \Delta, \varepsilon)$ and a certain class of Lévy processes on one of those involutive bialgebras.

We start with some general remarks.

Let $(\mathcal{C}, \Box)$ be a tensor category. Then we call an object $\mathcal{D}$ in $\mathcal{C}$ equipped with morphisms

$$\varepsilon : \mathcal{D} \to E, \qquad \Delta : \mathcal{D} \to \mathcal{D} \Box \mathcal{D}$$

a *dual semigroup in* $(\mathcal{C}, \Box)$, if the following diagrams commute.

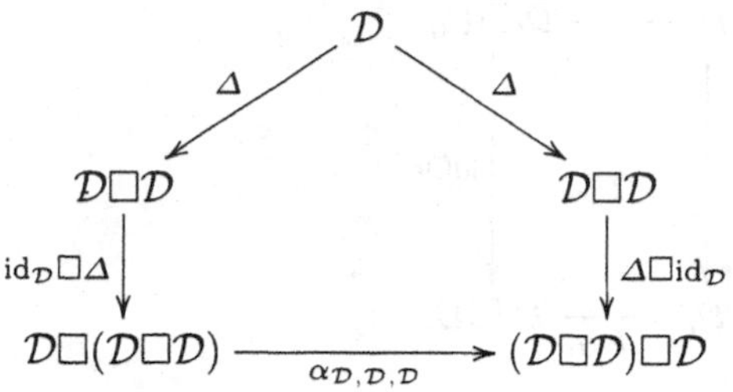

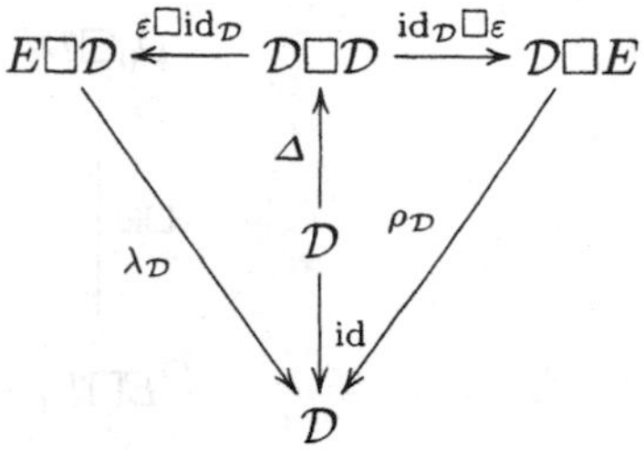

Proposition 4.6. *Let $\mathcal{D}$ be a dual semigroup in a tensor category and let $F : \mathcal{C} \to \mathfrak{Alg}$ be a cotensor functor with values in the category of unital algebras (equipped with the usual tensor product). Then $F(\mathcal{D})$ is a bialgebra with the counit $F_0 \circ F(\varepsilon)$ and the coproduct $F_2(\mathcal{D}, \mathcal{D}) \circ F(\Delta)$.*

Proof. We only prove the right half of the counit property. Applying F to $\lambda_\mathcal{D} \circ (\varepsilon\Box\mathrm{id}_\mathcal{D}) \circ \Delta = \mathrm{id}_\mathcal{D}$, we get $F(\lambda_\mathcal{D}) \circ F(\varepsilon\Box\mathrm{id}_\mathcal{D}) \circ F\Delta = \mathrm{id}_{F(\mathcal{D})}$. Using the naturality of F_2 and Diagram (3.3), we can extend this to the following commutative diagram,

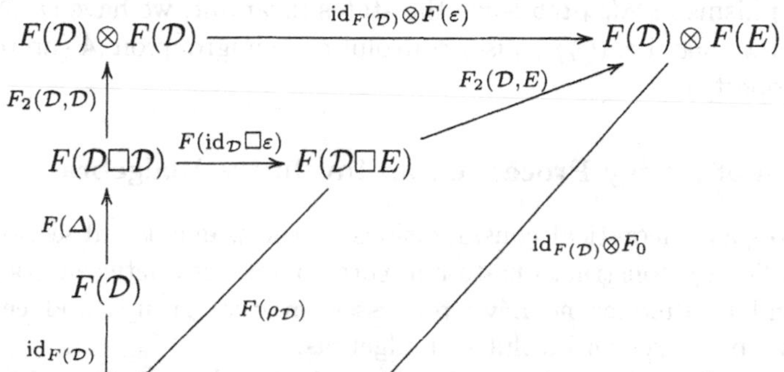

which proves the right counit property of $F(\mathcal{D})$. The proof of the left counit property is of course done by taking the mirror image of this diagram and replacing ρ by λ. The proof of the coassociativity requires a bigger diagram which makes use of (3.2). We leave it as an exercise for ambitious students. □

Assume now that we have a family $(\mathcal{D}_t)_{t\geq 0}$ of objects in $\mathcal{C}$ equipped with morphisms $\varepsilon : \mathcal{D}_0 \to E$ and $\delta_{st} : \mathcal{D}_{s+t} \to: \mathcal{D}_s\Box\mathcal{D}_t$ for $s, t \geq 0$ such that the following diagrams commute.

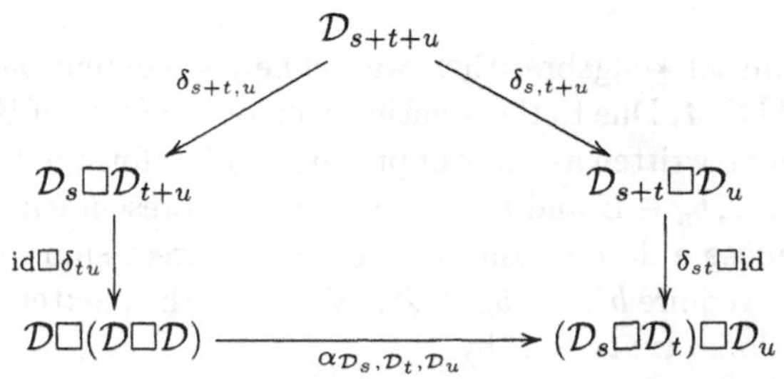

$$\begin{array}{ccccc} \mathcal{D}_0\Box\mathcal{D}_t & \xleftarrow{\delta_{0t}} & \mathcal{D}_t & \xrightarrow{\delta_{t0}} & \mathcal{D}_t\Box\mathcal{D}_0 \\ \downarrow{\scriptstyle \varepsilon\Box\mathrm{id}} & & \downarrow{\scriptstyle \mathrm{id}} & & \downarrow{\scriptstyle \mathrm{id}\Box\varepsilon} \\ E\Box\mathcal{D}_t & \xrightarrow[\lambda_{\mathcal{D}_t}]{} & \mathcal{D}_t & \xleftarrow[\rho_{\mathcal{D}_t}]{} & \mathcal{D}_t\Box E \end{array}$$

In the application we have in mind the objects $\mathcal{D}_t$ will be pairs consisting of a fixed dual semigroup $\mathcal{B}$ and a state φ_t on $\mathcal{B}$ that belongs to a convolution semigroup $(\varphi_t)_{t\geq 0}$ on $\mathcal{B}$. The morphisms δ_{st} and ε will be the coproduct and the counit of $\mathcal{B}$.

If there exists a cotensor functor $F : \mathcal{C} \to \mathfrak{AlgProb}$, $F(\mathcal{D}_t) = (\mathcal{A}_t, \varphi_t)$ such that the algebras $\mathrm{Alg}\big(F(\mathcal{D}_t)\big) = \mathcal{A}_t$ and the morphisms $F_2(\mathcal{D}_s, \mathcal{D}_t) \circ F(\delta_{st})$ are do not depend on s and t, then $\mathcal{A} = \mathrm{Alg}\big(F(\mathcal{D}_t)\big)$ is again a bialgebra with coproduct $\tilde{\Delta} = F_2(\mathcal{D}_s, \mathcal{D}_t) \circ F(\delta_{st})$ and the counit $\tilde{\varepsilon} = F_0 \circ F(\varepsilon)$, as in Proposition 4.6.

Since morphisms in $\mathfrak{AlgProb}$ leave the states invariant, we have $\varphi_s \otimes \varphi_t \circ \tilde{\Delta} = \varphi_{s+t}$ and $\varphi_0 = \tilde{\varepsilon}$, i.e. $(\varphi_t)_{t\geq 0}$ is a convolution semigroup on $\mathcal{A}$ (up to the continuity property).

Construction of a Lévy Process on an Involutive Bialgebra

After the category theoretical considerations of the previous subsection we shall now explicitely construct one-to-one correspondences between boolean, monotone, and anti-monotone Lévy processes on dual groups and certain classes of Lévy processes on involutive bialgebras.

Let $M = \{\mathbf{1}, p\}$ be the unital semigroup with two elements and the multiplication $p^2 = \mathbf{1}p = p\mathbf{1} = p$, $\mathbf{1}^2 = \mathbf{1}$. Its 'group algebra' $\mathbb{C}M = \mathrm{span}\,\{\mathbf{1}, p\}$ is an involutive bialgebra with comultiplication $\Delta(\mathbf{1}) = \mathbf{1} \otimes \mathbf{1}$, $\Delta(p) = p \otimes p$, counit $\varepsilon(\mathbf{1}) = \varepsilon(p) = 1$, and involution $\mathbf{1}^* = \mathbf{1}$, $p^* = p$. The involutive bialgebra $\mathbb{C}M$ was already used by Lenczweski [Len98, Len01] to give a tensor product construction for a large family of products of quantum probability spaces including the boolean and the free product and to define and study the additive convolutions associated to these products. As a unital $*$-algebra it is also used in Skeide's approach to boolean calculus, cf. [Ske01], where it is introduced as the unitization of $\mathbb{C}$. It also plays an important role in [Sch00, FS00].

Let $\mathcal{B}$ be a unital $*$-algebra, then we define its p-extension $\tilde{\mathcal{B}}$ as the free product $\tilde{\mathcal{B}} = \mathcal{B} \coprod \mathbb{C}M$. Due to the identification of the units of $\mathcal{B}$ and $\mathbb{C}M$, any element of $\tilde{\mathcal{B}}$ can be written as sums of products of the form $p^\alpha b_1 p b_2 p \cdots p b_n p^\omega$ with $n \in \mathbb{N}$, $b_1, \ldots, b_n \in \mathcal{B}$ and $\alpha, \omega = 0, 1$. This representation can be made unique, if we choose a decomposition of $\mathcal{B}$ into a direct sum of vector spaces $\mathcal{B} = \mathbb{C}\mathbf{1} \oplus \mathcal{V}^0$ and require $b_1, \ldots, b_n \in \mathcal{V}^0$. We define the p-extension $\tilde{\varphi} : \tilde{\mathcal{B}} \to \mathbb{C}$ of a unital functional $\varphi : \mathcal{B} \to \mathbb{C}$ by

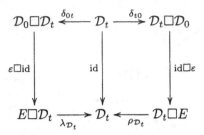

$$\tilde{\varphi}(p^\alpha b_1 p b_2 p \cdots p b_n p^\omega) = \varphi(b_1)\varphi(b_2)\cdots\varphi(b_n) \tag{4.4}$$

and $\tilde{\varphi}(p) = 1$. The p-extension does not depend on the decomposition $\mathcal{B} = \mathbb{C}\mathbf{1} \oplus \mathcal{V}^0$, since Equation (4.4) actually holds not only for $b_1, \ldots, b_n \in \mathcal{V}^0$, but also for $b_1, \ldots, b_n \in \mathcal{B}$.

If $\mathcal{B}_1, \ldots, \mathcal{B}_n$ are unital $*$-algebras that can be written as direct sums $\mathcal{B}_i = \mathbb{C}\mathbf{1} \oplus \mathcal{B}_i^0$ of $*$-algebras, then we can define unital $*$-algebra homomorphisms $I^{\mathrm{B}}_{k,\mathcal{B}_1,\ldots,\mathcal{B}_n}, I^{\mathrm{M}}_{k,\mathcal{B}_1,\ldots,\mathcal{B}_n}, I^{\mathrm{AM}}_{k,\mathcal{B}_1,\ldots,\mathcal{B}_n} : \mathcal{B}_k \to \tilde{\mathcal{B}}_1 \otimes \cdots \otimes \tilde{\mathcal{B}}_n$ for $k = 1, \ldots, n$ by

$$I^{\mathrm{B}}_{k,\mathcal{B}_1,\ldots,\mathcal{B}_n}(b) = \underbrace{p \otimes \cdots \otimes p}_{k-1 \text{ times}} \otimes b \otimes \underbrace{p \otimes \cdots \otimes p}_{n-k \text{ times}},$$

$$I^{\mathrm{M}}_{k,\mathcal{B}_1,\ldots,\mathcal{B}_n}(b) = \underbrace{\mathbf{1} \otimes \cdots \otimes \mathbf{1}}_{k-1 \text{ times}} \otimes b \otimes \underbrace{p \otimes \cdots \otimes p}_{n-k \text{ times}},$$

$$I^{\mathrm{AM}}_{k,\mathcal{B}_1,\ldots,\mathcal{B}_n}(b) = \underbrace{p \otimes \cdots \otimes p}_{k-1 \text{ times}} \otimes b \otimes \underbrace{\mathbf{1} \otimes \cdots \otimes \mathbf{1}}_{n-k \text{ times}},$$

for $b \in \mathcal{B}_k^0$.

Let $n \in \mathbb{N}$, $1 \leq k \leq n$, and denote the canonical inclusions of $\mathcal{B}_k$ into the kth factor of the free product $\coprod_{j=1}^n \mathcal{B}_j$ by i_k. Then, by the universal property, there exist unique unital $*$-algebra homomorphisms $R^\bullet_{\mathcal{B}_1,\ldots,\mathcal{B}_n} : \coprod_{k=1}^n \mathcal{B}_k \to \otimes_{k=1}^n \tilde{\mathcal{B}}_k$ such that

$$R^\bullet_{\mathcal{B}_1,\ldots,\mathcal{B}_n} \circ i_k = I^\bullet_{k,\mathcal{B}_1,\ldots,\mathcal{B}_n},$$

for $\bullet \in \{\mathrm{B}, \mathrm{M}, \mathrm{AM}\}$.

Proposition 4.7. *Let $(\mathcal{B}, \Delta, \varepsilon)$ be a dual semigroup. Then we have the following three involutive bialgebras $(\tilde{\mathcal{B}}, \overline{\Delta}_{\mathrm{B}}, \tilde{\varepsilon})$, $(\tilde{\mathcal{B}}, \overline{\Delta}_{\mathrm{M}}, \tilde{\varepsilon})$, and $(\tilde{\mathcal{B}}, \overline{\Delta}_{\mathrm{AM}}, \tilde{\varepsilon})$, where the comultiplications are defined by*

$$\overline{\Delta}_{\mathrm{B}} = R^{\mathrm{B}}_{\mathcal{B},\mathcal{B}} \circ \Delta,$$
$$\overline{\Delta}_{\mathrm{M}} = R^{\mathrm{M}}_{\mathcal{B},\mathcal{B}} \circ \Delta,$$
$$\overline{\Delta}_{\mathrm{AM}} = R^{\mathrm{AM}}_{\mathcal{B},\mathcal{B}} \circ \Delta,$$

on $\mathcal{B}$ and by

$$\overline{\Delta}_{\mathrm{B}}(p) = \overline{\Delta}_{\mathrm{M}}(p) = \overline{\Delta}_{\mathrm{AM}}(p) = p \otimes p$$

on $\mathbb{C}M$.

Remark 4.8. This is actually a direct consequence of Proposition 4.6. Below we give an explicit proof.

Proof. We will prove that $(\tilde{\mathcal{B}}, \overline{\Delta}_{\mathrm{B}}, \tilde{\varepsilon})$ is an involutive bialgebra, the proofs for $(\tilde{\mathcal{B}}, \overline{\Delta}_{\mathrm{M}}, \tilde{\varepsilon})$ and $(\tilde{\mathcal{B}}, \overline{\Delta}_{\mathrm{AM}}, \tilde{\varepsilon})$ are similar.

It is clear that $\overline{\Delta}_{\mathrm{B}} : \tilde{\mathcal{B}} \to \tilde{\mathcal{B}} \otimes \tilde{\mathcal{B}}$ and $\tilde{\varepsilon} : \tilde{\mathcal{B}} \to \mathbb{C}$ are unital $*$-algebra homomorphisms, so we only have to check the coassociativity and the counit property. That they are satisfied for p is also immediately clear. The proof for elements of $\mathcal{B}$ is similar to the proof of [Zha91, Theorem 4.2]. We get

$$\begin{aligned}(\overline{\Delta}_{\mathrm{B}} \otimes \mathrm{id}_{\tilde{\mathcal{B}}}) \circ \overline{\Delta}_{\mathrm{B}}\big|_{\mathcal{B}} &= R^{\mathrm{B}}_{\mathcal{B},\mathcal{B},\mathcal{B}} \circ \left(\Delta \coprod \mathrm{id}_{\mathcal{B}}\right) \circ \Delta \\ &= R^{\mathrm{B}}_{\mathcal{B},\mathcal{B},\mathcal{B}} \circ \left(\mathrm{id}_{\mathcal{B}} \coprod \Delta\right) \circ \Delta \\ &= (\mathrm{id}_{\tilde{\mathcal{B}}} \otimes \overline{\Delta}_{\mathrm{B}}) \circ \overline{\Delta}_{\mathrm{B}}\big|_{\mathcal{B}}\end{aligned}$$

and

$$\begin{aligned}(\tilde{\varepsilon} \otimes \mathrm{id}_{\tilde{\mathcal{B}}}) \circ \overline{\Delta}_{\mathrm{B}}\big|_{\mathcal{B}} &= (\tilde{\varepsilon} \otimes \mathrm{id}_{\tilde{\mathcal{B}}}) \circ R^{\mathrm{B}}_{\mathcal{B},\mathcal{B}} \circ \Delta \\ &= \left(\varepsilon \coprod \mathrm{id}_{\mathcal{B}}\right) \circ \Delta = \mathrm{id}_{\mathcal{B}} \\ &= \left(\mathrm{id}_{\mathcal{B}} \coprod \varepsilon\right) \circ \Delta \\ &= (\mathrm{id}_{\tilde{\mathcal{B}}} \otimes \tilde{\varepsilon}) \circ R^{\mathrm{B}}_{\mathcal{B},\mathcal{B}} \circ \Delta \\ &= (\mathrm{id}_{\tilde{\mathcal{B}}} \otimes \tilde{\varepsilon}) \circ \overline{\Delta}_{\mathrm{B}}\big|_{\mathcal{B}}.\end{aligned}$$

□

These three involutive bialgebras are important for us, because the boolean convolution (monotone convolution, anti-monotone convolution, respectively) of unital functionals on a dual semigroup $(\mathcal{B}, \Delta, \varepsilon)$ becomes the convolution with respect to the comultiplication $\overline{\Delta}_{\mathrm{B}}$ ($\overline{\Delta}_{\mathrm{M}}$, $\overline{\Delta}_{\mathrm{AM}}$, respectively) of their p-extensions on $\tilde{\mathcal{B}}$.

Proposition 4.9. *Let $(\mathcal{B}, \Delta, \varepsilon)$ be a dual semigroup and $\varphi_1, \varphi_2 : \mathcal{B} \to \mathbb{C}$ two unital functionals on $\mathcal{B}$. Then we have*

$$\begin{aligned}(\widetilde{\varphi_1 \diamond \varphi_2}) \circ \Delta &= (\tilde{\varphi}_1 \otimes \tilde{\varphi}_2) \circ \overline{\Delta}_{\mathrm{B}}, \\ (\widetilde{\varphi_1 \triangleright \varphi_2}) \circ \Delta &= (\tilde{\varphi}_1 \otimes \tilde{\varphi}_2) \circ \overline{\Delta}_{\mathrm{M}}, \\ (\widetilde{\varphi_1 \triangleleft \varphi_2}) \circ \Delta &= (\tilde{\varphi}_1 \otimes \tilde{\varphi}_2) \circ \overline{\Delta}_{\mathrm{AM}}.\end{aligned}$$

Proof. Let $b \in \mathcal{B}^0$. As an element of $\mathcal{B} \coprod \mathcal{B}$, $\Delta(b)$ can be written in the form $\Delta(b) = \sum_{\epsilon \in \mathbb{A}} b^\epsilon \in \bigoplus_{\epsilon \in \mathbb{A}} \mathcal{B}_\epsilon$. Only finitely many terms of this sum are non-zero. The individual summands are tensor products $b^\epsilon = b_1^\epsilon \otimes \cdots \otimes b_{|\epsilon|}^\epsilon$ and due to the counit property we have $b^\emptyset = 0$. Therefore we have

$$(\varphi_1 \diamond \varphi_2) \circ \Delta(b) = \sum_{\substack{\epsilon \in \mathbb{A} \\ \epsilon \neq \emptyset}} \prod_{k=1}^{|\epsilon|} \varphi_{\epsilon_k}(b_k^\epsilon).$$

For the right-hand-side, we get the same expression on $\mathcal{B}$,

$$
\begin{aligned}
(\tilde{\varphi}_1 \otimes \tilde{\varphi}_2) \circ \overline{\Delta}_{\mathrm{B}}(b) &= (\tilde{\varphi}_1 \otimes \tilde{\varphi}_2) \circ R^{\mathrm{B}}_{\mathcal{B},\mathcal{B}} \circ \Delta(b) \\
&= (\tilde{\varphi}_1 \otimes \tilde{\varphi}_2) \circ R^{\mathrm{B}}_{\mathcal{B},\mathcal{B}} \sum_{\epsilon \in \mathbb{A}} b_1^\epsilon \otimes \cdots \otimes b_{|\epsilon|}^\epsilon \\
&= \sum_{\substack{\epsilon \in \mathbb{A} \\ \epsilon_1 = 1}} \tilde{\varphi}_1(b_1^\epsilon p b_3^\epsilon \cdots) \tilde{\varphi}_2(p b_2^\epsilon p \cdots) \\
&\quad + \sum_{\substack{\epsilon \in \mathbb{A} \\ \epsilon_1 = 2}} \tilde{\varphi}_1(p b_2^\epsilon p \cdots) \tilde{\varphi}_2(b_1^\epsilon p b_3^\epsilon \cdots) \\
&= \sum_{\substack{\epsilon \in \mathbb{A} \\ \epsilon \neq \emptyset}} \prod_{k=1}^{|\epsilon|} \varphi_{\epsilon_k}(b_k^\epsilon).
\end{aligned}
$$

To conclude, observe

$$
\overline{\Delta}_{\mathrm{B}}(p^\alpha b_1 p \cdots p b_n p^\omega) = (p^\alpha \otimes p^\alpha) \overline{\Delta}_{\mathrm{B}}(b_1)(p \otimes p) \cdots (p \otimes p) \overline{\Delta}_{\mathrm{B}}(b_n)(p^\omega \otimes p^\omega)
$$

for all $b_1, \ldots, b_n \in \mathcal{B}$, $\alpha, \omega \in \{0,1\}$, and therefore

$$
(\tilde{\varphi}_1 \otimes \tilde{\varphi}_2) \circ \overline{\Delta}_{\mathrm{B}} = \widetilde{(\tilde{\varphi}_1 \otimes \tilde{\varphi}_2) \circ \overline{\Delta}_{\mathrm{B}}\big|_{\mathcal{B}}} = \widetilde{(\varphi_1 \diamond \varphi_2)} \circ \Delta.
$$

The proof for the monotone and anti-monotone product is similar. □

We can now state our first main result.

Theorem 4.10. *Let $(\mathcal{B}, \Delta, \varepsilon)$ be a dual semigroup. We have a one-to-one correspondence between boolean (monotone, anti-monotone, respectively) Lévy processes on the dual semigroup $(\mathcal{B}, \Delta, \varepsilon)$ and Lévy processes on the involutive bialgebra $(\tilde{\mathcal{B}}, \overline{\Delta}_{\mathrm{B}}, \tilde{\varepsilon})$ ($(\tilde{\mathcal{B}}, \overline{\Delta}_{\mathrm{M}}, \tilde{\varepsilon})$, $(\tilde{\mathcal{B}}, \overline{\Delta}_{\mathrm{AM}}, \tilde{\varepsilon})$, respectively), whose marginal distributions satisfy*

$$
\varphi_t(p^\alpha b_1 p \cdots p b_n p^\omega) = \varphi_t(b_1) \cdots \varphi_t(b_n) \tag{4.5}
$$

for all $t \geq 0$, $b_1, \ldots, b_n \in \mathcal{B}$, $\alpha, \omega \in \{0,1\}$.

Proof. Condition (4.5) says that the functionals φ_t on $\tilde{\mathcal{B}}$ are equal to the p-extension of their restriction to $\mathcal{B}$.

Let $\{j_{st}\}_{0 \leq s \leq t \leq T}$ be a boolean (monotone, anti-monotone, respectively) Lévy process on the dual semigroup $(\mathcal{B}, \Delta, \varepsilon)$ with convolution semigroup $\varphi_{t-s} = \Phi \circ j_{st}$. Then, by Proposition 4.9, their p-extensions $\{\tilde{\varphi}_t\}_{t \geq 0}$ form a convolution semigroup on the involutive bialgebra $(\tilde{\mathcal{B}}, \overline{\Delta}_{\mathrm{B}}, \tilde{\varepsilon})$ ($(\tilde{\mathcal{B}}, \overline{\Delta}_{\mathrm{M}}, \tilde{\varepsilon})$, $(\tilde{\mathcal{B}}, \overline{\Delta}_{\mathrm{AM}}, \tilde{\varepsilon})$, respectively). Thus there exists a unique (up to equivalence) Lévy process $\{\tilde{j}_{st}\}_{0 \leq s \leq t \leq T}$ on the involutive bialgebra $(\tilde{\mathcal{B}}, \overline{\Delta}_{\mathrm{B}}, \tilde{\varepsilon})$ ($(\tilde{\mathcal{B}}, \overline{\Delta}_{\mathrm{M}}, \tilde{\varepsilon})$, $(\tilde{\mathcal{B}}, \overline{\Delta}_{\mathrm{AM}}, \tilde{\varepsilon})$, respectively) with these marginal distribution.

Conversely, let $\{j_{st}\}_{0 \leq s \leq t \leq T}$ be a Lévy process on the involutive bialgebra $(\tilde{\mathcal{B}}, \overline{\Delta}_{\mathrm{B}}, \tilde{\varepsilon})$ ($(\tilde{\mathcal{B}}, \overline{\Delta}_{\mathrm{M}}, \tilde{\varepsilon})$, $(\tilde{\mathcal{B}}, \overline{\Delta}_{\mathrm{AM}}, \tilde{\varepsilon})$, respectively) with marginal distributions

$\{\varphi_t\}_{t\geq 0}$ and suppose that the functionals φ_t satisfy Equation (4.5). Then, by Proposition 4.9, their restrictions to $\mathcal{B}$ form a convolution semigroup on the dual semigroup $(\mathcal{B},\Delta,\varepsilon)$ with respect to the boolean (monotone, anti-monotone, respectively) convolution and therefore there exists a unique (up to equivalence) boolean (monotone, anti-monotone, respectively) Lévy process on the dual semigroup $(\mathcal{B},\Delta,\varepsilon)$ that has these marginal distributions.

The correspondence is one-to-one, because the p-extension establishes a bijection between unital functionals on $\mathcal{B}$ and unital functionals on $\tilde{\mathcal{B}}$ that satisfy Condition (4.5). Furthermore, a unital functional on $\mathcal{B}$ is positive if and only if its p-extension is positive on $\tilde{\mathcal{B}}$. □

We will now reformulate Equation (4.5) in terms of the generator of the process. Let $n\geq 1$, $b_1,\ldots,b_n\in\mathcal{B}^0=\ker\varepsilon$, $\alpha,\omega\in\{0,1\}$, then we have

$$\begin{aligned}\psi(p^\alpha b_1 p\cdots pb_n p^\omega) &= \lim_{t\searrow 0}\frac{1}{t}\big(\varphi_t(p^\alpha b_1 p\cdots pb_n p^\omega)-\tilde{\varepsilon}(p^\alpha b_1 p\cdots pb_n p^\omega)\big)\\ &= \lim_{t\searrow 0}\frac{1}{t}\big(\varphi_t(b_1)\cdots\varphi_t(b_n)-\varepsilon(b_1)\cdots\varepsilon(b_n)\big)\\ &= \sum_{k=1}^n \varepsilon(b_1)\cdots\varepsilon(b_{k-1})\psi(b_k)\varepsilon(b_{k+1})\cdots\varepsilon(b_n)\\ &= \begin{cases}\psi(b_1) & \text{if } n=1,\\ 0 & \text{if } n>1.\end{cases}\end{aligned}$$

Conversely, let $\{\varphi_t:\tilde{\mathcal{B}}\to\mathbb{C}\}_{t\geq 0}$ be a convolution semigroup on $(\tilde{\mathcal{B}},\overline{\Delta}_\bullet,\tilde{\varepsilon})$, $\bullet\in\{\mathrm{B},\mathrm{M},\mathrm{AM}\}$, whose generator $\psi:\tilde{\mathcal{B}}\to\mathbb{C}$ satisfies $\psi(\mathbf{1})=\psi(p)=0$ and

$$\psi(p^\alpha b_1 p\cdots pb_n p^\omega)=\begin{cases}\psi(b_1) & \text{if } n=1,\\ 0 & \text{if } n>1,\end{cases}\tag{4.6}$$

for all $n\geq 1$, $b_1,\ldots,b_n\in\mathcal{B}^0=\ker\varepsilon$, $\alpha,\omega\in\{0,1\}$. For $b_1,\ldots,b_n\in\mathcal{B}^0$, $\overline{\Delta}_\bullet(b_i)$ is of the form $\overline{\Delta}_\bullet(b_i)=b_i\otimes\mathbf{1}+\mathbf{1}\otimes b_i+\sum_{k=1}^{n_i}b_{i,k}^{(1)}\otimes b_{i,k}^{(2)}$, with $b_{i,k}^{(1)},b_{i,k}^{(2)}\in\ker\tilde{\varepsilon}$. By the fundamental theorem of coalgebras [Swe69] there exists a finite-dimensional subcoalgebra $\mathcal{C}\subseteq\tilde{\mathcal{B}}$ of $\tilde{\mathcal{B}}$ that contains all possible products of $\mathbf{1},b_i,b_{i,k_i}^{(1)},b_{i,k_i}^{(2)}$, $i=1,\ldots,n$, $k_i=1,\ldots,n_i$.

Then we have

$$\begin{aligned}&\varphi_{s+t}|_{\mathcal{C}}(p^\alpha b_1 p\cdots pb_n p^\omega)\\ &=(\varphi_s|_{\mathcal{C}}\otimes\varphi_t|_{\mathcal{C}})\big((p\alpha\otimes p^\alpha)\overline{\Delta}_\bullet(b_1)(p\otimes p)\cdots(p\otimes p)\overline{\Delta}_\bullet(b_n)(p^\omega\otimes p^\omega)\big)\end{aligned}$$

and, using (4.6), we find the differential equation

$$\begin{aligned}
&\dot{\varphi}_s|_{\mathcal{C}}\left(p^\alpha b_1 p \cdots p b_n p^\omega\right) \\
&= \sum_{i=1}^n \varphi_s|_{\mathcal{C}}\left(p^\alpha b_1 p \cdots b_{i-1} p \mathbf{1} p b_{i+1} p \cdots b_n p^\omega\right)\psi(b_i) \\
&+ \sum_{i=1}^n \sum_{k_i=1}^{n_i} \varphi|_{\mathcal{C}}\left(p^\alpha b_1 p \cdots b_{i-1} p b^{(1)}_{i,k_i} p b_{i+1} p \cdots b_n p^\omega\right)\psi(b^{(2)}_{i,k_i})
\end{aligned} \tag{4.7}$$

for $\{\varphi_t|_{\mathcal{C}}\}_{t\geq 0}$. This a linear inhomogeneous differential equation for a function with values in the finite-dimensional complex vector space $\mathcal{C}^*$ and it has a unique global solution for every initial value $\varphi_0|_{\mathcal{C}}$. Since we have

$$\begin{aligned}
\dot{\varphi}_s(b_i) &= (\varphi_s \otimes \psi)\left(b_i \otimes \mathbf{1} + \mathbf{1} \otimes b_i + \sum_{k=1}^{n_i} b^{(1)}_{i,k} \otimes b^{(2)}_{i,k}\right) \\
&= \psi(b_i) + \sum_{k_i=1}^{n_i} \varphi_s(b^{(1)}_{i,k_i})\psi(b^{(2)}_{i,k_i}),
\end{aligned}$$

we see that $\left\{ \widetilde{(\varphi_t|_{\mathcal{B}})}\Big|_{\mathcal{C}} \right\}_{t\geq 0}$ satisfies the differential equation (4.7). The initial values also agree,

$$\varphi_0(p^\alpha b_1 p \cdots p b_n p^\omega) = \tilde{\varepsilon}(p^\alpha b_1 p \cdots p b_n p^\omega) = \varepsilon(b_1)\cdots\varepsilon(b_n) = \varphi_0(b_1)\cdots\varphi_0(b_n)$$

and therefore it follows that $\{\varphi_t\}_{t\geq 0}$ satisfies Condition (4.5).

We have shown the following.

Lemma 4.11. *Let $\{\varphi_t : \tilde{\mathcal{B}} \to \mathbb{C}\}_{t\geq 0}$ be a convolution semigroup of unital functionals on the involutive bialgebra $(\tilde{\mathcal{B}}, \overline{\Delta}_\bullet, \tilde{\varepsilon})$, $\bullet \in \{\mathrm{B}, \mathrm{M}, \mathrm{AM}\}$, and let $\psi : \tilde{\mathcal{B}} \to \mathbb{C}$ be its infinitesimal generator.*

Then the functionals of the convolution semigroup $\{\varphi_t\}_{t\geq 0}$ satisfy (4.5) for all $t \geq 0$, if and only if its generator ψ satisfies (4.6).

For every linear functional $\psi : \mathcal{B} \to \mathbb{C}$ on $\mathcal{B}$ there exists only one unique functional $\hat{\psi} : \tilde{\mathcal{B}} \to \mathbb{C}$ with $\hat{\psi}|_{\mathcal{B}} = \psi$ that satisfies Condition (4.6). And since this functional $\hat{\psi}$ is hermitian and conditionally positive, if and only if ψ is hermitian and conditionally positive, we have shown the following.

Corollary 4.12. *We have a one-to-one correspondence between boolean Lévy processes, monotone Lévy processes, and anti-monotone Lévy processes on a dual semigroup $(\mathcal{B}, \Delta, \varepsilon)$ and generators, i.e. hermitian, conditionally positive, linear functionals $\psi : \mathcal{B} \to \mathbb{C}$ on $\mathcal{B}$ with $\psi(\mathbf{1}) = 0$.*

Another corollary of Theorem 4.10 is the Schoenberg correspondence for the boolean, monotone, and anti-monotone convolution.

Corollary 4.13. (Schoenberg correspondence) *Let $\{\varphi_t\}_{t\geq 0}$ be a convolution semigroup of unital functionals with respect to the tensor, boolean, monotone, or anti-monotone convolution on a dual semigroup $(\mathcal{B}, \Delta, \varepsilon)$ and let $\psi : \mathcal{B} \to \mathbb{C}$ be defined by*

$$\psi(b) = \lim_{t \searrow 0} \frac{1}{t}\big(\varphi_t(b) - \varepsilon(b)\big)$$

for $b \in \mathcal{B}$. Then the following statements are equivalent.

(i) φ_t is positive for all $t \geq 0$.
(ii) ψ is hermitian and conditionally positive.

We have now obtained a classification of boolean, monotone, and anti-monotone Lévy processes on a given dual semigroup in terms of a class of Lévy processes on a certain involutive bialgebra and in terms of their generators. In the next subsection we will see how to construct realizations.

Construction of Boolean, Monotone, and Anti-Monotone Lévy Processes

The following theorem gives us a way to construct realizations of boolean, monotone, and anti-monotone Lévy processes.

Theorem 4.14. *Let $\{k_{st}^{\mathrm{B}}\}_{0\leq s\leq t\leq T}$ ($\{k_{st}^{\mathrm{M}}\}_{0\leq s\leq t\leq T}, \{k_{st}^{\mathrm{AM}}\}_{0\leq s\leq t\leq T}$, respectively) be a boolean (monotone, anti-monotone, respectively) Lévy process with generator ψ on some dual semigroup $(\mathcal{B}, \Delta, \varepsilon)$. Denote the unique extension of $\psi : \mathcal{B} \to \mathbb{C}$ determined by Equation (4.6) by $\hat{\psi} : \tilde{\mathcal{B}} \to \mathbb{C}$.*

If $\{\tilde{\jmath}_{st}^{\mathrm{B}}\}_{0\leq s\leq t\leq T}$ ($\{\tilde{\jmath}_{st}^{\mathrm{M}}\}_{0\leq s\leq t\leq T}, \{\tilde{\jmath}_{st}^{\mathrm{AM}}\}_{0\leq s\leq t\leq T}$, respectively) is a Lévy process on the involutive bialgebra $(\tilde{\mathcal{B}}, \overline{\Delta}_{\mathrm{B}}, \tilde{\varepsilon})$ ($(\tilde{\mathcal{B}}, \overline{\Delta}_{\mathrm{M}}, \tilde{\varepsilon})$, $(\tilde{\mathcal{B}}, \overline{\Delta}_{\mathrm{AM}}, \tilde{\varepsilon})$, respectively), then the quantum stochastic process $\{\jmath_{st}^{\mathrm{B}}\}_{0\leq s\leq t\leq T}$ ($\{\jmath_{st}^{\mathrm{M}}\}_{0\leq s\leq t\leq T}$, $\{\jmath_{st}^{\mathrm{AM}}\}_{0\leq s\leq t\leq T}$, respectively) on $\mathcal{B}$ defined by

$$
\begin{aligned}
j_{st}^{\mathrm{B}}(\mathbf{1}) &= \mathrm{id}, & j_{st}^{\mathrm{B}}(b) &= \tilde{\jmath}_{0s}^{\mathrm{B}}(p)\tilde{\jmath}_{st}^{\mathrm{B}}(b)\tilde{\jmath}_{tT}^{\mathrm{B}}(p) & \text{for } b \in \mathcal{B}^0 = \ker \varepsilon,\\
j_{st}^{\mathrm{M}}(\mathbf{1}) &= \mathrm{id}, & j_{st}^{\mathrm{M}}(b) &= \tilde{\jmath}_{st}^{\mathrm{M}}(b)\tilde{\jmath}_{tT}^{\mathrm{M}}(p) & \text{for } b \in \mathcal{B}^0 = \ker \varepsilon,\\
j_{st}^{\mathrm{AM}}(\mathbf{1}) &= \mathrm{id}, & j_{st}^{\mathrm{AM}}(b) &= \tilde{\jmath}_{0s}^{\mathrm{AM}}(p)\tilde{\jmath}_{st}^{\mathrm{AM}}(b) & \text{for } b \in \mathcal{B}^0 = \ker \varepsilon,
\end{aligned}
$$

for $0 \leq s \leq t \leq T$, is a boolean (monotone, anti-monotone, respectively) Lévy process on the dual semigroup $(\mathcal{B}, \Delta, \varepsilon)$. Furthermore, if $\{\tilde{\jmath}_{st}^{\mathrm{B}}\}_{0\leq s\leq t\leq T}$ ($\{\tilde{\jmath}_{st}^{\mathrm{M}}\}_{0\leq s\leq t\leq T}, \{\tilde{\jmath}_{st}^{\mathrm{AM}}\}_{0\leq s\leq t\leq T}$, respectively) has generator $\hat{\psi}$, then $\{\jmath_{st}^{\mathrm{B}}\}_{0\leq s\leq t\leq T}$ ($\{\jmath_{st}^{\mathrm{M}}\}_{0\leq s\leq t\leq T}, \{\jmath_{st}^{\mathrm{AM}}\}_{0\leq s\leq t\leq T}$, respectively) is equivalent to $\{k_{st}^{\mathrm{B}}\}_{0\leq s\leq t\leq T}$ ($\{k_{st}^{\mathrm{M}}\}_{0\leq s\leq t\leq T}, \{k_{st}^{\mathrm{AM}}\}_{0\leq s\leq t\leq T}$, respectively).

Remark 4.15. Every Lévy process on an involutive bialgebra can be realized on boson Fock space as solution of quantum stochastic differential equations, see Theorem 1.15 or [Sch93, Theorem 2.5.3]. Therefore Theorem 4.14 implies that boolean, monotone, and anti-monotone Lévy processes can also always be realized on a boson Fock space. We will refer to the realizations obtained in this way as standard Fock realization.

It is natural to conjecture that monotone and anti-monotone Lévy processes can also be realized on their respective Fock spaces (see Subsection 4.3) as solutions of monotone or anti-monotone quantum stochastic differential equations, like this has been proved for the tensor case in [Sch93, Theorem 2.5.3] and discussed for free and boolean case in [Sch95b, BG01]. We will show in Subsection 4.3 that this is really possible.

Proof. $\{\tilde{\jmath}^{\bullet}_{st}\}_{0\le s\le t\le T}$ is a Lévy process on the involutive bialgebra $(\tilde{\mathcal{B}}, \overline{\Delta}_{\mathrm{B}}, \tilde{\varepsilon})$, $\bullet \in \{\mathrm{B}, \mathrm{M}, \mathrm{AM}\}$, and therefore, by the independence property of its increments, we have

$$\left[\tilde{\jmath}^{\bullet}_{st}(b_1), \tilde{\jmath}^{\bullet}_{s't'}(b_2)\right] = 0$$

for all $0 \le s \le t \le T$, $0 \le s' \le t' \le T$ with $]s,t[\cap]s',t'[= \emptyset$ and all $b_1, b_2 \in \tilde{\mathcal{B}}$. Using this property one immediately sees that the $j^{\bullet}_{st}$ are unital $*$-algebra homomorphisms. Using again the independence of the increments of $\{\tilde{\jmath}^{\bullet}_{st}\}_{0\le s\le t\le T}$ and the fact that its marginal distributions $\varphi^{\bullet}_{st} = \Phi \circ \tilde{\jmath}^{\bullet}_{0s}$, $0 \le s \le t \le T$, satisfy Equation (4.5), we get

$$\Phi\big(j^{\mathrm{B}}_{st}(b)\big) = \Phi\big(\tilde{\jmath}^{\mathrm{B}}_{0s}(p)\tilde{\jmath}^{\mathrm{B}}_{st}(b)\tilde{\jmath}^{\mathrm{B}}_{tT}(p)\big) = \Phi\big(\tilde{\jmath}^{\mathrm{B}}_{0s}(p)\big)\Phi\big(\tilde{\jmath}^{\mathrm{B}}_{st}(b)\big)\Phi\big(\tilde{\jmath}^{\mathrm{B}}_{tT}(p)\big) = \varphi^{\mathrm{B}}_{st}(b)$$

and similarly

$$\Phi\big(j^{\mathrm{M}}_{st}(b)\big) = \varphi^{\mathrm{M}}_{st}(b),$$
$$\Phi\big(j^{\mathrm{AM}}_{st}(b)\big) = \varphi^{\mathrm{AM}}_{st}(b),$$

for all $b \in \mathcal{B}^0$. Thus the marginal distributions of $\{j^{\bullet}_{st}\}_{0\le s\le t\le T}$ are simply the restrictions of the marginal distributions of $\{\tilde{\jmath}^{\bullet}_{st}\}_{0\le s\le t\le T}$. This proves the stationarity and the weak continuity of $\{j^{\bullet}_{st}\}_{0\le s\le t\le T}$, it only remains to show the increment property and the independence of the increments. We check these for the boolean case, the other two cases are similar. Let $b \in \mathcal{B}^0$ with $\Delta(b) = \sum_{\epsilon\in\mathbb{A}} b^{\epsilon}$, where $b^{\epsilon} = b^{\epsilon}_1 \otimes \cdots b^{\epsilon}_{\epsilon_{|\epsilon|}} \in \mathcal{B}_{\epsilon} = (\mathcal{B}^0)^{\otimes|\epsilon|}$, then we have

$$\overline{\Delta}_{\mathrm{B}}(b) = \sum_{\substack{\epsilon\in\mathbb{A}\\ \epsilon_1=1}} b^{\epsilon}_1 p b^{\epsilon}_3 \cdots \otimes p b^{\epsilon}_2 p \cdots + \sum_{\substack{\epsilon\in\mathbb{A}\\ \epsilon_1=2}} p b^{\epsilon}_2 p \cdots \otimes b^{\epsilon}_1 p b^{\epsilon}_3 \cdots \tag{4.8}$$

We set $j^{\mathrm{B}}_{st} = j_1$, $j^{\mathrm{B}}_{tu} = j_2$, and get

$$
\begin{aligned}
&m_{\mathcal{A}} \circ \left(j^{\mathrm{B}}_{st} \coprod j^{\mathrm{B}}_{tu} \right) \circ \Delta(b) \\
&= \sum_{\substack{\epsilon \in \mathbb{A} \\ \epsilon \neq \emptyset}} j_{\epsilon_1}(b^{\epsilon}_1) j_{\epsilon_2}(b^{\epsilon}_2) \cdots j_{\epsilon_{|\epsilon|}}(b^{\epsilon}_{|\epsilon|}) \\
&= \sum_{\substack{\epsilon \in \mathbb{A} \\ \epsilon_1 = 1}} \tilde{\jmath}^{\mathrm{B}}_{0s}(p) \tilde{\jmath}^{\mathrm{B}}_{st}(b^{\epsilon}_1) \tilde{\jmath}^{\mathrm{B}}_{tT}(p) \tilde{\jmath}^{\mathrm{B}}_{0t}(p) \tilde{\jmath}^{\mathrm{B}}_{tu}(b^{\epsilon}_2) \tilde{\jmath}^{\mathrm{B}}_{uT}(p) \cdots \tilde{\jmath}^{\mathrm{B}}_{0s}(p) \tilde{\jmath}^{\mathrm{B}}_{st}(b^{\epsilon}_{|\epsilon|}) \tilde{\jmath}^{\mathrm{B}}_{tT}(p) \\
&\quad + \sum_{\substack{\epsilon \in \mathbb{A} \\ \epsilon_1 = 2}} \tilde{\jmath}^{\mathrm{B}}_{0t}(p) \tilde{\jmath}^{\mathrm{B}}_{tu}(b^{\epsilon}_1) \tilde{\jmath}^{\mathrm{B}}_{uT}(p) \tilde{\jmath}^{\mathrm{B}}_{0s}(p) \tilde{\jmath}^{\mathrm{B}}_{st}(b^{\epsilon}_2) \tilde{\jmath}^{\mathrm{B}}_{tT}(p) \cdots \tilde{\jmath}^{\mathrm{B}}_{0t}(p) \tilde{\jmath}^{\mathrm{B}}_{tu}(b^{\epsilon}_{|\epsilon|}) \tilde{\jmath}^{\mathrm{B}}_{uT}(p) \\
&= \tilde{\jmath}^{\mathrm{B}}_{0s}(p) \left(\sum_{\substack{\epsilon \in \mathbb{A} \\ \epsilon_1 = 1}} \tilde{\jmath}^{\mathrm{B}}_{st}(b^{\epsilon}_1) \tilde{\jmath}^{\mathrm{B}}_{st}(p) \tilde{\jmath}^{\mathrm{B}}_{st}(b^{\epsilon}_3) \cdots \tilde{\jmath}^{\mathrm{B}}_{tu}(p) \tilde{\jmath}^{\mathrm{B}}_{tu}(b^{\epsilon}_2) \tilde{\jmath}^{\mathrm{B}}_{tu}(p) \cdots \right) \tilde{\jmath}^{\mathrm{B}}_{uT}(p) \\
&\quad + \tilde{\jmath}^{\mathrm{B}}_{0s}(p) \left(\sum_{\substack{\epsilon \in \mathbb{A} \\ \epsilon_1 = 2}} \tilde{\jmath}^{\mathrm{B}}_{st}(p) \tilde{\jmath}^{\mathrm{B}}_{st}(b^{\epsilon}_2) \tilde{\jmath}^{\mathrm{B}}_{st}(p) \cdots \tilde{\jmath}^{\mathrm{B}}_{tu}(b^{\epsilon}_1) \tilde{\jmath}^{\mathrm{B}}_{tu}(p) \tilde{\jmath}^{\mathrm{B}}_{tu}(b^{\epsilon}_3) \cdots \right) \tilde{\jmath}^{\mathrm{B}}_{uT}(p) \\
&= \tilde{\jmath}^{\mathrm{B}}_{0s}(p) \left(m_{\mathcal{A}} \circ (\tilde{\jmath}^{\mathrm{B}}_{st} \otimes \tilde{\jmath}^{\mathrm{B}}_{tu}) \circ \overline{\Delta}_{\mathrm{B}}(b) \right) \tilde{\jmath}^{\mathrm{B}}_{uT}(p) \\
&= \tilde{\jmath}^{\mathrm{B}}_{0s}(p) \tilde{\jmath}^{\mathrm{B}}_{su}(b) \tilde{\jmath}^{\mathrm{B}}_{uT}(p) = j^{\mathrm{B}}_{su}(b).
\end{aligned}
$$

For the boolean independence of the increments of $\{j^{\mathrm{B}}_{st}\}_{0 \le s \le t \le T}$, we have to check

$$
\Phi \circ m_{\mathcal{A}} \circ \left(j^{\mathrm{B}}_{s_1 t_1} \coprod \cdots \coprod j^{\mathrm{B}}_{s_n t_n} \right) = \varphi^{\mathrm{B}}_{s_1 t_1}|_{\mathcal{B}} \diamond \cdots \diamond \varphi^{\mathrm{B}}_{s_n t_n}|_{\mathcal{B}}
$$

for all $n \in \mathbb{N}$ and $0 \le s_1 \le t_1 \le s_2 \le \cdots \le t_n \le T$. Let, e.g., $n = 2$, and take an element of $\mathcal{B} \coprod \mathcal{B}$ of the form $i_1(a_1) i_2(b_1) \cdots i_n(b_n)$, with $a_1, \ldots, a_n, b_1, \ldots, b_n \in \mathcal{B}^0$. Then we have

$$
\begin{aligned}
&\Phi \circ m_{\mathcal{A}} \circ \left(j^{\mathrm{B}}_{s_1 t_1} \coprod j^{\mathrm{B}}_{s_2 t_2} \right) \left(i_1(a_1) i_2(b_1) \cdots i_n(b_n) \right) \\
&= \Phi \left(\tilde{\jmath}^{\mathrm{B}}_{0 s_1}(p) \tilde{\jmath}^{\mathrm{B}}_{s_1 t_1}(a_1) \tilde{\jmath}^{\mathrm{B}}_{t_1 T}(p) \tilde{\jmath}^{\mathrm{B}}_{0 s_2}(p) \tilde{\jmath}^{\mathrm{B}}_{s_2 t_2}(b_1) \tilde{\jmath}^{\mathrm{B}}_{t_2 T}(p) \cdots \tilde{\jmath}^{\mathrm{B}}_{0 s_2}(p) \tilde{\jmath}^{\mathrm{B}}_{s_2 t_2}(b_n) \tilde{\jmath}^{\mathrm{B}}_{t_2 T}(p) \right) \\
&= \Phi \left(\tilde{\jmath}^{\mathrm{B}}_{0 s_1}(p) \tilde{\jmath}^{\mathrm{B}}_{s_1 t_1}(a_1) \tilde{\jmath}^{\mathrm{B}}_{s_1 t_1}(p) \cdots \tilde{\jmath}^{\mathrm{B}}_{s_1 t_1}(a_n) \jmath^{\mathrm{B}}_{s_1 t_2}(p) \tilde{\jmath}^{\mathrm{B}}_{s_2 t_2}(b_1) \cdots \tilde{\jmath}^{\mathrm{B}}_{s_2 t_2}(b_n) \tilde{\jmath}^{\mathrm{B}}_{t_2 T}(p) \right) \\
&= \varphi^{\mathrm{B}}_{s_1 t_1}(a_1 p a_2 p \cdots p a_n) \varphi^{\mathrm{B}}_{s_2 t_2}(p b_1 p \cdots p b_n) = \prod_{j=1}^{n} \varphi^{\mathrm{B}}_{s_1 t_1}(a_j) \prod_{j=1}^{n} \varphi^{\mathrm{B}}_{s_2 t_2}(b_j) \\
&= \left(\varphi^{\mathrm{B}}_{s_1 t_1} \diamond \varphi^{\mathrm{B}}_{s_2 t_2} \right) \left(i_1(a_1) i_2(b_1) \cdots i_n(b_n) \right).
\end{aligned}
$$

The calculations for the other cases and general n are similar. □

For the actual construction of $\{\tilde{\jmath}^{\mathrm{B}}_{st}\}_{0 \le s \le t \le T}$ ($\{\tilde{\jmath}^{\mathrm{M}}_{st}\}_{0 \le s \le t \le T}$, $\{\tilde{\jmath}^{\mathrm{AM}}_{st}\}_{0 \le s \le t \le T}$, respectively) via quantum stochastic calculus, we need to know the Schürmann triple of $\hat{\psi}$.

Proposition 4.16. *Let $\mathcal{B}$ be a unital $*$-algebra, $\psi : \mathcal{B} \to \mathbb{C}$ a generator, i.e. a hermitian, conditionally positive linear functional with $\psi(\mathbf{1}) = 0$, and*

$\hat{\psi} : \tilde{\mathcal{B}} \to \mathbb{C}$ the extension of ψ to $\tilde{\mathcal{B}}$ given by Equation (4.6). If (ρ, η, ψ) is a Schürmann triple of ψ, then a Schürmann triple $(\hat{\rho}, \hat{\eta}, \hat{\psi})$ for $\hat{\psi}$ is given by

$$\begin{aligned} \hat{\rho}|_{\mathcal{B}} &= \rho, \quad \hat{\rho}(p) = 0, \\ \hat{\eta}|_{\mathcal{B}} &= \eta, \quad \hat{\eta}(p) = 0, \\ \hat{\psi}|_{\mathcal{B}} &= \psi, \quad \hat{\psi}(p) = 0, \end{aligned}$$

in particular, it can be defined on the same pre-Hilbert space as (ρ, η, ψ).

Proof. The restrictions of $\hat{\rho}$ and $\hat{\eta}$ to $\mathcal{B}$ have to be unitarily equivalent to ρ and η, respectively, since $\hat{\psi}|_{\mathcal{B}} = \psi$. We can calculate the norm of $\hat{\eta}(p)$ with Equation (1.3), we get

$$\hat{\psi}(p) = \hat{\psi}(p^2) = \tilde{\varepsilon}(p)\hat{\psi}(p) + \langle \hat{\eta}(p^*), \hat{\eta}(p) \rangle + \hat{\psi}(p)\tilde{\varepsilon}(p)$$

and therefore $||\hat{\eta}(p)||^2 = -\hat{\psi}(p) = 0$. From Equation (1.2) follows

$$\hat{\eta}(p^\alpha b_1 p b_2 p \cdots p b_n p^\omega) = \begin{cases} \eta(b_1) & \text{if } n = 1, \alpha = 0, \omega \in \{0,1\}, \\ 0 & \text{if } n > 1 \text{ or } \alpha = 1. \end{cases}$$

For the representation $\hat{\rho}$ we get

$$\hat{\rho}(p)\eta(b) = \hat{\eta}(pb) - \hat{\eta}(p)\varepsilon(b) = 0$$

for all $b \in \mathcal{B}$. □

The Lévy processes $\{\tilde{\jmath}^\bullet_{st}\}_{0 \le s \le t \le T}$ on the involutive bialgebras $(\tilde{\mathcal{B}}, \overline{\Delta}_\bullet, \tilde{\epsilon})$, $\bullet \in \{\mathrm{B}, \mathrm{M}, \mathrm{AM}\}$, with the generator $\hat{\psi}$ can now be constructed as solutions of the quantum stochastic differential equations

$$\tilde{\jmath}^\bullet_{st}(b) = \tilde{\varepsilon}(b)\mathrm{id} + \left(\int_s^t \tilde{\jmath}^\bullet_{s\tau} \otimes \mathrm{d}I_\tau \right) \overline{\Delta}_\bullet(b), \qquad \text{for all } b \in \tilde{\mathcal{B}},$$

where the integrator $\mathrm{d}I$ is given by

$$\mathrm{d}I_t(b) = \mathrm{d}\Lambda_t(\hat{\rho}(b) - \tilde{\varepsilon}(b)\mathrm{id}) + \mathrm{d}A^+_t(\hat{\eta}(b)) + \mathrm{d}A_t(\hat{\eta}(b^*)) + \hat{\psi}(b)\mathrm{d}t.$$

The element $p \in \tilde{\mathcal{B}}$ is group-like, i.e. $\overline{\Delta}_\bullet(p) = p \otimes p$, and mapped to zero by any Schürmann triple $(\hat{\rho}, \hat{\eta}, \hat{\psi})$ on $\tilde{\mathcal{B}}$ that is obtained by extending a Schürmann triple (ρ, η, ψ) on $\mathcal{B}$ as in Proposition 4.16. Therefore we can compute $\{\tilde{\jmath}^\bullet_{st}(p)\}_{0 \le s \le t \le T}$ without specifying $\bullet \in \{\mathrm{B}, \mathrm{M}, \mathrm{AM}\}$ or knowing the Schürmann triple (ρ, η, ψ).

Proposition 4.17. *Let $\{\tilde{\jmath}^\bullet_{st}\}_{0 \le s \le t \le T}$ be a Lévy process on $(\tilde{\mathcal{B}}, \overline{\Delta}_\bullet, \tilde{\epsilon})$, $\bullet \in \{\mathrm{B}, \mathrm{M}, \mathrm{AM}\}$, whose Schürmann triple $(\hat{\rho}, \hat{\eta}, \hat{\psi})$ is of the form given in Proposition 4.16. Denote by 0_{st} the projection from $L^2([0,T[, D)$ to $L^2([0,s[, D) \oplus L^2([t,T[, D) \subseteq L^2([0,T[, D)$,*

$$0_{st}f(\tau) = \begin{cases} f(\tau) & \text{if } \tau \notin [s,t[, \\ 0 & \text{if } \tau \in [s,t[, \end{cases}$$

Then

$$\tilde{j}^{\bullet}_{st}(p) = \Gamma(0_{st}) \quad \text{for all } 0 \leq s \leq t \leq T,$$

i.e. $\tilde{j}^{\bullet}_{st}(p)$ *is equal to the second quantization of* 0_{st} *for all* $0 \leq s \leq t \leq T$ *and* $\bullet \in \{\mathrm{B}, \mathrm{M}, \mathrm{AM}\}$.

Proof. This follows immediately from the quantum stochastic differential equation

$$\tilde{j}^{\bullet}_{st}(p) = \mathrm{id} - \int_s^t \tilde{j}^{\bullet}_{s\tau}(p)\mathrm{d}\Lambda_\tau(\mathrm{id}).$$

□

Boson Fock Space Realization of Boolean, Monotone, and Anti-Monotone Quantum Stochastic Calculus

For each of the independences treated in this chapter, we can define a Fock space with a creation, annihilation and conservation process, and develop a quantum stochastic calculus. For the monotone case, this was done in [Mur97, Lu97], for the boolean calculus see, e.g., [BGDS01] and the references therein.

Since the integrator processes of these calculi have independent and stationary increments, we can use our previous results to realize them on a boson Fock space. Furthermore, we can embed the corresponding Fock spaces into a boson Fock space and thus reduce the boolean, monotone, and anti-monotone quantum stochastic calculus to the quantum stochastic calculus on boson Fock space defined in [HP84] (but the integrands one obtains in the boolean or monotone case turn out to be not adapted in general). For the anti-monotone creation and annihilation process with one degree of freedom, this was already done in [Par99] (see also [Lie99]).

Let H be a Hilbert space. Its conjugate or dual is, as a set, equal to $\overline{H} = \{\overline{u} | u \in H\}$. The addition and scalar multiplication are defined by

$$\overline{u} + \overline{v} = \overline{u+v}, \qquad , z\overline{u} = \overline{\overline{z}u}, \qquad \text{for } u, v \in H, \quad z \in \mathbb{C}.$$

Then $V(H) = H \otimes \overline{H} \oplus \overline{H} \oplus H$ (algebraic tensor product and direct sum, no completion) is an involutive complex vector space with the involution

$$(v \otimes \overline{u} + \overline{x} + y)^* = u \otimes \overline{v} + \overline{y} + x, \qquad \text{for } u, v, x, y \in H.$$

We will also write $|u\rangle\langle v|$ for $u \otimes \overline{v}$. Let now $\mathcal{B}_H$ be the free unital $*$-algebra over $V(H)$. This algebra can be made into a dual semigroup, if we define the comultiplication and counit by

$$\Delta v = i_1(v) + i_2(v),$$

and $\varepsilon(v) = 0$ for $v \in V(H)$ and extend them as unital $*$-algebra homomorphisms. On this dual semigroup we can define the fundamental noises for all our independences. For the Schürmann triple we take the Hilbert space H, the representation ρ of $\mathcal{B}_H$ on H defined by

$$\rho(u) = \rho(\overline{u}) = 0, \quad \rho\big(|u\rangle\langle v|\big) : H \ni x \mapsto \langle v, x\rangle u \in H,$$

the cocycle $\eta : \mathcal{B}_H \to H$ with

$$\eta(u) = u, \quad \eta(\overline{u}) = \eta\big(|u\rangle\langle v|\big) = 0,$$

and the generator $\psi : \mathcal{B}_H \to \mathbb{C}$ with

$$\psi(\mathbf{1}) = \psi(u) = \psi(\overline{u}) = \psi\big(|u\rangle\langle v|\big) = 0,$$

for all $u, v \in H$.

A realization of the tensor Lévy process $\{j_{st}\}_{0\leq s\leq t}$ on the dual semigroup $(\mathcal{B}_H, \Delta, \varepsilon)$ with this Schürmann triple on the boson Fock space $\Gamma\big(L^2(\mathbb{R}_+, H)\big)$ is given by

$$j_{st}(u) = A^+_{st}(u), \quad j_{st}(\overline{u}) = A_{st}(u), \quad j_{st}\big(|u\rangle\langle v|\big) = \Lambda_{st}\big(|u\rangle\langle v|\big),$$

for all $0 \leq s \leq t \leq T$, $u, v \in H$.

Boolean Calculus

Let H be a Hilbert space. The boolean Fock space over $L^2([0,T[;H) \cong L^2([0,T]) \otimes H$ is defined as $\Gamma_{\mathrm{B}}\big(L^2([0,T[,H)\big) = \mathbb{C} \oplus L^2([0,T[,H)$. We will write the elements of $\Gamma_{\mathrm{B}}\big(L^2([0,T[,H)\big)$ as vectors

$$\begin{pmatrix}\lambda \\ f\end{pmatrix}$$

with $\lambda \in \mathbb{C}$ and $f \in L^2([0,T[,H)$. The boolean creation, annihilation, and conservation processes are defined as

$$A^{\mathrm{B}+}_{st}(u)\begin{pmatrix}\lambda \\ f\end{pmatrix} = \begin{pmatrix}0 \\ \lambda u \mathbf{1}_{[s,t[}\end{pmatrix},$$

$$A^{\mathrm{B}}_{st}(u)\begin{pmatrix}\lambda \\ f\end{pmatrix} = \begin{pmatrix}\int_s^t \langle u, f(\tau)\rangle \mathrm{d}\tau \\ 0\end{pmatrix},$$

$$\Lambda^{\mathrm{B}}_{st}\big(|u\rangle\langle v|\big)\begin{pmatrix}\lambda \\ f\end{pmatrix} = \begin{pmatrix}0 \\ \mathbf{1}_{[s,t[}(\cdot)\langle v, f(\cdot)\rangle u\end{pmatrix},$$

for $\lambda \in \mathbb{C}$, $f \in L^2([0,T[,H)$, $u, v \in H$. These operators define a boolean Lévy process $\{k^{\mathrm{B}}_{st}\}_{0\leq s\leq t\leq T}$ on the dual semigroup $(\mathcal{B}_H, \Delta, \varepsilon)$ with respect to the vacuum expectation, if we set

$$k_{st}^{\mathrm{B}}(u) = A_{st}^{\mathrm{B}+}(u), \quad k_{st}^{\mathrm{B}}(\overline{u}) = A_{st}^{\mathrm{B}}(u), \quad k_{st}^{\mathrm{B}}(|u\rangle\langle v|) = \Lambda_{st}^{\mathrm{B}}(|u\rangle\langle v|),$$

for all $0 \leq s \leq t \leq T$, $u, v \in H$, and extend the k_{st}^{B} as unital $*$-algebra homomorphisms to $\mathcal{B}_H$.

On the other hand, using Theorem 4.14 and Proposition 4.16, we can define a realization of the same Lévy process on a boson Fock space. Since the comultiplication $\overline{\Delta}_{\mathrm{B}}$ acts on elements of the involutive bialgebra $(\tilde{\mathcal{B}}_H, \overline{\Delta}_{\mathrm{B}}, \tilde{\varepsilon})$ as

$$\overline{\Delta}_{\mathrm{B}}(v) = v \otimes p + p \otimes v, \qquad \text{for } v \in V(H),$$

we have to solve the quantum stochastic differential equations

$$\begin{aligned}
\tilde{j}_{st}^{\mathrm{B}}(u) &= \int_s^t \Gamma(0_{s\tau})\mathrm{d}A_\tau^+(u) - \int_s^t \tilde{j}_{s\tau}^{\mathrm{B}}(u)\mathrm{d}\Lambda_\tau(\mathrm{id}_H), \\
\tilde{j}_{st}^{\mathrm{B}}(\overline{u}) &= \int_s^t \Gamma(0_{s\tau})\mathrm{d}A_\tau(u) - \int_s^t \tilde{j}_{s\tau}^{\mathrm{B}}(\overline{u})\mathrm{d}\Lambda_\tau(\mathrm{id}_H), \\
\tilde{j}_{st}^{\mathrm{B}}(|u\rangle\langle v|) &= \int_s^t \Gamma(0_{s\tau})\mathrm{d}\Lambda_\tau(|u\rangle\langle v|) - \int_s^t \tilde{j}_{s\tau}^{\mathrm{B}}(|u\rangle\langle v|)\mathrm{d}\Lambda_\tau(\mathrm{id}_H),
\end{aligned}$$

and set

$$\begin{aligned}
j_{st}^{\mathrm{B}}(u) &= \Gamma(0_{0s})\tilde{j}_{st}^{\mathrm{B}}(u)\Gamma(0_{tT}), \\
j_{st}^{\mathrm{B}}(\overline{u}) &= \Gamma(0_{0s})\tilde{j}_{st}^{\mathrm{B}}(\overline{u})\Gamma(0_{tT}), \\
j_{st}^{\mathrm{B}}(|u\rangle\langle v|) &= \Gamma(0_{0s})\tilde{j}_{st}^{\mathrm{B}}(|u\rangle\langle v|)\Gamma(0_{tT}),
\end{aligned}$$

These operators act on exponential vectors as

$$\begin{aligned}
j_{st}^{\mathrm{B}}(u)\mathcal{E}(f) &= u\mathbf{1}_{[s,t[}, \\
j_{st}^{\mathrm{B}}(\overline{u})\mathcal{E}(f) &= \int_s^t \langle u, f(\tau)\rangle \mathrm{d}\tau\,\Omega, \\
j_{st}^{\mathrm{B}}(|u\rangle\langle v|)\mathcal{E}(f) &= \mathbf{1}_{[s,t[}\langle v, f(\cdot)\rangle u,
\end{aligned}$$

for $0 \leq s \leq t \leq T$, $f \in L^2([0,T[)$, $u, v \in H$.

Since $\{k_{st}^{\mathrm{B}}\}_{0 \leq s \leq t \leq T}$ and $\{j_{st}^{\mathrm{B}}\}_{0 \leq s \leq t \leq T}$ are boolean Lévy processes on the dual semigroup $(\mathcal{B}_H, \Delta, \varepsilon)$ with the same generator, they are equivalent.

If we isometrically embed the boolean Fock space $\Gamma_{\mathrm{B}}(L^2([0,T[,H))$ into the boson Fock space $\Gamma(L^2([0,T[,H))$ in the natural way,

$$\theta_{\mathrm{B}} : \Gamma_{\mathrm{B}}(L^2([0,T[,H)) \to \Gamma(L^2([0,T[,H)), \qquad \theta_{\mathrm{B}}\begin{pmatrix}\lambda \\ f\end{pmatrix} = \lambda\Omega + f,$$

for $\lambda \in \mathbb{C}$, $f \in L^2([0,T[,H)$, then we have

$$k_{st}^{\mathrm{B}}(b) = \theta_{\mathrm{B}}^* j_{st}^{\mathrm{B}}(b)\theta_{\mathrm{B}}$$

for all $b \in \mathcal{B}$.

Anti-Monotone Calculus

We will treat the anti-monotone calculus first, because it leads to simpler quantum stochastic differential equations. The monotone calculus can then be constructed using time-reversal, cf. Lemma 4.4.

We can construct the monotone and the anti-monotone calculus on the same Fock space. Let

$$\mathbb{T}_n = \{(t_1,\ldots,t_n)|0 \le t_1 \le t_2 \le \cdots \le t_n \le T\} \subseteq [0,T[^n \subseteq \mathbb{R}^n,$$

then the monotone and anti-monotone Fock space $\Gamma_{\rm M}\big(L^2([0,T[,H)\big)$ over $L^2([0,T[,H)$ can be defined as

$$\Gamma_{\rm M}\big(L^2([0,T[,H)\big) = \mathbb{C}\Omega \oplus \bigoplus_{n=1}^{\infty} L^2(\mathbb{T}_n, H^{\overline{\otimes} n}),$$

where where $H^{\overline{\otimes} n}$ denotes the n-fold Hilbert space tensor product of H and the measure on $\mathbb{T}_n$ is the restriction of the Lebesgue measure on $\mathbb{R}^n$ to $\mathbb{T}_n$. Since $\mathbb{T}_n \subseteq [0,T[^n$, we can interpret $f_1 \otimes \cdots \otimes f_n \in L^2([0,T[,H)^{\overline{\otimes} n} \cong L^2([0,T[^n, H^{\overline{\otimes} n})$ also as an element of $L^2(\mathbb{T}_n, H^{\overline{\otimes} n})$ (by restriction).

The anti-monotone creation, annihilation, and conservation operator are defined by

$$\begin{aligned}
&A_{st}^{\rm AM+}(u) f_1 \otimes \cdots \otimes f_n(t_1,\ldots,t_{n+1}) \\
&= \mathbf{1}_{[s,t[}(t_1) u \otimes f_1 \otimes \cdots \otimes f_n(t_2,\ldots,t_{n+1}) \\
&A_{st}^{\rm AM}(u) f_1 \otimes \cdots \otimes f_n(t_1,\ldots,t_{n-1}) \\
&= \int_s^{\min(t,t_1)} \langle u, f_1(\tau)\rangle {\rm d}\tau f_2 \otimes \cdots \otimes f_n(t_1,\ldots,t_{n-1}) \\
&\Lambda_{st}^{\rm AM}\big(|u\rangle\langle v|\big) f_1 \otimes \cdots \otimes f_n(t_1,\ldots,t_n) \\
&= \mathbf{1}_{[s,t[}(t_1)\langle v, f_1(t_1)\rangle u \otimes f_2 \otimes \cdots \otimes f_n(t_2,\ldots,t_n),
\end{aligned}$$

for $0 \le s \le t \le T$, $0 \le t_1 \le t_2 \le \cdots \le t_n \le t_{n+1} \le T$, $u,v \in H$.

These operators define an anti-monotone Lévy process $\{k_{st}^{\rm AM}\}_{0\le s\le t\le T}$ on the dual semigroup $\mathcal{B}$ with respect to the vacuum expectation, if we set

$$k_{st}^{\rm AM}(u) = A_{st}^{\rm AM+}(u), \quad k_{st}^{\rm AM}(\overline{u}) = A_{st}^{\rm AM}(u), \quad k_{st}^{\rm AM}\big(|u\rangle\langle v|\big) = \Lambda_{st}^{\rm AM}\big(|u\rangle\langle v|\big),$$

for all $0 \le s \le t \le T$, $u,v \in H$, and extend the $k_{st}^{\rm AM}$ as unital $*$-algebra homomorphisms to $\mathcal{B}$.

We can define a realization of the same Lévy process on a boson Fock space with Theorem 4.14. The anti-monotone annihilation operators $j_{st}^{\rm AM}(\overline{u})$, $u \in H$, obtained this way act on exponential vectors as

$$j_{st}^{\rm AM}(u)\mathcal{E}(f) = u\mathbf{1}_{[s,t[}(\cdot) \otimes_s \mathcal{E}(0_{0\cdot} f), \qquad f \in L^2([0,T[,H),$$

and the anti-monotone creation operators are given by $j_{st}^{\mathrm{AM}}(u) = j_{st}^{\mathrm{AM}}(\overline{u})^*$, $u \in H$. On symmetric simple tensors $f_1 \otimes \cdots \otimes f_n \in L^2([0,T[, H^{\overline{\otimes} n})$ they act as

$$\begin{aligned} & j_{st}^{\mathrm{AM}}(u) f_1 \otimes \cdots \otimes f_n(t_1, \ldots, t_{n+1}) \\ = & f(t_1) \otimes \cdots \otimes f_{k-1}(t_{k-1}) \otimes u \mathbf{1}_{[s,t[}(t_k) \otimes f_{k+1}(t_{k+1}) \otimes \cdots \otimes f_n(t_n) \end{aligned}$$

where k has to be chosen such that $t_k = \min\{t_1, \ldots, t_{n+1}\}$.

Since $\{k_{st}^{\mathrm{AM}}\}_{0 \leq s \leq t \leq T}$ and $\{j_{st}^{\mathrm{AM}}\}_{0 \leq s \leq t \leq T}$ are boolean Lévy processes on the dual semigroup $\mathcal{B}$ with the same generator, they are equivalent.

A unitary map $\theta_{\mathrm{M}} : \Gamma_{\mathrm{M}}\big(L^2([0,T[, H)\big) \to \Gamma\big(L^2([0,T[, H)\big)$ can be defined by extending functions on $\mathbb{T}_n$ to symmetric functions on $[0,T[^n$ and dividing them by $\sqrt{n!}$. The adjoint $\theta_{\mathrm{M}}^* : \Gamma\big(L^2([0,T[, H)\big) \to \Gamma_{\mathrm{M}}\big(L^2([0,T[, H)\big)$ of θ_{M} acts on simple tensors $f_1 \otimes \cdots \otimes f_n \in L^2([0,T[, H)^{\overline{\otimes} n} \cong L^2([0,T[^n, H^{\overline{\otimes} n})$ as restriction to $\mathbb{T}_n$ and multiplication by $\sqrt{n!}$, i.e.

$$\theta_{\mathrm{M}}^* f_1 \otimes \cdots \otimes f_n(t_1, \ldots, t_n) = \sqrt{n!} f_1(t_1) \otimes \cdots \otimes f_n(t_n),$$

for all $f_1, \ldots, f_n \in L^2([0,T[, H)$, $(t_1, \ldots, t_n) \in \mathbb{T}_n$.

This isomorphism intertwines between $\{k_{st}^{\mathrm{AM}}\}_{0 \leq s \leq t \leq T}$ and $\{j_{st}^{\mathrm{AM}}\}_{0 \leq s \leq t \leq T}$, we have

$$k_{st}^{\mathrm{AM}}(b) = \theta_{\mathrm{M}}^* j_{st}^{\mathrm{AM}}(b) \theta_{\mathrm{M}}$$

for all $0 \leq s \leq t \leq T$ and $b \in \mathcal{B}_H$.

Monotone Calculus

The monotone creation, annihilation, and conservation operator on the monotone Fock space $\Gamma_{\mathrm{M}}\big(L^2([0,T[, H)\big)$ can be defined by

$$\begin{aligned} & A_{st}^{\mathrm{M}+}(u) f_1 \otimes \cdots \otimes f_n(t_1, \ldots, t_{n+1}) \\ = & f_1 \otimes \cdots \otimes f_n(t_1, \ldots, t_n) \otimes \mathbf{1}_{[s,t[}(t_{n+1}) u \\ & A_{st}^{\mathrm{AM}}(u) f_1 \otimes \cdots \otimes f_n(t_1, \ldots, t_{n-1}) \\ = & \int_{\max(s, t_{n-1})}^{t} \langle u, f_n(\tau) \rangle \mathrm{d}\tau f_1 \otimes \cdots \otimes f_{n-1}(t_1, \ldots, t_{n-1}) \\ & \Lambda_{st}^{\mathrm{AM}}\big(|u\rangle\langle v|\big) f_1 \otimes \cdots \otimes f_n(t_1, \ldots, t_n) \\ = & f_1 \otimes \cdots \otimes f_{n-1}(t_1, \ldots, t_{n-1}) \mathbf{1}_{[s,t[}(t_n) \langle v, f_n(t_n) \rangle u, \end{aligned}$$

for $0 \leq s \leq t \leq T$, $u, v \in H$. These operators define a monotone Lévy process $\{k_{st}^{\mathrm{M}}\}_{0 \leq s \leq t \leq T}$ on the dual semigroup $\mathcal{B}$ with respect to the vacuum expectation, if we set

$$k_{st}^{\mathrm{M}}(u) = A_{st}^{\mathrm{M}+}(u), \quad k_{st}^{\mathrm{M}}(\overline{u}) = A_{st}^{\mathrm{M}}(u), \quad k_{st}^{\mathrm{M}}\big(|u\rangle\langle v|\big) = \Lambda_{st}^{\mathrm{M}}\big(|u\rangle\langle v|\big),$$

for all $0 \leq s \leq t \leq T$, $u, v \in H$, and extend the k_{st}^{M} as unital $*$-algebra homomorphisms to $\mathcal{B}$.

Define a time-reversal $R : \Gamma_M\big(L^2([0,T[,H)\big) \to \Gamma_M\big(L^2([0,T[,H)\big)$ for the monotone Fock space by $R\Omega = \Omega$ and

$$Rf_1 \otimes \cdots \otimes f_n(t_1,\ldots,t_n) = f_n(T-t_n) \otimes \cdots \otimes f_1(T-t_1),$$

for $(t_1,\ldots,t_n) \in \mathbb{T}_n$, $f,\ldots,f_n \in L^2(\mathbb{T}_n)$. The time-reversal R is unitary and satisfies $R^2 = \mathrm{id}_{\Gamma_{\mathrm{M}}(L^2([0,T[;H))}$. It intertwines between the monotone and anti-monotone noise on the monotone Fock space, i.e. we have

$$k^{\mathrm{AM}}_{st}(b) = Rk^{\mathrm{M}}_{T-t,T-s}(b)R$$

for all $0 \le s \le t \le T$, $b \in \mathcal{B}_H$. On the boson Fock space we have to consider $R_{\mathrm{M}} = \theta_{\mathrm{M}} R \theta^*_{\mathrm{M}} : \Gamma\big(L^2([0,T[,H)\big) \to \Gamma\big(L^2([0,T[,H)\big)$. This map is again unitary and satisfies also $R^2_{\mathrm{M}} = \mathrm{id}$. It follows that the realization $\{j^{\mathrm{M}}_{st}\}_{0\le s\le t\le T}$ of $\{k^{\mathrm{M}}_{st}\}_{0\le s\le t\le T}$ on boson Fock space can be defined via

$$\begin{aligned}
j^{\mathrm{M}}_{st}(u) &= \int_s^t \mathrm{d}\tilde{A}^+_\tau(u)\Gamma(0_{\tau T}),\\
j^{\mathrm{M}}_{st}(\overline{u}) &= \int_s^t \mathrm{d}\tilde{A}_\tau(u)\Gamma(0_{\tau T}),\\
j^{\mathrm{M}}_{st}\big(|u\rangle\langle v|\big) &= \int_s^t \mathrm{d}\tilde{\Lambda}_\tau\big(|u\rangle\langle v|\big)\Gamma(0_{\tau T}),
\end{aligned}$$

where the integrals are *backward* quantum stochastic integrals.

Remark 4.18. Taking $H = \mathbb{C}$ and comparing these equations with [Sch93, Section 4.3], one recognizes that our realization of the monotone creation and annihilation process on the boson Fock space can be written as

$$\begin{aligned}
\theta_{\mathrm{M}} A^{\mathrm{M}+}_{st}(1)\theta^*_{\mathrm{M}} &= j^{\mathrm{M}}_{st}(1) = X^*_{st}\Gamma(0_{tT}),\\
\theta_{\mathrm{M}} A^{\mathrm{M}}_{st}(1)\theta^*_{\mathrm{M}} &= j^{\mathrm{M}}_{st}(\overline{1}) = X_{st}\Gamma(0_{tT}),
\end{aligned}$$

where $\{(X^*_{st}, X_{st})\}_{0\le s\le t\le T}$ is the quantum Azéma martingale [Par90, Sch91a] with parameter $q = 0$, cf. Subsection I.1.5. Note that here 1 denotes the unit of $H = \mathbb{C}$, not the unit of $\mathcal{B}_{\mathbb{C}}$.

Realization of boolean, monotone, and anti-monotone Lévy process on boolean, monotone, and anti-monotone Fock spaces

Free and boolean Lévy processes on dual semigroups can be realized as solutions of free or boolean quantum stochastic equations on the free or boolean Fock space, see e.g. [Sch95b]. A full proof of this fact is still missing, because it would require a generalization of their calculi to unbounded coefficients, but for a large class of examples this has been shown in [BG01, Section 6.5] for the boolean case. For dual semigroups that are generated by primitive elements (i.e. $\Delta(v) = i_1(v) + 1_2(v)$) it is sufficient to determine the operators $j_{0t}(v)$, which have additive free or boolean increments. It turns out that they can always be represented as a linear combination of the corresponding creators, annihilators, conservation operators and time (which contains the projection $\Gamma(0_{0T})$ to the vacuum in the boolean case), cf. [GSS92, BG01].

We will sketch, how one can show the same for monotone and anti-monotone Lévy processes on dual semigroups.

We can write the fundamental integrators of the anti-monotone calculus on the monotone Fock space $\Gamma_{\mathrm{M}}\big(L^2([0,t[,H)\big)$ as

$$\begin{aligned} \mathrm{d}A_t^{\mathrm{AM}+}(u) &= \theta_{\mathrm{M}}^*\Gamma(0_{0t})\mathrm{d}A_t^+(u)\theta_{\mathrm{M}}, \\ \mathrm{d}A_t^{\mathrm{AM}}(u) &= \theta_{\mathrm{M}}^*\Gamma(0_{0t})\mathrm{d}A_t(u)\theta_{\mathrm{M}}, \\ \mathrm{d}\Lambda_t^{\mathrm{AM}}\big(|u\rangle\langle v|\big) &= \theta_{\mathrm{M}}^*\Gamma(0_{0t})\mathrm{d}\Lambda_t\big(|u\rangle\langle v|\big)\theta_{\mathrm{M}}, \end{aligned}$$

where $\theta_{\mathrm{M}} : \Gamma_{\mathrm{M}}\big(L^2([0,t[,H)\big) \to \Gamma\big(L^2([0,t[,H)\big)$ is the unitary isomorphism introduced in 4.3. Anti-monotone stochastic integrals can be defined using this isomorphism. We call an operator process $\{X_t\}_{0\le t\le T}$ on the monotone Fock space anti-monotonically adapted, if $\{\theta_{\mathrm{M}}^* X_t \theta_{\mathrm{M}}\}_{0\le t\le T}$ is adapted on the boson Fock space $\Gamma\big(L^2([0,t[,H)\big)$ and define the integral by

$$\int_0^T X_t \mathrm{d}I_t^{\mathrm{AM}} := \theta_{\mathrm{M}} \left(\int_0^T \theta_{\mathrm{M}}^* X_t \theta_{\mathrm{M}} \mathrm{d}I_t \right) \theta_{\mathrm{M}}^*$$

for

$$\begin{aligned} \mathrm{d}I_t^{\mathrm{AM}} &= \mathrm{d}\Lambda_t^{\mathrm{AM}}\big(|x\rangle\langle y|\big) + \mathrm{d}A_t^{\mathrm{AM}+}(u) + \mathrm{d}A_t^{\mathrm{AM}}(v), \\ \mathrm{d}I_t &= \Gamma(0_{0t})\Big(\mathrm{d}\Lambda_t\big(|x\rangle\langle y|\big) + \mathrm{d}A_t^{\mathrm{AM}+}(u) + \mathrm{d}A_t^{\mathrm{AM}}(v)\Big), \end{aligned}$$

for $x, y, u, v \in H$. In this way all the domains, kernels, etc., defined in [Sch93, Chapter 2] can be translated to the monotone Fock space.

Using the form of the comultiplication of $(\tilde{\mathcal{B}}, \overline{\Delta}_{\mathrm{AM}}, \tilde{\varepsilon})$, the quantum stochastic equation for the Lévy process on the involutive bialgebra $(\tilde{\mathcal{B}}, \overline{\Delta}_{\mathrm{AM}}, \tilde{\varepsilon})$ that we associated to an anti-monotone Lévy process on the dual semigroup $(\mathcal{B}, \Delta, \varepsilon)$ in Theorem 4.10, and Theorem 4.14, one can now derive a representation theorem for anti-monotone Lévy processes on dual semigroups.

To state our result we need the free product $\coprod^0$ without unification of units. This is the coproduct in the category of all $*$-algebras (not necessarily unital). The two free products $\coprod$ and $\coprod^0$ are related by

$$(\mathbb{C}\mathbf{1} \oplus \mathcal{A}) \coprod (\mathbb{C}\mathbf{1} \oplus \mathcal{B}) \cong \mathbb{C}\mathbf{1} \oplus \left(\mathcal{A} \coprod\nolimits^0 \mathcal{B} \right).$$

We will use the notation $\Gamma_{\mathrm{M}}(0_{st}) = \theta_{\mathrm{M}}^*\Gamma(0_{st})\theta_{\mathrm{M}}$, $0 \le s \le t \le T$.

Theorem 4.19. *Let $(\mathcal{B}, \Delta, \varepsilon)$ be a dual semigroup and let (ρ, η, ψ) be a Schürmann triple on $\mathcal{B}$ over some pre-Hilbert space D. Then the anti-monotone stochastic differential equations*

$$j_{st}(b) = \int_s^t \left(j_{s\tau} \coprod^0 \mathrm{d}I_\tau^{\mathrm{AM}} \right) \circ \Delta(b), \quad \textit{for } b \in \mathcal{B}^0 = \ker \varepsilon, \tag{4.9}$$

with

$$\mathrm{d}I_\tau^{\mathrm{AM}}(b) = \mathrm{d}\Lambda_t^{\mathrm{AM}}(\rho(b)) + \mathrm{d}A_t^{\mathrm{AM}+}(\eta(b)) + \mathrm{d}A_t^{\mathrm{AM}}(\eta(b^*)) + \psi(b)\Gamma_{\mathrm{M}}(0_{0\tau})\mathrm{d}t,$$

have solutions (unique in $\theta_{\mathrm{M}}^* \underline{\mathcal{A}}_D \theta_{\mathrm{M}}$). If we set $j_{st}(\mathbf{1}_\mathcal{B}) = \mathrm{id}$, then $\{j_{st}\}_{0 \le s \le t \le T}$ is an anti-monotone Lévy process on the dual semigroup $(\mathcal{B}, \Delta, \varepsilon)$ with respect to the vacuum state. Furthermore, any anti-monotone Lévy process on the dual semigroup $(\mathcal{B}, \Delta, \varepsilon)$ with generator ψ is equivalent to $\{j_{st}\}_{0 \le s \le t \le T}$.

Remark 4.20. Let $b \in \mathcal{B}^0$, $\Delta(b) = \sum_{\epsilon \in \mathbb{A}} b^\epsilon$, $b^\epsilon \in \mathcal{B}_\epsilon$, then Equation (4.9) has to be interpreted as

$$j_{st}(b) = \sum_{\substack{\epsilon \in \mathbb{A} \\ \epsilon_1 = 1, \epsilon \ne (1)}} \int_s^t j_{s\tau}(b_1^\epsilon)\mathrm{d}I_\tau^{\mathrm{AM}}(b_2^\epsilon) j_{s\tau}(b_3^\epsilon) \cdots$$

$$+ \sum_{\substack{\epsilon \in \mathbb{A} \\ \epsilon_1 = 2}} \int_s^t \mathrm{d}I_\tau^{\mathrm{AM}}(b_1^\epsilon) j_{s\tau}(b_2^\epsilon) \mathrm{d}I_\tau^{\mathrm{AM}}(b_3^\epsilon) \cdots,$$

see also [Sch95b]. This equation can be simplified using the relation

$$\mathrm{d}I_t^{\mathrm{AM}}(b_1) X_t \mathrm{d}I_t^{\mathrm{AM}}(b_2) = \langle \Omega, X_t \Omega \rangle \left(\mathrm{d}I_t^{\mathrm{AM}}(b_1) \bullet \mathrm{d}I_t^{\mathrm{AM}}(b_2) \right)$$

for $b_1, b_2 \in \mathcal{B}^0$ and anti-monotonically adapted operator processes $\{X_t\}_{0 \le t \le T}$, where the product '$\bullet$' is defined by the anti-monotone Itô table

$\bullet$	$\mathrm{d}A^{\mathrm{AM}+}(u_1)$	$\mathrm{d}\Lambda^{\mathrm{AM}}(\lvert x_1 \rangle\langle y_1 \rvert)$	$\mathrm{d}A^{\mathrm{AM}}(v_1)$	$\mathrm{d}t$
$\mathrm{d}A^{\mathrm{AM}+}(u_2)$	0	0	0	0
$\mathrm{d}\Lambda^{\mathrm{AM}}(\lvert x_2 \rangle\langle y_2 \rvert)$	$\langle y_2, u_1 \rangle \mathrm{d}A^{\mathrm{AM}+}(x_2)$	$\langle y_2, x_1 \rangle \mathrm{d}\Lambda^{\mathrm{AM}}(\lvert x_2 \rangle\langle y_1 \rvert)$	0	0
$\mathrm{d}A^{\mathrm{AM}}(v_2)$	$\langle v_2, u_1 \rangle \Gamma_{\mathrm{M}}(0_{0t})\mathrm{d}t$	$\langle v_2, x_1 \rangle \mathrm{d}A^{\mathrm{AM}}(y_1)$	0	0
$\mathrm{d}t$	0	0	0	0

for $u_i, v_i, x_i, y_i \in D$, $i = 1, 2$.

One can check that $\mathrm{d}I_t^{\mathrm{AM}}$ is actually a homomorphism on $\mathcal{B}^0$ for the Itô product, i.e.

$$\mathrm{d}I_t^{\mathrm{AM}}(b_1) \bullet \mathrm{d}I_t^{\mathrm{AM}}(b_2) = \mathrm{d}I_t^{\mathrm{AM}}(b_1 b_2),$$

for all $b_1, b_2 \in \mathcal{B}^0$.

Using the time-reversal R defined in 4.3, we also get a realization of monotone Lévy processes on the monotone Fock space as solutions of backward monotone stochastic differential equations.

It follows also that operator processes with monotonically or anti-monotonically independent additive increments can be written as linear combination of the four fundamental noises, where the time process has to be taken as $T_{st}^{\mathrm{AM}} = \int_s^t \Gamma_{\mathrm{M}}(0_{0\tau})\mathrm{d}\tau$, $0 \le s \le t \le T$, for the anti-monotone case and $T_{st}^{\mathrm{M}} = \int_s^t \Gamma_{\mathrm{M}}(0_{\tau T})\mathrm{d}\tau$ for the monotone case.

References

[AFS02] L. Accardi, U. Franz, and M. Skeide. Renormalized squares of white noise and other non-Gaussian noises as Lévy processes on real Lie algebras. *Comm. Math. Phys.*, 228(1):123–150, 2002.

[Ans02] M. Anshelevich. Itô formula for free stochastic integrals. *J. Funct. Anal.*, 188(1):292–315, 2002.

[Ans03] M. Anshelevich. Free martingale polynomials. *J. Funct. Anal.*, 201(1):228–261, 2003.

[App05] D. Applebaum. Lectures on classical Lévy process in Euclidean spaces and groups. In [QIIP-I], pp. 1–98, 2005.

[ASW88] L. Accardi, M. Schürmann, and W.v. Waldenfels. Quantum independent increment processes on superalgebras. *Math. Z.*, 198:451–477, 1988.

[Bel98] V.P. Belavkin. On quantum Itô algebras. *Math. Phys. Lett.*, 7:1–16, 1998.

[BG01] A. Ben Ghorbal. *Fondements algébrique des probabilités quantiques et calcul stochastique sur l'espace de Fock booléen.* PhD thesis, Université Henri Poincaré-Nancy 1, 2001.

[BGDS01] A. Ben Ghorbal and M. Schürmann. Quantum stochastic calculus on Boolean Fock space. *Infin. Dimens. Anal. Quantum Probab. Relat. Top.*, Vol. 7, No. 4, pp. 631–650, 2004.

[BGS99] A. Ben Ghorbal and M. Schürmann. On the algebraic foundations of a non-commutative probability theory. Prépublication 99/17, Institut E. Cartan, Nancy, 1999.

[BGS02] A. Ben Ghorbal and M. Schürmann. Non-commutative notions of stochastic independence. *Math. Proc. Cambridge Philos. Soc.*, 133(3):531–561, 2002.

[BH96] G.M. Bergman and A.O. Hausknecht. *Co-groups and co-rings in categories of associative rings*, volume 45 of *Mathematical Surveys and Monographs.* American Mathematical Society, Providence, RI, 1996.

[Bha01] B. V. Rajarama Bhat. Cocycles of CCR flows. *Mem. Amer. Math. Soc.*, 149(709), 2001.

[Bha05] B.V.R. Bhat. Dilations, cocycles, and product systems. In [QIIP-I], pp. 273–291

[Bia98] P. Biane. Processes with free increments. *Math. Z.*, 227(1):143–174, 1998.

[BNT02a] O. E. Barndorff-Nielsen and S. Thorbjørnsen. Lévy laws in free probability. *Proc. Natl. Acad. Sci. USA*, 99(26):16568–16575 (electronic), 2002.

[BNT02b] O. E. Barndorff-Nielsen and S. Thorbjørnsen. Self-decomposability and Lévy processes in free probability. *Bernoulli*, 8(3):323–366, 2002.

[BNT05] O.E. Barndorff-Nielsen and S. Thorbjørnsen. On the roles of classical and free lévy processes in theory and applications. In this volume [QIIP-II], 2005.

[Eme89] M. Emery. On the Azéma martingales. In *Séminaire de Probabilités XXIII*, volume 1372 of *Lecture Notes in Math.* Springer-Verlag, Berlin, 1989.

[Fra00] U. Franz. Lévy processes on quantum groups. In *Probability on algebraic structures (Gainesville, FL, 1999)*, volume 261 of *Contemp. Math.*, pages 161–179. Amer. Math. Soc., Providence, RI, 2000.

[Fra01] U. Franz. Monotone independence is associative. *Infin. Dimens. Anal. Quantum Probab. Relat. Top.*, 4(3):401–407, 2001.

[Fra03a] U. Franz. Lévy processes on real Lie algebras. First Sino-German Conference on Stochastic Analysis (A Satellite Conference of ICM 2002), Beijing, China, August 29 - September 3, 2002, 2003.

[Fra03b] U. Franz. Unification of boolean, monotone, anti-monotone, and tensor independence and Lévy process. *Math. Z.*, 243(4):779–816, 2003.

[FS99] U. Franz and R. Schott. *Stochastic Processes and Operator Calculus on Quantum Groups.* Kluwer Academic Publishers, Dordrecht, 1999.

[FS00] U. Franz and M. Schürmann. Lévy processes on quantum hypergroups. In *Infinite dimensional harmonic analysis (Kyoto, 1999)*, pages 93–114. Gräbner, Altendorf, 2000.

[FS03] U. Franz and M. Skeide, 2003. in preparation.

[GKS76] V. Gorini, A. Kossakowski, and E. C. G. Sudarshan. Completely positive dynamical semigroups of N-level systems. *J. Mathematical Phys.*, 17(5):821–825, 1976.

[GSS92] P. Glockner, M. Schürmann, and R. Speicher. Realization of free white noise. *Arch. Math.*, 58:407–416, 1992.

[Gui72] A. Guichardet. *Symmetric Hilbert spaces and related topics*, volume 261 of *Lecture Notes in Math.* Springer-Verlag, Berlin, 1972.

[GvW89] P. Glockner and W. von Waldenfels. The relations of the noncommutative coefficient algebra of the unitary group. In *Quantum probability and applications, IV (Rome, 1987)*, volume 1396 of *Lecture Notes in Math.*, pages 182–220. Springer, Berlin, 1989.

[Hol01] A.S. Holevo. *Statistical structure of quantum theory*, volume 67 of *Lecture Notes in Physics. Monographs.* Springer-Verlag, Berlin, 2001.

[HP84] R. L. Hudson and K. R. Parthasarathy. Quantum Ito's formula and stochastic evolutions. *Comm. Math. Phys.*, 93(3):301–323, 1984.

[HP86] R. L. Hudson and K. R. Parthasarathy. Unification of fermion and boson stochastic calculus. *Comm. Math. Phys.*, 104(3):457–470, 1986.

[Len98] R. Lenczewski. Unification of independence in quantum probability. *Infin. Dimens. Anal. Quantum Probab. Relat. Top.*, 1(3):383–405, 1998.

[Len01] Romuald Lenczewski. Filtered random variables, bialgebras, and convolutions. *J. Math. Phys.*, 42(12):5876–5903, 2001.

[Lie99] V. Liebscher. On a central limit theorem for monotone noise. *Infin. Dimens. Anal. Quantum Probab. Relat. Top.*, 2(1):155–167, 1999.

[Lin76] G. Lindblad. On the generators of quantum dynamical semigroups. *Comm. Math. Phys.*, 48(2):119–130, 1976.

[Lin05] J.M. Lindsay. Quantum stochastic analysis – an introduction. In: [QIIP-I], pp. 181–271, 2005.

[Lu97] Y. G. Lu. An interacting free Fock space and the arcsine law. *Probab. Math. Statist.*, 17(1):149–166, 1997.

[Mac98] S. MacLane. *Categories for the working mathematician*, volume 5 of *Graduate texts in mathematics.* Springer-Verlag, Berlin, 2 edition, 1998.

[Mey95] P.-A. Meyer. *Quantum Probability for Probabilists*, volume 1538 of *Lecture Notes in Math.* Springer-Verlag, Berlin, 2nd edition, 1995.

[Mur97] N. Muraki. Noncommutative Brownian motion in monotone Fock space. *Comm. Math. Phys.*, 183(3):557–570, 1997.

[Mur03] N. Muraki. The five independences as natural products. *Inf. Dim. Anal., quant. probab. and rel. fields*, 6(3):337-371, 2003.

[Mur02] N. Muraki. The five independences as quasi-universal products. *Inf. Dim. Anal., quant. probab. and rel. fields*, 5(1):113–134, 2002.

[Par90] K.R. Parthasarathy. Azéma martingales and quantum stochastic calculus. In R.R. Bahadur, editor, *Proc. R.C. Bose Memorial Symposium*, pages 551–569. Wiley Eastern, 1990.

[Par99] K.R. Parthasarathy. A Boson Fock space realization of arcsine Brownian motion. *Sankhyā Ser. A*, 61(3):305–311, 1999.

[PS72] K.R. Parthasarathy and K. Schmidt. *Positive definite kernels, continuous tensor products, and central limit theorems of probability theory*, volume 272 of *Lecture Notes in Math.* Springer-Verlag, Berlin, 1972.

[PS98] K. R. Parthasarathy and V. S. Sunder. Exponentials of indicator functions are total in the boson Fock space $\Gamma(L^2[0,1])$. In *Quantum probability communications*, QP-PQ, X, pages 281–284. World Sci. Publishing, River Edge, NJ, 1998.

[QIIP-I] D. Applebaum, B.V.R. Bhat, J. Kustermans, J.M. Lindsay. *Quantum Independent Increment Processes I: From Classical Probability to Quantum Stochastic Calculus* U. Franz, M. Schürmann (eds.), Lecture Notes in Math., Vol. 1865, Springer, 2005.

[QIIP-II] O.E. Barndorff-Nielsen, U. Franz, R. Gohm, B. Kümmerer, S. Thorbjørnsen. *Quantum Independent Increment Processes II: Structure of Quantum Lévy Processes, Classical Probability and Physics*, U. Franz, M. Schürmann (eds.), Lecture Notes in Math., Vol. 1866, Springer, 2005.

[Sch90] M. Schürmann. Noncommutative stochastic processes with independent and stationary increments satisfy quantum stochastic differential equations. *Probab. Theory Related Fields*, 84(4):473–490, 1990.

[Sch91a] M. Schürmann. The Azéma martingales as components of quantum independent increment processes. In J. Azéma, P.A. Meyer, and M. Yor, editors, *Séminaire de Probabilités XXV*, volume 1485 of *Lecture Notes in Math.* Springer-Verlag, 1991.

[Sch91b] Michael Schürmann. Quantum stochastic processes with independent additive increments. *J. Multivariate Anal.*, 38(1):15–35, 1991.

[Sch93] M. Schürmann. *White Noise on Bialgebras*, volume 1544 of *Lecture Notes in Math.* Springer-Verlag, Berlin, 1993.

[Sch95a] M. Schürmann. Direct sums of tensor products and non-commutative independence. *J. Funct. Anal.*, 1995.

[Sch95b] M. Schürmann. Non-commutative probability on algebraic structures. In H. Heyer, editor, *Proceedings of XI Oberwolfach Conference on Probability Measures on Groups and Related Structures*, pages 332–356. World Scientific, 1995.

[Sch97] M. Schürmann. Cours de DEA. Université Louis Pasteur, Strasbourg, 1997.

[Sch00] M. Schürmann. Operator processes majorizing their quadratic variation. *Infin. Dimens. Anal. Quantum Probab. Relat. Top*, 3(1):99–120, 2000.

[Ske00] M. Skeide. Indicator functions of intervals are totalizing in the symmetric Fock space $\Gamma(L^2(\mathbb{R}_+))$. In L. Accardi, H.-H. Kuo, N. Obata, K. Saito, Si Si, and L. Streit, editors, *Trends in Contemporary Infinite Dimensional Analysis and Quantum Probability.* Istituto Italiano di Cultura, Kyoto, 2000.

[Ske01] M. Skeide. Hilbert modules and applications in quantum probability. Habilitation thesis, 2001.

[Spe97] R. Speicher. Universal products. In D. Voiculescu, editor, *Free probability theory. Papers from a workshop on random matrices and operator algebra free products, Toronto, Canada, March 1995*, volume 12 of *Fields Inst. Commun.*, pages 257–266. American Mathematical Society, Providence, RI, 1997.

[Str00] R. F. Streater. Classical and quantum probability. *J. Math. Phys.*, 41(6):3556–3603, 2000.

[Swe69] M. E. Sweedler. *Hopf Algebras.* Benjamin, New York, 1969.

[VDN92] D. Voiculescu, K. Dykema, and A. Nica. *Free Random Variables.* AMS, 1992.

[Voi87] D. Voiculescu. Dual algebraic structures on operator algebras related to free products. *J. Oper. Theory*, 17:85–98, 1987.

[Voi90] D. Voiculescu. Noncommutative random variables and spectral problems in free product C^*-algebras. *Rocky Mountain J. Math.*, 20(2):263–283, 1990.

[Wal73] W.v. Waldenfels. An approach to the theory of pressure broadening of spectral lines. In M. Behara, K. Krickeberg, and J. Wolfowitz, editors, *Probability and Information Theory II*, volume 296 of *Lecture Notes in Math.* Springer-Verlag, Berlin, 1973.

[Wal84] W.v. Waldenfels. Ito solution of the linear quantum stochastic differential equation describing light emission and absorption. In *Quantum probability and applications to the quantum theory of irreversible processes, Proc. int. Workshop, Villa Mondragone/Italy 1982*, volume 1055 of *Lecture Notes in Math.*, pages 384–411. Springer-Verlag, 1984.

[Zha91] J.J. Zhang. H-algebras. *Adv. Math.*, 89(2):144–191, 1991.

Quantum Markov Processes and Applications in Physics

Burkhard Kümmerer

Fachbereich Mathematik
Technische Universität Darmstadt
Schloßgartenstraße 7
64289 Darmstadt, Germany
kuemmerer@mathematik.tu-darmstadt.de

B. Kümmerer: *Quantum Markov Processes and Applications in Physics*,
Lect. Notes Math. **1866**, 259–330 (2006)
www.springerlink.com

Introduction

In this course we discuss aspects of the theory of stationary quantum Markov processes.

By 'processes' we mean stochastic processes; hence, ideas of probability theory are central to our discussions. The attribute *'Markov'* indicates that we are mainly concerned with forms of stochastic behaviour where the (probabilities of) future states depend on the present state, but beyond this the behaviour in the past has no further influence on the future behaviour of the process.

The attribute *'quantum'* refers to the fact that we want to include stochastic behaviour of quantum systems into our considerations; this does not mean, however, that we discuss quantum systems exclusively. While quantum systems are described in the language of Hilbert spaces and operators, classical systems are modelled by phase spaces and functions on phase spaces. A mathematical language which allows a unified description of both types of systems is provided by the theory of operator algebras. This is the language we shall use throughout these lectures. Noncommutativity of such an algebra corresponds to quantum features of the system while classical systems are modelled by commutative algebras. The price paid for this generality lies in the abstractness of the mathematical theory of operator algebras. We seek to compensate its abstractness by giving a detailed description of two particular physical systems, a spin-$\frac{1}{2}$-particle in a stochastic magnetic field (Chapter 6) and the micro-maser (Chapter 7).

Finally, the attribute *'stationary'* indicates that we are mainly interested in a stochastic behaviour which possesses a distinguished stationary state, often referred to as an equilibrium distribution or equilibrium state . This does not mean, that we usually find the system in such a stationary state, but in a number of cases an initial state will converge to a stationary state if we wait long enough. The mere existence of a stationary state as a reference state has a number of pleasant mathematical consequences. First it allows, classically speaking, to work on a fixed measure space, which does not depend on the initial state of the process and does not change in time. In the operator algebraic description this is reflected by the fact that the mathematics can be done within the framework of von Neumann algebras, frequently equipped with a faithful normal reference state. They can be viewed as non-commutative versions of spaces of the type $L^\infty(\Omega, \Sigma, \mu)$. A second useful consequence of stationarity is the fact that the time evolution of such a process can be implemented by a group of automorphisms on the underlying von Neumann algebra of observables, leaving the reference state fixed. This relates stationary processes to stationary dynamical systems, in particular to their ergodic theory. From this point of view a stationary stochastic process is simply a dynamical system, given by a group of automorphisms with a stationary state on a von Neumann algebra, where the action on a distinguished subalgebra – the time zero algebra – is of particular interest. As an example of the fruitfulness of this point of view we discuss in Chapter 4 a scattering theory for Markov processes. The existence of stationary states is again fundamental in our discussion of the ergodic theory of repeated measurement in the final Chapter 10.

Needless to say that many important stochastic processes are not stationary, like the paradigmatic process of Brownian motion. However, even here stationarity is present, as Brownian motion belongs to the class of processes with stationary independent increments. Many efforts have been spent on employing the stationarity of its increments to the theory of Brownian motion. The approach of Hida in [Hid] is a famous example: The basic idea is to consider Brownian motion as a function of its stationary increment process, white noise, and early developments of quantum stochastic calculus on Fock space can be considered as an extension of this approach. Recent developments of these ideas can be found in the present two volumes.

We end with a brief guide through the contents of these lectures: A first part (Chapters 1–3) introduces and discusses basic notions which are needed for the following discussion of stationary quantum Markov processes. In particular, we introduce a special class of such Markov processes in Chapter 3. It will play a prominent role in the following parts of these lectures. The second part (Chapter 4) looks at this class of stationary Markov processes from the point of view of scattering theory. In a third part (Chapters 5–8) we show that such Markov processes do naturally occur in the description of certain physical systems. The final part (Chapters 8–10) discusses a different type of stochastic processes which describe repeated measurement. The aim is to discuss the ergodic properties of such processes.

Parts of these notes are adaptions and revised versions from texts of two earlier summer schools in Grenoble [Kü3] and Dresden [Kü4].

Acknowledgements: It is a pleasure to thank Uwe Franz and Michael Schürman for their hospitality not only during this summer school but at many occasions during the past few years. Particular thanks go to Uwe Franz for his patience with these notes. I would like to thank Florian Haag und Nadiem Sissouno for their help during the final proof-reading. Above all I would like to thank Hans Maassen. Large parts of the material included in these notes result from our collaboration in friendship over many years.

1 Quantum Mechanics

Our first aim is to introduce quantum Markov processes. In order to do this we start by giving a mathematical description of quantum mechanics. This frame will be extended in the next section in such a way that it also incorporates the description of classical systems.

1.1 The Axioms of Quantum Mechanics

Following the ideas of J.v. Neumann [JvN] quantum mechanics can be axiomatized as follows:
To a physical system there corresponds a Hilbert space $\mathcal{H}$ such that

1. *Pure states* of this system are described by unit vectors in $\mathcal{H}$ (determined up to a phase).
2. *Observables* of this system are described by (possibly unbounded) self-adjoint operators on $\mathcal{H}$.
3. If the system is in a state described by the unit vector $\xi \in \mathcal{H}$ then the *measurement* of an observable described by a self-adjoint operator X yields the *expectation value* $\mathbb{E}(X) = \langle X\xi, \xi \rangle$.

4. If an observable is described by the self-adjoint operator X on $\mathcal{H}$ then the observable obtained from it by changing the scale of the measurement apparatus via a measurable function f is described by the operator $f(X)$. Here, $f(X)$ is obtained from X by use of the spectral theorem (cf. Section 1.3).

   If f is a bounded function then $f(X)$ is a bounded operator; therefore, from a theoretical point of view working with bounded operators suffices.

From these axioms one can deduce large parts of the quantum mechanical formalism (cf. the discussion in Section 1.3). Determining $\mathcal{H}$, X, and ξ, however, is a different problem which is not touched in these axioms.

1.2 An Example: Two–Level Systems

In order to have a concrete example in mind consider a quantum mechanical two–level system like a spin–$\frac{1}{2}$–particle. The corresponding Hilbert space is the two-dimensional Hilbert space $\mathcal{H} = \mathbb{C}^2$ and a standard set of observables is given by the self-adjoint matrices

$$\sigma_x = \begin{pmatrix} 0 & 1 \\ 1 & 0 \end{pmatrix}, \quad \sigma_y = \begin{pmatrix} 0 & -i \\ i & 0 \end{pmatrix}, \quad \sigma_z = \begin{pmatrix} 1 & 0 \\ 0 & -1 \end{pmatrix}$$

which may be interpreted as describing the measurement of polarization in x, y, and z-direction, respectively.

Every self-adjoint matrix is a unique real linear combination of $\mathbb{1}, \sigma_x, \sigma_y, \sigma_z$ and such a matrix

$$\Phi = \alpha \cdot \mathbb{1} + x \cdot \sigma_x + y \cdot \sigma_y + z \cdot \sigma_z = \begin{pmatrix} \alpha + z & x - iy \\ x + iy & \alpha - z \end{pmatrix}$$

is a density matrix of a mixed state iff, by definition, $\Phi \geq 0$ and $tr(\Phi) = 1$, hence iff $\alpha = \frac{1}{2}$ and $x^2 + y^2 + z^2 \leq \frac{1}{4}$.

Thus the convex set of mixed states can be identified with a (full) ball in $\mathbb{R}^3$ (of radius $\frac{1}{2}$ in our parametrization) and the pure states of the system correspond to the extreme points, i.e. to the points on the surface of this ball.

1.3 How Quantum Mechanics is Related to Classical Probability

The formalism of quantum mechanics is not as different from the formalism of classical probability as it might seem at a first glance. The link between both of them is established by the *spectral theorem* (cf. [RS]):

If X is a self-adjoint operator on a separable Hilbert space then there exist

- a probability space (Ω, Σ, μ),
- a real-valued random variable $Y : \Omega \to \mathbb{R}$,
- a unitary $u : \mathcal{H} \to L^2(\Omega, \Sigma, \mu)$,

such that $uXu^* = M_Y$, where M_Y is the operator acting on $L^2(\Omega, \Sigma, \mu)$ by pointwise multiplication with Y.

If follows that the *spectrum* $\sigma(X)$ of X is equal to $\sigma(M_Y)$, hence it is given by the essential range of the random variable Y. The function Y can be composed with any further real or complex function f which is defined on the (essential) range of Y, hence on the spectrum of X. Therefore we can also define the operator

$$f(X) := u^* \cdot M_{f \circ Y} \cdot u$$

for any such function f.

It thus appears that a self-adjoint operator can be identified with a real-valued random variable. There is only one problem: Two self-adjoint operators may not be equivalent to multiplication operators on the same probability space with the same intertwining unitary u. Indeed, a family of self-adjoint operators on $\mathcal{H}$ admits a simultaneous realization by multiplication operators on one probability space if and only if they commute. It is only at this point, the occurrence of non-commuting self-adjoint operators, where quantum mechanics separates from classical probability.

As long as only one self-adjoint operator is involved, we can proceed further as in classical probability:

A state $\xi \in \mathcal{H}$ induces a probability measure μ_ξ on the spectrum $\sigma(X) \subseteq \mathbb{R}$ which is uniquely characterized by the property

$$\langle f(X)\xi, \xi \rangle = \int_{\mathbb{R}} f(\lambda)\, d\mu_\xi(\lambda)$$

for all bounded measurable functions f on $\mathbb{R}$. The measure μ_ξ is called the *spectral measure* of X with respect to ξ but it may also be viewed as the distribution of X:

The function $u\xi \in L^2(\Omega, \Sigma, \mu)$ is a unit vector, therefore, its squared pointwise absolute value $|u\xi|^2$ is, with respect to μ, the density of a probability measure on (Ω, Σ) and μ_ξ is the distribution of Y with respect to this probability measure.

The quantum mechanical interpretation of μ_ξ is given in the next statement.

Proposition 1.1. *A measurement of an observable X on a system in a state ξ gives a value in $\sigma(X)$ and the probability distribution of these values is given by μ_ξ.*

This result can be deduced from the axioms in Section 1.1 as follows: Let $f := \chi := \chi_{\sigma(X)^C}$ be the characteristic function of the complement of $\sigma(X)$. By Axiom 4 a measurement of $\chi(X)$ yields a value 0 or 1. Therefore, the probability that this measurement gives the value 1 is equal to the expectation of this measurement, hence equal to

$$\langle \chi(X)\xi, \xi \rangle = \langle 0\xi, \xi \rangle = 0 \,.$$

It follows that a measurement of $\chi(X)$ gives 0, hence measuring X gives a value in $\sigma(X)$. More generally, if $A \subseteq \sigma(X)$ then the probability for obtaining from a measurement of X a value in A is the probability to obtain the value 1 in a measurement of $\chi_A(X)$ (again we used the fourth axiom), which is given by

$$\langle \chi_A(X)\xi, \xi \rangle = \int_{\mathbb{R}} \chi_A \mathrm{d}\mu_\xi = \mu_\xi(A) \,.$$

The above proof could have been condensed. But in its present form it shows more clearly the role played by the fourth axiom.

Corollary 1.2. *A measurement of an observable X on a system in a state ξ gives a value in a subset $A \subseteq \sigma(X)$ with certainty iff $1 = \mu_\xi(A) = \langle \chi_A(X)\xi, \xi \rangle$, hence if and only if $\chi_A(X)\xi = \xi$. This means, that ξ is an eigenvector with eigenvalue 1 of the spectral projection $\chi_A(X)$ of X.*

It follows that after a measurement of the observable X, if it resulted in a value in $A \subseteq \sigma(X)$, the state of the system has changed to a vector in $\chi_A(X)\mathcal{H}$. The reason is that an immediate second measurement of X should now give a value in A with certainty.

In such a manner one can now proceed further deducing, step by step, the formalism of quantum mechanics from these axioms.

2 Unified Description of Classical and Quantum Systems

In this second chapter we extend the mathematical model in such a way that it allows to describe classical systems and quantum systems simultaneously. Additional motivation is given in [KüMa2].

2.1 Probability Spaces

Observables

In the above formulation of the second axiom of quantum mechanics we have been a bit vague: We left open how many self-adjoint operators correspond to physical observables. We are now going to use this freedom:

Axiom 2, improved version. There is a $*$-algebra $\mathcal{A}$ of bounded operators on $\mathcal{H}$ such that the (bounded) observables of the system are described by the self-adjoint operators in $\mathcal{A}$.

Here the word $*$*-algebra* means: If $x, y \in \mathcal{A}$, then also $x + y$, λx ($\lambda \in \mathbb{C}$), $x \cdot y$, and the adjont x^* are elements of $\mathcal{A}$. In the literature the adjoint of x is sometimes denoted by $x^\dagger$.

$\mathcal{A}$ is called the *algebra of observables* of the system. For simplicity we assume that $\mathcal{A}$ contains the identity $\mathbb{1}$. For mathematical convenience $\mathcal{A}$ is usually assumed to be closed either in the norm – it is then called a C^* – *algebra* – or in the strong operator topology – in this case it is called a *von Neumann algebra* or W^* *-algebra*.

In a truly quantum situation with only finitely many degrees of freedom one would choose $\mathcal{A} = \mathcal{B}(\mathcal{H})$, the algebra of all bounded operators on $\mathcal{H}$. Indeed, von Neumann in his formulation of quantum mechanics assumed this explicitly. This assumption is known as his *irreducibility axiom* .

On the other hand, if (Ω, Σ, μ) is a probability space then bounded real-valued random variables (the classical pendant to observables in quantum mechanics) are functions in $L^\infty(\Omega, \Sigma, \mu)$ and any such function can be viewed as a bounded multiplication operator on $L^2(\Omega, \Sigma, \mu)$. Therefore, classical systems correspond to (subalgebras of) algebras of the type $L^\infty(\Omega, \Sigma, \mu)$, which are now viewed as algebras of multiplication operators. Moreover, it is a non-trivial fact (cf. [Tak2]) that any commutative von Neumann algebra is isomorphic to some $L^\infty(\Omega, \Sigma, \mu)$. Therefore, it is safe to say that classical systems correspond to commutative algebras of observables. If we do not think in probabilistic terms but in terms of classical mechanics then Ω becomes the phase space of the system and the first choice for μ is the Liouville measure on Ω.

States

The next problem is to find a unified description of quantum mechanical states on the one hand and classical probability measures on the other. The idea is that both give rise to expectation values of observables. Moreover, they are uniquely determined by the collection of all expectation values. Thus, we will axiomatize the notion of an expectation value.

Starting again with quantum mechanics a state given by a unit vector $\xi \in \mathcal{H}$ gives rise to the *expectation functional*

$$\varphi_\xi : \mathcal{B}(\mathcal{H}) \ni x \mapsto \langle x\xi, \xi \rangle \in \mathbb{C}\,.$$

The functional φ_ξ is linear, positive ($\varphi_\xi(x) \geq 0$ if $x \geq 0$) and normalized ($\varphi_\xi(\mathbb{1}) = 1$). More generally, if ρ is a density matrix on $\mathcal{H}$, then

$$\varphi_\rho : \mathcal{B}(\mathcal{H}) \ni x \mapsto tr(\rho\, x) \in \mathbb{C}$$

still enjoys the same properties. (A *density matrix* or *density operator* ρ on $\mathcal{H}$ is a positive operator ρ such that $tr(\rho) = 1$ where tr denotes the trace.)

On the other hand, if (Ω, Σ, μ) is a classical probability space, then the probability measure μ gives rise to the expectation functional

$$\varphi_\mu : L^\infty(\Omega, \Sigma, \mu) \ni f \mapsto \mathbb{E}(f) = \int_\Omega f d\mu \in \mathbb{C}\,.$$

Again, φ_μ is a linear, positive, and normalized functional on $L^\infty(\Omega, \Sigma, \mu)$. This leads to the following notions.

Definition 2.1. *A state on an algebra $\mathcal{A}$ of observables is a positive normalized linear functional*

$$\varphi : \mathcal{A} \to \mathbb{C}.$$

If φ is a state on $\mathcal{A}$ then the pair $(\mathcal{A}, \varphi)$ is called a probability space.

Instead of calling φ a 'state' one could call it a 'probability measure' as well, but the term 'state' has become common. In order to avoid confusion with classical probability spaces, a pair $(\mathcal{A}, \varphi)$ is sometimes called *quantum probability space* or *non-commutative probability space*, despite the fact that it may describe a classical system and $\mathcal{A}$ may be commutative. Finally we note that under certain continuity conditions a state on $\mathcal{B}(\mathcal{H})$ is induced by a density matrix and a state on $L^\infty(\Omega, \Sigma, \mu)$ comes from a probability measure on (Ω, Σ) (see below).

2.2 From the Vocabulary of Operator Algebras

As might become clear from the above, the language of *operator algebras* is appropriate when a unified mathematical description of classical systems and quantum systems is needed. For convenience we review some basic notions from the vocabulary of operator algebras. For further information we refer to the books on this subject like [Tak2].

As mentioned above operator algebras can be viewed as *-algebras of bounded operators on some Hilbert space $\mathcal{H}$, closed either in the operator norm (*C*-algebra*) or in the strong operator topology (*von Neumann algebra*). Here, operators $(x_i)_{i\in I} \subseteq B(\mathcal{H})$ converge to an operator $x \in B(\mathcal{H})$ in the *strong operator topology* if $(x_i(\xi))_{i\in I}$ converges to $x(\xi)$ for every vector $\xi \in \mathcal{H}$. Therefore, strong operator convergence is weaker than convergence in the operator norm. It follows that von Neumann algebras are also C*-algebras. But for many purposes convergence in the operator norm is too strong while most C*-algebras are not closed in the strong operator topology. Conversely, von Neumann algebras are 'very large' when considered as C*-algebras. There is also an abstract characterization of C*-algebras as Banach *-algebras for which $\|x^*x\| = \|x\|^2$ for all elements x (the usefulness of this condition is by far not obvious). Von Neumann algebras are abstractly characterized as C*-algebras which have, as a Banach space, a predual.

A typical example of a commutative C*-algebra is $C(K)$, the algebra of continuous functions on a compact space K, and every commutative C*-algebra with an identity is isomorphic to an algebra of this type. A typical example of a commutative von Neumann algebra is $L^\infty(\Omega, \Sigma, \mu)$ (here (Ω, Σ, μ) should be a localizable measure space) and every commutative von Neumann algebra is isomorphic to an algebra of this type. The algebras M_n of $n \times n$-matrices and, more generally, the algebra $B(\mathcal{H})$ of all bounded operators on a Hilbert space $\mathcal{H}$ are C*-algebras and von Neumann algebras. On the other hand the algebra of all compact operators on $\mathcal{H}$ is only a C*-algebra whenever $\mathcal{H}$ is not finite dimensional. Other C*-algebras which are interesting from the point of view of physics are the C*-algebras of the canonical commutation relations (CCR) and of the canonical anticommutation relations (CAR) (cf. [EvLe]).

Elements x with $x = x^*$ are called *self-adjoint* as they are represented by self-adjoint operators. It is less obvious that elements of the form x^*x should be called *positive*. If y is an operator on some Hilbert space then by the spectral theorem y is positive semidefinite if and only if $y = x^*x$ for some operator x. But is not so easy to see that also for an abstract C*-algebra this leads to the right notion of positivity.

As motivated above a *state* on a C*-algebra $\mathcal{A}$ is abstractly defined as a linear functional $\varphi : \mathcal{A} \to \mathbb{C}$ which is positive (in view of the above this means that $\varphi(x^*x) \geq 0$ for all $x \in \mathcal{A}$) and normalized, i.e. $\|\varphi\| = 1$. If $\mathcal{A}$ has an identity and φ is already positive then $\|\varphi\| = 1$ whenever $\varphi(\mathbb{1}) = 1$. A state is thus an element of the Banach space dual of a C*-algebra $\mathcal{A}$. If $\mathcal{A}$ is a von Neumann algebra and φ is not only in the dual but in the predual of $\mathcal{A}$ then it is called a *normal* state. There are various characterizations of normal states by continuity or order continuity properties. For the moment it is enough to know that a state φ on a commutative von Neumann algebra $L^\infty(\Omega, \Sigma, \mu)$ is normal if and only if there is a 'density' function $f_\varphi \in L^1(\Omega, \Sigma, \mu)$ such that $\varphi(g) = \int_\Omega f_\varphi g \mathrm{d}\mu$ for all $g \in L^\infty(\Omega, \Sigma, \mu)$. A state φ on the von Neumann algebra $B(\mathcal{H})$ is normal if and only if there is a density matrix ρ_φ on $\mathcal{H}$ such that $\varphi(x) = tr(\rho_\varphi \cdot x)$ for all $x \in B(\mathcal{H})$.

The mathematical duality between states and observables has its counterpart in the description of time evolutions of quantum systems: By their very nature time evolutions are transformations on the space of (normal) states. The Banach space adjoint of such a transformation is a transformation on the dual space of observables. In the language of physics a description of time evolutions on the states is referred to as the *Schrödinger picture* while the *Heisenberg picture* refers to a description on the space of observables. These two descriptions are dual to each other and they are equivalent from a theoretical point of view. But spaces of observables have a richer algebraic structure (e.g., operators can be multiplied). Therefore, working in the Heisenberg picture can be of great mathematical advantage, although a discussion in the Schrödinger picture is closer to intuition.

3 Towards Markov Processes

In this chapter we discuss, step by step, the notions which will finally lead to the definition of a Markov process in the operator algebraic language.

3.1 Random Variables and Stochastic Processes

We are looking for a definition of a Markov process which covers the classical and the quantum case. We already saw that in this general context there is no state space Ω_0 available such that the system could jump between the points of Ω_0. Even if we generalized points of Ω_0 to pure states on an algebra $\mathcal{A}_0$ of observables then a state given by a density matrix can not be interpreted in a unique way as a probability measure on the pure states (the state space of M_2, cf. 1.2, demonstrates this problem drastically). Consequently, there is no direct way to talk about transition probabilities and transition operators in this general context and we will introduce transition operators only much later via conditional expectations.

Instead we proceed with defining random variables first. Unfortunately, the notion of a general random variable seems to be the most abstract and unaccessible notion of quantum probability.

From the foregoing it should be clear that a real-valued random variable is a self-adjoint operator in $\mathcal{A}$. But what would happen if one wanted to consider random variables having other state spaces? For example, when studying the behaviour of a two–level system one wants to consider polarization in all space directions simultaneously. In classical probability it is enough to change from $\Omega_0 = \mathbb{R}$ to more general versions of Ω_0 like $\Omega_0 = \mathbb{R}^3$. Now we need an algebraic description of Ω_0 and this is obtained as follows ([AFL]).

If $X : (\Omega, \Sigma, \mu) \mapsto \Omega_0$ is a random variable and $f : \Omega_0 \to \mathbb{C}$ is a measurable function then

$$i_X(f) := f \circ X : (\Omega, \Sigma, \mu) \to \mathbb{C}$$

is measurable. Moreover, $f \mapsto i_X(f)$ is a $*$–homomorphism from the algebra $\mathcal{A}_0$ of all bounded measurable $\mathbb{C}$-valued functions on Ω_0 into $\mathcal{A} := L^\infty(\Omega, \Sigma, \mu)$ with $i_X(\mathbb{1}) = \mathbb{1}$. ($*$*-homomorphism* means that i_X preserves addition, multiplication by scalars, multiplication, and involution which is complex conjugation in this case).

We are allowing now $\mathcal{A}_0$ and $\mathcal{A}$ to be non-commutative algebras of observables. For the first part of our discussion they could be any *-algebras of operators on a Hilbert space. Later in our discussion we have to require that they are C*-algebras or even von Neumann algebras. We thus arrive at the following definition.

Definition 3.1. *([AFL]) A* random variable *on $\mathcal{A}$ with values in $\mathcal{A}_0$ is an identity preserving $*$-homomorphism*

$$i : \mathcal{A}_0 \mapsto \mathcal{A} .$$

It may be confusing that the arrow seems to point into the wrong direction, but this comes from the fact that our description is dual to the classical formulation. Nevertheless our definition describes an influence of $\mathcal{A}$ onto $\mathcal{A}_0$:

If the 'world' $\mathcal{A}$ is in a certain state φ then i induces the state $\varphi \circ i$ on $\mathcal{A}_0$ given by $\mathcal{A}_0 \ni x \mapsto \varphi(i(x)) \in \mathbb{C}$. If i comes from a classical random variable X as above then $\varphi \circ i$ is the state induced by the distribution of X hence it can be called the *distribution* of i also in the general case.

From now on we equip $\mathcal{A}$ with a state φ thus obtaining a probability space $(\mathcal{A}, \varphi)$. Once having defined the notion of a random variable the definition of a stochastic process is obvious:

Definition 3.2. *A* stochastic process *indexed by a time parameter in* $\mathbb{T}$ *is a family*

$$i_t : \mathcal{A}_0 \to (\mathcal{A}, \varphi) \quad , \quad t \in \mathbb{T} ,$$

of random variables. Such a process will also be denoted by $(\mathcal{A}, \varphi, (i_t)_{t\in\mathbb{T}}; \mathcal{A}_0)$.

Stationary stochastic processes are of particular importance in classical probability. In the spirit of our reformulations of classical concepts the following generalizes this notion.

Definition 3.3. *A stochastic process* $(i_t)_{t\in\mathbb{T}} : \mathcal{A}_0 \to (\mathcal{A}, \varphi)$ *is called* stationary *if for all* $s \geq 0$

$$\varphi(i_{t_1}(x_1) \cdot \ldots \cdot i_{t_n}(x_n)) = \varphi(i_{t_1+s}(x_1) \cdot \ldots \cdot i_{t_n+s}(x_n))$$

with $n \in \mathbb{N}$, $x_1, \ldots, x_n \in \mathcal{A}_0$, $t_1, \ldots, t_n \in \mathbb{T}$ *arbitrarily.*

As in the classical situation this means that multiple time correlations depend only on time differences. It should also be noted that it is not sufficient to require the above identity only for ordered times $t_1 \leq t_2 \leq \ldots \leq t_n$.

Finally, if a classical stochastic process is represented on the space of its paths then time translation is induced by the time shift on the path space. This is turned into the following definition:

Definition 3.4. *A process* $(i_t)_{t\in\mathbb{T}} : \mathcal{A}_0 \to (\mathcal{A}, \varphi)$ *admits a* time translation *if there are* $*$*-homomorphisms* $\alpha_t : \mathcal{A} \to \mathcal{A}$ $(t \in \mathbb{T})$ *such that*

i) $\alpha_{s+t} = \alpha_s \circ \alpha_t$ *for all* $s, t \in \mathbb{T}$
ii) $i_t = \alpha_t \circ i_0$ *for all* $t \in \mathbb{T}$.

In this case we may also denote the process $(\mathcal{A}, \varphi, (i_t)_{t\in\mathbb{T}}; \mathcal{A}_0)$ *by* $(\mathcal{A}, \varphi, (\alpha_t)_{t\in\mathbb{T}}; \mathcal{A}_0)$.

In most cases, in particular if the process is stationary, such a time translation exists. In the stationary case, it leaves the state φ invariant.

3.2 Conditional Expectations

Before we can formulate a Markov property for a stochastic process we should talk about conditional expectations. The idea is as in the classical framework:

One is starting with a probability space (Ω, Σ, μ) which describes our knowledge about the system in the following sense: We expect an event $A \in \Sigma$ to occur with probability $\mu(A)$. Now assume that we obtain some additional information on the probabilities of the events in a σ-subalgebra $\Sigma_0 \subseteq \Sigma$. Their probabilities are now given by a new probability measure ν on (Ω, Σ_0). It leads to improved – conditional – probabilities for all events of Σ given by a probability measure $\tilde{\nu}$ on (Ω, Σ) which extends ν on (Ω, Σ_0). (Since ν is absolutely continuous with respect to the restriction of μ to Σ_0, it has a Σ_0-measurable density f by the Radon Nikodym theorem, and one can put $d\tilde{\nu} = f d\mu$.)

Similarly, we now start with a (quantum) probability space $(\mathcal{A}, \varphi)$. If we perform a measurement of a self-adjoint observable $x \in \mathcal{A}$ we expect the value $\varphi(x)$. Assume again that we gained some additional information about the expectation values of the observables in a subalgebra $\mathcal{A}_0$ (for example by an observation): Now we expect a value $\psi(x)$ for the outcome of a measurement of $x \in \mathcal{A}_0$ where ψ is a new state on $\mathcal{A}_0$. As above this should change our expectation for all measurements on $\mathcal{A}$ in an appropriate way, expressed by a state $\tilde{\psi}$ on $\mathcal{A}$. Unfortunately, there is no general Radon Nikodym theorem for states on operator algebras which gives all the desired properties. Thus we have to proceed more carefully.

Mathematically speaking we should have an extension map Q assigning to each state ψ on $\mathcal{A}_0$ a state $\tilde{\psi} = Q(\psi)$ on $\mathcal{A}$; the map should thus satisfy $Q(\psi)(x) = \psi(x)$ for all $x \in \mathcal{A}_0$. Moreover, if $\psi(x) = \varphi(x)$ for all $x \in \mathcal{A}_0$, that is if there is no additional information, then the state φ should remain unchanged, hence we should require $Q(\psi) = \varphi$ in this case. If we require, in addition, that Q is an affine map $(Q(\lambda\psi_1 + (1-\lambda)\psi_2) = \lambda Q(\psi_1) + (1-\lambda)Q(\psi_2)$ for states ψ_1 and ψ_2 on $\mathcal{A}_0$ and $0 \leq \lambda \leq 1)$ and has a certain continuity property (weak *-continuous if $\mathcal{A}_0$ and $\mathcal{A}$ are C*–algebras) then one can easily show that there exists a unique linear map $P : \mathcal{A} \to \mathcal{A}$ such that $P(\mathcal{A}) = \mathcal{A}_0$, $P^2 = P$, and $||P|| \leq 1$, which has the property $Q(\psi(x)) = \psi(P(x))$ for all states ψ on $\mathcal{A}_0$ and $x \in \mathcal{A}$: Up to identification of $\mathcal{A}_0$ with a subalgebra of $\mathcal{A}$ the map P is the adjoint of Q. The passage from Q to P means to change from a state picture (Schrödinger picture) into the dual observable picture (Heisenberg picture). If $\mathcal{A}_0$ and $\mathcal{A}$ are C*–algebras then such a map P is called a *projection of norm one* and it automatically enjoys further properties: P maps positive elements of $\mathcal{A}$ into positive elements and it has the *module property*

$$P(axb) = aP(x)b$$

for $a, b \in \mathcal{A}_0$, $x \in \mathcal{A}$ ([Tak2]). Therefore, such a map P is called a *conditional expectation* from $\mathcal{A}$ onto $\mathcal{A}_0$.

From the property $\varphi(P(x)) = \varphi(x)$ for all $x \in \mathcal{A}$ it follows that there is at most one such projection P. Indeed, with respect to the scalar product $< x, y >_\varphi := \varphi(y^* x)$ induced by φ on $\mathcal{A}$ the map P becomes an orthogonal projection. Therefore, we will talk about *the* conditional expectation $P : (\mathcal{A}, \varphi) \to \mathcal{A}_0$ ... if it exists.

Typical examples for conditional expectations are conditional expectations on commutative algebras (on commutative von Neumann algebras they always exist by the Radon Nikodym theorem) and *conditional expectations of tensor type*: If $\mathcal{A}_0$ and $\mathcal{C}$ are C*-algebras and ψ is a state on $\mathcal{C}$ then

$$P_\psi : \mathcal{A}_0 \otimes \mathcal{C} \ni x \otimes y \mapsto \psi(y) \cdot x \otimes \mathbb{1}$$

extends to a conditional expectation from the (minimal) tensor product $\mathcal{A} := \mathcal{A}_0 \otimes \mathcal{C}$ onto $\mathcal{A}_0 \otimes \mathbb{1}$ (cf. [Tak2]. If $\mathcal{A}_0$ and $\mathcal{C}$ are von Neumann algebras and ψ is a normal state on $\mathcal{C}$ then P_ψ can be further extended to a conditional expectation which is defined on the larger 'von Neumann algebra tensor product' of $\mathcal{A}_0$ and $\mathcal{C}$ ([Tak2]). Sometimes it is convenient to identify $\mathcal{A}_0$ with the subalgebra $\mathcal{A}_0 \otimes \mathbb{1}$ of $\mathcal{A}_0 \otimes \mathcal{C}$ and to call the map defined by $\mathcal{A}_0 \otimes \mathcal{C} \ni x \otimes y \mapsto \psi(y)x \in \mathcal{A}_0$ a conditional expectation, too. From its definition it is clear that P_ψ leaves every state $\varphi_0 \otimes \psi$ invariant where φ_0 is any state on $\mathcal{A}_0$.

In general, the existence of a conditional expectation from $(\mathcal{A}, \varphi)$ onto a subalgebra $\mathcal{A}_0$ is a difficult problem and in many cases it simply does not exist: Equip $\mathcal{A} = M_2$ with a state φ which is induced from the density matrix $\begin{pmatrix} \lambda & 0 \\ 0 & 1-\lambda \end{pmatrix}$ $(0 \le \lambda \le 1)$. Then the conditional expectation P from (M_2, φ) onto

$$\mathcal{A}_0 = \left\{ \begin{pmatrix} a & 0 \\ 0 & b \end{pmatrix} : a, b \in \mathbb{C} \right\}$$

does exist while the conditional expectation from (M_2, φ) onto the commutative subalgebra

$$\mathcal{A}_0 = \left\{ \begin{pmatrix} a & b \\ b & a \end{pmatrix} : a, b \in \mathbb{C} \right\}$$

does not exist (we still insist on the invariance of φ) whenever $\lambda \neq \frac{1}{2}$.

There is a beautiful theorem due to M. Takesaki ([Tak1]) which solves the problem of existence of conditional expectations in great generality. Since we will not need this theorem explicitly we refer for it to the literature. It suffices to note that requiring the existence of a conditional expectation can be a strong condition. On the other hand, from a probabilistic point of view it can nevertheless make sense to require its existence as we have seen above.

With the help of conditional expectations we can define transition operators:

Definition 3.5. *Suppose* $i_1, i_2 : \mathcal{A}_0 \to (\mathcal{A}, \varphi)$ *are two random variables such that* i_1 *is injective and thus can be inverted on its range. If the conditional* *expectation* $P : (\mathcal{A}, \varphi) \to i_1(\mathcal{A}_0)$ *exists then the operator* $T : \mathcal{A}_0 \to \mathcal{A}_0$ *defined by*

$$T(x) := i_1^{-1} P(i_2(x))$$

for $x \in \mathcal{A}_1$ *is called a* transition operator.

If the random variables i_1 and i_2 are random variables of a stochastic process at times t_1 and t_2 ($t_1 < t_2$) then T describes the transitions from time t_1 to time t_2.

3.3 Markov Processes

Using conditional expectations we can now formulate a Markov property which generalizes the Markov property for classical processes:

Let $(i_t)_{t \in \mathbb{T}} : \mathcal{A}_0 \to (\mathcal{A}, \varphi)$ be a stochastic process. For $I \subseteq \mathbb{T}$ we denote by $\mathcal{A}_I$ the subalgebra of $\mathcal{A}$ generated by $\{i_t(x) : x \in \mathcal{A}_0, t \in I\}$. In particular, subalgebras $\mathcal{A}_{t]}$ and $\mathcal{A}_{[t}$ are defined as in the classical context. A subalgebra $\mathcal{A}_I$ generalizes the algebra of functions on a classical probability space which are measurable with respect to the σ-subalgebra generated by the random variables at times $t \in I$.

Definition 3.6. *The process* $(i_t)_{t \in \mathbb{T}}$ *is a Markov process if for all* $t \in \mathbb{T}$ *the conditional expectation*

$$P_{t]} : (\mathcal{A}, \varphi) \to \mathcal{A}_{t]}$$

exists and

$$\text{for all } x \in \mathcal{A}_{[t} \text{ we have } P_{t]}(x) \in i_t(\mathcal{A}_0)\,.$$

If, in particular, the conditional expectation $P_t : (\mathcal{A}, \varphi) \to i_t(\mathcal{A}_0)$ exists, then this requirement is equivalent to $P_{t]}(x) = P_t(x)$ for all $x \in \mathcal{A}_{[t}$. This parallels the classical definition.

Clearly, a definition without requiring the existence of conditional expectations is more general and one can imagine several generalizations of the above definition. On the other hand the existence of $P_0 : (\mathcal{A}, \varphi) \to i_0(\mathcal{A}_0) = \mathcal{A}_{\{0\}}$ allows us to define transition operators as above: Assume again, as is the case in most situations, that i_0 is injective. Then $i_0(\mathcal{A}_0)$ is an isomorphic image of $\mathcal{A}_0$ in $\mathcal{A}$ on which i_0 can be inverted. Thus we can define the *transition operator* T_t by

$$T_t : \mathcal{A}_0 \to \mathcal{A}_0 : x \mapsto i_0^{-1} P_0 i_t(x)\,.$$

From its definition it is clear that T_t is an identity preserving (completely) positive operator, as it is the composition of such operators. Moreover, it generalizes the classical transition operators and the Markov property again implies the semigroup law

$$T_{s+t} = T_s \cdot T_t \text{ for } s,t \geq 0$$

while $T_0 = \mathbb{1}$ is obvious from the definition. The derivation of the semigroup law from the Markov property is sometimes called the *quantum regression theorem*, although in the present context it is an easy exercise.

In the classical case we have a converse: Any such semigroup comes from a Markov process which, in addition, is essentially uniquely determined by the semigroup. It is a natural question whether this extends to the general context. Unfortunately, it does not. But there is one good news: For a semigroup on the algebra M_n of complex $n \times n$–matrices there does exist a Markov process which can be constructed on Fock space (cf. Sect. 9.3). For details we refer to [Par]. However, this Markov process is not uniquely determined by its semigroup as we will see in Sect. 6.3. Moreover, if the semigroup $(T_t)_{t\geq 0}$ on $\mathcal{A}_0$ admits a stationary state φ_0, that is, $\varphi_0(T_t(x)) = \varphi_0(x)$ for $x \in \mathcal{A}_0$, $t \geq 0$, then one should expect that it comes from a stationary Markov process as it is the case for classical processes. But here we run into severe problems. They are basically due to the fact that in a truly quantum situation interesting joint distributions – states on tensor products of algebras – do not admit conditional expectations. As an illustration of this kind of problem consider the following situation.

Consider $\mathcal{A}_0 = M_n$, $2 \leq n \leq \infty$. Such an algebra $\mathcal{A}_0$ describes a truly quantum mechanical system. Moreover, consider any random variable $i : \mathcal{A}_0 \to (\mathcal{A}, \varphi)$.

Proposition 3.7. *The algebra $\mathcal{A}$ decomposes as*

$$\begin{aligned} \mathcal{A} &\simeq M_n \otimes \mathcal{C} \text{ for some algebra } \mathcal{C} \text{, such that} \\ i(x) &= x \otimes \mathbb{1} \quad \text{for all } x \in \mathcal{A}_0 = M_n \, . \end{aligned}$$

Proof: Put $\mathcal{C} := \{y \in \mathcal{A} : i(x) \cdot y = y \cdot i(x) \text{ for all } x \in \mathcal{A}_0\}$.

Moreover, the existence of a conditional expectation forces the state φ to split, too:

Proposition 3.8. *If the conditional expectation*

$$P : (\mathcal{A}, \varphi) \to i(\mathcal{A}_0) = M_n \otimes \mathbb{1}$$

exists then there is a state ψ on $\mathcal{C}$ such that

$$\varphi = \varphi_0 \otimes \psi$$

i.e., $\varphi(x \otimes y) = \varphi_0(x) \cdot \psi(y)$ for $x \in \mathcal{A}_0$, $y \in \mathcal{C}$ with $\varphi_0(x) := \varphi(x \otimes \mathbb{1})$. It follows that

$$P(x \otimes y) = \psi(y) \cdot x \otimes \mathbb{1} \, ,$$

hence P is a conditional expectation of tensor type (cf. Sect. 3.2).

Again, the proof is easy: From the module property of P it follows that P maps the relative commutant $\mathbb{1} \otimes \mathcal{C}$ of $i(\mathcal{A}_0)$ into the center of M_n, hence onto the multiples of $\mathbb{1}$; thus P on $\mathbb{1} \otimes \mathcal{C}$ defines a state ψ on $\mathcal{C}$.

Therefore, if $\mathcal{A}_0 = M_n$ then the existence of the conditional expectation $P : (\mathcal{A}, \varphi) \to \mathcal{A}_0$ forces the state to split into a product state hence the state can not represent a non-trivial joint distribution.

3.4 A Construction Scheme for Markov Processes

The discussion in the previous section seems to indicate that there are no interesting Markov processes in the truly quantum context: On the one hand we would like to have a conditional expectation onto the time zero algebra $\mathcal{A}_0$ of the process, on the other hand, if $\mathcal{A}_0 = M_n$, this condition forces the state to split into a tensor product and this prevents the state from representing an interesting joint distribution. Nevertheless, there is a way to bypass this problem. This approach to stationary Markov processes was initiated in ([Kü2]). It avoids the above problem by putting the information about the relationship between different times into the dynamics rather than into the state:

We freely use the language introduced in the previous sections. We note that the following construction can be carried out on different levels: If the algebras are merely *-algebras of operators then the tensor products are meant to be algebraic tensor products. If we work in the category of C*-algebras then we use the minimal tensor product of C*-algebras (cf. [Tak2]). In most cases, by stationarity, we can even turn to the closures in the strong operator topology and work in the category of von Neumann algebras. Then all algebras are von Neumann algebras, the states are assumed to be normal states, and the tensor products are tensor products of von Neumann algebras (cf. [Tak2]). In many cases we may even assume that the states are faithful: If a normal state is stationary for some automorphism on a von Neumann algebra then its support projection, too, is invariant under this automorphism and we may consider the restriction of the whole process to the part where the state is faithful. In particular, when the state is faithful on the initial algebra $\mathcal{A}_0$ (see below), then all states can be assumed to be faithful. On the other hand, as long as we work on an purely algebraic level or on a C*-algebraic level, the following construction makes sense even if we refrain from all stationarity assumptions.

We start with the probability space $(\mathcal{A}_0, \varphi_0)$ for the time–zero-algebra of the Markov process to be constructed. Given any further probability space $(\mathcal{C}_0, \psi_0)$ then we can form their tensor product

$$(\mathcal{A}_0, \varphi_0) \otimes (\mathcal{C}_0, \psi_0) := (\mathcal{A}_0 \otimes \mathcal{C}_0, \varphi_0 \otimes \psi_0) ,$$

where $\mathcal{A}_0 \otimes \mathcal{C}_0$ is the tensor product of $\mathcal{A}_0$ and $\mathcal{C}_0$ and $\varphi_0 \otimes \psi_0$ is the product state on $\mathcal{A}_0 \otimes \mathcal{C}_0$ determined by $\varphi_0 \otimes \psi_0(x \otimes y) = \varphi_0(x) \cdot \psi_0(y)$ for $x \in \mathcal{A}_0$,

$y \in \mathcal{C}_0$. Finally, let α_1 be any automorphism of $(\mathcal{A}_0, \varphi_0) \otimes (\mathcal{C}_0, \psi_0)$ that means that α_1 is an automorphism of the algebra $\mathcal{A}_0 \otimes \mathcal{C}_0$ which leaves the state $\varphi \otimes \psi$ invariant. From these ingredients we now construct a stationary Markov process:

There is also an infinite tensor product of probability spaces. In particular, we can form the infinite tensor product $\bigotimes_{\mathbb{Z}}(\mathcal{C}_0, \psi_0)$: The algebra $\bigotimes_{\mathbb{Z}} \mathcal{C}_0$ is the closed linear span of elements of the form $\cdots \otimes \mathbb{1} \otimes x_{-n} \otimes \cdots \otimes x_n \otimes \mathbb{1} \otimes \cdots$ and the state on such elements is defined as $\psi_0(x_{-n}) \cdot \ldots \cdot \psi_0(x_n)$ for $x_i \in \mathcal{C}_0$, $n \in \mathbb{N}$, $-n \leq i \leq n$. Then $\bigotimes_{\mathbb{Z}}(\mathcal{C}_0, \psi_0)$ is again a probability space which we denote by $(\mathcal{C}, \psi)$. Moreover, the tensor right shift on the elementary tensors extends to an automorphism S of $(\mathcal{C}, \psi)$.

We now form the probability space

$$(\mathcal{A}, \varphi) := (\mathcal{A}_0, \varphi_0) \otimes (\mathcal{C}, \psi) = (\mathcal{A}_0, \varphi_0) \otimes \Big(\bigotimes_{\mathbb{Z}}(\mathcal{C}_0, \psi_0)\Big)$$

and identify $(\mathcal{A}_0, \varphi_0) \otimes (\mathcal{C}_0, \psi_0)$ with a subalgebra of $(\mathcal{A}, \varphi)$ by identifying $(\mathcal{C}_0, \psi_0)$ with the zero factor ($n = 0$) of $\bigotimes_{\mathbb{Z}}(\mathcal{C}_0, \psi_0)$. Thus, by letting it act as the identity on all other factors of $\bigotimes_{\mathbb{Z}}(\mathcal{C}_0, \psi_0)$, we can trivially extend α_1 from an automorphism of $(\mathcal{A}_0, \varphi_0) \otimes (\mathcal{C}_0, \psi_0)$ to an automorphism of $(\mathcal{A}, \varphi)$. This extension is still denoted by α_1. Similarly, S is extended to the automorphism $Id \otimes S$ of $(\mathcal{A}, \varphi) = (\mathcal{A}_0, \varphi_0) \otimes (\mathcal{C}, \psi)$, acting as the identity on $\mathcal{A}_0 \otimes \mathbb{1} \subseteq \mathcal{A}$. Finally, we define the automorphism

$$\alpha := \alpha_1 \circ (Id \otimes S) .$$

This construction may be summarized in the following picture:

$$\begin{array}{ccc} & \left.\begin{array}{c} (\mathcal{A}_0, \varphi_0) \\ \otimes \\ \cdots \otimes (\mathcal{C}_0, \psi_0) \otimes (\mathcal{C}_0, \psi_0) \end{array}\right\} \alpha_1 & \otimes (\mathcal{C}_0, \psi_0) \otimes \cdots \\ & \xrightarrow{\qquad\qquad S \qquad\qquad} & \end{array}$$

The identification of $\mathcal{A}_0$ with the subalgebra $\mathcal{A}_0 \otimes \mathbb{1}$ of $\mathcal{A}$ gives rise to a random variable $i_0 : \mathcal{A}_0 \to (\mathcal{A}, \varphi)$. From i_0 we obtain random variables i_n for $n \in \mathbb{Z}$ by $i_n := \alpha^n \circ i_0$. Thus we obtain a stochastic process $(i_n)_{n \in \mathbb{Z}}$ which admits a time translation α. This process is stationary (α_1 as well as S preserve the state φ) and the conditional expectation $P_0 : (\mathcal{A}, \varphi) \to \mathcal{A}_0$ exists (cf. Sect. 3.2).

Theorem 3.9. *The above stochastic process $(\mathcal{A}, \varphi, (\alpha^n)_{n \in \mathbb{Z}}; \mathcal{A}_0)$ is a stationary Markov process.*

The proof is by inspection: By stationarity it is enough to show that for all x in the future algebra $\mathcal{A}_{[0}$ we have $P_{0]}(x) \in \mathcal{A}_0$. But the algebra $\mathcal{A}_{[0}$ is obviously contained in

$$\begin{array}{c} (\mathcal{A}_0, \varphi_0) \\ \otimes \\ \cdots \otimes \mathbb{1} \otimes (\mathcal{C}_0, \psi_0) \otimes (\mathcal{C}_0, \psi_0) \otimes \cdots \end{array}$$

while the past $\mathcal{A}_{0]}$ is contained in

$$\begin{array}{c} (\mathcal{A}_0, \varphi_0) \\ \otimes \\ \cdots \otimes (\mathcal{C}_0, \psi_0) \otimes \quad \mathbb{1} \quad \otimes \mathbb{1} \otimes \cdots \end{array}$$

Discussion

This construction can also be carried out in the special case, where all algebras are commutative. It then gives a construction scheme for classical Markov processes, which is different from its canonical realization on the space of its paths. It is not difficult to show that every classical discrete time stationary Markov process can be obtained in this way. However, this process may not be minimal, i.e., $\mathcal{A}_{\mathbb{Z}}$ may be strictly contained in $\mathcal{A}$.

Given the initial algebra $(\mathcal{A}_0, \varphi_0)$ then a Markov process as above is determined by the probability space $(\mathcal{C}_0, \psi_0)$ and the automorphism α_1. In particular, the transition operator can be computed from $T(x) = P_0 \circ \alpha_1(x \otimes \mathbb{1})$ for $x \in \mathcal{A}_0$. It generates the semigroup $(T^n)_{n \in \mathbb{N}}$ of transition operators on $(\mathcal{A}_0, \varphi_0)$ (cf. Section 3.3). By construction the state φ_0 is stationary, i.e., $\varphi_0 \circ T = \varphi_0$.

Conversely, given a transition operator T of $(\mathcal{A}_0, \varphi_0)$ with φ_0 stationary, if one wants to construct a corresponding stationary Markov process, then it is enough to find $(\mathcal{C}_0, \psi_0)$ and α_1 as above. This makes the problem easier compared to the original problem of guessing the whole Markov process, but it is by no means trivial. In fact, given T, there is no universal scheme for finding $(\mathcal{C}_0, \psi_0)$ and α_1, and there are some deep mathematical problems associated with their existence. On the other hand, if one refrains from the stationarity requirements then the Stinespring representation easily leads to constructions of the above type (cf. Section 10.3).

We finally remark that for $\mathcal{A}_0 = M_n$ this form of a Markov process is typical and even, in a sense, necessary. In fact there are theorems which show that if $\mathcal{A}_0 = M_n$ then an arbitrary Markov process has a structure similar to the one above: It is always a *coupling of $\mathcal{A}_0$ to a shift system.* The meaning of this will be made more precise in the next chapter. Further information can be found in [Kü3].

3.5 Dilations

The relation between a Markov process with time translations $(\alpha_t)_t$ on $(\mathcal{A}, \varphi)$ and its semigroup $(T_t)_t$ of transition operators on $\mathcal{A}_0$ can be brought into the form of a diagram:

$$\begin{array}{ccc} \mathcal{A}_0 & \xrightarrow{T_t} & \mathcal{A}_0 \\ {\scriptstyle i_0}\downarrow & & \uparrow{\scriptstyle P_0} \\ (\mathcal{A},\varphi) & \xrightarrow[\alpha_t]{} & (\mathcal{A},\varphi) \end{array}$$

This diagram commutes for all $t \geq 0$.

From this point of view the Markovian time evolution $(\alpha_t)_t$ can be understood as an extension of the *irreversible* time evolution $(T_t)_t$ on $\mathcal{A}_0$ to an evolution of *-homomorphisms or even *-automorphisms on the large algebra $\mathcal{A}$. Such an extension is referred to as a *dilation* of $(T_t)_t$ to $(\alpha_t)_t$. The paradigmatic dilation theory is the theory of unitary dilations of contraction semigroups on Hilbert spaces, defined by the commuting diagram

$$\begin{array}{ccc} \mathcal{H}_0 & \xrightarrow{T_t} & \mathcal{H}_0 \\ {\scriptstyle i_0}\downarrow & & \uparrow{\scriptstyle P_0} \\ \mathcal{H} & \xrightarrow[U_t]{} & \mathcal{H} \end{array}$$

Here $(T_t)_{t\geq 0}$ is a semigroup of contractions on a Hilbert space $\mathcal{H}_0$, $(U_t)_t$ is a unitary group on a Hilbert space $\mathcal{H}$, $i_0 : \mathcal{H}_0 \to \mathcal{H}$ is an isometric embedding, and P_0 is the Hilbert space adjoint of i_0, which may be identified with the orthogonal projection from $\mathcal{H}$ onto $\mathcal{H}_0$. The diagram has to commute for all $t \geq 0$.

There is an extensive literature on unitary dilations starting with the pioneering books [SzNF] and [LaPh]. It turned out to be fruitful to look at Markov processes and open systems from the point of view of dilations, like for example in [EvLe] and [Kü2]. In fact, the next chapter on scattering is a demonstration of this: P.D. Lax and R. S. Phillips based their approach to scattering theory in [LaPh] on unitary dilations and our original idea in [KüMa3] was to transfer some of their ideas to the theory of operator algebraic Markov processes. Meanwhile this transfer has found various interesting applications. One is to the preparation of quantum states which is discussed in Chapter 7.

There is a deeper reason why the understanding of unitary dilations can be helpful for the understanding of Markov processes as the following section will show.

3.6 Dilations from the Point of View of Categories

The relation between the above two types of dilations can be brought beyond the level of an intuitive feeling of similarity. For simplicity we discuss the case of a discrete time parameter only:

Consider a category whose objects form a class $\mathcal{O}$. For any two objects $O_1, O_2 \in \mathcal{O}$ denote by $\mathcal{M}(O_1, O_2)$ the morphisms from O_1 to O_2. By $Id_O \in \mathcal{M}(O, O)$ denote the identity morphism of an object $O \in \mathcal{O}$, which is characterized by $Id_O \circ T = T$ for all $T \in \mathcal{M}(A, O)$ and $S \circ Id_O = S$ for all $S \in \mathcal{M}(O,B)$ where A and B are any further objects in $\mathcal{O}$. Finally, a morphism $T \in \mathcal{M}(O,O)$ is called an *automorphism* of O if there exists a morphism $T^{-1} \in \mathcal{M}(O,O)$ such that $T^{-1} \circ T = Id_O = T \circ T^{-1}$.

Now we can formulate the general concept of a dilation (cf. [Kü2]):

Definition 3.10. *Given $T \in \mathcal{M}(O,O)$ for some object $O \in \mathcal{O}$ then we call a quadruple $(\hat{O}, \hat{T}; i, P)$ a dilation of (O,T) if $\hat{T} \in \mathcal{M}(\hat{O},\hat{O})$ is an automorphism of $\hat{O}$ and $i \in \mathcal{M}(O,\hat{O})$ and $P \in \mathcal{M}(\hat{O},O)$ are morphisms such that the diagram*

$$\begin{array}{ccc} O & \xrightarrow{T^n} & O \\ {\scriptstyle i}\downarrow & & \uparrow{\scriptstyle P} \\ \hat{O} & \xrightarrow[\hat{T}^n]{} & \hat{O} \end{array}$$

commutes for all $n \in \mathbb{N}_0$. Here we adopt the convention $T^0 = Id_O$ for any morphism $T \in \mathcal{M}(O,O)$.

For the special case $n = 0$ the commutativity of the dilation diagram implies $P \circ i = Id_O$. Hence $(i \circ P)^2 = i \circ P \circ i \circ P = i \circ Id_O \circ P = i \circ P$, i.e., $i \circ P \in \mathcal{M}(\hat{O},\hat{O})$ is an idempotent morphism.

Now we can specialize to the case where the objects of the category are Hilbert spaces and the morphisms are contractions between Hilbert spaces. In this category automorphisms are unitaries while idempotent morphisms are orthogonal projections. Therefore, if $\mathcal{H}_0$ is some Hilbert space, $T \in \mathcal{M}(\mathcal{H}_0,\mathcal{H}_0)$ is a contraction, and $(\mathcal{H}, U; i_0, P_0)$ is a dilation of $(\mathcal{H}_0, T)$, then U is unitary, $i_0 : \mathcal{H}_0 \to \mathcal{H}$ is an isometry, and the orthogonal projection $i_0 \circ P_0$ projects onto the subspace $i_0(\mathcal{H}_0) \subseteq \mathcal{H}$. We thus retain the definition of a unitary dilation.

On the other hand we can specialize to the category whose objects are probability spaces $(\mathcal{A}, \varphi)$ where $\mathcal{A}$ is a von Neumann algebra and φ is a faithful normal state on $\mathcal{A}$. As morphisms between two such objects $(\mathcal{A}, \varphi)$ and $(\mathcal{B}, \psi)$ we consider completely positive operators $T : \mathcal{A} \to \mathcal{B}$ which are identity preserving, i.e., $T(\mathbb{1}_\mathcal{A}) = \mathbb{1}_\mathcal{B}$, and respect the states, i.e., $\psi \circ T = \varphi$. (For further information on completely positive operators we refer to Chapter 8). In this category an automorphism of $(\mathcal{A}, \varphi)$ is a *-automorphism of $\mathcal{A}$ which leaves the state φ fixed. Moreover, an idempotent morphism P of $(\mathcal{A}, \varphi)$ turns out to be a conditional expectation onto a von Neumann subalgebra $\mathcal{A}_0$ of $\mathcal{A}$ [KüNa]. Therefore, if T is a morphism of a probability space $(\mathcal{A}_0, \varphi_0)$ and $(\mathcal{A}, \varphi, \alpha; i_0, P_0)$ is a dilation of $(\mathcal{A}_0, \varphi_0, T)$ (we omit the additional brackets around probability spaces) then $i : \mathcal{A}_0 \to \mathcal{A}$ is an injective *-homomorphism, hence a random variable, $P_0 \circ i_0$ is the conditional expectation from $(\mathcal{A}, \varphi)$ onto $i_0(\mathcal{A}_0)$, and $(\mathcal{A}, \varphi, (\alpha^n)_{n\in\mathbb{Z}}; i_0(\mathcal{A}_0))$ is a stationary stochastic process with $(\alpha^n)_{n\in\mathbb{Z}}$ as its time translation and $(T^n)_{n\in\mathbb{N}_0}$ as its transition operators. In particular, we have obtained a dilation as in the foregoing Section 3.5. Depending on the situation it can simplify notation to identify $\mathcal{A}_0$ with the subalgebra $i_0(\mathcal{A}_0) \subseteq \mathcal{A}$ and we will freely do so, whenever it seems to be convenient.

This discussion shows that unitary dilations and stationary Markov processes are just two realizations of the general concept of a dilation. In fact, the relation between those two realizations is even closer: Between the two categories above there are functors in both directions which, in particular, carry dilations into dilations:

The GNS-construction associates with a probability space $(\mathcal{A}, \varphi)$ a Hilbert space $\mathcal{H}_\varphi$ which is obtained from completing $\mathcal{A}$ with respect to the scalar product $<x, y>_\varphi := \varphi(y^*x)$ for $x, y \in \mathcal{A}$. A morphism $T : (\mathcal{A}, \varphi) \to (\mathcal{B}, \psi)$ is turned into a contraction $T_{\varphi,\psi} : \mathcal{H}_\varphi \to \mathcal{H}_\psi$, as follows from the Cauchy-Schwarz inequality for completely positive operators (cf. Chapter 8). Thus the GNS-construction turns a dilation of $(\mathcal{A}, \varphi, T)$ into a unitary dilation of $(\mathcal{H}_\varphi, T_{\varphi,\varphi})$. However this functorial relation is of minor interest, since in general this unitary dilation is far from being unique.

There are, however, several interesting functors into the other direction. We sketch only briefly some of them:

Given a Hilbert space $\mathcal{H}$ there is, up to stochastic equivalence, a unique family of real valued centered Gaussian random variables $\{X(\xi) : \xi \in \mathcal{H}\}$ on some probability space (Ω, Σ, μ), such that $\mathcal{H} \ni \xi \to X(\xi)$ is linear and $\mathbb{E}(X(\xi) \cdot X(\eta)) = <\xi, \eta>$ for $\xi, \eta \in \mathcal{H}$. Assuming that the σ-algebra Σ is already generated by the random variables $\{X(\xi) : \xi \in \mathcal{H}\}$ we obtain an object $(\mathcal{A}, \varphi)$ with $\mathcal{A} = L^\infty(\Omega, \Sigma, \mu)$ and $\varphi(f) = \int_\Omega f d\mu$ for $f \in \mathcal{A}$. Moreover, consider two Hilbert spaces $\mathcal{H}$ and $\mathcal{K}$ leading, as above, to two families of Gaussian random variables $\{X(\xi) : \xi \in \mathcal{H}\}$ and $\{Y(\eta) : \eta \in \mathcal{K}\}$ on probability spaces $(\Omega_1, \Sigma_1, \mu_1)$ and $(\Omega_2, \Sigma_2, \mu_2)$, respectively. It follows from the theory of Gaussian random variables (cf. [Hid]) that to a contraction $T : \mathcal{H} \to \mathcal{K}$ there is canonically associated a positive identity preserving operator $\tilde{T} : L^1(\Omega_1, \Sigma_1, \mu_1) \to L^1(\Omega_2, \Sigma_2, \mu_2)$ with $\tilde{T}(X(\xi)) = Y(T\xi)$ $(\xi \in \mathcal{H})$ which maps $L^\infty(\Omega_1, \Sigma_1, \mu_1)$ into $L^\infty(\Omega_2, \Sigma_2, \mu_2)$. It thus leads to a morphism $T : (\mathcal{A}, \varphi) \to (\mathcal{B}, \psi)$ with $\mathcal{A} := L^\infty(\Omega_1, \Sigma_1, \mu_1)$, $\varphi(f) := \int_{\Omega_1} f \, d\mu_1$ for $f \in \mathcal{A}$, and $\mathcal{B} := L^\infty(\Omega_2, \Sigma_2, \mu_2)$, $\psi(g) := \int_{\Omega_2} g \, d\mu_2$ for $g \in \mathcal{B}$. Therefore, this 'Gaussian functor' carries unitary dilations into classical Gaussian Markov processes, usually called Ornstein-Uhlenbeck processes.

Similarly, there are functors carrying Hilbert spaces into non-commutative probability spaces. The best known of these functors come from the theory of canonical commutation relations (CCR) and from canonical anticommutation relations (CAR). In both cases, fixing an 'inverse temperature' $\beta > 0$, to a Hilbert space $\mathcal{H}$ there is associated a von Neumann algebra $\mathcal{A}$ of canonical commutation relations or anticommutation relations, respectively, which is equipped with a faithful normal state φ_β, called the equilibrium state at inverse temperature β (for the CCR case this functor is used in our discussion in Section 4.6). Again, contractions between Hilbert spaces are carried into morphisms between the corresponding probability spaces. Hence unitary dilations are carried into non-commutative stationary Markov processes. For details we refer to [EvLe] and [Eva]. An extension of these functors to the case of q-commutation relations has been studied in [BKS].

In order to provide a unified language for all these situations we make the following definition.

Definition 3.11. *Consider a functor which carries Hilbert spaces as objects into probability spaces of the form* $(\mathcal{A}, \varphi)$ *with* $\mathcal{A}$ *a von Neumann algebra and* φ *a faithful normal state on* $\mathcal{A}$*, and which carries contractions between Hilbert spaces into morphisms between such probability spaces. Such a functor is called a* functor of white noise, *if, in addition, the trivial zero-dimensional Hilbert space is carried into the trivial one-dimensional von Neumann algebra* $\mathbb{C}\mathbb{1}$ *and if families of contractions between Hilbert spaces which converge in the strong operator topology are carried into morphisms which converge in the pointwise strong operator topology.*

The name *functor of white noise* will become in Section 4.3. From the above discussion it is already clear that unitaries are carried into automorphisms while orthogonal projections are carried into conditional expectations ([KüNa]). In particular, subspaces of a Hilbert space correspond to subalgebras of the corresponding von Neumann algebra. Moreover, orthogonal subspaces correspond to independent subalgebras in the sense described in Section 4.3. The functor is called *minimal* if the algebra corresponding to some Hilbert space $\mathcal{H}$ is algebraically generated by the subalgebras corresponding to Hilbert subspaces of $\mathcal{H}$ which generate $\mathcal{H}$ linearly. The continuity assumption could be omitted but it assures that, in particular, strongly continuous unitary groups are carried into pointwise weak*-continuous groups of automorphisms. Finally, we will see in the next section that a unitary dilation is carried into a stationary Markov process by any such functor.

All functors mentioned above are minimal functors of white noise.

4 Scattering for Markov Processes

The Markov processes constructed in Section 3.4 above have a particular structure which we call "coupling to white noise". The part $(\mathcal{C}, \psi, S)$ is a non-commutative Bernoulli shift, i.e., a white noise in discrete time, to which the system algebra $\mathcal{A}_0$ is coupled via the automorphism α_1. Thus the evolution α of the whole Markov process may be considered as a perturbation of the white noise evolution S by the coupling α_1. By means of scattering theory we can compare the evolution α with the "free evolution" S. The operator algebraic part of the following material is taken from [KüMa3] to which we refer for further details and proofs.

4.1 On the Geometry of Unitary Dilations

Before entering into the operator algebraic discussion it may be useful to have a more detailed look at the geometry of unitary dilations. On the one hand this shows that the particular structure of the Markov processes constructed in Section 3.4 is more natural than it might seem at a first glance. On the other hand these considerations will motivate the operator algebraic discussions to come.

It should be clear from the above discussion about categories that the Hilbert space analogue of a two-sided stationary stochastic process with time translation in discrete time is given by a triple $(\mathcal{H}, U; \mathcal{H}_0)$ where $\mathcal{H}$ is a Hilbert space, $U : \mathcal{H} \to \mathcal{H}$ is a unitary and $\mathcal{H}_0 \subseteq \mathcal{H}$ is a distinguished subspace. This subspace describes the 'time zero' part, $U^n\mathcal{H}_0$ the 'time n part' of this process. If $P_0 : \mathcal{H} \to \mathcal{H}_0$ denotes the orthogonal projection from $\mathcal{H}$ onto $\mathcal{H}_0$ then the operators $T_n : \mathcal{H}_0 \to \mathcal{H}_0$ with $T_n := P_0 U^n P_0$, $n \in \mathbb{Z}$, are the Hilbert space versions of the transition operators of a stochastic process. In general, the family $(T_n)_{n\in\mathbb{N}_0}$ will not form a semigroup, i.e., T_n may well be different from T_1^n for $n \geq 2$. Still, the process $(\mathcal{H}, U; \mathcal{H}_0)$ may be called a unitary dilation of $(\mathcal{H}_0, (T_n)_{n\in\mathbb{Z}})$, which now means that the diagram

$$\begin{array}{ccc} \mathcal{H}_0 & \xrightarrow{T_n} & \mathcal{H}_0 \\ {\scriptstyle i_0}\downarrow & & \uparrow{\scriptstyle P_0} \\ \mathcal{H} & \underset{U^n}{\longrightarrow} & \mathcal{H} \end{array}$$

commutes for all $n \in \mathbb{Z}$. Here we identify $\mathcal{H}_0$ via the isometry i_0 with a subspace of $\mathcal{H}$. The following theorem characterizes the families $(T_n)_{n\in\mathbb{Z}}$ of operators on $\mathcal{H}_0$ which allow a unitary dilation in the sense above:

Theorem 4.1. *[SzNF] For a family $(T_n)_{n\in\mathbb{Z}}$ of contractions of $\mathcal{H}_0$ the following conditions are equivalent:*

a) $(\mathcal{H}_0, (T_n)_{n\in\mathbb{Z}})$ has a unitary dilation.
b) $T_0 = \mathbb{1}_{\mathcal{H}_0}$ and the family $(T_n)_{n\in\mathbb{Z}}$ is positive definite *, i.e., for all $n \in \mathbb{N}$ and for all choices of vectors $\xi_1, \ldots, \xi_n \in \mathcal{H}_0$:*

$$\sum_{i,j=1}^{n} < T_{i-j}\, \xi_i, \xi_j > \; \geq \; 0 \, .$$

Moreover, if the unitary dilation is minimal, *i.e., if $\mathcal{H}$ is the closed linear span of $\{U^n\xi : \xi \in \mathcal{H}_0,\ n \in \mathbb{Z}\}$, then the unitary dilation is uniquely determined up to unitary equivalence.*

If $T : \mathcal{H}_0 \to \mathcal{H}_0$ is a contraction and if we define $T_n := T^n$ for $n \geq 0$ and $T_n := (T^{-n})^*$ for $n < 0$ then this family $(T_n)_{n\in\mathbb{Z}}$ is positive definite and thus it has a unitary dilation $(\mathcal{H}, U; \mathcal{H}_0)$ (cf. [SzNF]). In slight abuse of language we call $(\mathcal{H}, U; \mathcal{H}_0)$ a unitary dilation of $(\mathcal{H}_0, T)$ also in this case.

In order to understand the geometry of such a unitary dilation we define for a general triple $(\mathcal{H}, U; \mathcal{H}_0)$ as above and for any subset $I \subseteq \mathbb{Z}$ the subspace

$\mathcal{H}_I$ as the closed linear span of $\{U^n\xi : \xi \in \mathcal{H}_0, n \in I\}$ and $P_I : \mathcal{H} \to \mathcal{H}_I$ as the orthogonal projection from $\mathcal{H}$ onto $\mathcal{H}_I$. For simplicity we denote $\mathcal{H}_{\{n\}}$ by $\mathcal{H}_n$ and $P_{\{n\}}$ by P_n for $n \in \mathbb{Z}$, too.

The following observation describes the geometry of a unitary dilation of $(\mathcal{H}_0, T)$:

Proposition 4.2. *For a unitary dilation $(\mathcal{H}, U; \mathcal{H}_0)$ of a positive definite family $(T_n)_{n\in\mathbb{Z}}$ the following conditions are equivalent:*

a) $(\mathcal{H}, U; \mathcal{H}_0)$ is a unitary dilation of a semigroup, i.e., $T_n = T_1^n$ for $n \in \mathbb{N}$.
b) For all $\xi \in \mathcal{H}_0$ and for all $n, m \in \mathbb{N}$: $U^m P_0^\perp U^n \xi$ is orthogonal to $\mathcal{H}_0$.
c) For all $\xi \in \mathcal{H}_{[0,\infty[}$ we have $P_{]-\infty,0]}(\xi) = P_0(\xi)$.

Here, $P_0^\perp$ denotes the orthogonal projection $\mathbb{1} - P_0$ onto the orthogonal complement $\mathcal{H}_0^\perp$ of $\mathcal{H}_0$. Condition b) can be roughly rephrased by saying that the part of the vector $U^n\xi$ which is orthogonal to $\mathcal{H}_0$, i.e., which 'has left' $\mathcal{H}_0$, will stay orthogonal to $\mathcal{H}_0$ at all later times, too. We therefore refer to this condition as the 'they never come back principle'. Condition c) is the linear version of the Markov property as formulated in Section 3.3.

Proof: Given $\xi \in \mathcal{H}_0$ and $n, m \geq 0$ we obtain

$$\begin{aligned}
T_{n+m}\xi &= P_0 U^{n+m}\xi = P_0 U^n U^m \xi = P_0 U^n (P_0 + P_0^\perp) U^m \xi \\
&= P_0 U^n P_0 U^m \xi + P_0 U^n P_0^\perp U^m \xi \\
&= T_n T_m \xi + P_0 U^n P_0^\perp U^m \xi \,.
\end{aligned}$$

Thus $T_{n+m} = T_n T_m$ if and only if $P_0 U^n P_0^\perp U^m \xi = 0$ for all $\xi \in \mathcal{H}_0$, which proves the equivalence of a) and b).

In order to prove the implication $b) \Rightarrow c)$ decompose $\eta := U^n\xi$ with $\xi \in \mathcal{H}_0$, $n \geq 0$, as

$$\eta = P_0\eta + P_0^\perp \eta \,.$$

By assumption, we have for all $\zeta \in \mathcal{H}_0$:

$$0 = \langle U^m P_0^\perp \eta, \zeta \rangle = \langle P_0^\perp \eta, U^{-m}\zeta \rangle \,,$$

hence $P_0^\perp \eta$ is orthogonal to $\mathcal{H}_{]-\infty,0]}$ as this holds for all $m \geq 0$; it follows that

$$P_{]-\infty,0]}\eta = P_{]-\infty,0]} P_0 \eta + P_{]-\infty,0]} P_0^\perp \eta = P_0 \eta \,.$$

Since the set of these vectors η is total in $\mathcal{H}_{[0,\infty[}$ the assertion holds for all $\eta \in \mathcal{H}_{[0,\infty[}$.

Finally, in order to deduce condition a) from condition c) we 'apply' U^n to condition c) and find

$$P_{]-\infty,n]}\xi = P_n \xi$$

for all $\xi \in \mathcal{H}_{[n,\infty[}$. Therefore, we obtain for $\xi \in \mathcal{H}_0$ and $n, m \geq 0$:

$$\begin{aligned} T_{n+m}\xi &= P_0 U^{n+m}\xi = P_0 U^n U^m \xi \\ &= P_0 P_{]-\infty,n]} U^n U^m \xi = P_0 P_n U^n U^m \xi \\ &= P_0 U^n P_0 U^m \xi = P_0 U^n T_m \xi \\ &= T_n T_m \xi. \end{aligned}$$

□

It should be noted that the above result and its proof hold in continuous time as well.

Corollary 4.3. *A (minimal) functor of white noise carries a unitary dilation of a semigroup into a stationary Markov process.*

The proof is immediate from the above condition c) as such a functor translates the linear Markov property into the Markov property as defined in Section 3.3. Moreover, it is clear that such a functor carries the semigroup of the unitary dilation into the semigroup of transition operators of the corresponding Markov process. Finally, we remark that an Ornstein-Uhlenbeck process is obtained by applying the Gaussian functor as above to a unitary dilation.

The above geometric characterization of unitary dilations of semigroups can be used in order to guess such a unitary dilation: Start with a contraction $T : \mathcal{H}_0 \to \mathcal{H}_0$ and assume that $(\mathcal{H}, U; \mathcal{H}_0)$ is a unitary dilation of $(\mathcal{H}_0, T)$. First of all the unitary U has to compensate the defect by which T differs from a unitary. This defect can be determined as follows: Given $\xi \in \mathcal{H}_0$ we obtain

$$\begin{aligned} \|U\xi\|^2 - \|T\xi\|^2 &= <\xi,\xi> - <T\xi,T\xi> = <\xi,\xi> - <T^*T\xi,\xi> \\ &= <\mathbb{1} - T^*T\xi,\xi> \\ &= \|\sqrt{\mathbb{1} - T^*T}\xi\|^2 . \end{aligned}$$

Therefore,

$$\begin{pmatrix} T \\ \\ \sqrt{\mathbb{1} - T^*T} \end{pmatrix} : \mathcal{H}_0 \mapsto \begin{matrix} \mathcal{H}_0 \\ \oplus \\ \mathcal{H}_0 \end{matrix}$$

is an isometry. (We write operators on direct sums of copies of $\mathcal{H}_0$ as block matrices with entries from $B(\mathcal{H}_0)$.)

The easiest way to complete this isometry in order to obtain a unitary is by putting

$$U_1 := \begin{pmatrix} T & -\sqrt{\mathbb{1} - TT^*} \\ \\ \sqrt{\mathbb{1} - T^*T} & T^* \end{pmatrix} \quad \text{on} \quad \begin{matrix} \mathcal{H}_0 \\ \oplus \\ \mathcal{H}_0 \end{matrix}$$

A short computation is necessary in order to show that $T\sqrt{\mathbb{1} - T^*T} = \sqrt{\mathbb{1} - TT^*}\, T$, hence U_1 is indeed a unitary. Identifying the original copy of $\mathcal{H}_0$

with the upper component of this direct sum we obviously have $P_0 U_1 P_0 = T$. On the other hand, if T was not already an isometry then $P_0 U_1^2 P_0$ would differ from T^2. The reason is that in this case the 'they never come back principle' from the above proposition is obviously violated. In order to get it satisfied we need to take care that elements after once having left the upper copy of $\mathcal{H}_0$ and hence having arrived at the lower copy of $\mathcal{H}_0$ are not brought back into the upper copy of $\mathcal{H}_0$, in other words, they have to be brought away. The easiest way to accomplish this is just to shift away these elements. But also the elements having been shifted away are not allowed to come back, so they have to be shifted further. Continuing this way of reasoning and also taking care of negative times one finally arrives at a unitary dilation which has a structure analogously to the one of the Markov process in Section 3.4: Put

$$\mathcal{H} := \mathcal{H}_0 \oplus \Big(\bigoplus_{\mathbb{Z}} \mathcal{H}_0\Big) = \mathcal{H}_0 \oplus l^2(\mathbb{Z}; \mathcal{H}_0) .$$

Let U_1 act on $\mathcal{H}_0 \oplus \mathcal{H}_0^0$ where $\mathcal{H}_0^0$ denotes the zero'th summand of $\bigoplus_{\mathbb{Z}} \mathcal{H}_0$, and extend U_1 trivially to a unitary on all of $\mathcal{H}$ by letting it act as the identity on the other summands. Denote by S the right shift on $\bigoplus_{\mathbb{Z}} \mathcal{H}_0 = l^2(\mathbb{Z}; \mathcal{H}_0)$ and extend it trivially to a unitary by letting it act as the identity on the summand $\mathcal{H}_0 \oplus 0$. Finally, put $U := U_1 \circ S$ and define $i_0 : \mathcal{H}_0 \ni \xi \mapsto \xi \oplus 0 \in \mathcal{H}$, where the 0 is the zero in $l^2(\mathbb{Z}, \mathcal{H}_0)$, and put $P_0 := i^*$. This construction may be summarized by the following picture:

$$\begin{array}{c} \mathcal{H}_0 \\ \oplus \\ \cdots \oplus \mathcal{H}_0 \oplus \mathcal{H}_0 \end{array}\Bigg\}\, U_1 \quad \oplus \mathcal{H}_0 \oplus \cdots$$
$$\xrightarrow[\hspace{8em}]{}$$
$$S$$

By the above reasoning it is clear that $(\mathcal{H}, U; i_0, P_0)$ is a unitary dilation of $(\mathcal{H}_0, T)$. In general, this unitary dilation will not be minimal, but this can easily be corrected: Put $\mathcal{L} := \overline{\sqrt{\mathbb{1} - TT^*}\mathcal{H}_0}$ and $\mathcal{K} := \overline{\sqrt{\mathbb{1} - T^*T}\mathcal{H}_0}$ where the bar denotes the closure. If we substitute in the above picture the copies of $\mathcal{H}_0$ by $\mathcal{L}$ for $n \geq 0$ and by $\mathcal{K}$ for $n < 0$ so that the whole space $\mathcal{H}$ is now of the form

$$\begin{array}{c} \mathcal{H}_0 \\ \oplus \\ \cdots \oplus \mathcal{K} \oplus \ \mathcal{L} \ \oplus \mathcal{L} \oplus \cdots \end{array}$$

then the unitary U as a whole is still well defined on this space and the dilation will be minimal. For more details on the structure of unitary dilations of semigroups in discrete and in continuous time we refer to [KüS1].

4.2 Scattering for Unitary Dilations

In the above situation the unitary U might be considered as a perturbation of the free evolution S, which is a shift, by the local perturbation U_1. This is a simple example of the situation which is discussed in the Lax-Phillips approach to scattering theory in [LaPh]. One way to compare the evolutions U and S is to consider the wave operator

$$\Phi_- := \lim_{n\to\infty} S^{-n}U^n ,$$

if it exists. On $\xi \in \mathcal{H}_{[0,\infty[} \cap \mathcal{H}_0^\perp$ we have $U\xi = S\xi$, hence $\Phi_-\xi = \xi$ for such ξ. From this observation it is almost immediate to conclude that

$$\lim_{n\to\infty} S^{-n}U^n i_0(\xi)$$

exists for $\xi \in \mathcal{H}_0$ if and only if $\lim_{n\to\infty} T^n\xi$ exists. From this one easily derives the following result:

Proposition 4.4. *In the above situation the following conditions are equivalent:*

a) $\Phi_- := \lim_{n\to\infty} S^{-n}U^n$ *exists in the strong operator topology and* $\Phi_-(\mathcal{H}) \subseteq \mathcal{H}_0^\perp$.

b) $\lim_{n\to\infty} T^n = 0$ *in the strong operator topology.*

If this is the case then $\Phi_- U = S|_{\mathcal{H}_0^\perp}\Phi_-$. Since $S|_{\mathcal{H}_0^\perp}$ is a shift, it follows, in particular, that U is unitarily equivalent to a shift.

The following sections intend to develop an analogous approach for Markov processes. They give a review of some of the results obtained in [KüMa3].

4.3 Markov Processes as Couplings to White Noise

For the following discussion we assume that all algebras are von Neumann algebras and all states are faithful and normal.

Independence

On a probability space $(\mathcal{A}, \varphi)$ we frequently will consider the topology induced by the norm $\|x\|_\varphi^2 := \varphi(x^*x)$, which on bounded sets of $\mathcal{A}$ agrees with the $s(\mathcal{A}, \mathcal{A}_*)$ topology or the strong operator topology ($\mathcal{A}_*$ denotes the predual of the von Neumann algebra $\mathcal{A}$).

Definition 4.5. *Given* $(\mathcal{A}, \varphi)$ *then two von Neumann subalgebras* $\mathcal{A}_1$ *and* $\mathcal{A}_2$ *of* $\mathcal{A}$ *are* independent *subalgebras of* $(\mathcal{A}, \varphi)$ *or* independent with respect to φ, *if there exist conditional expectations* P_1 *and* P_2 *from* $(\mathcal{A}, \varphi)$ *onto* $\mathcal{A}_1$ *and* $\mathcal{A}_2$, *respectively, and if*

$$\varphi(x_1x_2) = \varphi(x_1)\varphi(x_2)$$

for any elements $x_1 \in \mathcal{A}_1$, $x_2 \in \mathcal{A}_2$.

Independence of subalgebras may be considered as an algebraic analogue to orthogonality of subspaces in Hilbert space theory. Indeed, it is a short exercise to prove that a functor of white noise as discussed in Section 3.5 will always turn orthogonal subspaces of a Hilbert space into independent subalgebras.

The typical example of independence is the situation where

$$(\mathcal{A}, \varphi) = (\mathcal{A}_1 \otimes \mathcal{A}_2, \varphi_1 \otimes \varphi_2) ;$$

then $\mathcal{A}_1 \otimes \mathbb{1}$ and $\mathbb{1} \otimes \mathcal{A}_2$ are independent.

There are, however, very different examples of independence. Another example is obtained by taking $\mathcal{A}$ as the II_1-factor of the free group with two generators a and b, equipped with the trace, and $\mathcal{A}_1$ and $\mathcal{A}_2$ as the commutative subalgebras generated by the unitaries U_a and U_b, respectively, representing the generators a and b. In this case $\mathcal{A}_1$ and $\mathcal{A}_2$ are called freely independent. Other examples of independence are studied in [BKS], [KüMa2]. A more detailed discussion of independence is contained in [Kü3] and in [KüMa2].

White Noise

Roughly speaking *white noise* means that we have a stochastic process where subalgebras for disjoint times are independent. In continuous time, however, we cannot have a continuous time evolution on the one hand and independent subalgebras of observables for each individual time $t \in \mathbb{R}$ on the other hand. Therefore, in continuous time the notion of a stochastic process is too restrictive for our purpose and we have to consider subalgebras for time intervalls instead of for individual times. This is the idea behind the following definition. It should be interpreted as our version of white noise as a generalized stationary stochastic process as it is formulated for the classical case in [Hid].

Definition 4.6. *A (non-commutative) white noise in time $\mathbb{T} = \mathbb{Z}$ or $\mathbb{T} = \mathbb{R}$ is a quadruple $(\mathcal{C}, \psi, S_t; \mathcal{C}_{[0,t]})$ where $(\mathcal{C}, \psi)$ is a probability space, $(S_t)_{t \in \mathbb{T}}$ is a group of automorphisms of $(\mathcal{C}, \psi)$, pointwise weak*-continuous in the case $\mathbb{T} = \mathbb{R}$, and for each $t \in \mathbb{T}$, $t \geq 0$, $\mathcal{C}_{[0,t]}$ is a von Neumann subalgebra of $\mathcal{C}$ such that*

(i) $\mathcal{C}$ is generated by the subalgebras $\{ S_s(\mathcal{C}_{[0,t]}) \mid t \geq 0, s \in \mathbb{T} \}$;
(ii) $\mathcal{C}_{[0,s+t]}$ is generated by $\mathcal{C}_{[0,s]}$ and $S_s(\mathcal{C}_{[0,t]})$, $(s,t \geq 0)$;
(iii) $\mathcal{C}_{[0,s]}$ and $S_r(\mathcal{C}_{[0,t]})$ are independent subalgebras of $(\mathcal{C}, \psi)$ whenever $s,t \geq 0$ and $r > s$.

In such a situation we can define the algebras $\mathcal{C}_{[s,t]} := S_s(\mathcal{C}_{[0,t-s]})$ whenever $s \leq t$. Then subalgebras associated with disjoint time intervals are independent. For an open interval I we denote by $\mathcal{C}_I$ the union of all subalgebras $\mathcal{C}_J$ with the interval $J \subset I$ closed.

Classical examples in discrete time are provided by Bernoulli systems with n states in $X := \{1, \cdots, n\}$ and probability distribution $\mu := \{\lambda_1, \cdots, \lambda_n\}$ on X. Define $\mathcal{C} := L^\infty(X^{\mathbb{Z}}, \mu^{\mathbb{Z}})$, denote by S the map on $\mathcal{C}$ which is induced by the coordinate left shift on $X^{\mathbb{Z}}$, and define $\mathcal{C}_{[0,t]}$ as the set of all functions in $\mathcal{C}$ which depend only on the time interval $[0, t]$. Then $(\mathcal{C}, \psi, S_t; \mathcal{C}_{[0,t]})$ is a white noise in the sense of the above definition.

This example is canonically generalised to the algebraic and non-commutative setting: one starts with some non-commutative probability space $(\mathcal{C}_0, \psi_0)$, defines $(\mathcal{C}, \psi)$ as the infinite tensor product $\bigotimes_{\mathbb{Z}}(\mathcal{C}_0, \psi_0)$ with respect to the infinite product state $\bigotimes_{\mathbb{Z}} \psi_0$, S as the tensor right shift on $\mathcal{C}$, and $\mathcal{C}_{[0,t]}$ as the subalgebra generated by operators of the form $\cdots \mathbb{1} \otimes \mathbb{1} \otimes x_0 \otimes x_1 \otimes \cdots \otimes x_t \otimes \mathbb{1} \otimes \cdots$ in $\mathcal{C}$. Then $(\mathcal{C}, \psi, S_t; \mathcal{C}_{[0,t]})$ is a white noise. If $\mathcal{C}_0$ is commutative and finite dimensional then this example reduces to the previous one.

Other non-commutative examples can be constructed by using other forms of independence, cf., e.g., [Kü3], [KüMa2].

As examples in continuous time one has, as the continuous analogue of a Bernoulli system, classical white noise as it is discussed in [Hid]. Non-commutative Boson white noise on the CCR algebra may be considered as the continuous analogue of a non-commutative Bernoulli shift. Similarly, there is the non-commutative Fermi white noise on the CAR algebra. Again, more examples can be provided, such as free white noise and q-white noise [BKS].

In our algebraic context, white noise will play the same role which is played by the two-sided Hilbert space shift systems on $L^2(\mathbb{R}; \mathcal{N})$ or $l^2(\mathbb{Z}; \mathcal{N})$ in the Hilbert space context, where $\mathcal{N}$ is some auxiliary Hilbert space (cf. [SzNF], [LaPh]). Indeed, any minimal functor of white noise will carry such a Hilbert space shift system into a white noise in the sense of our definition. In particular, Gaussian white noise as it is discussed in [Hid] is obtained by applying the Gaussian functor to the Hilbert space shift system $L^2(\mathbb{R})$, equipped with the right translations. This explains the name 'functor of white noise' we have chosen for such a functor.

Couplings to White Noise

Consider a two-sided stochastic process $(\mathcal{A}, \varphi, (\alpha_t)_{t\in\mathbb{T}}; \mathcal{A}_0)$ indexed by time $\mathbb{T} = \mathbb{Z}$ or $\mathbb{R}$. For short we simply write $(\mathcal{A}, \varphi, \alpha_t; \mathcal{A}_0)$ for such a process. We assume that the conditional expectation $P_0 : (\mathcal{A}, \varphi) \to \mathcal{A}_0$ exists. It follows from [Tak1] that also the conditional expectations $P_I : (\mathcal{A}, \varphi) \to \mathcal{A}_I$ exist for any time interval I.

The following definition axiomatizes a type of Markov process of which the Markov processes constructed above are paradigmatic examples.

Definition 4.7. *A stationary process $(\mathcal{A}, \varphi, \alpha_t; \mathcal{A}_0)$ is a coupling to white noise if there exists a von Neumann subalgebra $\mathcal{C}$ of $\mathcal{A}$ and a (weak*-continuous) group of automorphisms $(S_t)_{t\in\mathbb{T}}$ of $(\mathcal{A}, \varphi)$ such that*

(i) $\mathcal{A}$ *is generated by* $\mathcal{A}_0$ *and* $\mathcal{C}$ *;*
(ii) $\mathcal{A}_0$ *and* $\mathcal{C}$ *are independent subalgebras of* $(\mathcal{A}, \varphi)$*;*
(iii) There exist subalgebras $\mathcal{C}_{[0,t]}$, $t \geq 0$, *of* $\mathcal{C}$ *such that* $(\mathcal{C}, S_t|_\mathcal{C}, \varphi|_\mathcal{C}; \mathcal{C}_{[0,t]})$ *is a white noise and* $S_t|_{\mathcal{A}_0}$ *is the identity;*
iv) For all $t \geq 0$ *the map* α_t *coincides with* S_t *on* $\mathcal{C}_{[0,\infty)}$ *and on* $\mathcal{C}_{(-\infty,-t)}$, *whereas* α_t *maps* $\mathcal{A}_0 \vee \mathcal{C}_{[-t,0]}$ *into* $\mathcal{A}_0 \vee \mathcal{C}_{[0,t]}$ *;*
(v) $\mathcal{A}_{[0,t]} \subset \mathcal{A}_0 \vee \mathcal{C}_{[0,t]}$.

Here $\mathcal{A} \vee \mathcal{B}$ denotes the von Neumann subalgebra generated by von Neumann subalgebras $\mathcal{A}$ and $\mathcal{B}$.

It it obvious that the Markov processes constructed in Section 3.4 give examples of couplings to white noise. Examples of independence other than tensor products lead to other examples of couplings to white noise. Indeed, whenever we apply a minimal functor of white noise to a unitary dilation as described in Section 4.1 then the result will be a coupling to white noise. This is the reason why we work with these abstract notions of couplings to white noise. It is easy to see that whenever a stationary process is a coupling to white noise in the above sense then it will be a Markov process.

In such a situation we define the *coupling operators* $C_t := \alpha_t \circ S_{-t}$ for $t \geq 0$. So $\alpha_t = C_t \circ S_t$ and $(C_t)_{t\geq 0}$ can be extended to a cocycle of the automorphism group S_t and we consider $(\alpha_t)_{t\in\mathbb{T}}$ as a perturbation of $(S_t)_{t\in\mathbb{T}}$. Our requirements imply that $C_t|_{\mathcal{C}_{[t,\infty)}} = \mathrm{Id}$ and $C_t|_{\mathcal{C}_{(-\infty,0)}} = \mathrm{Id}$ for $t \geq 0$.

There is a physical interpretation of the above coupling structure which provides a motivation for its study. The subalgebra $\mathcal{A}_0$ of $\mathcal{A}$ may be interpreted as the algebra of observables of an open system, e.g, a radiating atom, while $\mathcal{C}$ contains the observables of the surroundings (e.g., the electromagnetic field) with which the open system interacts. Then S_t naturally describes the free evolution of the surroundings, and α_t that of the coupled system. Later in these lectures we will discuss examples of such physical systems.

4.4 Scattering

Let us from now on assume that $(\mathcal{A}, \varphi, \alpha_t; \mathcal{A}_0)$ is a Markov process which has the structure of a coupling to the white noise $(\mathcal{C}, \psi, S_t; \mathcal{C}_{[0,t]})$. We are interested in the question, under what conditions every element of $\mathcal{A}$ eventually ends up in the outgoing noise algebra $\mathcal{C}_{[0,\infty)}$. In scattering theory, this property is called *asymptotic completeness* .

In the physical interpretation of quantum optics this means that any observable of the atom or molecule can eventually be measured by observing the emitted radiation alone. Another example will be discussed in Chapter 7.

We start by defining the von Neumann subalgebra $\mathcal{A}_{\text{out}}$ of those elements in $\mathcal{A}$ which eventually end up in $\mathcal{C}_{[0,\infty)}$:

$$\mathcal{A}_{\text{out}} := \overline{\bigcup_{t\geq 0} \alpha_{-t}(\mathcal{C}_{[0,\infty)})}.$$

The closure refers to the $\|\cdot\|_\varphi$-norm. Let Q denote the conditional expectation from $(\mathcal{A}, \varphi)$ onto the outgoing noise algebra $\mathcal{C}_{[0,\infty)}$.

Lemma 4.8. *For $x \in \mathcal{A}$ the following conditions are equivalent:*

a) $x \in \mathcal{A}_{\text{out}}$.
b) $\lim_{t\to\infty} \|Q \circ \alpha_t(x)\|_\varphi = \|x\|_\varphi$.
c) $\|\cdot\|_\varphi$ -$\lim_{t\to\infty} S_{-t} \circ \alpha_t(x)$ *exists and lies in* $\mathcal{C}$.

*If these conditions hold, then the limit in (c) defines an isometric *-homomorphism* $\Phi_- : \mathcal{A}_{\text{out}} \to \mathcal{C}$.

Lemma 4.9. *For all $x \in \mathcal{C}$ the limit* $\|\cdot\|_\varphi$ -$\lim_{t\to\infty} \alpha_{-t} \circ S_t(x) =: \Omega_-(x)$ *exists and* $\Phi_-\Omega_- = \mathrm{Id}_\mathcal{C}$. *In particular,* $\Phi_- : \mathcal{A}_{\text{out}} \to \mathcal{C}$ *is an isomorphism.*

In scattering theory the operators Ω_- and Φ_-, and the related operators $\Omega_+ := \lim_{t\to\infty} \alpha_t \circ S_{-t}$ and $\Phi_+ := S_t \circ \alpha_{-t}$ (taken as strong operator limits in the $\|\cdot\|_\varphi$ - norm) are known as the *Møller operators* or *wave operators* ([LaPh]) associated to the evolutions $(S_t)_{t\in\mathbb{T}}$ and $(\alpha_t)_{t\in\mathbb{T}}$. The basic result is the following.

Theorem 4.10. *[KüMa3] For a stationary process which is a coupling to white noise the following conditions are equivalent:*

a) $\mathcal{A} = \mathcal{A}_{\text{out}}$.
b) For all $x \in \mathcal{A}_0$ we have $\lim_{t\to\infty} \|Q \circ \alpha_t(x)\|_\varphi = \|x\|_\varphi$.
c) The process has an outgoing translation representation, i.e., there exists an isomorphism $j \colon (\mathcal{A}, \varphi) \to (\mathcal{C}, \psi)$ *with* $j|_{\mathcal{C}_{[0,\infty)}} = \mathrm{Id}$ *such that* $S_t \circ j = j \circ \alpha_t$.

A stationary Markov process which is a coupling to white noise and satisfies these conditions will be called *asymptotically complete.*

4.5 Criteria for Asymptotic Completeness

In this section we shall formulate concrete criteria for the asymptotic completeness of a stationary Markov process coupled to white noise.

As before, let Q denote the conditional expectation from $(\mathcal{A}, \varphi)$ onto the outgoing noise algebra $\mathcal{C}_{[0,\infty)}$, and put $Q^\perp := \mathrm{Id}_\mathcal{A} - Q$. For $t \geq 0$, let Z_t denote the compression $Q^\perp \alpha_t Q^\perp$ of the coupled evolution to the orthogonal complement of the outgoing noise.

Lemma 4.11. *$(Z_t)_{t\geq 0}$ is a semigroup, i.e., for all $s, t \geq 0$,*

$$Z_{s+t} = Z_s \circ Z_t.$$

Now, let us note that for $a \in \mathcal{A}_0$

$$Z_t(a) = Q^\perp \alpha_t Q^\perp(a) = Q^\perp \alpha_t\big(a - \varphi(a) \cdot \mathbb{1}\big) = Q^\perp \alpha_t(a),$$

so that

$$\|Z_t(a)\|_\varphi^2 = \|a\|_\varphi^2 - \|Q\alpha_t(a)\|_\varphi^2.$$

Hence, by the above theorem asymptotic completeness is equivalent to the condition that for all $a \in \mathcal{A}_0$

$$\|Z_t(a)\|_\varphi \longrightarrow 0 \qquad \text{as} \qquad t \longrightarrow \infty\,.$$

In what follows concrete criteria are given to test this property of Z_t in the case of finite dimensional $\mathcal{A}_0$ and a tensor product structure of the coupling to white noise.

Theorem 4.12. *[KüMa3] Let $(\mathcal{A}, \varphi, \alpha_t; \mathcal{A}_0)$ be a Markov process with a finite dimensional algebra $\mathcal{A}_0$, and assume that this process is a tensor product coupling to a white noise $(\mathcal{C}, \psi, S)$. Let $Q^\perp$ and Z_t be as described above, and let $e_1, e_2, \ldots, e_n$ be an orthonormal basis of $\mathcal{A}_0$ with respect to the scalar product induced by φ on $\mathcal{A}_0$. Then the following conditions are equivalent:*

a) $\mathcal{A} = \mathcal{A}_{\text{out}}$.
b) For all $a \in \mathcal{A}_0$, $\lim_{t\to\infty} \|Z_t(a)\|_\varphi = 0$.
c) For all nonzero $a \in \mathcal{A}_0$ there exists $t \geq 0$ such that $\|Z_t(a)\|_\varphi < \|a\|_\varphi$.
d) For some $t \geq 0$, the n-tuple $\{\, Q \circ \alpha_t(e_j) \mid j = 1, 2, \cdots n \,\}$ is linearly independent.
e) For some $\varepsilon \geq 0$, $t \geq 0$, and all $x \in \mathcal{A}_{[0,\infty)}$,

$$\|Z_t x\|_\varphi \leq (1 - \varepsilon)\|x\|_\varphi.$$

4.6 Asymptotic Completeness in Quantum Stochastic Calculus

As a first application to a physical model we consider the coupling of a finite dimensional matrix algebra to Bose noise. This is a satisfactory physical model for an atom or molecule in the electromagnetic field, provided that the widths of its spectral lines are small when compared to the frequencies of the radiation the particle is exposed to. In [RoMa] this model was used to calculate the nontrivial physical phenomenon known as the 'dynamical Stark effect', namely the splitting of a fluorescence line into three parts with specified height and width ratios, when the atom is subjected to extremely strong, almost resonant radiation. The effect was calculated against a thermal radiation background, which is needed in order to ensure faithfulness of the state on the noise algebra. In the limit where the temperature of this background radiation tends to zero, the results agreed with those in the physics literature, both theoretical [Mol] and experimental [SSH].

The model mentioned above falls into the class of Markov chains with a finite dimensional algebra $\mathcal{A}_0$ driven by Bose noise, as described briefly below. In this section, we cast criterion (c) for asymptotic completeness of the above theorem into a manageable form for these Markov processes.

Although the main emphasis in these notes is put on discrete time, in the following we freely use notions from quantum stochastic calculus. Some additional information on Lindblad generators and stochastic differential equations may be found in Sect. 9.3. For a complete discussion we refer to [KüMa3].

For $\mathcal{A}_0$ we take the algebra M_n of all complex $n \times n$ matrices, on which a faithful state φ_0 is given by

$$\varphi_0(x) := \operatorname{tr}(\rho x).$$

Here, ρ is a diagonal matrix with strictly positive diagonal elements summing up to 1. The modular group of $(\mathcal{A}_0, \varphi_0)$ is given by

$$\sigma_t(x) := \rho^{-it} x \rho^{it}.$$

We shall couple the system $(\mathcal{A}_0, \varphi_0)$ to Bose noise (cf. [Par], [ApH], [LiMa]). Let $\mathcal{C}$ denote the Weyl algebra over an m-fold direct sum of copies of $L^2(\mathbb{R})$, on which the state ψ is given by

$$\psi(W(f_1 \oplus f_2 \oplus \cdots \oplus f_m)) := \exp\left(-\tfrac{1}{2}\sum_{j=1}^{m} \coth(\tfrac{1}{2}\beta_j)\|f_j\|^2\right).$$

The probability space $(\mathcal{C}, \psi)$ describes a noise source consisting of m channels which contain thermal radiation at inverse temperatures $\beta_1, \beta_2, \cdots, \beta_m$. Let the free time evolution S_t on $\mathcal{C}$ be induced by the right shift on the functions $f_1, f_2, \cdots, f_m \in L^2(\mathbb{R})$. The GNS representation of $(\mathcal{C}, \psi)$ lives on the $2m$-th tensor power of the Boson Fock space over $L^2(\mathbb{R})$ (cf. [Par]), where annihilation operators $A_j(t)$, $(j = 1, \cdots, m)$ are defined by

$$A_j(t) := (\mathbb{1} \otimes \mathbb{1}) \otimes \cdots \otimes (c_j^- A(t) \otimes \mathbb{1} - c_j^+ \mathbb{1} \otimes A(t)^*) \otimes \cdots \otimes (\mathbb{1} \otimes \mathbb{1}).$$

The operator is in the j-th position and the constants c_j^+ and c_j^- are given by

$$c_j^+ := \sqrt{\frac{e^{\beta_j}}{e^{\beta_j} + 1}}, \qquad c_j^- := \sqrt{\frac{1}{e^{\beta_j} + 1}}.$$

In [LiMa], Section 9, Markov processes $(\mathcal{A}, \varphi, \alpha_t; \mathcal{A}_0)$ are constructed by coupling to these Bose noise channels. They are of the following form.

$$\begin{aligned} \mathcal{A} &:= \mathcal{A}_0 \otimes \mathcal{C} \\ \varphi &:= \varphi_0 \otimes \psi \qquad\qquad \text{with } P_0(x \otimes y) := \psi(y)x; \\ \alpha_t(a) &:= u_t^*(\mathrm{Id} \otimes S_t)(a)u_t, \quad (t \geq 0); \qquad \alpha_t := (\alpha_{-t})^{-1}, \quad (t < 0), \end{aligned}$$

where u_t is the solution of the quantum stochastic differential equation

$$du_t = \left(\sum_{j=1}^{m} \left(v_j \otimes dA_j^*(t) - v_j^* \otimes dA_j(t) - \tfrac{1}{2}(c_j^+ v_j^* v_j + c_j^- v_j v_j^*) \otimes \mathbb{1} \cdot dt \right) \right.$$
$$\left. +(ih \otimes \mathbb{1}) \cdot dt \right) u_t,$$

with initial condition $u_0 = \mathbb{1}$. The semigroup of transition operators on $(\mathcal{A}_0, \varphi_0)$ associated to this Markov process is given by

$$P_0 \circ \alpha_t(a) =: T_t(a) = e^{-tL}(a)$$

for $a \in \mathcal{A}_0$, where the infinitesimal generator $L : \mathcal{A}_0 \to \mathcal{A}_0$ is given by

$$L(a) = i[h,a] - \tfrac{1}{2} \sum_{j=1}^{m} \left(c_j^+ (v_j^* v_j a - 2v_j^* a v_j + a v_j^* v_j) + c_j^- (v_j v_j^* a - 2 v_j a v_j^* + a v_j v_j^*) \right).$$

Here $v_j \in \mathcal{A}_0 = M_n$ must be eigenvectors of the modular group σ_t of $(\mathcal{A}_0, \varphi_0)$ and h must be fixed under σ_t.

Now, the key observation in [LiMa] and [RoMa] which we need here is the following. Let L_j^ε be the operator $x \mapsto [v_j^\varepsilon, x]$ on $\mathcal{A}_0$.

Observation. If Q is the projection onto the future noise algebra $\mathcal{C}_{[0,\infty)}$, then

$$\|Q\alpha_t(x \otimes \mathbb{1})\|^2$$
$$= \sum_{k=0}^{\infty} \sum_{j \in \{1,\cdots,m\}^k} \sum_{\varepsilon \in \{-1,1\}^k} c_{j(1)}^{\varepsilon(1)} \cdots c_{j(k)}^{\varepsilon(k)}$$
$$\int_{0 \le s_1 \le \cdots \le s_k \le t} \left| \varphi \left(T_{t-s_k} L_{j(k)}^{\varepsilon(k)} T_{s_k - s_{k-1}} \cdots T_{s_2 - s_1} L_{j(1)}^{\varepsilon(1)} T_{s_1}(x) \right) \right|^2 ds_1 \cdots ds_k.$$

Together with the above theorem this leads to the following results concerning asymptotic completeness.

Proposition 4.13. *The system* $(\mathcal{A}, \varphi, \alpha_t; \mathcal{A}_0)$ *described above is asymptotically complete if and only if for all nonzero* $x \in M_n$ *there are* $t > 0$, $k \in \mathbb{N}$, *and* $s_1, s_2, \cdots, s_k$ *satisfying* $0 \le s_1 \le \cdots \le s_k \le t$, $j(1), \cdots, j(k) \in \{1, \cdots m\}$ *and* $\varepsilon \in \{-1,1\}^m$ *such that*

$$\varphi \left(T_{t-s_k} L_{j(k)}^{\varepsilon(k)} \cdots T_{s_2 - s_1} L_{j(1)}^{\varepsilon(1)} T_{s_1}(x) \right) \neq 0.$$

In particular, if φ_0 is a trace, i.e. $\rho = \frac{1}{n}\mathbb{1}$ in the above, then $\varphi_0 \circ T_t = \varphi$ and $\varphi_0 \circ L_j^\varepsilon = 0$, so that the system can never be asymptotically complete for $n \ge 2$. This agrees with the general idea that a tracial state φ should correspond to noise at infinite temperature, i.e., to classical noise [KüMa1]. Obviously, if $\mathcal{C}$ is commutative there can be no isomorphism j between $\mathcal{C}$ and $\mathcal{C} \otimes M_n$.

Corollary 4.14. *A sufficient condition for* $(\mathcal{A}, \varphi, \alpha_t; \mathcal{A}_0)$ *to be asymptotically complete is that for all* $x \in M_n$ *there exists* $k \in \mathbb{N}$, $j \in \{1, 2, \cdots, m\}^k$, *and* $\varepsilon \in \{-1, 1\}^k$ *such that*

$$\varphi\left(L_{j(k)}^{\varepsilon(k)} \cdots L_{j(1)}^{\varepsilon(1)}\right) \neq 0.$$

In particular, the Wigner-Weisskopf atom treated in [RoMa] is asymptotically complete.

5 Markov Processes in the Physics Literature

In this chapter we compare our approach to Markov processes developed in the first three chapters with other ways of describing Markovian behaviour in the physics literature.

5.1 Open Systems

First, we compare our formalism of quantum probability with a standard discussion of open quantum systems as it can be found in a typical book on quantum optics. We will find that these approaches can be easily translated into each other. The main difference is that the discussion of open systems in physics usually uses the Schrödinger picture while we work in the Heisenberg picture which is dual to it. The linking idea is that a random variable i identifies $\mathcal{A}_0$ with the observables of an open subsystem of $(\mathcal{A}, \varphi)$.

Being more specific the description of an open system usually starts with a Hilbert space

$$\mathcal{H} = \mathcal{H}_s \otimes \mathcal{H}_b \ .$$

The total Hilbert space $\mathcal{H}$ decomposes into a Hilbert space $\mathcal{H}_s$ for the open subsystem and a Hilbert space $\mathcal{H}_b$ for the rest of the system which is usually considered as a bath .

Correspondingly, the total Hamiltonian decomposes as

$$\mathbb{H} = \mathbb{H}_s + \mathbb{H}_b + \mathbb{H}_{int} \ ,$$

more precisely,

$$\mathbb{H} = \mathbb{H}_s \otimes \mathbb{1} + \mathbb{1} \otimes \mathbb{H}_b + \mathbb{H}_{int}$$

where $\mathbb{H}_s$ is the free Hamiltonian of the system, $\mathbb{H}_b$ is the free Hamiltonian of the bath and $\mathbb{H}_{int}$ stands for the interaction Hamiltonian.

At the beginning, at time $t = 0$, the bath is usually assumed to be in an equilibrium state. Hence its state is given by a density operator ρ_b on $\mathbb{H}_b$ which commutes with $\mathbb{H}_b$: $[\rho_b, \mathbb{H}_b] = 0$.

Next, one can frequently find a sentence similar to "if the open system is in a state ρ_s then the composed system is in the state $\rho_s \otimes \rho_b$". The mapping $\rho_s \mapsto \rho_s \otimes \rho_b$ from states of the open system into states of the composed system is dual to a conditional expectation.

Indeed, if we denote by $\mathcal{A}_0$ the algebra $B(\mathcal{H}_s)$ and by $\mathcal{C}$ the algebra $B(\mathcal{H}_b)$ and if ψ_b on $\mathcal{C}$ is the state induced by ρ_b that is $\psi_b(y) = tr_b(\rho_b \cdot y)$ for $y \in \mathcal{C}$, then the mapping

$$\mathcal{A}_0 \otimes \mathcal{C} \ni x \otimes y \mapsto \psi_b(y) \cdot x \otimes \mathbb{1}$$

extends to a conditional expectation of tensor type $P = P_{\psi_b}$ from $\mathcal{A}_0 \otimes \mathcal{C}$ to $\mathcal{A}_0 \otimes \mathbb{1}$ such that

$$tr_s(\rho_s(P(x \otimes y))) = tr(\rho_s \otimes \rho_b \cdot x \otimes y)$$

where we identified $\mathcal{A}_0 \otimes \mathbb{1}$ with $\mathcal{A}_0$. This duality is an example of the type of duality discussed in Sect. 2.2.

A further step in discussing open systems is the introduction of the *partial trace* over the bath: If the state of the composed system is described by a density operator ρ on $\mathcal{H}_s \otimes \mathcal{H}_b$ (which, in general, will not split into a tensor product of density operators) then the corresponding state of the open system is given by the partial trace $tr_b(\rho)$ of ρ over $\mathcal{H}_b$. The partial trace on a tensor product $\rho = \rho_1 \otimes \rho_2$ of density matrices ρ_1 on $\mathcal{H}_s$ and ρ_2 on $\mathcal{H}_b$ is defined as

$$tr_b(\rho) = tr_b(\rho_1 \otimes \rho_2) = tr_b(\rho_2) \cdot \rho_1$$

and is extended to general ρ by linearity and continuity. It thus has the property

$$tr(\rho \cdot x \otimes \mathbb{1}) = tr_s(tr_b(\rho) \cdot x)$$

for all $x \in \mathcal{A}_0$, that is x on $\mathcal{H}_s$, and is therefore dual to the random variable

$$i : B(\mathcal{H}_s) \ni x \mapsto x \otimes \mathbb{1} \in B(\mathcal{H}_s) \otimes B(\mathcal{H}_b)\,.$$

The time evolution in the Schrödinger picture is given by $\rho \mapsto u_t \rho u_t^*$ with $u_t = \mathrm{e}^{\mathrm{i}\mathbb{H}t}$. Dual to it is the time evolution

$$x \mapsto u_t^* x u_t$$

in the Heisenberg picture which can be viewed as a time translation α_t of a stochastic process $(i_t)_t$ with $i_t(x) := \alpha_t \circ i(x)$.

Finally, the *reduced time evolution* on the states of the open system maps an initial state ρ_s of this system into

$$\rho_s(t) := tr_b(u_t \cdot \rho_s \otimes \rho_b \cdot u_t^*)\,.$$

Thus the map $\rho_s \mapsto \rho_s(t)$ is the composition of the maps $\rho_s \mapsto \rho_s \otimes \rho_b$, $\rho \mapsto u_t \rho u_t^*$, and $\rho \mapsto tr_b(\rho)$. Hence it is dual to the composition of the maps i, α_t, and P, that is to

$$T_t : \mathcal{A}_0 \mapsto \mathcal{A}_0 : x \mapsto P \circ \alpha_t \circ i(x) = P(i_t(x))$$

which is a transition operator of this stochastic process.

In almost all realistic models this stochastic process will not have the Markov property. Nevertheless, in order to make the model accessible to computations one frequently performs a so–called 'Markovian limit'. Mathematically this turns this process into a kind of Markov process. Physically, it changes the system in such a way that the dynamics of the heat bath looses its memory. Hence its time evolution would become a kind of white noise. In many cases it is not possible to perform such a limit rigorously on the whole system. In important cases one can show that at least the reduced dynamics of the open system converges to a semigroup (e.g. when performing a weak coupling limit cf. [Dav2]). Sometimes one already starts with the white noise dynamics of a heat bath and changes only the coupling (singular coupling limit cf. [KüS1]).

5.2 Phase Space Methods

In the physics literature on quantum optics one can frequently find a different approach to quantum stochastic processes: if the system under observation is mathematically equivalent to a system of one or several quantum harmonic oscillators – as it is the case for one or several modes of the quantized electromagnetic field – then phase space representations are available for the density matrices of the system. The most prominent of these representations are the P–representation, the Wigner–representation, and the Q–representation (there exist other such representations and even representations for states of other quantum systems). The idea is to represent a state by a density function, a measure, or a distribution on the phase space of the corresponding classical physical system. These density functions are interpreted as classical probability distributions although they are not always positive. This provides a tool to take advantage of ideas of classical probability:

If $(T_t)_{t\geq 0}$ on $\mathcal{A}_0$ is a semigroup of transition operators it induces a time evolution $\rho \mapsto \rho_t$ on the density operators and thus on the corresponding densities on phase space.

With a bit of luck this evolution can be treated as if it were the evolution of probabilities of a classical Markov process and the machinery of partial differential equations can be brought into play (cf. also our remarks in Section 9.1). It should be noted, however, that a phase space representation does not inherit all properties from the quantum Markov process. It is a description of Markovian behaviour on the level of a phenomenological description. But it can not be used to obtain a representation of the quantum Markov process on the space of its paths.

5.3 Markov Processes with Creation and Annihilation Operators

In the physics literature a Markov process of an open quantum system as in Sect. 5.1 is frequently given by certain families $(A_t^*)_t$ and $(A_t)_t$ of creation and annihilation operators. The relation to the above description is the following: If the open system has an algebra $\mathcal{A}_0$ of observables which contains an annihilation operator A_0 then a Markovian time evolution α_t of the composed system applies, in particular, to A_0 and gives an operator A_t. Sometimes the operators $(A_t)_t$ can be obtained by solving a quantum stochastic differential equation (cf. Sect. 9.3).

6 An Example on M_2

In this section we discuss Markov processes of the type discussed in Section 3.4 for the simplest non-commutative case. They have a physical interpretation in terms of a spin-$\frac{1}{2}$-particle in a stochastic magnetic field. More information on this example can be found in [Kül]. A continuous time version of this example is discussed in [KüS2].

6.1 The Example

We put $\mathcal{A}_0 := M_2$ and $\varphi_0 := tr$, the tracial state on M_2.

If $(\mathcal{C}_0, \psi_0)$ is any probability space then the algebra $M_2 \otimes \mathcal{C}$ is canonically isomorphic to the algebra $M_2(\mathcal{C})$ of 2×2-matrices with entries in $\mathcal{C}$: The element

$$\begin{pmatrix} x_{11} & x_{12} \\ x_{21} & x_{22} \end{pmatrix} \otimes \mathbb{1} \ \in \ M_2 \otimes \mathcal{C}$$

corresponds to

$$\begin{pmatrix} x_{11} \cdot \mathbb{1} & x_{12} \cdot \mathbb{1} \\ x_{21} \cdot \mathbb{1} & x_{22} \cdot \mathbb{1} \end{pmatrix} \ \in \ M_2(\mathcal{C}) \,,$$

while the element

$$\mathbb{1} \otimes c \ \in \ M_2 \otimes \mathcal{C} \quad (c \in \mathcal{C})$$

corresponds to

$$\begin{pmatrix} c & 0 \\ 0 & c \end{pmatrix} \ \in \ M_2(\mathcal{C}) \,.$$

Accordingly, the state $tr \otimes \psi$ on $M_2 \otimes \mathcal{C}$ is identified with

$$M_2(\mathcal{C}) \ni \begin{pmatrix} c_{11} & c_{12} \\ c_{21} & c_{22} \end{pmatrix} \mapsto \tfrac{1}{2}(\psi(c_{11}) + \psi(c_{22}))$$

on $M_2(\mathcal{C})$, and the conditional expectation P_0 from $(M_2 \otimes \mathcal{C}, tr \otimes \psi)$ onto $M_2 \otimes \mathbb{1}$ reads as

$$M_2(\mathcal{C}) \ni \begin{pmatrix} c_{11} & c_{12} \\ c_{21} & c_{22} \end{pmatrix} \mapsto \begin{pmatrix} \psi\,(c_{11}) & \psi\,(c_{12}) \\ \psi\,(c_{21}) & \psi\,(c_{22}) \end{pmatrix} \in M_2$$

when we identify $M_2 \otimes \mathbb{1}$ with M_2 itself.

In Sect. 3.4 we saw that whenever we have a non-commutative probability space $(\mathcal{C}_0, \psi_0)$ and an automorphism α_1 of $(M_2 \otimes \mathcal{C}_0, tr \otimes \psi_0)$, then we can extend this to a stationary Markov process. We begin with the simplest possible choice for $(\mathcal{C}_0, \psi_0)$: put $\Omega_0 := \{-1, 1\}$ and consider the probability measure μ_0 on Ω_0 given by $\mu_0(\{-1\}) = \frac{1}{2} = \mu_0(\{1\})$. The algebra $\mathcal{C}_0 := L^\infty(\Omega_0, \mu_0)$ is just $\mathbb{C}^2$ and the probability measure μ_0 induces the state ψ_0 on $\mathcal{C}_0$ which is given by $\psi_0(f) = \frac{1}{2}f(-1) + \frac{1}{2}f(1)$ for a vector $f \in \mathcal{C}_0$.

In this special case there is yet another picture for the algebra $M_2 \otimes \mathcal{C}_0 = M_2 \otimes \mathbb{C}^2$. It can be canonically identified with the direct sum $M_2 \oplus M_2$ in the following way. When elements of $M_2 \otimes \mathcal{C}_0 = M_2(\mathcal{C}_0)$ are written as 2×2-matrices with entries f_{ij} in $\mathcal{C}_0 = L^\infty(\Omega_0, \mu_0)$, then an isomorphism is given by

$$M_2(\mathcal{C}_0) \to M_2 \oplus M_2 : \begin{pmatrix} f_{11} & f_{12} \\ f_{21} & f_{22} \end{pmatrix} \mapsto \begin{pmatrix} f_{11}(-1) & f_{12}(-1) \\ f_{21}(-1) & f_{22}(-1) \end{pmatrix} \oplus \begin{pmatrix} f_{11}(1) & f_{12}(1) \\ f_{21}(1) & f_{22}(1) \end{pmatrix}.$$

Finally, we need to define an automorphism α_1. We introduce the following notation: a unitary u in an algebra $\mathcal{A}$ induces an *inner automorphism* $Ad\,u : \mathcal{A} \to \mathcal{A}, x \mapsto u^* \cdot x \cdot u$. For any real number ω we define the unitary $w_\omega := \begin{pmatrix} 1 & 0 \\ 0 & e^{i\omega} \end{pmatrix} \in M_2$. It induces the inner automorphism

$$Ad\,w_\omega : M_2 \to M_2, \quad \begin{pmatrix} x_{11} & x_{12} \\ x_{21} & x_{22} \end{pmatrix} \mapsto \begin{pmatrix} x_{11} & x_{12}e^{i\omega} \\ x_{21}e^{-i\omega} & x_{22} \end{pmatrix}.$$

Now, for some fixed ω define the unitary $u := w_{-\omega} \oplus w_\omega \in M_2 \oplus M_2 = M_2 \otimes \mathcal{C}_0$. It induces the automorphism $\alpha_1 := Ad\,u$ which is given by $Ad\,w_{-\omega} \oplus Ad\,w_\omega$ on $M_2 \oplus M_2$.

To these ingredients there corresponds a stationary Markov process as in Sect. 3.4. From the above identifications it can be immediately verified that the corresponding one–step transition operator is given by

$$T : M_2 \to M_2, \quad x = \begin{pmatrix} x_{11} & x_{12} \\ x_{21} & x_{22} \end{pmatrix} \mapsto P_0 \circ \alpha_1(x \otimes \mathbb{1}) = \begin{pmatrix} x_{11} & x_{12}\rho \\ x_{21}\rho & x_{22} \end{pmatrix}$$

where $\rho = \frac{1}{2}(e^{i\omega} + e^{-i\omega}) = \cos(\omega)$.

6.2 A Physical Interpretation: Spins in a Stochastic Magnetic Field

We now show that this Markov process has a natural physical interpretation: it can be viewed as the description of a spin-$\frac{1}{2}$-particle in a stochastic magnetic field. This system is at the basis of nuclear magnetic resonance.

Spin Relaxation

We interpret the matrices σ_x, σ_y, and σ_z in M_2 as observables of (multiples of) the spin component of a spin-$\frac{1}{2}$-particle in the x-, y-, and z-directions, respectively (cf. Sect. 1.2).

If a probe of many spin-$\frac{1}{2}$-particles is brought into an irregular magnetic field in the z-direction, one finds that the behaviour in time of this probe is described by the semigroup of operators on M_2 given by

$$T_t : M_2 \to M_2 : x = \begin{pmatrix} x_{11} & x_{12} \\ x_{21} & x_{22} \end{pmatrix} \mapsto \begin{pmatrix} x_{11} & x_{12} \cdot \mathrm{e}^{-\frac{1}{2}\overline{\lambda} t} \\ x_{21} \cdot \mathrm{e}^{-\frac{1}{2}\lambda t} & x_{22} \end{pmatrix},$$

where the real part of λ is larger than zero.

When we restrict to discrete time steps and assume λ to be real (in physical terms this means that we change to the interaction picture), then this semigroup reduces to the powers of the single transition operator

$$T : M_2 \to M_2 : x = \begin{pmatrix} x_{11} & x_{12} \\ x_{21} & x_{22} \end{pmatrix} \mapsto \begin{pmatrix} x_{11} & \rho \cdot x_{12} \\ \rho \cdot x_{21} & x_{22} \end{pmatrix}$$

for some ρ, $0 \leq \rho < 1$. This is just the operator, for which we constructed the Markov process in the previous section. We see that polarization in the z-direction remains unaffected, while polarization in the x-direction and y-direction dissipates to zero. We want to see whether our Markov process gives a reasonable physical explanation for the observed relaxation.

A Spin$-\frac{1}{2}-$Particle in a Magnetic Field

A spin-$\frac{1}{2}$-particle in a magnetic field B in the z-direction is described by the Hamiltonian $\mathbb{H} = \frac{1}{2}\frac{e}{m}B \cdot \sigma_z = \frac{1}{2}\omega \cdot \sigma_z$, where e is the electric charge and m the mass of the particle. ω is called the Larmor–frequency. The time evolution, given by $\mathrm{e}^{-\mathrm{i}\mathbb{H}t}$, describes a rotation of the spin–particle around the z-axis with this frequency:

$$Ad\,\mathrm{e}^{-\mathrm{i}\mathbb{H}t}\left(\begin{pmatrix} x_{11} & x_{12} \\ x_{21} & x_{22} \end{pmatrix}\right) = \begin{pmatrix} x_{11} & \mathrm{e}^{\mathrm{i}\omega t} x_{12} \\ \mathrm{e}^{-\mathrm{i}\omega t} x_{21} & x_{22} \end{pmatrix}.$$

Since we are discussing the situation for discrete time steps, we consider the unitary

$$\overline{w}_\omega := \mathrm{e}^{-\mathrm{i}\mathbb{H}} = \begin{pmatrix} \mathrm{e}^{-\mathrm{i}\omega/2} & 0 \\ 0 & \mathrm{e}^{\mathrm{i}\omega/2} \end{pmatrix}.$$

It describes the effect of the time evolution after one time unit in a field of strength B. Note that $Ad\,\overline{w}_\omega = Ad\,w_\omega$ with $w_\omega = \begin{pmatrix} 1 & 0 \\ 0 & \mathrm{e}^{\mathrm{i}\omega} \end{pmatrix}$ as in Sect. 6.1.

A Spin$-\frac{1}{2}-$Particle in a Magnetic Field with Two Possible Values

Imagine now that the magnetic field is constant during one time unit, that it always has the same absolute value $|B|$ such that $\cos\omega = \rho$, but that it points into +z-direction and –z-direction with equal probability $\frac{1}{2}$. Representing the two possible states of the field by the points in $\Omega_0 = \{+1, -1\}$, then the magnetic field is described by the probability space $(\Omega_0, \mu_0) = (\{+1, -1\}, (\frac{1}{2}, \frac{1}{2}))$ as in the previous section. The algebraic description of this magnetic field leads to $(\mathcal{C}_0, \psi_0)$ where $\mathcal{C}_0$ is the two-dimensional commutative algebra $\mathbb{C}^2$, considered as the algebra of functions on the two points of Ω_0, while ψ_0 is the state on $\mathcal{C}_0$ which is induced by the probability measure μ_0.

The spin-$\frac{1}{2}$-particle is described by the algebra of observables $\mathcal{A}_0 = M_2$ and assuming that we know nothing about its polarization, then its state is appropriately given by the tracial state tr on M_2 (this state is also called the "chaotic state").

Therefore, the system which is composed of a spin-$\frac{1}{2}$-particle and of a magnetic field with two possible values, has $M_2 \otimes \mathcal{C}_0$ as its algebra of observables. We use the identification of this algebra with the algebra $M_2 \oplus M_2$ as it was described in Section 6.1.

The point $-1 \in \Omega_0$ corresponds to the field in –z-direction. Therefore, the first summand of $M_2 \oplus M_2$ corresponds to the spin-$\frac{1}{2}$-particle in the field in –z-direction and the time evolution on this summand is thus given by $Ad\,\overline{w}_{-\omega} = Ad\,w_{-\omega}$. On the second summand it is accordingly given by $Ad\,\overline{w}_{\omega} = Ad\,w_{\omega}$. Therefore, the time evolution of the whole composed system is given by the automorphism $\alpha_1 = Ad\,w_{-\omega} \oplus Ad\,w_{\omega}$ on $(M_2 \otimes \mathcal{C}_0, tr \otimes \psi_0)$. We thus have all the ingredients which we needed in Section 3.4 in order to construct a Markov process.

A Spin$-\frac{1}{2}-$Particle in a Stochastic Magnetic Field

What is the interpretation of the whole Markov process? As in Section 3.4, denote by $(\mathcal{C}, \psi)$ the infinite tensor product of copies of $(\mathcal{C}_0, \psi_0)$, and denote by S the tensor right shift on it. Then $(\mathcal{C}, \psi)$ is the algebraic description of the classical probability space (Ω, μ) whose points are two-sided infinite sequences of -1's and 1's, equipped with the product measure constructed from $\mu_0 = (\frac{1}{2}, \frac{1}{2})$. The tensor right shift S is induced from the left shift on these sequences. Therefore, $(\mathcal{C}, \psi, S; \mathcal{C}_0)$ is the algebraic description of the classical Bernoulli–process, which describes, for example, the tossing of a coin, or the behaviour of a stochastic magnetic field with two possible values, $+B$ or $-B$, which are chosen according to the outcomes of the coin toss: $(\mathcal{C}, \psi, S)$ is the mathematical model of such a stochastic magnetic field. Its time zero-component is coupled to the spin-$\frac{1}{2}$-particle via the interaction–automorphism α_1. Finally, the Markov process as a whole describes the spin-$\frac{1}{2}$-particle which is interacting with this surrounding stochastic magnetic field.

This is precisely how one explains the spin relaxation T: The algebra M_2 of spin observables represents a large ensemble of many spin-$\frac{1}{2}$-particles. Assume, for example, that at time zero they all point in the x-direction. So one measures a macroscopic magnetic moment in this direction. Now they feel the above stochastic magnetic field in z-direction. In one time unit, half of the ensemble feels a field in –z-direction and starts to rotate around the z-axis, say clockwise; the other half feels a field in +z-direction and starts to rotate counterclockwise. Therefore, the polarization of the single spins goes out of phase and the overall polarization in x-direction after one time step reduces by a factor ρ. Alltogether, the change of polarization is appropriately described by T. After another time unit, cards are shuffled again: two other halfs of particles, stochastically independent of the previous ones, feel the magnetic fields in –z-direction and +z-direction, respectively. The overall effect in polarization is now given by T^2, and so on. This description of the behaviour of the particles in the stochastic magnetic field is precisely reflected by the structure of our Markov process.

6.3 Further Discussion of the Example

The idea behind the construction of our example in Sect. 6.1 depended on writing the transition operator T as a convex combination of the two automorphisms $Ad\,w_{-\omega}$ and $Ad\,w_{\omega}$. This idea can be generalized. In fact, whenever a transition operator of a probability space $(\mathcal{A}_0, \varphi_0)$ is a convex combination of automorphisms of $(\mathcal{A}_0, \varphi_0)$ or even a convex integral of such automorphisms, a Markov process can be constructed in a similar way ([Kü2]). There is even a generalization to continuous time of this idea, which is worked out in ([KüMa1]).

We do not want to enter into such generality here. But it is worth going at least one step further in this direction. Obviously, there are many more ways of writing T as a convex combination of automorphisms of M_2: let μ_0 be any probability measure on the intervall $[-\pi, \pi]$ such that $\int_{-\pi}^{\pi} e^{i\omega} d\mu_0(\omega) = \rho$. Obviously, there are many such probability measures. When we identify the intervall $[-\pi, \pi]$ canonically with the unit circle in the complex plane and μ_0 with a probability measure on it, this simply means that the barycenter of μ_0 is ρ. Then it is clear that $T = \int_{-\pi}^{\pi} Ad\,w_{\omega} d\mu_0(\omega)$, i.e., T is a convex integral of automorphisms of the type $Ad\,w_{\omega}$. To any such representation of T there correspond $(\mathcal{C}_0, \psi_0)$ and α_1 as follows. Put $\mathcal{C}_0 := L^{\infty}([-\pi, \pi], \mu_0)$ and let ψ_0 be the state on $\mathcal{C}_0$ induced by μ_0. The function $[-\pi, \pi] \ni \omega \mapsto e^{i\omega}$ defines a unitary v in $\mathcal{C}_0$. It gives rise to a unitary $u := \begin{pmatrix} 1\!\!1 & 0 \\ 0 & v \end{pmatrix} \in M_2(\mathcal{C}_0) \cong M_2 \otimes \mathcal{C}_0$ and thus to an automorphism $\alpha_1 := Ad\,u$ of $(M_2 \otimes \mathcal{C}_0, tr \otimes \psi_0)$. Our example of Sect. 6.1 is retained when choosing $\mu_0 := \frac{1}{2}\delta_{-\omega} + \delta_{\omega}$, where δ_x denotes the Dirac measure at point x (obviously, it was no restriction to assume $\omega \in [-\pi, \pi]$).

In this way for any such μ we obtain a Markov process for the same transition operator T. By computing the classical dynamical entropy of the commutative part of these processes one sees that there are uncountably many non-equivalent Markov processes of this type. This is in sharp contrast to the classical theory of Markov processes: up to stochastic equivalence a classical Markov process is uniquely determined by its semigroup of transition operators. On the other hand, our discussion of the physical interpretation in the previous section shows that these different Markov processes are not artificial, but they correspond to different physical situations: The probability measure μ_0 on the points ω appears as a probability measure on the possible values of the magnetic fields. It was rather artificial when we first assumed that the field B can only attain two different values of equal absolute value. In general, we can describe any stochastic magnetic field in the z-direction as long as it has no memory in time.

There are even non-commutative Markov processes for a classical transition operator which are contained in these examples: The algebra M_2 contains the two-dimensional commutative subalgebra generated by the observable σ_x, and the whole Markov–process can be restricted to the subalgebra generated by the translates of this observable. This gives a Markov process with values in the two-dimensional subalgebra $\mathbb{C}^2$, which still is non-commutative for certain choices of μ_0. Thus we also have non-commutative processes for a classical transition matrix. Details may be found in [Kü2].

7 The Micro-Maser as a Quantum Markov Process

The micro-maser experiment as it is carried through by H. Walther [VBWW] turns out to be another experimental realization of a quantum Markov process with all the structure described in Section 3.4. It turns out that the scattering theory for such processes leads to some suggestions on how to use a micro-maser for the preparation of interesting quantum states. In the following we give a description of this recent considerations. For details we refer to [WBKM] for the results on the micro-maser, to [KüMa3] for the mathematical background on general scattering theory, and to [Haa] for the asymptotic completeness of this system. For the physics of this experiment we refer to [VBWW].

7.1 The Experiment

In the micro-maser experiment a beam of isolated Rubidium atoms is prepared. The atoms of this beam are prepared in highly exited Rydberg states and for the following only two of these states are relevant. Therefore we may consider the atoms as quantum mechanical two-level systems. Thus the algebra of observables for a single atom is the algebra M_2 of 2×2-matrices. The atoms with a fixed velocity are singled out and sent through a micro-wave cavity which has small holes on both sides for the atoms to pass through this cavity. During their passage through the cavity the atoms interact with one mode of the electromagnetic field in this cavity which is in tune with the energy difference of the two levels of these atoms. One mode of the electromagnetic field is described mathematically as a quantum harmonic oscillator. Hence its algebra of observable is given by $B(\mathcal{H})$ where $\mathcal{H} = L^2(\mathbb{R})$ or $\mathcal{H} = l^2(\mathbb{N})$, depending on whether we work in the position representation or in the energy representation. The atomic beam is weak enough so there is at most one atom inside the cavity at a time and since the atoms all come with the same velocity there is a fixed time for the interaction between atom and field for each of these atoms. To simplify the discussion further we assume that the time between the passage through the cavity of two successive atoms is always the same. So there is a time unit such that one atom passes during one time unit. This is not realistic but due to the particular form of the model (cf. below) the free evolution of the field commutes with the interaction evolution and can be handled separately. Therefore it is easy to turn from this description to a more realistic description afterwards where the arrival times of atoms in the cavity have, for example, a Poissonian distribution.

For the moment we do not specify the algebras and the interaction involved and obtain the following scheme of description for the experiment: φ stands for the state of the field mode and $(\rho_i)_i$ denote the states of the successive atoms. For the following discussion it will be convenient to describe states by their density matrices.

Micro-
Wave-   isolated Rubidium atoms
Cavity   in Rydberg-states

φ

$\cdots \; \rho_{-1} \; \otimes \; \rho_0 \; \otimes \; \rho_1 \; \otimes \; \rho_2 \; \cdots$

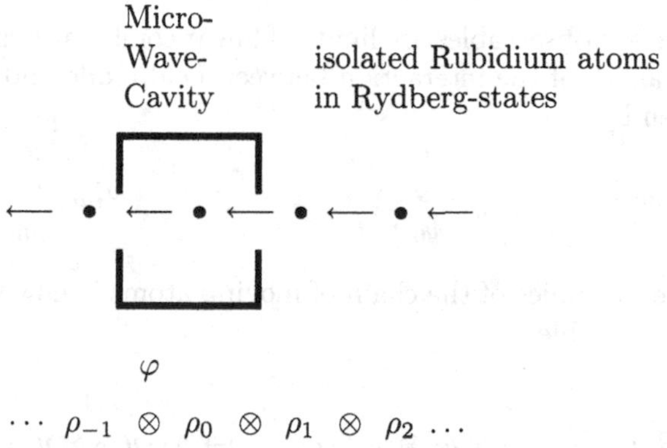

7.2 The Micro-Maser Realizes a Quantum Markov Process

We consider the time evolution in the interaction picture. For one time step the time evolution naturally decomposes into two parts. One part describes the interaction between a passing atom and the field, the other part describes the moving atoms.

Consider one atom which is passing through the cavity during one time step. Assuming that before the passage the cavity was in a state φ and the atom was in a state ρ then the state of the system consising of field mode and atom is now given by $u_{int} \cdot \varphi \otimes \rho \cdot u_{int}^*$ where $u_{int} = e^{i\mathbb{H}t_0}$, $\mathbb{H}$ is the Hamiltonian, and t_0 is the interaction time given by the time an atom needs to pass through the cavity.

The other part of the time evolution describes the moving atoms. For one time unit it is the tensor right shift in the tensor product of states of the flying atoms. Thus the time evolution for one step of the whole system might be written in the following suggestive way:

$$\begin{array}{ccccc} & & \varphi & & \\ & & u_{int} \otimes u_{int}^* & & \\ \text{tensor left shift } (\cdots \rho_{-1} & \otimes & \rho_0 & \otimes & \rho_1 \otimes \rho_2 \cdots) \end{array}$$

We continue to use this suggestive picture for our description. Then a description of this system in the Heisenberg picture looks as follows: If $x \in B(\mathcal{H})$ is an observable of the field mode and $(y_i)_i \in M_2$ are observables of the atoms then a typical observable of the whole systems is given by

$$\begin{array}{ccc} x & & B(\mathcal{H}) \\ \otimes & \in & \otimes \\ \cdots y_{-1} \otimes y_0 \otimes y_1 \cdots & & \cdots M_2 \otimes M_2 \otimes M_2 \cdots \end{array}$$

and arbitrary observables are limits of linear combinations of such observables. The dynamics of the interaction between field mode and one passing atom is now given by

$$\alpha_{int} : \quad \begin{array}{c} x \\ \otimes \\ y_0 \end{array} \quad \mapsto \quad u_{int}^* \cdot \begin{array}{c} x \\ \otimes \\ y_0 \end{array} \cdot u_{int}$$

while the dynamics of the chain of moving atoms is now the tensor right shift on the observables:

$$S : \quad \cdots y_{-1} \otimes y_0 \otimes y_1 \otimes y_2 \cdots \mapsto \cdots y_{-2} \otimes y_{-1} \otimes y_0 \otimes y_1 \cdots$$

Therefore, the complete dynamics for one time step is given by $\alpha := \alpha_{int} \cdot S$ and can be written as

$$\begin{array}{ccc} & \left.\begin{array}{c} B(\mathcal{H}) \\ \otimes \\ M_2 \end{array}\right\} \alpha_{int} & \\ \cdots \otimes M_2 \otimes & & \otimes M_2 \otimes \cdots \end{array}$$

$$\xrightarrow[S]{\hspace{3cm}}$$

We see that the dynamics of this systems is a realization of the dynamics of a quantum Markov process of the type as discussed in Sect. 3.4.

7.3 The Jaynes–Cummings Interaction

Before further investigating this Markov process we need to be more specific on the nature of the interaction between field mode and two-level atoms. In the micro-maser regime it is a good approximation to assume that the interaction is described by the Jaynes–Cummings model: On the Hilbert space $l^2(\mathbb{N}) \otimes \mathbb{C}^2$ of field mode and atom we can use the simplified Hamiltonian given by

$$\begin{aligned}
\mathbb{H} &= \hbar\omega_F a^* a \otimes \mathbb{1} + \mathbb{1} \otimes \frac{\hbar}{2}\omega_A \sigma_z + g\hbar(a + a^*) \otimes (\sigma_+ + \sigma_-) \\
&\rightsquigarrow \hbar\omega_F a^* a \otimes \mathbb{1} + \mathbb{1} \otimes \frac{\hbar}{2}\omega_A \sigma_z + g\hbar(a \otimes \sigma_+ + a^* \otimes \sigma_-) \\
&\rightsquigarrow \hbar\omega\ a^* a \otimes \mathbb{1} + \mathbb{1} \otimes \frac{\hbar}{2}\omega\ \sigma_z + g\hbar(a \otimes \sigma_+ + a^* \otimes \sigma_-) \,.
\end{aligned}$$

Here the first line is the original Hamiltonian of a field–atom interaction where ω_F is the frequency of the field mode, ω_A is the frequency for the transition between the two levels of our atoms, and g is the coupling constant. In the second line this Hamiltonian is simplified by the rotating wave approximation and in the third line we further assume $\omega_F = \omega_A =: \omega$. The operators σ_+ and σ_- are the raising and lowering operators of a two-level system. The Hamiltonian generates the unitary group

$$U(t) = \mathrm{e}^{-\frac{i}{\hbar}\mathbb{H}t}$$

and we put $u_{int} := U(t_0)$ where t_0 is the interaction time needed for one atom to pass through the cavity.

We denote by $|n\rangle \otimes |\downarrow\rangle$ and $|n\rangle \otimes |\uparrow\rangle$ the canonical basis vectors of the Hilbert space where $|n\rangle$ denotes the n-th eigenstate of the harmonic oscillator and $|\uparrow\rangle$ and $|\downarrow\rangle$ are the two eigenstates of the two-level atom. The Hilbert space decomposes into subspaces which are invariant under the Hamiltonian and the time evolution:

Denote by $\mathcal{H}_0$ the one-dimensional subspace spanned by $|0\rangle \otimes |\downarrow\rangle$; then the restriction of $\mathbb{H}$ to $\mathcal{H}_0$ is given by $\mathbb{H}_0 = 0$. Hence the restriction of $U(t)$ to $\mathcal{H}_0$ is $U_0(t) = 1$. For $k \in \mathbb{N}$ denote by $\mathcal{H}_k$ the two-dimensional subspace spanned by the vectors $|k\rangle \otimes |\downarrow\rangle$ and $|k-1\rangle \otimes |\uparrow\rangle$. Then the restriction of $\mathbb{H}$ to $\mathcal{H}_k$ is given by

$$\mathbb{H}_k = \hbar \cdot \begin{pmatrix} \omega k & g\sqrt{k} \\ g\sqrt{k} & \omega k \end{pmatrix}$$

and hence the restriction of $U(t)$ to $\mathcal{H}_k$ is

$$U_k(t) = \mathrm{e}^{\mathrm{i}\omega kt} \begin{pmatrix} \cos g\sqrt{k}t & -\mathrm{i}\sin g\sqrt{k}t \\ -\mathrm{i}\sin g\sqrt{k}t & \cos g\sqrt{k}t \end{pmatrix} .$$

Finally, if for some inverse temperatur β, $0 < \beta < \infty$, φ_β and ψ_β are the equilibrium states for the free Hamiltonian of the field mode and of the two-level-atom, respectively, then $\varphi_\beta \otimes \psi_\beta$ is invariant under the full time evolution generated by the Jaynes–Cummings interaction Hamiltonian $\mathbb{H}$ from above. Therefore, $\alpha_1 := \alpha_{int} := Ad\, u_{int}$ on $B(\mathcal{H}) \otimes M_2$ leaves this state invariant and the dynamics of the micro-maser is the dynamics of a full stationary Markov process $(\mathcal{A}, \varphi, \alpha_t; \mathcal{A}_0)$ as discussed in Sect. 3.4: Put

$$(\mathcal{A}, \varphi) := (B(\mathcal{H}), \varphi_\beta) \otimes (\bigotimes_{\mathbb{Z}} (M_2, \psi_\beta)) ,$$

$\alpha_t := \alpha^t$ for $t \in \mathbb{Z}$ with $\alpha := \alpha_{int} \circ S$, and $\mathcal{A}_0 := B(\mathcal{H})$.

7.4 Asymptotic Completeness and Preparation of Quantum States

The long-term behaviour of this system depends very much on whether or not a so-called *trapped state condition* is fulfilled. That means that for some $k \in \mathbb{N}$ the constant $g\sqrt{k}t_0$ is an integer multiple $n\pi$ of π for some $n \in \mathbb{N}$. In this case the transition

$$|k-1\rangle \otimes |\uparrow\rangle \longleftrightarrow |k\rangle \otimes |\downarrow\rangle$$

is blocked. Therefore, if the initial state of the micro-maser has a density matrix with non-zero entries only in the upper left $k-1 \times k-1$ corner then the atoms, in whichever state they are, will not be able to create a state in the micro-maser with more than $k-1$ photons. This has been used [VBWW] to prepare two-photon number states experimentally: the initial state of the field mode is the vacuum, the two-level atoms are in the upper state $|\uparrow\rangle$ and the interaction time is chosen such that the transition from two to three photons is blocked. This forces the field-mode into the two-photon number state.

On the other hand, if no trapped state condition is fulfilled and all transitions are possible then the state of the field-mode can be controlled by the states of the passing atoms [WBKM]. The mathematical reason is the following theorem:

Theorem 7.1. *If no trapped state condition is fulfilled then for every inverse temperature $\beta > 0$ the Markov process $(\mathcal{A}, \varphi, \alpha_t; \mathcal{A}_0)$ as above, which describes the time evolution of the micro-maser, is asymptotically complete.*

A proof is worked out in [Haa].

For convenience we recall from Chapter 4 that a Markov process as in Section 3.4 is *asymptotically complete* if for all $x \in \mathcal{A}$

$$\begin{aligned} &\Phi_-(x) := \lim_{n\to\infty} S^{-n}\alpha^n(x) \quad \text{exists strongly} \\ \text{and} \quad &\Phi_-(x) \in \mathbb{1} \otimes \mathcal{C}\,. \end{aligned}$$

Moreover, as was noted in Chapter 4, it suffices if this condition is satisfied for all $x \in \mathcal{A}_0$. For $x \in \mathcal{A}_0$, however, we find that

$$\begin{aligned} \alpha^n(x \otimes \mathbb{1}) &= u_n^* \cdot x \otimes \mathbb{1} \cdot u_n \\ u_n &:= S^{n-1}(u_{int}) \cdot S^{n-2}(u_{int}) \cdot \ldots \cdot S(U_{int}) \cdot u_{int} \end{aligned}$$

and asymptotic completeness roughly means that for $x \in \mathcal{A}_0$ and for very large $n \in \mathbb{N}$ there exists $x_{out}^n \in \mathcal{C}$ such that

$$\alpha^n(x \otimes \mathbb{1}) = u_n^* \cdot x \otimes \mathbb{1} \cdot u_n \approx \mathbb{1} \otimes x_{out}^n\,.$$

We translate this into the Schrödinger picture and, for a moment, we use again density matrices for the description of states. Then we find that if such a Markov process is asymptotically complete then for any density matrix φ_n of $\mathcal{A}_0$ and large $n \in \mathbb{N}$ we can find a density matrix ρ_0 of $\mathcal{C}$ such that

$$u_n \cdot \varphi_0 \otimes \rho_0 \cdot u_n^* \approx \varphi_n \otimes \rho'$$

for some density matrix ρ' of $\mathcal{C}$ and the choice of ρ_0 is independent of the initial state φ_0 on $\mathcal{A}_0$. This means that if we want to prepare a state φ_n on $\mathcal{A}_0$ (in our case of the field mode) then even without knowing the initial state φ_0 of $\mathcal{A}_0$ we can prepare an initial state ρ_0 on $\mathcal{C}$ such that the state $\varphi_0 \otimes \rho_0$ evolves after n time steps, at least up to some ε, into the state φ_n on $\mathcal{A}_0$ and some other state ρ' of $\mathcal{C}$ which, however, is not entangled with $\mathcal{A}_0$.

This intuition can be made precise as follows: For simplicity we use discrete time and assume that $(\mathcal{A}, \varphi, \alpha; \mathcal{A}_0)$ is a Markov process which is a coupling to a white noise $(\mathcal{C}, \psi, S; \mathcal{C}_{[0,n]})$.

Definition 7.2. *We say that a normal state φ_∞ on $\mathcal{A}_0$* can be prepared *if there is a sequence ψ_n of normal states on $\mathcal{C}$ such that for all $x \in \mathcal{A}_0$ and all normal initial states θ on $\mathcal{A}_0$*

$$\lim_{n\to\infty} \theta \otimes \psi_n \circ \alpha^n(x \otimes \mathbb{1}) = \varphi_\infty(x)\,.$$

It turns out that for systems like the micro-maser this condition is even equivalent to asymptotic completeness:

Theorem 7.3. *If the Markov process $(\mathcal{A}, \varphi, \alpha; \mathcal{A}_0)$ is of the form as considered in Section 3.4 and if, in addition, the initial algebra $\mathcal{A}_0$ is finite dimensional or isomorphic to $B(\mathcal{H})$ for some Hilbert space $\mathcal{H}$ then the following conditions are equivalent:*

a) The Markov process $(\mathcal{A}, \varphi, \alpha; \mathcal{A}_0)$ is asymptotically complete.
b) Every normal state on $\mathcal{A}_0$ can be prepared.

A proof of this result is contained in [Haa]. This theorem is also the key for proving the above theorem on the asymptotic completeness of the micro-maser.

Therefore, from a mathematical point of view it is possible to prepare an arbitrary state of the field-mode with arbitrary accuracy by sending suitably prepared atoms through the cavity. This raises the question whether also from a physical point of view states of the micro-maser can be prepared by this method. This question has been investigated in [WBKM], [Wel]. The results show that already with a small number of atoms one can prepare interesting states of the field mode with a very high fidelity. Details can be found in [WBKM]. As an illustration we give a concrete example: If the field mode is initially in the vacuum $|0\rangle$ and one wants to prepare the two-photon number state $|2\rangle$ with 4 incoming atoms then by choosing an optimal interaction time t_{int} one can prepare the state $|2\rangle$ with a fidelity of 99.87% if the four atoms are prepared in the state

$$\begin{aligned} |\psi_0\rangle = \; & \sqrt{0.867}|\uparrow\rangle|\uparrow\rangle|\downarrow\rangle|\downarrow\rangle \\ & +\sqrt{0.069}|\uparrow\rangle|\downarrow\rangle|\uparrow\rangle|\downarrow\rangle \\ & -\sqrt{0.052}|\downarrow\rangle|\uparrow\rangle|\uparrow\rangle|\downarrow\rangle \\ & +\sqrt{0.005}|\uparrow\rangle|\downarrow\rangle|\downarrow\rangle|\uparrow\rangle \\ & -\sqrt{0.004}|\downarrow\rangle|\uparrow\rangle|\downarrow\rangle|\uparrow\rangle \\ & +\sqrt{0.003}|\downarrow\rangle|\downarrow\rangle|\uparrow\rangle|\uparrow\rangle \; . \end{aligned}$$

8 Completely Positive Operators

8.1 Complete Positivity

After the discussion of some specific examples from physics we now come back to discussing the general theory. A physical system is again described by its algebra $\mathcal{A}$ of observables. We assume that $\mathcal{A}$ is, at least, a C^*–algebra of operators on some Hilbert space and we can always assume that $\mathbb{1} \in \mathcal{A}$. A normalized positive linear state functional $\varphi : \mathcal{A} \to \mathbb{C}$ is interpreted either as a physical state of the system or as a probability measure.

All time evolutions and other 'operations' which we have considered so far had the property of carrying states into states. This was necessary in order to be consistent with their physical or probabilistic interpretation. In the Heisenberg picture these 'operations' are described by operators on algebras of operators. In order to avoid such an accumulation of 'operators' we talk synonymously about maps. Given two C*-algebras $\mathcal{A}$ and $\mathcal{B}$ then it is obvious that for a map $T : \mathcal{A} \to \mathcal{B}$ the following two conditions are equivalent:

a) T is state preserving: for every state φ on $\mathcal{B}$ the functional

$$\varphi \circ T : \mathcal{A} \ni x \mapsto \varphi(T(x))$$

on $\mathcal{A}$ is a state, too.

b) T is positive and identity preserving: $T(x) \geq 0$ for $x \in \mathcal{A}$, $x \geq 0$, and $T(\mathbb{1}) = \mathbb{1}$.

Indeed, all maps which we have considered so far had this property. A closer inspection, however, shows that these maps satisfy an even stronger notion of positivity called *complete positivity.*

Definition 8.1. *A map $T : \mathcal{A} \to \mathcal{B}$ is n–*positive *if*

$$T \otimes Id_n : \mathcal{A} \otimes M_n \to \mathcal{B} \otimes M_n : x \otimes y \mapsto T(x) \otimes y$$

is positive. It is completely positive *if T is n–positive for all $n \in \mathbb{N}$.*

Elements of $\mathcal{A} \otimes M_n$ may be represented as $n \times n$–matrices with entries from $\mathcal{A}$. In this representation the operator $T \otimes Id_n$ appears as the map which carries such an $n \times n$–matrix $(x_{ij})_{i,j}$ into $(T(x_{ij}))_{i,j}$ with $x_{ij} \in \mathcal{A}$. Thus T is n-positive if such non-negative $n \times n$-matrices are mapped again into non-negative $n \times n$-matrices.

From the definition it is clear that 1–positivity is just positivity and $(n+1)$–positivity implies n–positivity: in the above matrix representation elements of $\mathcal{A} \otimes M_n$ can be identified with $n \times n$–matrices in the upper left corner of all $(n+1) \times (n+1)$–matrices in $\mathcal{A} \otimes M_{n+1}$.

It is a non–trivial theorem that for commutative $\mathcal{A}$ or commutative $\mathcal{B}$ positivity already implies complete positivity (cf. [Tak2], IV. 3). If $\mathcal{A}$ and $\mathcal{B}$ are both non-commutative algebras, this is no longer true. The simplest (and typical) example is the transposition on the (complex) 2×2–matices M_2. The map

$$M_2 \ni \begin{pmatrix} a\, b \\ c\, d \end{pmatrix} \mapsto \begin{pmatrix} a\, c \\ b\, d \end{pmatrix} \in M_2$$

is positive but not 2–positive hence not completely positive. From this example one can proceed further to show that for all n there are maps which are n–positive but not $(n+1)$–positive. It is true, however, that on M_n n–positivity already implies complete positivity.

It is an important property of 2–positive and hence of completely positive maps that they satisfy a Schwarz–type inequality:

$$\|T\| T(x^*x) \geq T(x)^* T(x)$$

for $x \in \mathcal{A}$ (the property $T(x^*) = T(x)^*$ follows from positivity).

It can be shown that *-homomorphisms and conditional expectations are automatically completely positive. All maps which we have considered so far are either of these types or are compositions of such maps, like transition operators. Hence they are all completely positive. This is the mathematical reason why we have only met completely positive operators.

One could wonder, however, whether there is also a physical reason for this fact.

8.2 Interpretation of Complete Positivity

In the introduction to this paragraph we argued that time evolutions should be described by positive identity preserving maps. Now suppose that T is such a time evolution on a system $\mathcal{A}$ and that S is a time evolution of a different system $\mathcal{B}$. Even if these systems have nothing to do with each other we can consider them – if only in our minds – as parts of the composed system $\mathcal{A} \otimes \mathcal{B}$ whose time evolution should then be given by $T \otimes S$ – there is no interaction. Being the time evolution of a physical system the operator $T \otimes S$, too, should be positive and identity preserving. This, however, is not automatic: already for the simple case $\mathcal{B} = M_2$ and $S = Id$ there are counter-examples as mentioned above. This is the place where complete positivity comes into play. With this stronger notion of positivity we can avoid the above problem.

Indeed, if $T : \mathcal{A}_1 \to \mathcal{A}_2$ and $S : \mathcal{B}_1 \to \mathcal{B}_2$ are completely positive operators then $T \otimes S$ can be defined uniquely on the minimal tensor product $\mathcal{A}_1 \otimes \mathcal{B}_1$ and it becomes again a completely positive operator from $\mathcal{A}_1 \otimes \mathcal{B}_1$ into $\mathcal{A}_2 \otimes \mathcal{B}_2$. It suffices to require that T preserves its positivity property when tensored with the maps Id on M_n. Then T can be tensored with any other map having this property and the composed system still has the right positivity property: Complete positivity is stable under forming tensor products. Indeed, this holds not only for C*-tensor products, but also for tensor products in the category of von Neumann algebras as well. For these theorems and related results we refer to the literature, for example ([Tak2], IV. 4 and IV. 5).

8.3 Representations of Completely Positive Operators

The fundamental theorem behind almost all results on complete positivity is Stinespring's famous representation theorem for completely positive maps. Consider a map $T : \mathcal{A} \to \mathcal{B}$. Since $\mathcal{B}$ is an operator algebra it is contained in $B(\mathcal{H})$ for some Hilbert space $\mathcal{H}$ and it is no restriction to assume that T is a map $T : \mathcal{A} \to B(\mathcal{H})$.

Theorem 8.2. *(Stinespring 1955, cf. [Tak2]). For a map $T : \mathcal{A} \to B(\mathcal{H})$ the following conditions are equivalent:*

a) T is completely positive.
b) There is a further Hilbert space $\mathcal{K}$, a representation $\pi : \mathcal{A} \to B(\mathcal{K})$ and a bounded linear map $v : \mathcal{H} \to \mathcal{K}$ such that

$$T(x) = v^*\pi(x)v$$

for all $x \in \mathcal{A}$. If $T(\mathbb{1}) = \mathbb{1}$ then v is an isometry.

The triple $(\mathcal{K}, \pi, v)$ is called a *Stinespring representation* for T. If it is *minimal* that is, the linear span of $\{\pi(x)v\xi \ , \ \xi \in \mathcal{H} \ , \ x \in \mathcal{A}\}$ is dense in $\mathcal{K}$, then the Stinespring representation is *unique up to unitary equivalence.*

From Stinespring's theorem it is easy to derive the following *concrete representation* for completely positive operators on M_n.

Theorem 8.3. *For $T : M_n \to M_n$ the following conditions are equivalent:*

a) T is completely positive.
b) There are elements $a_1, \dots, a_k \in M_n$ for some k such that

$$T(x) = \sum_{i=1}^{k} a_i^* x a_i \ .$$

Clearly, T is identity preserving if and only if $\sum_{i=1}^{k} a_i^* a_i = \mathbb{1}$.

Such decompositions of completely positive operators are omnipresent whenever completely positive operators occur in a physical context. It is important to note that such a decomposition is by no means uniquely determined by T (see below). In a physical context different decompositions rather correspond to different physical situations (cf. the discussion in Sect. 6.3; cf. also Sect. 10.2).

The following basic facts can be derived from Stinespring's theorem without much difficulty:

A concrete representation $T(x) = \sum_{i=1}^{k} a_i^* x a_i$ for T can always be chosen such that $\{a_1, a_2, \dots, a_k\} \subseteq M_n$ is linearly independent, in particular, $k \le n^2$. We call such a representation *minimal.* The cardinality k of a minimal representation of T is uniquely determined by T, i.e., two minimal representations of T have the same cardinality. Finally, all minimal representations can be characterized by the following result.

Proposition 8.4. *Let $T(x) = \sum_{i=1}^{k} a_i^* x a_i$ and $S(x) = \sum_{j=1}^{l} b_j^* x b_j$ be two minimal representations of completely positive operators S and T on M_n. The following conditions are equivalent:*

a) $S = T$.

b) $k = l$ and there is a unitary $k \times k$ –matrix $\Lambda = (\lambda_{ij})_{i,j}$ such that

$$a_i = \sum_{j=1}^{k} \lambda_{ij}\, b_j\ .$$

The results on concrete representations have an obvious generalization to the case $n = \infty$. Then infinite sums may occur, but they must converge in the strong operator topology on $B(\mathcal{H})$.

9 Semigroups of Completely Positive Operators and Lindblad Generators

9.1 Generators of Lindblad Form

In Section 3.3 we saw that to each Markov process there is always associated a semigroup of completely positive transition operators on the initial algebra $\mathcal{A}_0$. If time is continuous then in all cases of physical interest this semigroup $(T_t)_{t\geq 0}$ will be strongly continuous. According to the general theory of one-parameter semigroups (cf. [Dav2]) the semigroup has a generator L such that

$$\frac{\mathrm{d}}{\mathrm{d}t} T_t(x) = L(T_t(x))$$

for all x in the domain of L, which is formally written as $T_t = \mathrm{e}^{Lt}$. In the case of a classical Markov process with values in $\mathbb{R}^n$ one can say much more. Typically, L has the form of a partial differential operator of second order of a very specific form like

$$Lf(x) = \sum_i a_i(x) \frac{\partial}{\partial x_i} f(x) + \sum_{i,j} \frac{1}{2}\, b_{ij}(x)\, \frac{\partial^2}{\partial x_i \partial x_j} f(x) + \int_{\mathbb{R}^n} f(y) \mathrm{d}w(y)$$

for f a twice continuously differentiable function on $\mathbb{R}^n$ and suitable functions a_i, b_{ij} and a measure $w(\cdot, t)$.

It is natural to wonder whether a similar characterization of generators can be given in the non-commutative case. This turns out to be a difficult problem and much research on this problem remains to be done. A first breakthrough was obtained in a celebrated paper by G. Lindblad [Lin] in 1976 and at the same time, for the finite dimensional case, in [GKS].

Theorem 9.1. *Let $(T_t)_{t\geq 0}$ be a semigroup of completely positive identity preserving operators on M_n with generator L.*

Then there is a completely positive operator $M : M_n \to M_n$ and a self-adjoint element $h \in M_n$ such that

$$L(x) = \mathrm{i}[h,x] + M(x) - \frac{1}{2}(M(\mathbb{1})x + xM(\mathbb{1})).$$

where, as usual, $[h,x]$ *stands for the commutator* $hx - xh$*. Conversely, every operator* L *of this form generates a semigroup of completely positive identity preserving operators.*

Since we know that every such M has a concrete representation as

$$M(x) = \sum_i a_i^* x a_i$$

we obtain for L the representation

$$L(x) = \mathrm{i}[h,x] + \sum_i a_i^* x a_i - \frac{1}{2}(a_i^* a_i x + x a_i^* a_i)$$

This representation is usually called the *Lindblad form* of the generator.

Lindblad was able to prove this result for norm-continuous semigroups on $B(\mathcal{H})$ for infinite dimensional $\mathcal{H}$. In this situation L is still a bounded operator. If one wants to treat the general case of strongly continuous semigroups on $B(\mathcal{H})$ then one has to take into account, for example, infinite unbounded sums of bounded and unbounded operators a_i. Until today no general characterization of such generators is available, which would generalize the representation of L as a second order differential operator as indicated above. Nevertheless, Lindblad's characterization seems to be 'philosophically true' as in most cases of physical interest unbounded generators also appear to be in Lindblad form. Typically, the operators a_i are creation and annihilation operators.

9.2 Interpretation of Generators of Lindblad Form

The relation between a generator in Lindblad form and the above partial differential operator is not so obvious. The following observation might clarify their relation. For an extended discussion we refer to [KüMa1].

For $h \in M_n$ consider the operator D on M_n given by

$$D : x \mapsto \mathrm{i}[h,x] = \mathrm{i}(hx - xh) \quad (x \in M_n)\ .$$

Then

$$D(xy) = D(x) \cdot y + x \cdot D(y)$$

Hence D is a derivation.

In Lindblad's theorem h is self-adjoint and in this case D is a real derivation (i.e. $D(x^*) = D(x)^*$) and generates the time evolution $x \mapsto \mathrm{e}^{+iht} x \mathrm{e}^{-iht}$ which is implemented by the unitary group $(\mathrm{e}^{iht})_{t\in\mathbb{R}}$. Therefore, for self-adjoint h the term $x \mapsto \mathrm{i}[h,x]$ is a 'quantum derivative' of first order and corresponds to a drift term.

For the second derivative we obtain after a short computation

$$\begin{aligned} D^2(x) &= \mathrm{i}[h, \mathrm{i}[h, x]] \\ &= 2(hxh - \tfrac{1}{2}(h^2x + xh^2)) \ . \end{aligned}$$

This resembles the second part of a generator in Lindblad form. It shows that for self-adjoint a the term

$$axa - \frac{1}{2}(a^2x + xa^2)$$

is a second derivative and thus generates a quantum diffusion.

On the other hand for $a = u$ unitary the term $a^*xa - \frac{1}{2}(a^*ax + xa^*a)$ turns into $u^*xu - x$ which generates a jump process: If we define the *jump operator* $J(x) := u^*xu$ and

$$L(x) := J(x) - x = (J - Id)(x) \quad \text{then}$$

$$\begin{aligned} \mathrm{e}^{Lt} &= \mathrm{e}^{(J-Id)t} = \mathrm{e}^{-t} \cdot \mathrm{e}^{Jt} \\ &= \textstyle\sum_{n=0}^{\infty} \mathrm{e}^{-t} \frac{t^n}{n!} J^n \ . \end{aligned}$$

This is a Poissonian convex combination of the jumps $\{J^n \ , \ n \in \mathbb{N}\}$. Therefore, terms of this type correspond to classical jump processes.

In general a generator of Lindblad type $L = \sum_i a_i^* x a_i - \frac{1}{2}(a_i^* a_i x + a_i^* a_i x)$ can not be decomposed into summands with a_i self-adjoint and a_i unitary thus there are more general types of transitions. The cases which allow decompositions of this special type have been characterized and investigated in [KüMa1]. Roughly speaking a time evolution with such a generator can be interpreted as the time evolution of an open quantum system under the influence of a classical noise.

In the context of quantum trajectories decompositions of Lindblad type play an important role. They are closely related to *unravellings* of the time evolution T_t (cf., e.g., [Car], [KüMa4], [KüMa5]).

9.3 A Brief Look at Quantum Stochastic Differential Equations

We already mentioned that for a semigroup $(T_t)_{t \geq 0}$ of transition operators on a general initial algebra $\mathcal{A}_0$ there is no canonical procedure which leads to an analogue of the canonical representation of a classical Markov process on the space of its paths. For $\mathcal{A}_0 = M_n$, however, quantum stochastic calculus allows to construct a stochastic process which is almost a Markov process in the sense of our definition. But in most cases stationarity is not preserved by this construction.

Consider $T_t = \mathrm{e}^{Lt}$ on M_n and assume, for simplicity only, that the generator L has the simple Lindblad form

$$L(x) = \mathrm{i}[h, x] + b^*xb - \frac{1}{2}(b^*bx + xb^*b) .$$

Let $\mathcal{F}(L^2(\mathbb{R}))$ denote the symmetric Fock space of $L^2(\mathbb{R})$. For a test function $f \in L^2(\mathbb{R})$ there exist the creation operator $A^*(f)$ and annihilation operator $A(f)$ as unbounded operators on $\mathcal{F}(L^2(\mathbb{R}))$. For $f = \chi_{[0,t]}$, the characteristic function of the interval $[0, t] \subseteq \mathbb{R}$, the operators $A^*(f)$ and $A(f)$ are usually denoted by A_t^* (or $A_t^\dagger$) and A_t, respectively. It is known that the operators $B_t := A_t^* + A_t$ on $\mathcal{F}(L^2(\mathbb{R})), t \geq 0$, give a representation of classical Brownian motion by a commuting family of self-adjoint operators on $\mathcal{F}(L^2(\mathbb{R}))$ (cf. the discussion in Sect. 1.3). Starting from this observation R. Hudson and K.R. Parthasaraty have extended the classical Itô–calculus of stochastic integration with respect to Brownian motion to more general situations on symmetric Fock space. An account of this theory is given in [Par].

In particular, one can give a rigorous meaning to the stochastic differential equation

$$du_t = u_t \left(bdA_t^* + b^*dA_t + (ih - \frac{1}{2}b^*b)dt \right)$$

where bdA_t^* stands for $b \otimes dA_t^*$ on $\mathbb{C}^n \otimes \mathcal{F}(L^2(\mathbb{R}))$ and similarly for b^*dA_t, while $ih - \frac{1}{2}b^*b$ stands for $(ih - \frac{1}{2}b^*b) \otimes \mathbb{1}$ on $\mathbb{C}^n \otimes \mathcal{F}(L^2(\mathbb{R}))$. It can be shown that the solution exits, is unique, and is given by a family $(u_t)_{t\geq 0}$ of unitaries on $\mathbb{C}^n \otimes \mathcal{F}(L^2(\mathbb{R}))$ with $u_0 = \mathbb{1}$.

This leads to a stochastic process with random variables

$$i_t : M_n \ni x \mapsto u_t^* \cdot x \otimes \mathbb{1} \cdot u_t \in M_n \otimes \mathcal{B}(\mathcal{F}(L^2(\mathbb{R})))$$

which can, indeed, be viewed as a Markov process with transition operators $(T_t)_{t\geq 0}$. This construction can be applied to all semigroups of completely positive identity preserving operators on M_n and to many such semigroups on $B(\mathcal{H})$ for infinite dimensional $\mathcal{H}$.

10 Repeated Measurement and its Ergodic Theory

We already mentioned that in a physical context completely positive operators occur frequently in a particular concrete representation and that such a representation may carry additional physical information. In this chapter we discuss such a situation of particular importance: The state of a quantum system under the influence of a measurement. The state change of the system is described by a completely positive operator and depending on the particular observable to be measured this operator is decomposed into a concrete representation. After the discussion of a single measurement we turn to the situation where such a measurement is performed repeatedly as it is the case in the micro-maser example. We describe some recent results on the ergodic theory of the outcomes of a repeated measurement as well as of the state changes caused by it.

10.1 Measurement According to von Neumann

Consider a system described by its algebra $\mathcal{A}$ of observables which is in a state φ. In the typical quantum case $\mathcal{A}$ will be $B(\mathcal{H})$ and φ will be given by a density matrix ρ on $\mathcal{H}$. Continuing our discussion in Section 1.1 we consider the measurement of an observable given by a self-adjoint operator X on $\mathcal{H}$. For simplicity we assume that the spectrum $\sigma(X)$ is finite so that X has a spectral decomposition of the form $X = \sum_i \lambda_i p_i$ with $\sigma(X) = \{\lambda_1, \dots \lambda_n\}$ and orthogonal projections $p_1, p_2, \dots, p_n$ with $\Sigma_i p_i = \mathbb{1}$. According to the laws of quantum mechanics the spectrum $\sigma(X)$ is the set of possible outcomes of this measurement (cf. Sect. 1.1). The probability of measuring the value $\lambda_i \in \sigma(X)$ is given by

$$\varphi(p_i) \ = tr(\rho p_i)$$

and if this probability is different from zero then after such a measurement the state of the system has changed to the state

$$\varphi_i : x \mapsto \frac{\varphi(p_i x p_i)}{\varphi(p_i)}$$

with density matrix

$$\frac{p_i \rho p_i}{tr(p_i \rho)} .$$

It will be convenient to denote the state φ_i also by

$$\varphi_i = \frac{\varphi(p_i \cdot p_i)}{\varphi(p_i)} ,$$

leaving a dot where the argument x has to be inserted.

The spectral measure $\sigma(X) \ni \lambda_i \mapsto \varphi(p_i)$ defines a probability measure μ_{φ_0} on the set $\Omega_0 := \sigma(X)$ of possible outcomes. If we perform the measurement of X, but we ignore its outcome (this is sometimes called "measurement with deliberate ignorance") then the initial state φ has changed to the state φ_i with probability $\varphi(p_i)$. Therefore, the state of the system after such a measurement in ignorance of its outcome is adequately described by the state

$$\varphi_X := \Sigma_i \varphi(p_i) \cdot \varphi_i = \Sigma_i \varphi(p_i \cdot p_i) .$$

(Here it is no longer necessary to single out the cases with probability $\varphi(p_i) = 0$.)

Turning to the dual description in the Heisenberg picture an element $x \in \mathcal{A}$ changes as

$$x \mapsto \frac{p_i x p_i}{\varphi(p_i)}$$

if λ_i was measured. A measurement with deliberate ignorance is described by

$$x \mapsto \sum_i p_i x p_i$$

which is a conditional expectation of $\mathcal{A}$ onto the subalgebra $\{\sum_i p_i x p_i, x \in \mathcal{A}\}$.

10.2 Indirect Measurement According to K. Kraus

In many cases the observables of a system are not directly accessible to an observation or an observation would lead to an undesired destruction of the system as is typically the case if measuring photons.

In such a situation one obtains information on the state φ of the system by coupling the system to another system – a measurement apparatus – and reading off the value of an observable of the measurement apparatus. A mathematical description of such measurements was first given by K. Kraus [Kra].

As a typical example for such an indirect measurement consider the micromaser experiment which was discussed from a different point of view in Chapter 7. The system to be measured is the mode of the electromagnetic field inside the cavity with $\mathcal{A} = B(\mathcal{H})$ as its algebra of observables. It is initially in a state φ. A two-level atom sent through the cavity can be viewed as a measurement apparatus: If the atom is initially prepared in a state ψ on $\mathcal{C} = M_2$, it is then sent through the cavity where it can interact with the field mode, and it is measured after it has left the cavity, then this gives a typical example of such an indirect measurement.

Similarly, in general such a measurement procedure can be decomposed into the following steps:

α) Couple the system $\mathcal{A}$ in its initial state φ to another system – the measurement apparatus – with observable algebra $\mathcal{C}$, which is initially in a state ψ.

β) For a certain time t_0 the composed system evolves according to a dynamic $(\alpha_t)_t$. In the Heisenberg picture, $(\alpha_t)_{t\in\mathbb{R}}$ is a group of automorphisms of $\mathcal{A} \otimes \mathcal{C}$. After the interaction time t_0 the overall change of the system is given by $T_{int} := \alpha_{t_0}$.

γ) Now an observable $X = \sum_i \lambda_i p_i \in \mathcal{C}$ is measured and changes the state of the composed system accordingly.

δ) The new state of $\mathcal{A}$ is finally obtained by restricting the new state of the composed system to the operators in $\mathcal{A}$.

Mathematically each step corresponds to a map on states and the whole measurement is obtained by composing those four maps (on infinite dimensional algebras all states are assumed to be normal):

α) The measurement apparatus is assumed to be initially in a fixed state ψ. Therefore, in the Schrödinger picture, coupling $\mathcal{A}$ to $\mathcal{C}$ corresponds to the map $\varphi \mapsto \varphi \otimes \psi$ of states on $\mathcal{A}$ into states on $\mathcal{A} \otimes \mathcal{C}$. We already saw in Sect. 5.1 that dual to this map is the conditional expectation of tensor type

$$P_\psi : \mathcal{A} \otimes \mathcal{C} \mapsto \mathcal{A} : x \otimes y \mapsto \psi(y) \cdot x$$

which thus describes this step in the Heisenberg picture (again we identify $\mathcal{A}$ with the subalgebra $\mathcal{A} \otimes \mathbb{1}$ of $\mathcal{A} \otimes \mathcal{C}$ so that we may still call P_ψ a conditional expectation).

β) The time evolution of $\mathcal{A} \otimes \mathcal{C}$ during the interaction time t_0 is given by an automophism T_{int} on $\mathcal{A} \otimes \mathcal{C}$. It changes any state χ on $\mathcal{A} \otimes \mathcal{C}$ into $\chi \circ T_{int}$.

γ) A measurement of $X = \sum_i \lambda_i p_i \in \mathcal{C}$ changes a state χ on $\mathcal{A} \otimes \mathcal{C}$ into the state $\frac{\chi(\mathbb{1} \otimes p_i \; \cdot \; \mathbb{1} \otimes p_i)}{\chi(\mathbb{1} \otimes p_i)}$ and this happens with probability $\chi(\mathbb{1} \otimes p_i)$. It is convenient to consider this state change together with its probability. This can be described by the non-normalized but linear map

$$\chi \mapsto \chi(\mathbb{1} \otimes p_i \; \cdot \; \mathbb{1} \otimes p_i) \, .$$

Dual to this is the map

$$\mathcal{A} \otimes \mathcal{C} \ni z \mapsto \mathbb{1} \otimes p_i \cdot z \cdot \mathbb{1} \otimes p_i$$

which thus describes the unnormalized state change due to a measurement with outcome λ_i in the Heisenberg picture.
When turning from a measurement with outcome λ_i to a measurement with deliberate ignorance then the difference between the normalized and the unnormalized description will disappear.

δ) This final step maps a state χ on the composed system $\mathcal{A} \otimes \mathcal{C}$ to the state

$$\chi|_\mathcal{A} : \mathcal{A} \ni x \mapsto \chi(x \otimes \mathbb{1}) \, .$$

The density matrix of $\chi|_\mathcal{A}$ is obtained from the density matrix of χ by a partial trace over $\mathcal{C}$. As we already saw in Sect. 5.1 a description of this step in the dual Heisenberg picture is given by the map

$$\mathcal{A} \ni x \mapsto x \otimes \mathbb{1} \in \mathcal{A} \otimes \mathcal{C} \, .$$

By composing all four maps in the Schrödinger picture and in the Heisenberg picture we obtain

$$\begin{array}{ccccccccc}
\mathcal{A} & \alpha) & \mathcal{A} \otimes \mathcal{C} & \beta) & \mathcal{A} \otimes \mathcal{C} & \gamma) & \mathcal{A} \otimes \mathcal{C} & \delta) & \mathcal{A} \\
\varphi & \longrightarrow & \varphi \otimes \psi & \longrightarrow & \varphi \otimes \psi \circ T_{int} & \longrightarrow & \varphi_i & \longrightarrow & \varphi_i|_\mathcal{A} \\
P_\psi T_{int}(x \otimes p_i) & \longleftarrow & T_{int}(x \otimes p_i) & \longleftarrow & x \otimes p_i & \longleftarrow & x \otimes \mathbb{1} & \longleftarrow & x
\end{array}$$

with $\varphi_i := \varphi \otimes \psi \circ T_{int}(\mathbb{1} \otimes p_i \cdot \mathbb{1} \otimes p_i)$.

Altogether, the operator

$$T_i : \mathcal{A} \to \mathcal{A} : x \mapsto P_\psi T_{int}(x \otimes p_i)$$

describes, in the Heisenberg picture, the non-normalized change of states in such a measurement if the i-th value λ_i is the outcome. The probability for this to happen can be computed from the previous section as

$$\begin{aligned}\varphi \otimes \psi \circ T_{int}(\mathbb{1} \otimes p_i) &= \varphi \otimes \psi(\, T_{int}(\mathbb{1} \otimes p_i)\,) \\ &= \varphi(\, P_\psi T_{int}(\mathbb{1} \otimes p_i)\,) \\ &= \varphi(\, T_i(\mathbb{1})\,).\end{aligned}$$

When performing such a measurement but deliberately ignoring its outcome the change of the system is described (in the Heisenberg picture) by

$$T = \sum_i T_i\,.$$

Since the operators T_i were unnormalized we do not need to weight them with their probabilities. The operator T can be computed more explicitly: For $x \in \mathcal{A}$ we obtain

$$T(x) = \sum_i P_\psi T_{int}(x \otimes p_i) = P_\psi T_{int}(x \otimes \mathbb{1})$$

since $\sum_i p_i = \mathbb{1}$.

From their construction it is clear that all operators T and T_i are completely positive and, in addition, T is identity preserving that is $T(\mathbb{1}) = \mathbb{1}$. It should be noted that T does no longer depend on the particular observable $X \in \mathcal{C}$, but only on the interaction T_{int} and the initial state ψ of the apparatus $\mathcal{C}$. The particular decomposition of T reflects the particular choice of X.

10.3 Measurement of a Quantum System and Concrete Representations of Completely Positive Operators

Once again consider a 'true quantum situation' where $\mathcal{A}$ is given by the algebra M_n of all $n \times n$–matrices and $\mathcal{C}$ is given by M_m for some m. Assume further that we perform a kind of 'perfect measurement' : In order to draw a maximal amount of information from such a measurement the spectral projection p_i should be minimal hence 1–dimensional and the initial state ψ of the measurement apparatus should be a pure state. It then follows that there are operators $a_i \in \mathcal{A} = M_n$, $1 \leq i \leq m$, such that

$$\begin{aligned}T_i(x) &= a_i^* x a_i \qquad \text{and thus} \\ T(x) &= \textstyle\sum_i a_i^* x a_i\,.\end{aligned}$$

Indeed, every automophism T_{int} of $M_n \otimes M_m$ is implemented by a unitary $u \in M_n \otimes M_m$ such that $T_{int}(z) = Ad\,u(z) = u^* z u$ for $z \in M_n \otimes M_m$. Since $M_n \otimes M_m$ can be identified with $M_m(M_n)$, the algebra of $m \times m$–matrices with entries from M_n, the unitary u can be written as an $m \times m$ matrix

$$u = \begin{pmatrix} & & \\ & u_{ij} & \\ & & \end{pmatrix}_{m \times m}$$

with entries $u_{ij} \in M_n$, $1 \leq i, j \leq m$.

Moreover, the pure state ψ on M_m is a vector state induced by a unit vector $\begin{pmatrix} \psi_1 \\ \vdots \\ \psi_m \end{pmatrix} \in \mathbb{C}^m$ while p_i projects onto the 1–dimensional subspace spanned by a unit vector $\begin{pmatrix} \xi_1^i \\ \vdots \\ \xi_m^i \end{pmatrix} \in \mathbb{C}^m$.

A short computation shows that $T(x) = \sum_i T_i(x)$ where

$$\begin{aligned} T_i(x) &= P_\psi T_{int}(x \otimes p_i) = P_\psi(u^* \cdot x \otimes p_i \cdot u) \\ &= a_i^* x a_i \end{aligned}$$

with

$$a_i = (\overline{\xi}_1^i, \ldots, \overline{\xi}_m^i) \cdot \begin{pmatrix} & & \\ & u_{ij} & \\ & & \end{pmatrix} \begin{pmatrix} \psi_1 \\ \vdots \\ \psi_m \end{pmatrix}.$$

Summing up, a completely positive operator T with $T(\mathbb{1}) = \mathbb{1}$ describes the state change of a system in the Heisenberg picture due to a measurement with deliberate ignorance. It depends only on the coupling of the system to a measurement apparatus and on the initial state of the apparatus. The measurement of a specific observable $X = \sum_i \lambda_i p_i$ leads to a decomposition $T = \sum_i T_i$ where T_i describes the (non-normalized) change of states if the the outcome λ_i has occurred. The probability of this is given by $\psi(T_i(\mathbb{1}))$.

In the special case of a perfect quantum measurement the operators T_i are of the form $T_i(x) = a_i^* x a_i$ and the probability of an outcome λ_i is given by $\varphi(a_i^* a_i)$.

Conversely, a concrete representation $T(x) = \sum_i a_i^* x a_i$ for $T : M_n \to M_n$ with $T(\mathbb{1}) = \mathbb{1}$ may always be interpreted as coming from such a measurement: Since $T(\mathbb{1}) = \mathbb{1}$ the map

$$v := \begin{pmatrix} a_1 \\ \vdots \\ a_m \end{pmatrix} \quad \text{from } \mathbb{C}^n \text{ into } \mathbb{C}^n \otimes \mathbb{C}^m = \mathbb{C}^n \oplus \ldots \oplus \mathbb{C}^n$$

is an isometry and $T(x) = v^* \cdot x \otimes \mathbb{1} \cdot v$ is a Stinespring representation of T. Construct any unitary $u \in M_n \otimes M_m = M_m(M_n)$ which has $v = \begin{pmatrix} a_1 \\ \vdots \\ a_m \end{pmatrix}$ in its first column (there are many such unitaries) and put $\tilde{\psi} := \begin{pmatrix} 1 \\ 0 \\ \vdots \\ 0 \end{pmatrix} \in \mathbb{C}^m$ which induces the pure state ψ on M_n. Then

$$P_\psi(u^* \cdot x \otimes \mathbb{1} \cdot u) = v^* \cdot x \otimes \mathbb{1} \cdot v = T(x) .$$

Finally, with the orthogonal projection p_i onto the 1–dimensional subspace spanned by the i-th canonical basis vector $\begin{pmatrix} 0 \\ \vdots \\ 1 \\ 0 \\ \vdots \\ 0 \end{pmatrix}$ with 1 as the i-th entry, we obtain

$$P_\psi(u^* \cdot x \otimes p_i \cdot u) = a_i^* x a_i .$$

10.4 Repeated Measurement

Consider now the case where we repeat such a measurement infinitely often. At each time step we couple the system in its present state to the same measurement apparatus which is always prepared in the same initial state. We perform a measurement, thereby changing the state of the system, we then decouple the system from the apparatus, perform the measurement on the apparatus, and start the whole procedure again. Once more the micro–maser can serve as a perfect illustration of such a procedure: Continuing the discussion in Section 10.2 one is now sending many identically prepared atoms through the cavity, one after the other, and measuring their states after they have left the cavity.

For a mathematical description we continue the discussion in the previous section: Each single measurement can have an outcome i in a (finite) set Ω_0 (the particular eigenvalues play no further role thus it is enough just to index the possible outcomes). For simplicity assume that we perform a perfect quantum measurement. Then it is described by a completely positive identity preserving operator T on an algebra M_n ($n \in \mathbb{N}$ or $n = \infty$) with a concrete representation $T(x) = \sum_{i \in \Omega_0} a_i^* x a_i$.

A trajectory of the outcomes of a repeated measurement will be an element in

$$\Omega := \Omega_0^{\mathbb{N}} = \{(\omega_1, \omega_2, \ldots) : \omega_i \in \Omega_0\} .$$

Given the system is initially in a state φ then the probability of measuring $i_1 \in \Omega_0$ at the first measurement is $\varphi(a_{i_1}^* a_{i_1})$ and in this case its state changes to

$$\frac{\varphi(a_{i_1}^* \cdot a_{i_1})}{\varphi(a_{i_1}^* a_{i_1})} .$$

Therefore, the probability of measuring now $i_2 \in \Omega_0$ in a second measurement is given by $\varphi(a_{i_1}^* a_{i_2}^* a_{i_2} a_{i_1})$ and in this case the state changes further to

$$\frac{\varphi(a_{i_1}^* a_{i_2}^* \cdot a_{i_2} a_{i_1})}{\varphi(a_{i_1}^* a_{i_2}^* a_{i_2} a_{i_1})} .$$

Similarly, the probability of obtaining a sequence of outcomes $(i_1, \ldots, i_n) \in \Omega_0^n = \Omega_0 \times \ldots \times \Omega_0$ is given by

$$\mathbb{P}_\varphi^n((i_1, i_2, \ldots, i_n)) := \varphi(a_{i_1}^* a_{i_2}^* \cdot \ldots \cdot a_{i_n}^* a_{i_n} \cdot \ldots \cdot a_{i_2} a_{i_1})$$

which defines a probability measure $\mathbb{P}_\varphi^n$ on Ω_0^n.

The identity $\sum_{i \in \Omega_0} a_i^* a_i = T(\mathbb{1}) = \mathbb{1}$ immediately implies the compatibility condition

$$\mathbb{P}_\varphi^{n+1}((i_1, i_2, \ldots, i_n) \times \Omega_0) = \mathbb{P}_\varphi^n((i_1, \ldots, i_n)) .$$

Therefore, there is a unique probability measure $\mathbb{P}_\varphi$ on Ω defined on the σ-algebra Σ generated by cylinder sets

$$\Lambda_{i_1, \ldots, i_n} := \{\omega \in \Omega : \omega_1 = i_1, \ldots, \omega_n = i_n\}$$

such that

$$\mathbb{P}_\varphi(\Lambda_{i_1, \ldots, i_n}) = \mathbb{P}_\varphi^n((i_1, \ldots, i_n)) .$$

The measure $\mathbb{P}_\varphi$ contains all information on this repeated measurement: For every $A \in \Sigma$ the probability of measuring a trajectory in A is given by $\mathbb{P}_\varphi(A)$.

10.5 Ergodic Theorems for Repeated Measurements

Denote by σ the time shift on Ω that is $\sigma((\omega_1, \omega_2, \omega_3, \ldots)) = (\omega_2, \omega_3, \omega_4, \ldots)$. Then a short computation shows that

$$\mathbb{P}_\varphi(\sigma^{-1}(A)) = \mathbb{P}_{\varphi \circ T}(A)$$

for all sets $A \in \Sigma$. In particular, if φ is *stationary* for T, that is $\varphi \circ T = \varphi$, then $\mathbb{P}_\varphi$ is stationary for σ on Ω. This allows to use methods of classical ergodic theory for the analysis of trajectories for repeated quantum measurements. Indeed, what follows is an extension of Birkhoff's pointwise ergodic theorem to this situation.

Theorem 10.1. *Ergodic Theorem ([KüMa4]) If*

$$\lim_{n\to\infty} \frac{1}{N} \sum_{n=0}^{N-1} \varphi \circ T^n = \varphi_0$$

for all states φ then for any initial state φ and for any set $A \in \Sigma$ which is time invariant, that is $\sigma^{-1}(A) = A$, we have either $\mathbb{P}_\varphi(A) = 0$ or $\mathbb{P}_\varphi(A) = 1$.

We illustrate this theorem by an application: How likely is it to find during such a repeated measurement a certain sequence of outcomes $(i_1, \ldots, i_n) \in \Omega_0^n$? If the initial state is a T–invariant state φ_0 then the probability of finding this sequence as outcome of the measurements $k, k+1, \ldots k+n-1$ is the same as the probability for finding it for the first n measurements. In both cases it is given by $\varphi_0(a_{i_1}^* \ldots a_{i_n}^* a_{i_n} \ldots a_{i_1})$. However, it is also true that this probability is identical to the relative frequency of occurences of this sequence in an arbitrary individual trajectory:

Corollary 10.2. *For any initial state φ and for $(i_1, \ldots i_n) \in \Omega_0^n$*

$$\lim_{N\to\infty} \frac{1}{N} |\{j : j < N \ \ \text{and} \ \ \omega_{j+1} = i_1, \ldots, \omega_{j+n} = i_n\}|$$

$$= \varphi_0(a_{i_1}^* \cdot \ldots \cdot a_{i_n}^* a_{i_n} \cdot \ldots \cdot a_{i_1})$$

for $\mathbb{P}_\varphi$ - almost all paths $\omega \in \Omega_0^{\mathbb{N}}$.

Similarly, all kind of statistical information can be drawn from the observation of a single trajectory of the repeated measurement process: correlations can be measured as autocorrelations. This was tacitly assumed at many places in the literature but it has not been proven up to now. For proofs and further discussions we refer to [KüMa4], where the continuous time versions of the above results are treated.

If a sequence of n measurements has led to a sequence of outcomes $(i_1, \ldots, i_n) \in \Omega_0^n$ then the operator

$$T_{i_1 i_2 \ldots i_n} : x \mapsto a_{i_1}^* \ldots a_{i_n}^* x a_{i_n} \ldots a_{i_1}$$

describes the change of the system in the Heisenberg picture under this measurement, multiplied by the probability of this particular outcomes to occur. Similarly, to any subset $A \subseteq \Omega_0^n$ we associate the operator

$$T_A^n := \sum_{\omega \in \Omega_0^n} T_\omega \ .$$

In particular, $T_{\Omega_0^n} = T^n$.

For subsets $A \subseteq \Omega_0^n$ and $B \subseteq \Omega_0^m$ the set $A \times B$ may be naturally identified with a subset of $\Omega_0^n \times \Omega_0^m = \Omega_0^{n+m}$, and from the definition of T_A^n we obtain

$$T^{n+m}_{A\times B} = T^n_A \circ T^m_B .$$

Therefore, the operators $\{T^n_A : n \in \mathbb{N}, A \subseteq \Omega_0^n\}$ form a discrete time version of the type of quantum stochastic processes which have been considered in [Dav1] for the description of quantum counting processes.

Also for this type of quantum stochastic processes we could prove a pointwise ergodic theorem [KüMa5]. It concerns not only the outcomes of a repeated measurement but the quantum trajectories of the system itself which is being repeatedly measured.

Theorem 10.3. *[KüMa5] Under the same assumptions as in the above ergodic theorem*

$$\lim_{n\to\infty} \frac{1}{N} \sum_{n=1}^{N} \frac{\varphi(a^*_{i_1} \dots a^*_{i_n} \cdot a_{i_n} \dots a_{i_1})}{\varphi(a^*_{i_1} \dots a^*_{i_n} a_{i_n} \dots a_{i_1})} = \varphi_0$$

for any initial state φ and $\omega = (i_1, i_2, \dots)$ $\mathbb{P}_\varphi$ – almost surely.

The continuous time version of this theorem has been discussed and proven in [KüMa5]. We continue to discuss the discrete time version hoping that this shows the ideas of reasoning more clearly. In order to simplify notation we put

$$M_i\psi := \psi(a_i^* \cdot a_i)$$

for any state ψ. Thus $\sum_{i\in\Omega_0} M_i\psi = \psi \circ T$.

Given the initial state φ and $\omega \in \Omega$ we define

$$\Theta_n(\omega) := \frac{M_{\omega_n} \cdot \ldots \cdot M_{\omega_1}\varphi}{\|M_{\omega_n} \cdot \ldots \cdot M_{\omega_1}\varphi\|} = \frac{\varphi(a^*_{\omega_1} \dots a^*_{\omega_n} \cdot \; a_{\omega_n} \dots a_{\omega_1})}{\varphi(a^*_{\omega_1} \dots a^*_{\omega_n} a_{\omega_n} \dots a_{\omega_1})}$$

whenever $\|M_{\omega_n} \cdot \ldots \cdot M_{\omega_1}\varphi\| \neq 0$. By the definition of $\mathbb{P}_\varphi$ the maps $\Theta_n(\omega)$ are well-defined random variables on $(\Omega, \mathbb{P}_\varphi)$ with values in the states of $\mathcal{A}$. Putting $\Theta_0(\omega) := \varphi$ for $\omega \in \Omega$ we thus obtain a stochastic processs $(\Theta_n)_{n\geq 0}$ taking values in the state space of $\mathcal{A}$. A path of this process is also called a *quantum trajectory* . In this sense decompositions as $T(x) = \Sigma_i a_i^* x a_i$ define quantum trajectories.

Using these notions we can formulate a slightly more general version of the above theorem as follows.

Theorem 10.4. *For any initial state φ the pathwise time average*

$$\lim_{N\to\infty} \frac{1}{N} \sum_{n=0}^{N-1} \Theta_n(\omega)$$

exists for $\mathbb{P}_\varphi$-almost every $\omega \in \Omega$. The limit defines a random variable Θ_∞ taking values in the stationary states.

If, in particular, there is a unique stationary state φ_0 with $\varphi_0 \circ T = \varphi_0$ then

$$\lim_{N\to\infty} \frac{1}{N} \sum_{n=0}^{N-1} \Theta_n(\omega) = \varphi_0$$

$\mathbb{P}_\varphi$*-almost surely.*

Quantum trajectories are extensively used in the numerical simulation of irreversible behaviour of open quantum systems, in particular, for computing their equilibrium states (cf. [Car]). The theorem above shows that for purposes like this it is not necessary to perform multiple simulations and determine their sample average. Instead, it is enough to do a simulation along a single path only.

Proof: Since $\mathcal{A} = M_n$ is finite dimensional and $\|T\| = 1$ the operator T is *mean ergodic* , i.e.,

$$P := \lim_{N\to\infty} \frac{1}{N} \sum_{n=0}^{N-1} T^n$$

exists and P is the projection onto the set of fixed elements. It follows that $PT = TP = P$. For more information on ergodic theory we refer to [Kre] and [KüNa].

By Σ_n we denote the σ–subalgebra on Ω generated by the process $(\Theta_k)_{k\geq 0}$ up to time n. Thus Σ_n is generated by the cylinder sets $\{\Lambda_{i_1,\ldots,i_n}$, $(i_1,\ldots,i_n) \in \Omega_0^n\}$. As usual, $\mathbb{E}(X|\Sigma_n)$ denotes the conditional expectation of a random variable X on Ω with respect to Σ_n.

Evaluating the random variables Θ_n, $n \geq 0$, with values in the state space of $\mathcal{A}$ on an element $x \in \mathcal{A}$ we obtain scalar–valued random variables $\Theta_n^x : \Omega \ni \omega \mapsto \Theta_n(\omega)(x)$, $n \geq 0$. Whenever it is convenient we write also $\Theta_n(x)$ for Θ_n^x. For the following arguments we fix an arbitrary element $x \in \mathcal{A}$.

Key observation: On $\mathbb{P}_\varphi$-almost all $\omega \in \Omega$ we obtain

$$\begin{aligned} \mathbb{E}(\Theta_{n+1}(x)|\Sigma_n)(\omega) &= \textstyle\sum_{i\in\Omega_0} \|M_i\Theta_n(\omega)\| \cdot \frac{M_i\Theta_n(\omega)(x)}{\|M_i\Theta_n(\omega)\|} \\ &= \textstyle\sum_{i\in\Omega_0} M_i\Theta_n(\omega)(x) \qquad\qquad (*) \\ &= \Theta_n(\omega)(Tx)\ . \end{aligned}$$

Step 1: Define random variables

$$V_n := \Theta_{n+1}(x) - \Theta_n(Tx)\ , \quad n \geq 0\ ,$$

on $(\Omega, \mathbb{P}_\varphi)$. In order to simplify notation we now omit the argument $\omega \in \Omega$. The random variable V_n is Σ_{n+1}– measurable and $\mathbb{E}(V_n|\Sigma_n) = 0$ by $(*)$. Therefore, the process $(V_n)_{n\geq 0}$ consists of pairwise uncorrelated random variables, hence the process $(Y_n)_{n\geq 0}$ with

$$Y_n := \sum_{j=1}^{n} \frac{1}{j} V_j$$

is a martingale.

From $\mathbb{E}(V_j^2) \leq 4 \cdot \|x\|^2$ we infer $\mathbb{E}(Y_n^2) \leq 4 \cdot \|x\|^2 \cdot \frac{\pi^2}{6}$, hence $(Y_n)_{n \geq 1}$ is uniformly bounded in $L^1(\Omega, \mathbb{P}_\varphi)$. Thus, by the martingale convergence theorem (cf. [Dur]),

$$\lim_{n \to \infty} \sum_{j=1}^{n} \frac{1}{j} V_j =: Y_\infty$$

exists $\mathbb{P}_\varphi$–almost surely. Applying Kronecker's Lemma (cf. [Dur]), it follows that

$$\frac{1}{N} \sum_{j=0}^{N-1} V_j \underset{N \to \infty}{\longrightarrow} 0 \quad \mathbb{P}_\varphi\text{–almost surely,}$$

i.e.,

$$\frac{1}{N} \sum_{j=0}^{N-1} \Big(\Theta_{j+1}(x) - \Theta_j(Tx) \Big) \underset{N \to \infty}{\longrightarrow} 0 \quad \mathbb{P}_\varphi\text{–almost surely,}$$

hence

$$\frac{1}{N} \sum_{j=0}^{N-1} \Big(\Theta_j(x) - \Theta_j(Tx) \Big) \underset{N \to \infty}{\longrightarrow} 0 \quad \mathbb{P}_\varphi\text{–almost surely,}$$

since the last sum differs from the foregoing only by two summands which can be neglected when N becomes large. Applying T it follows that

$$\frac{1}{N} \sum_{j=0}^{N-1} \Big(\Theta_j(Tx) - \Theta_j(T^2 x) \Big) \underset{N \to \infty}{\longrightarrow} 0 \quad \mathbb{P}_\varphi\text{–almost surely,}$$

and by adding this to the foregoing expression we obtain

$$\frac{1}{N} \sum_{j=0}^{N-1} \Big(\Theta_j(x) - \Theta_j(T^2 x) \Big) \underset{N \to \infty}{\longrightarrow} 0 \quad \mathbb{P}_\varphi\text{–almost surely.}$$

By the same argument we see

$$\frac{1}{N} \sum_{j=0}^{N-1} \Big(\Theta_j(x) - \Theta_j(T^l x) \Big) \underset{N \to \infty}{\longrightarrow} 0 \quad \mathbb{P}_\varphi\text{–almost surely for all } l \in \mathbb{N}$$

and averaging this over the first m values of l yields

$$\frac{1}{N}\sum_{j=0}^{N-1}\left(\Theta_j(x) - \frac{1}{m}\sum_{l=0}^{m-1}\Theta_j(T^l x)\right) \underset{N\to\infty}{\longrightarrow} 0 \quad \mathbb{P}_\varphi\text{–almost surely for } m \in \mathbb{N}\,.$$

We may exchange the limits $N \to \infty$ and $m \to \infty$ and finally obtain

$$\frac{1}{N}\sum_{j=0}^{N-1}\left(\Theta_j(x) - \Theta_j(Px)\right) \underset{N\to\infty}{\longrightarrow} 0 \quad \mathbb{P}_\varphi\text{–almost surely.} \qquad (**)$$

Step 2: From the above key observation $(*)$ we obtain

$$\mathbb{E}(\Theta_{n+1}(Px)|\Sigma_n) = \Theta_n(TPx) = \Theta_n(Px)\,,$$

hence the process $(\Theta_n(Px))_{n\geq 0}$, too, is a uniformly bounded martingale which converges to a random variable Θ^x_∞ $\mathbb{P}_\varphi$–almost surely on Ω. By $(**)$ the averages of the difference $(\Theta_j(x) - \Theta_j(Px))_{j\geq 0}$ converge to zero, hence

$$\lim_{N\to\infty}\frac{1}{N}\sum_{j=0}^{N-1}\Theta_j(x) = \Theta^x_\infty \qquad \mathbb{P}_\varphi\text{– almost surely on } \Omega\,.$$

This holds for all $x \in \mathcal{A}$, hence the averages

$$\frac{1}{N}\sum_{j=0}^{N-1}\Theta_j$$

converge to some random variable Θ_∞ with values in the state space of $\mathcal{A}$ $\mathbb{P}_\varphi$–almost surely.

Finally, since $PTx = Tx$ for $x \in \mathcal{A}$, we obtain

$$\begin{aligned}\Theta_\infty(Tx) &= \lim_{n\to\infty}\Theta_n(PTx) = \lim_{n\to\infty}\Theta_n(Px)\\ &= \Theta_\infty(x)\,,\end{aligned}$$

hence Θ_∞ takes values in the stationary states.

□

If a quantum trajectory starts in a pure state φ it will clearly stay in the pure states for all times. However, our computer simulations showed that even if initially starting with a mixed state there was a tendency for the state to "purify" along a trajectory. There is an obvious exception: If T is decomposed into a convex combination of automorphisms, i.e., if the operators a_i are multiples of unitaries for all $i \in \Omega_0$ then a mixed state φ will never purify since all states along the trajectory will stay being unitarily equivalent to φ. In a sense this is the only exception:

For a state ψ on $\mathcal{A} = M_n$ we denote by ρ_ψ the corresponding density matrix such that $\psi(x) = tr(\rho_\psi \cdot x)$ where, as usual, tr denotes the trace on $\mathcal{A} = M_n$.

Definition 10.5. *A quantum trajectory* $(\Theta_n(\omega))_{n\geq 0}$ *purifies, if*

$$\lim_{n\to\infty} tr(\rho^2_{\Theta_n(\omega)}) = 1 \ .$$

Theorem 10.6. *[MaKü] The quantum trajectories* $(\Theta_n(\omega))_{n\geq 0}$, $\omega \in \Omega$, *purify* $\mathbb{P}_\varphi$*-almost surely or there exists a projection* $p \in \mathcal{A} = M_n$ *with dim p* ≥ 2, *such that* $pa_i^* a_i p = \lambda_i p$ *for all* $i \in \Omega_0$ *and* $\lambda_i \geq 0$.

Corollary 10.7. *On* $\mathcal{A} = M_2$ *quantum trajectories purify* $\mathbb{P}_\varphi$*-almost surely or* $a_i = \lambda_i u_i$ *for* $\lambda_i \in \mathbb{C}$ *and* $u_i \in M_2$ *unitary for all* $i \in \Omega_0$, *i.e., T is decomposed into a convex combination of automorphisms.*

References

[AFL] L.Accardi, F. Frigerio, J.T. Lewis: Quantum stochastic processes. Publ. RIMS 18 (1982), 97 - 133.

[ApH] D. Applebaum, R.L. Hudson: Fermion Itô's formula and stochastic evolutions. Commun. Math. Phys. 96 (1984), 473.

[BKS] M. Bożejko, B. Kümmerer, R. Speicher: q-Gaussian processes: non-commutative and classical aspects. Commun. Math. Phys. 185 (1997), 129 - 154.

[Car] H. J. Carmichael: *An Open Systems Approach to Quantum Optics.* Springer Verlag, Berlin 1993.

[Dav1] E.B. Davies: *Quantum Theory of Open Systems.* Academic Press, London 1976.

[Dav2] E. B. Davies: *One Parameter Semigroups* Academic Press, London 1980.

[Dur] R. Durett: *Probability: Theory and Examples.* Duxbury Press, Belmont 1996.

[Eva] D. E. Evans: Completely positive quasi-free maps on the CAR algebra. Commun. Math. Phys. 70 (1979), 53-68.

[EvLe] D. Evans, J.T. Lewis: *Dilations of Irreversible Evolutions in Algebraic Quantum Theory.* Comm. Dublin Inst. Adv. Stud. Ser A 24, 1977.

[GKS] V. Gorini, A. Kossakowski, E.C.G. Sudarshan: Completely positive dynamical semigroups of n-level systems, J. Math. Phys. 17 (1976), 821 - 825.

[Haa] F. Haag: Asymptotik von Quanten-Markov-Halbgruppen und Quanten-Markov-Prozessen, Dissertation, Darmstadt 2005.

[Hid] T. Hida: *Brownian motion.* Springer-Verlag, Berlin 1980.

[Kra] K. Kraus: General state changes in quantum theory. Ann. Phys. 64 (1971), 311 - 335.

[Kre] U. Krengel: *Ergodic Theorems.* Walter de Gruyter, Berlin-New York 1985.

[Kü1] B. Kümmerer: Examples of Markov dilations over the 2 × 2-matrices. In *Quantum Probability and Applications I*, Lecture Notes in Mathematics 1055, Springer-Verlag, Berlin-Heidelberg-New York-Tokyo 1984, 228 - 244.

[Kü2] B. Kümmerer: Markov dilations on W*-algebras. Journ. Funct. Anal. 63 (1985), 139 - 177.

[Kü3] B. Kümmerer: Stationary processes in quantum probability. *Quantum Probability Communications XI.* World Scientific 2003, 273 - 304.

[Kü4] B. Kümmerer: Quantum Markov processes. In *Coherent Evolution in Noisy Environments*, A. Buchleitner, K Hornberger (Eds.), Springer Lecture Notes in Physics 611 (2002), 139 - 198.

[KüMa1] B. Kümmerer, H. Maassen: The essentially commutative dilations of dynamical semigroups on M_n. Commun. Math. Phys. 109 (1987), 1 - 22.

[KüMa2] B. Kümmerer, H. Maassen: Elements of quantum probability. In *Quantum Probability Communications X*, World Scientific 1998, 73 - 100.

[KüMa3] B. Kümmerer, H. Maassen: A scattering theory for Markov chains. Infinite Dimensional Analysis, Quantum Probability and Related Topics, Vol. 3, No. 1 (2000), 161 - 176.

[KüMa4] B. Kümmerer, H. Maassen: An ergodic theorem for quantum counting processes. J. Phys. A: Math Gen. 36 (2003), 2155 - 2161.

[KüMa5] B. Kümmerer, H. Maassen: A pathwise ergodic theorem for quantum trajectories. J. Phys. A: Math. Gen. 37 (2004) 11889-11896.

[KüNa] B. Kümmerer, R.J. Nagel: Mean ergodic semigroups on W*-Algebras. Acta Sci. Math. 41 (1979), 151-159.

[KüS1] B. Kümmerer, W. Schröder: A new construction of unitary dilations: singular coupling to white noise. In *Quantum Probability and Applications II*, (L. Accardi, W. von Waldenfels, eds.) Springer, Berlin 1985, 332–347 (1985).

[KüS2] B. Kümmerer, W. Schröder: A Markov dilation of a non-quasifree Bloch evolution. Comm. Math. Phys. 90 (1983), 251-262.

[LaPh] P.D. Lax, R.S. Phillips: *Scattering Theory*. Academic Press, New York 1967.

[Lin] G. Lindblad: On the generators of quantum dynamical semigroups, Commun. Math. Phys. 48 (1976), 119 - 130.

[LiMa] J.M. Lindsay, H. Maassen: Stochastic calculus for quantum Brownian motion of non-minimal variance. In: *Mark Kac seminar on probability and physics*, Syllabus 1987–1992. CWI Syllabus 32 (1992), Amsterdam.

[MaKü] H. Maassen, B. Kümmerer: Purification of quantum trajectories, quant-ph/0505084, to appear in IMS Lecture Notes-Monograph Series.

[Mol] B.R. Mollow: Power spectrum of light scattered by two-level systems. Phys. Rev. 188 (1969), 1969–1975.

[JvN] John von Neumann: *Mathematische Grundlagen der Quantenmechanik*. Springer, Berlin 1932, 1968.

[Par] K.R. Parthasarathy: *An Introduction to Quantum Stochastic Calculus*. Birkhäuser Verlag, Basel 1992.

[RoMa] P. Robinson, H. Maassen: Quantum stochastic calculus and the dynamical Stark effect. Reports Math. Phys. 30 (1991), 185–203.

[RS] M. Reed, B. Simon: *Methods of Modern Mathematical Physics. I: Functional Analysis*. Academic Press, New York 1972.

[SSH] F. Schuda, C.R. Stroud, M. Hercher: Observation of resonant Stark effect at optical frequencies. Journ. Phys. B7 (1974), 198.

[SzNF] B. Sz.-Nagy, C. Foias: *Harmonic Analysis of Operators on Hilbert Space*. North Holland, Amsterdam 1970.

[Tak1] M. Takesaki: Conditional expectations in von Neumann algebras. J. Funct. Anal 9 (1971), 306 - 321.

[Tak2] M. Takesaki: *Theory of Operator Algebras I*. Springer, New York 1979.

[VBWW] B.T.H. Varcoe, S. Battke, M. Weidinger, H. Walther: Preparing pure photon number states of the radiation field. Nature 403 (2000), 743 - 746.

[WBKM] T. Wellens, A. Buchleitner and B. Kümmerer, H. Maassen: Quantum state preparation via asymptotic completeness. Phys. Rev. Letters 85 (2000), 3361.

[Wel] Thomas Wellens: Entanglement and Control of Quantum States. Dissertation, München 2002.

Index

Lecture Notes in Mathematics

For information about earlier volumes please contact your bookseller or Springer LNM Online archive: springerlink.com

Vol. 1674: G. Klaas, C. R. Leedham-Green, W. Plesken, Linear Pro-*p*-Groups of Finite Width (1997)
Vol. 1675: J. E. Yukich, Probability Theory of Classical Euclidean Optimization Problems (1998)
Vol. 1676: P. Cembranos, J. Mendoza, Banach Spaces of Vector-Valued Functions (1997)
Vol. 1677: N. Proskurin, Cubic Metaplectic Forms and Theta Functions (1998)
Vol. 1678: O. Krupková, The Geometry of Ordinary Variational Equations (1997)
Vol. 1679: K.-G. Grosse-Erdmann, The Blocking Technique. Weighted Mean Operators and Hardy's Inequality (1998)
Vol. 1680: K.-Z. Li, F. Oort, Moduli of Supersingular Abelian Varieties (1998)
Vol. 1681: G. J. Wirsching, The Dynamical System Generated by the 3n+1 Function (1998)
Vol. 1682: H.-D. Alber, Materials with Memory (1998)
Vol. 1683: A. Pomp, The Boundary-Domain Integral Method for Elliptic Systems (1998)
Vol. 1684: C. A. Berenstein, P. F. Ebenfelt, S. G. Gindikin, S. Helgason, A. E. Tumanov, Integral Geometry, Radon Transforms and Complex Analysis. Firenze, 1996. Editors: E. Casadio Tarabusi, M. A. Picardello, G. Zampieri (1998)
Vol. 1685: S. König, A. Zimmermann, Derived Equivalences for Group Rings (1998)
Vol. 1686: J. Azéma, M. Émery, M. Ledoux, M. Yor (Eds.), Séminaire de Probabilités XXXII (1998)
Vol. 1687: F. Bornemann, Homogenization in Time of Singularly Perturbed Mechanical Systems (1998)
Vol. 1688: S. Assing, W. Schmidt, Continuous Strong Markov Processes in Dimension One (1998)
Vol. 1689: W. Fulton, P. Pragacz, Schubert Varieties and Degeneracy Loci (1998)
Vol. 1690: M. T. Barlow, D. Nualart, Lectures on Probability Theory and Statistics. Editor: P. Bernard (1998)
Vol. 1691: R. Bezrukavnikov, M. Finkelberg, V. Schechtman, Factorizable Sheaves and Quantum Groups (1998)
Vol. 1692: T. M. W. Eyre, Quantum Stochastic Calculus and Representations of Lie Superalgebras (1998)
Vol. 1694: A. Braides, Approximation of Free-Discontinuity Problems (1998)
Vol. 1695: D. J. Hartfiel, Markov Set-Chains (1998)
Vol. 1696: E. Bouscaren (Ed.): Model Theory and Algebraic Geometry (1998)
Vol. 1697: B. Cockburn, C. Johnson, C.-W. Shu, E. Tadmor, Advanced Numerical Approximation of Nonlinear Hyperbolic Equations. Cetraro, Italy, 1997. Editor: A. Quarteroni (1998)
Vol. 1698: M. Bhattacharjee, D. Macpherson, R. G. Möller, P. Neumann, Notes on Infinite Permutation Groups (1998)
Vol. 1699: A. Inoue,Tomita-Takesaki Theory in Algebras of Unbounded Operators (1998)
Vol. 1700: W. A. Woyczyński, Burgers-KPZ Turbulence (1998)
Vol. 1701: Ti-Jun Xiao, J. Liang, The Cauchy Problem of Higher Order Abstract Differential Equations (1998)
Vol. 1702: J. Ma, J. Yong, Forward-Backward Stochastic Differential Equations and Their Applications (1999)
Vol. 1703: R. M. Dudley, R. Norvaiša, Differentiability of Six Operators on Nonsmooth Functions and p-Variation (1999)
Vol. 1704: H. Tamanoi, Elliptic Genera and Vertex Operator Super-Algebras (1999)
Vol. 1705: I. Nikolaev, E. Zhuzhoma, Flows in 2-dimensional Manifolds (1999)
Vol. 1706: S. Yu. Pilyugin, Shadowing in Dynamical Systems (1999)
Vol. 1707: R. Pytlak, Numerical Methods for Optimal Control Problems with State Constraints (1999)
Vol. 1708: K. Zuo, Representations of Fundamental Groups of Algebraic Varieties (1999)
Vol. 1709: J. Azéma, M. Émery, M. Ledoux, M. Yor (Eds.), Séminaire de Probabilités XXXIII (1999)
Vol. 1710: M. Koecher, The Minnesota Notes on Jordan Algebras and Their Applications (1999)
Vol. 1711: W. Ricker, Operator Algebras Generated by Commuting Projećtions: A Vector Measure Approach (1999)
Vol. 1712: N. Schwartz, J. J. Madden, Semi-algebraic Function Rings and Reflectors of Partially Ordered Rings (1999)
Vol. 1713: F. Bethuel, G. Huisken, S. Müller, K. Steffen, Calculus of Variations and Geometric Evolution Problems. Cetraro, 1996. Editors: S. Hildebrandt, M. Struwe (1999)
Vol. 1714: O. Diekmann, R. Durrett, K. P. Hadeler, P. K. Maini, H. L. Smith, Mathematics Inspired by Biology. Martina Franca, 1997. Editors: V. Capasso, O. Diekmann (1999)
Vol. 1715: N. V. Krylov, M. Röckner, J. Zabczyk, Stochastic PDE's and Kolmogorov Equations in Infinite Dimensions. Cetraro, 1998. Editor: G. Da Prato (1999)
Vol. 1716: J. Coates, R. Greenberg, K. A. Ribet, K. Rubin, Arithmetic Theory of Elliptic Curves. Cetraro, 1997. Editor: C. Viola (1999)
Vol. 1717: J. Bertoin, F. Martinelli, Y. Peres, Lectures on Probability Theory and Statistics. Saint-Flour, 1997. Editor: P. Bernard (1999)
Vol. 1718: A. Eberle, Uniqueness and Non-Uniqueness of Semigroups Generated by Singular Diffusion Operators (1999)
Vol. 1719: K. R. Meyer, Periodic Solutions of the N-Body Problem (1999)
Vol. 1720: D. Elworthy, Y. Le Jan, X-M. Li, On the Geometry of Diffusion Operators and Stochastic Flows (1999)
Vol. 1721: A. Iarrobino, V. Kanev, Power Sums, Gorenstein Algebras, and Determinantal Loci (1999)

Vol. 1722: R. McCutcheon, Elemental Methods in Ergodic Ramsey Theory (1999)
Vol. 1723: J. P. Croisille, C. Lebeau, Diffraction by an Immersed Elastic Wedge (1999)
Vol. 1724: V. N. Kolokoltsov, Semiclassical Analysis for Diffusions and Stochastic Processes (2000)
Vol. 1725: D. A. Wolf-Gladrow, Lattice-Gas Cellular Automata and Lattice Boltzmann Models (2000)
Vol. 1726: V. Marić, Regular Variation and Differential Equations (2000)
Vol. 1727: P. Kravanja M. Van Barel, Computing the Zeros of Analytic Functions (2000)
Vol. 1728: K. Gatermann Computer Algebra Methods for Equivariant Dynamical Systems (2000)
Vol. 1729: J. Azéma, M. Émery, M. Ledoux, M. Yor (Eds.) Séminaire de Probabilités XXXIV (2000)
Vol. 1730: S. Graf, H. Luschgy, Foundations of Quantization for Probability Distributions (2000)
Vol. 1731: T. Hsu, Quilts: Central Extensions, Braid Actions, and Finite Groups (2000)
Vol. 1732: K. Keller, Invariant Factors, Julia Equivalences and the (Abstract) Mandelbrot Set (2000)
Vol. 1733: K. Ritter, Average-Case Analysis of Numerical Problems (2000)
Vol. 1734: M. Espedal, A. Fasano, A. Mikelić, Filtration in Porous Media and Industrial Applications. Cetraro 1998. Editor: A. Fasano. 2000.
Vol. 1735: D. Yafaev, Scattering Theory: Some Old and New Problems (2000)
Vol. 1736: B. O. Turesson, Nonlinear Potential Theory and Weighted Sobolev Spaces (2000)
Vol. 1737: S. Wakabayashi, Classical Microlocal Analysis in the Space of Hyperfunctions (2000)
Vol. 1738: M. Émery, A. Nemirovski, D. Voiculescu, Lectures on Probability Theory and Statistics (2000)
Vol. 1739: R. Burkard, P. Deuflhard, A. Jameson, J.-L. Lions, G. Strang, Computational Mathematics Driven by Industrial Problems. Martina Franca, 1999. Editors: V. Capasso, H. Engl, J. Periaux (2000)
Vol. 1740: B. Kawohl, O. Pironneau, L. Tartar, J.-P. Zolesio, Optimal Shape Design. Tróia, Portugal 1999. Editors: A. Cellina, A. Ornelas (2000)
Vol. 1741: E. Lombardi, Oscillatory Integrals and Phenomena Beyond all Algebraic Orders (2000)
Vol. 1742: A. Unterberger, Quantization and Non-holomorphic Modular Forms (2000)
Vol. 1743: L. Habermann, Riemannian Metrics of Constant Mass and Moduli Spaces of Conformal Structures (2000)
Vol. 1744: M. Kunze, Non-Smooth Dynamical Systems (2000)
Vol. 1745: V. D. Milman, G. Schechtman (Eds.), Geometric Aspects of Functional Analysis. Israel Seminar 1999-2000 (2000)
Vol. 1746: A. Degtyarev, I. Itenberg, V. Kharlamov, Real Enriques Surfaces (2000)
Vol. 1747: L. W. Christensen, Gorenstein Dimensions (2000)
Vol. 1748: M. Ruzicka, Electrorheological Fluids: Modeling and Mathematical Theory (2001)
Vol. 1749: M. Fuchs, G. Seregin, Variational Methods for Problems from Plasticity Theory and for Generalized Newtonian Fluids (2001)
Vol. 1750: B. Conrad, Grothendieck Duality and Base Change (2001)
Vol. 1751: N. J. Cutland, Loeb Measures in Practice: Recent Advances (2001)
Vol. 1752: Y. V. Nesterenko, P. Philippon, Introduction to Algebraic Independence Theory (2001)
Vol. 1753: A. I. Bobenko, U. Eitner, Painlevé Equations in the Differential Geometry of Surfaces (2001)
Vol. 1754: W. Bertram, The Geometry of Jordan and Lie Structures (2001)
Vol. 1755: J. Azéma, M. Émery, M. Ledoux, M. Yor (Eds.), Séminaire de Probabilités XXXV (2001)
Vol. 1756: P. E. Zhidkov, Korteweg de Vries and Nonlinear Schrödinger Equations: Qualitative Theory (2001)
Vol. 1757: R. R. Phelps, Lectures on Choquet's Theorem (2001)
Vol. 1758: N. Monod, Continuous Bounded Cohomology of Locally Compact Groups (2001)
Vol. 1759: Y. Abe, K. Kopfermann, Toroidal Groups (2001)
Vol. 1760: D. Filipović, Consistency Problems for Heath-Jarrow-Morton Interest Rate Models (2001)
Vol. 1761: C. Adelmann, The Decomposition of Primes in Torsion Point Fields (2001)
Vol. 1762: S. Cerrai, Second Order PDE's in Finite and Infinite Dimension (2001)
Vol. 1763: J.-L. Loday, A. Frabetti, F. Chapoton, F. Goichot, Dialgebras and Related Operads (2001)
Vol. 1764: A. Cannas da Silva, Lectures on Symplectic Geometry (2001)
Vol. 1765: T. Kerler, V. V. Lyubashenko, Non-Semisimple Topological Quantum Field Theories for 3-Manifolds with Corners (2001)
Vol. 1766: H. Hennion, L. Hervé, Limit Theorems for Markov Chains and Stochastic Properties of Dynamical Systems by Quasi-Compactness (2001)
Vol. 1767: J. Xiao, Holomorphic Q Classes (2001)
Vol. 1768: M.J. Pflaum, Analytic and Geometric Study of Stratified Spaces (2001)
Vol. 1769: M. Alberich-Carramiñana, Geometry of the Plane Cremona Maps (2002)
Vol. 1770: H. Gluesing-Luerssen, Linear Delay-Differential Systems with Commensurate Delays: An Algebraic Approach (2002)
Vol. 1771: M. Émery, M. Yor (Eds.), Séminaire de Probabilités 1967-1980. A Selection in Martingale Theory (2002)
Vol. 1772: F. Burstall, D. Ferus, K. Leschke, F. Pedit, U. Pinkall, Conformal Geometry of Surfaces in S^4 (2002)
Vol. 1773: Z. Arad, M. Muzychuk, Standard Integral Table Algebras Generated by a Non-real Element of Small Degree (2002)
Vol. 1774: V. Runde, Lectures on Amenability (2002)
Vol. 1775: W. H. Meeks, A. Ros, H. Rosenberg, The Global Theory of Minimal Surfaces in Flat Spaces. Martina Franca 1999. Editor: G. P. Pirola (2002)
Vol. 1776: K. Behrend, C. Gomez, V. Tarasov, G. Tian, Quantum Comohology. Cetraro 1997. Editors: P. de Bartolomeis, B. Dubrovin, C. Reina (2002)
Vol. 1777: E. García-Río, D. N. Kupeli, R. Vázquez-Lorenzo, Osserman Manifolds in Semi-Riemannian Geometry (2002)
Vol. 1778: H. Kiechle, Theory of K-Loops (2002)
Vol. 1779: I. Chueshov, Monotone Random Systems (2002)
Vol. 1780: J. H. Bruinier, Borcherds Products on O(2,1) and Chern Classes of Heegner Divisors (2002)
Vol. 1781: E. Bolthausen, E. Perkins, A. van der Vaart, Lectures on Probability Theory and Statistics. Ecole d' Eté de Probabilités de Saint-Flour XXIX-1999. Editor: P. Bernard (2002)

Vol. 1782: C.-H. Chu, A. T.-M. Lau, Harmonic Functions on Groups and Fourier Algebras (2002)
Vol. 1783: L. Grüne, Asymptotic Behavior of Dynamical and Control Systems under Perturbation and Discretization (2002)
Vol. 1784: L.H. Eliasson, S. B. Kuksin, S. Marmi, J.-C. Yoccoz, Dynamical Systems and Small Divisors. Cetraro, Italy 1998. Editors: S. Marmi, J.-C. Yoccoz (2002)
Vol. 1785: J. Arias de Reyna, Pointwise Convergence of Fourier Series (2002)
Vol. 1786: S. D. Cutkosky, Monomialization of Morphisms from 3-Folds to Surfaces (2002)
Vol. 1787: S. Caenepeel, G. Militaru, S. Zhu, Frobenius and Separable Functors for Generalized Module Categories and Nonlinear Equations (2002)
Vol. 1788: A. Vasil'ev, Moduli of Families of Curves for Conformal and Quasiconformal Mappings (2002)
Vol. 1789: Y. Sommerhäuser, Yetter-Drinfel'd Hopf algebras over groups of prime order (2002)
Vol. 1790: X. Zhan, Matrix Inequalities (2002)
Vol. 1791: M. Knebusch, D. Zhang, Manis Valuations and Prüfer Extensions I: A new Chapter in Commutative Algebra (2002)
Vol. 1792: D. D. Ang, R. Gorenflo, V. K. Le, D. D. Trong, Moment Theory and Some Inverse Problems in Potential Theory and Heat Conduction (2002)
Vol. 1793: J. Cortés Monforte, Geometric, Control and Numerical Aspects of Nonholonomic Systems (2002)
Vol. 1794: N. Pytheas Fogg, Substitution in Dynamics, Arithmetics and Combinatorics. Editors: V. Berthé, S. Ferenczi, C. Mauduit, A. Siegel (2002)
Vol. 1795: H. Li, Filtered-Graded Transfer in Using Noncommutative Gröbner Bases (2002)
Vol. 1796: J.M. Melenk, hp-Finite Element Methods for Singular Perturbations (2002)
Vol. 1797: B. Schmidt, Characters and Cyclotomic Fields in Finite Geometry (2002)
Vol. 1798: W.M. Oliva, Geometric Mechanics (2002)
Vol. 1799: H. Pajot, Analytic Capacity, Rectifiability, Menger Curvature and the Cauchy Integral (2002)
Vol. 1800: O. Gabber, L. Ramero, Almost Ring Theory (2003)
Vol. 1801: J. Azéma, M. Émery, M. Ledoux, M. Yor (Eds.), Séminaire de Probabilités XXXVI (2003)
Vol. 1802: V. Capasso, E. Merzbach, B.G. Ivanoff, M. Dozzi, R. Dalang, T. Mountford, Topics in Spatial Stochastic Processes. Martina Franca, Italy 2001. Editor: E. Merzbach (2003)
Vol. 1803: G. Dolzmann, Variational Methods for Crystalline Microstructure – Analysis and Computation (2003)
Vol. 1804: I. Cherednik, Ya. Markov, R. Howe, G. Lusztig, Iwahori-Hecke Algebras and their Representation Theory. Martina Franca, Italy 1999. Editors: V. Baldoni, D. Barbasch (2003)
Vol. 1805: F. Cao, Geometric Curve Evolution and Image Processing (2003)
Vol. 1806: H. Broer, I. Hoveijn. G. Lunther, G. Vegter, Bifurcations in Hamiltonian Systems. Computing Singularities by Gröbner Bases (2003)
Vol. 1807: V. D. Milman, G. Schechtman (Eds.), Geometric Aspects of Functional Analysis. Israel Seminar 2000-2002 (2003)
Vol. 1808: W. Schindler, Measures with Symmetry Properties (2003)
Vol. 1809: O. Steinbach, Stability Estimates for Hybrid Coupled Domain Decomposition Methods (2003)
Vol. 1810: J. Wengenroth, Derived Functors in Functional Analysis (2003)
Vol. 1811: J. Stevens, Deformations of Singularities (2003)
Vol. 1812: L. Ambrosio, K. Deckelnick, G. Dziuk, M. Mimura, V. A. Solonnikov, H. M. Soner, Mathematical Aspects of Evolving Interfaces. Madeira, Funchal, Portugal 2000. Editors: P. Colli, J. F. Rodrigues (2003)
Vol. 1813: L. Ambrosio, L. A. Caffarelli, Y. Brenier, G. Buttazzo, C. Villani, Optimal Transportation and its Applications. Martina Franca, Italy 2001. Editors: L. A. Caffarelli, S. Salsa (2003)
Vol. 1814: P. Bank, F. Baudoin, H. Föllmer, L.C.G. Rogers, M. Soner, N. Touzi, Paris-Princeton Lectures on Mathematical Finance 2002 (2003)
Vol. 1815: A. M. Vershik (Ed.), Asymptotic Combinatorics with Applications to Mathematical Physics. St. Petersburg, Russia 2001 (2003)
Vol. 1816: S. Albeverio, W. Schachermayer, M. Talagrand, Lectures on Probability Theory and Statistics. Ecole d'Eté de Probabilités de Saint-Flour XXX-2000. Editor: P. Bernard (2003)
Vol. 1817: E. Koelink, W. Van Assche(Eds.), Orthogonal Polynomials and Special Functions. Leuven 2002 (2003)
Vol. 1818: M. Bildhauer, Convex Variational Problems with Linear, nearly Linear and/or Anisotropic Growth Conditions (2003)
Vol. 1819: D. Masser, Yu. V. Nesterenko, H. P. Schlickewei, W. M. Schmidt, M. Waldschmidt, Diophantine Approximation. Cetraro, Italy 2000. Editors: F. Amoroso, U. Zannier (2003)
Vol. 1820: F. Hiai, H. Kosaki, Means of Hilbert Space Operators (2003)
Vol. 1821: S. Teufel, Adiabatic Perturbation Theory in Quantum Dynamics (2003)
Vol. 1822: S.-N. Chow, R. Conti, R. Johnson, J. Mallet-Paret, R. Nussbaum, Dynamical Systems. Cetraro, Italy 2000. Editors: J. W. Macki, P. Zecca (2003)
Vol. 1823: A. M. Anile, W. Allegretto, C. Ringhofer, Mathematical Problems in Semiconductor Physics. Cetraro, Italy 1998. Editor: A. M. Anile (2003)
Vol. 1824: J. A. Navarro González, J. B. Sancho de Salas, $\mathscr{C}^\infty$ – Differentiable Spaces (2003)
Vol. 1825: J. H. Bramble, A. Cohen, W. Dahmen, Multiscale Problems and Methods in Numerical Simulations, Martina Franca, Italy 2001. Editor: C. Canuto (2003)
Vol. 1826: K. Dohmen, Improved Bonferroni Inequalities via Abstract Tubes. Inequalities and Identities of Inclusion-Exclusion Type. VIII, 113 p, 2003.
Vol. 1827: K. M. Pilgrim, Combinations of Complex Dynamical Systems. IX, 118 p, 2003.
Vol. 1828: D. J. Green, Gröbner Bases and the Computation of Group Cohomology. XII, 138 p, 2003.
Vol. 1829: E. Altman, B. Gaujal, A. Hordijk, Discrete-Event Control of Stochastic Networks: Multimodularity and Regularity. XIV, 313 p, 2003.
Vol. 1830: M. I. Gil', Operator Functions and Localization of Spectra. XIV, 256 p, 2003.
Vol. 1831: A. Connes, J. Cuntz, E. Guentner, N. Higson, J. E. Kaminker, Noncommutative Geometry, Martina Franca, Italy 2002. Editors: S. Doplicher, L. Longo (2004)
Vol. 1832: J. Azéma, M. Émery, M. Ledoux, M. Yor (Eds.), Séminaire de Probabilités XXXVII (2003)
Vol. 1833: D.-Q. Jiang, M. Qian, M.-P. Qian, Mathematical Theory of Nonequilibrium Steady States. On the Frontier of Probability and Dynamical Systems. IX, 280 p, 2004.

Vol. 1834: Yo. Yomdin, G. Comte, Tame Geometry with Application in Smooth Analysis. VIII, 186 p, 2004.
Vol. 1835: O.T. Izhboldin, B. Kahn, N.A. Karpenko, A. Vishik, Geometric Methods in the Algebraic Theory of Quadratic Forms. Summer School, Lens, 2000. Editor: J.-P. Tignol (2004)
Vol. 1836: C. Năstăsescu, F. Van Oystaeyen, Methods of Graded Rings. XIII, 304 p, 2004.
Vol. 1837: S. Tavaré, O. Zeitouni, Lectures on Probability Theory and Statistics. Ecole d'Eté de Probabilités de Saint-Flour XXXI-2001. Editor: J. Picard (2004)
Vol. 1838: A.J. Ganesh, N.W. O'Connell, D.J. Wischik, Big Queues. XII, 254 p, 2004.
Vol. 1839: R. Gohm, Noncommutative Stationary Processes. VIII, 170 p, 2004.
Vol. 1840: B. Tsirelson, W. Werner, Lectures on Probability Theory and Statistics. Ecole d'Eté de Probabilités de Saint-Flour XXXII-2002. Editor: J. Picard (2004)
Vol. 1841: W. Reichel, Uniqueness Theorems for Variational Problems by the Method of Transformation Groups (2004)
Vol. 1842: T. Johnsen, A.L. Knutsen, K3 Projective Models in Scrolls (2004)
Vol. 1843: B. Jefferies, Spectral Properties of Noncommuting Operators (2004)
Vol. 1844: K.F. Siburg, The Principle of Least Action in Geometry and Dynamics (2004)
Vol. 1845: Min Ho Lee, Mixed Automorphic Forms, Torus Bundles, and Jacobi Forms (2004)
Vol. 1846: H. Ammari, H. Kang, Reconstruction of Small Inhomogeneities from Boundary Measurements (2004)
Vol. 1847: T.R. Bielecki, T. Björk, M. Jeanblanc, M. Rutkowski, J.A. Scheinkman, W. Xiong, Paris-Princeton Lectures on Mathematical Finance 2003 (2004)
Vol. 1848: M. Abate, J. E. Fornaess, X. Huang, J. P. Rosay, A. Tumanov, Real Methods in Complex and CR Geometry, Martina Franca, Italy 2002. Editors: D. Zaitsev, G. Zampieri (2004)
Vol. 1849: Martin L. Brown, Heegner Modules and Elliptic Curves (2004)
Vol. 1850: V. D. Milman, G. Schechtman (Eds.), Geometric Aspects of Functional Analysis. Israel Seminar 2002-2003 (2004)
Vol. 1851: O. Catoni, Statistical Learning Theory and Stochastic Optimization (2004)
Vol. 1852: A.S. Kechris, B.D. Miller, Topics in Orbit Equivalence (2004)
Vol. 1853: Ch. Favre, M. Jonsson, The Valuative Tree (2004)
Vol. 1854: O. Saeki, Topology of Singular Fibers of Differential Maps (2004)
Vol. 1855: G. Da Prato, P.C. Kunstmann, I. Lasiecka, A. Lunardi, R. Schnaubelt, L. Weis, Functional Analytic Methods for Evolution Equations. Editors: M. Iannelli, R. Nagel, S. Piazzera (2004)
Vol. 1856: K. Back, T.R. Bielecki, C. Hipp, S. Peng, W. Schachermayer, Stochastic Methods in Finance, Bressanone/Brixen, Italy, 2003. Editors: M. Fritelli, W. Runggaldier (2004)
Vol. 1857: M. Émery, M. Ledoux, M. Yor (Eds.), Séminaire de Probabilités XXXVIII (2005)
Vol. 1858: A.S. Cherny, H.-J. Engelbert, Singular Stochastic Differential Equations (2005)
Vol. 1859: E. Letellier, Fourier Transforms of Invariant Functions on Finite Reductive Lie Algebras (2005)
Vol. 1860: A. Borisyuk, G.B. Ermentrout, A. Friedman, D. Terman, Tutorials in Mathematical Biosciences I. Mathematical Neurosciences (2005)
Vol. 1861: G. Benettin, J. Henrard, S. Kuksin, Hamiltonian Dynamics – Theory and Applications, Cetraro, Italy, 1999. Editor: A. Giorgilli (2005)
Vol. 1862: B. Helffer, F. Nier, Hypoelliptic Estimates and Spectral Theory for Fokker-Planck Operators and Witten Laplacians (2005)
Vol. 1863: H. Fürh, Abstract Harmonic Analysis of Continuous Wavelet Transforms (2005)
Vol. 1864: K. Efstathiou, Metamorphoses of Hamiltonian Systems with Symmetries (2005)
Vol. 1865: D. Applebaum, B.V. R. Bhat, J. Kustermans, J. M. Lindsay, Quantum Independent Increment Processes I. From Classical Probability to Quantum Stochastic Calculus. Editors: M. Schürmann, U. Franz (2005)
Vol. 1866: O.E. Barndorff-Nielsen, U. Franz, R. Gohm, B. Kümmerer, S. Thorbjønsen, Quantum Independent Increment Processes II. Structure of Quantum Lévy Processes, Classical Probability, and Physics. Editors: M. Schürmann, U. Franz, (2005)

Recent Reprints and New Editions

Vol. 1200: V. D. Milman, G. Schechtman (Eds.), Asymptotic Theory of Finite Dimensional Normed Spaces. 1986. – Corrected Second Printing (2001)
Vol. 1471: M. Courtieu, A.A. Panchishkin, Non-Archimedean L-Functions and Arithmetical Siegel Modular Forms. – Second Edition (2003)
Vol. 1618: G. Pisier, Similarity Problems and Completely Bounded Maps. 1995 – Second, Expanded Edition (2001)
Vol. 1629: J.D. Moore, Lectures on Seiberg-Witten Invariants. 1997 – Second Edition (2001)
Vol. 1638: P. Vanhaecke, Integrable Systems in the realm of Algebraic Geometry. 1996 – Second Edition (2001)
Vol. 1702: J. Ma, J. Yong, Forward-Backward Stochastic Differential Equations and their Applications. 1999. – Corrected 3rd printing (2005)

4. Manuscripts should in general be submitted in English. Final manuscripts should contain at least 100 pages of mathematical text and should always include
   - a general table of contents;
   - an informative introduction, with adequate motivation and perhaps some historical remarks: it should be accessible to a reader not intimately familiar with the topic treated;
   - a global subject index: as a rule this is genuinely helpful for the reader.

   Lecture Notes volumes are, as a rule, printed digitally from the authors' files. We strongly recommend that all contributions in a volume be written in the same LaTeX version, preferably LaTeX2e. To ensure best results, authors are asked to use the LaTeX2e style files available from Springer's web-server at

   ftp://ftp.springer.de/pub/tex/latex/mathegl/mono.zip (for monographs) and
   ftp://ftp.springer.de/pub/tex/latex/mathegl/mult.zip (for summer schools/tutorials).

   Additional technical instructions, if necessary, are available on request from:

   lnm@springer-sbm.com.

5. Careful preparation of the manuscripts will help keep production time short besides ensuring satisfactory appearance of the finished book in print and online. After acceptance of the manuscript authors will be asked to prepare the final LaTeX source files (and also the corresponding dvi-, pdf- or zipped ps-file) together with the final printout made from these files. The LaTeX source files are essential for producing the full-text online version of the book. For the existing online volumes of LNM see:
   http://www.springerlink.com/openurl.asp?genre=journal&issn=0075-8434.

   The actual production of a Lecture Notes volume takes approximately 8 weeks.

6. Volume editors receive a total of 50 free copies of their volume to be shared with the authors, but no royalties. They and the authors are entitled to a discount of 33.3 % on the price of Springer books purchased for their personal use, if ordering directly from Springer.

7. Commitment to publish is made by letter of intent rather than by signing a formal contract. Springer-Verlag secures the copyright for each volume. Authors are free to reuse material contained in their LNM volumes in later publications: A brief written (or e-mail) request for formal permission is sufficient.

Addresses:

Professor J.-M. Morel, CMLA,
École Normale Supérieure de Cachan,
61 Avenue du Président Wilson, 94235 Cachan Cedex, France
E-mail: Jean-Michel.Morel@cmla.ens-cachan.fr

Professor F. Takens, Mathematisch Instituut,
Rijksuniversiteit Groningen, Postbus 800,
9700 AV Groningen, The Netherlands
E-mail: F.Takens@math.rug.nl

Professor B. Teissier, Institut Mathématique de Jussieu,
UMR 7586 du CNRS, Équipe "Géométrie et Dynamique",
175 rue du Chevaleret, 75013 Paris, France
E-mail: teissier@math.jussieu.fr

For the "Mathematical Biosciences Subseries" of LNM :
Professor P. K. Maini, Center for Mathematical Biology,
Mathematical Institute, 24-29 St Giles,
Oxford OX1 3LP, UK
E-mail : maini@maths.ox.ac.uk

Springer, Mathematics Editorial I, Tiergartenstr. 17,
69121 Heidelberg, Germany,
Tel.: +49 (6221) 487-8410
Fax: +49 (6221) 487-8355
E-mail: lnm@springer-sbm.com

Printing: Krips bv, Meppel
Binding: Stürtz, Würzburg